Engineering
Thermodynamics
with Worked Examples

Second Edition

Engineering Thermodynamics
with Worked Examples

Second Edition

Nihal E Wijeysundera

World Scientific

NEW JERSEY • LONDON • SINGAPORE • BEIJING • SHANGHAI • HONG KONG • TAIPEI • CHENNAI • TOKYO

Published by

World Scientific Publishing Co. Pte. Ltd.

5 Toh Tuck Link, Singapore 596224

USA office: 27 Warren Street, Suite 401-402, Hackensack, NJ 07601

UK office: 57 Shelton Street, Covent Garden, London WC2H 9HE

British Library Cataloguing-in-Publication Data
A catalogue record for this book is available from the British Library.

ENGINEERING THERMODYNAMICS WITH WORKED EXAMPLES
2nd Edition

ISBN 978-981-3148-07-9
ISBN 978-981-3148-08-6 (pbk)

Printed in Singapore

Preface

This textbook is intended mainly for use in undergraduate engineering courses in thermodynamics, and it covers all the topics that are traditionally taught in such courses. A novel feature of the book is the inclusion of a series of worked examples in each chapter. These examples are carefully chosen to expose students to diverse applications of engineering thermodynamics. In particular, care has been taken not to repeat the same type of example, with different numerical values, which would only increase the number of problems presented. Each worked example is designed to illustrate the application of an important concept introduced in the chapter to a practical situation. At the end of each chapter there are an additional series of problems for which numerical answers are provided. For the instructor, this book should provide a useful source for problems, to be included in tutorials and illustrations in courses.

The first two chapters on systems and thermodynamic properties provide the groundwork for applying the laws of thermodynamics. The third chapter on work interactions, builds on the concepts studied in basic mechanics courses. Chapters 4 and 5 present the application of the first law to closed and open systems respectively. The second law is developed in chapter 6, following the historical route that uses heat engines. The property, entropy, emerges as a consequence of the second law in chapter 7. The analysis of open systems, using the first law and the second law, is presented in chapter 8. The analysis of vapor power cycles, gas power cycles, and refrigeration systems are included in chapters 9, 10 and 11 respectively. Chapter 12 deals with non-reactive mixtures, which have important applications in air conditioning. The emphasis in chapter 13 on reactive mixtures is on combustion processes.

In this second edition of the book I have included two new chapters. Chapter 14 deals with thermodynamic property relations which are important in the development of thermodynamic property-data tables. Worked examples are used to illustrate the computation procedure to obtain the values of the derived properties from the measured properties.

Presented in chapter 15 is an introduction to the statistical interpretation of entropy. Although the computation of entropy production is emphasized in most of the earlier chapters, its physical meaning can be further clarified using the microscopic view point of matter. In a limited way, chapter 15 could also provide some essential background to the study of statistical thermodynamics.

Due to the interconnectedness of the topics in thermodynamics, a majority of the worked examples in earlier chapters 1, 2 and 3 could have been extended to the application of the first and second laws in later chapters, thereby generating additional worked examples. I have avoided doing this, which at first sight might appear as a lost opportunity. However, I have deliberately left this useful avenue, for extending the examples, open to the user of the book.

I was fortunate to have had the opportunity to teach a number of courses in thermodynamics, energy conversion, refrigeration and air conditioning, at the Department of Mechanical Engineering, National University of Singapore (NUS). The notes developed for these courses provided the framework and much of the material for this book. I am thankful to my colleagues in the energy and bio-thermal division at NUS, with whom I shared the teaching of these courses, for many valuable discussions on the applications of thermodynamics.

I wish to thank Professor Terry Hollands who willingly reviewed an early draft of the book. I benefited much from his insightful comments and suggestions on the content, and presentation of the subject matter. I am thankful to Dr. Raisul Islam who helped produce the illustrations in the first three chapters of the book. Thanks are due to my sons, Duminda and Harindra, and my daughters-in-law, Sindhu and Sophia, for their encouragement and help in different ways.

Finally, my heartfelt thanks are given to my wife Kamani for her constant encouragement and generous support towards the completion of this project. This book is dedicated to her.

N.E. Wijeysundera

Contents

Chapter 1

Thermodynamic Systems and Properties

In this chapter we shall introduce some of the important concepts and definitions that provide the basic framework for the study of engineering thermodynamics. The logical development and generalization of the first and second laws of thermodynamics in later chapters of the book rely on these concepts and definitions.

1.1 Thermodynamics

Thermodynamics deals with energy and matter, their transformations and the interactions between them. Since all engineering systems involve energy and matter, the applications of thermodynamics are very broad. Mechanical and chemical engineering are two of the traditional branches of engineering where applications of thermodynamics abound. Although the laws of thermodynamics are universally valid, their presentation and the mode of development could vary according to the engineering discipline.

1.2 Systems and Surroundings

For the analysis of physical situations involving matter and energy, it is useful to isolate regions or entities on which attention may be focused with a view to applying the laws of thermodynamics. These regions or entities are called *thermodynamic systems*. A real or imaginary *boundary* is drawn to physically demarcate and clearly identify the system. The boundary may be fixed or deformable. Everything that lies outside the boundary constitutes the *surroundings*. Systems can be further categorized based on the type of matter and energy interactions between them and their surroundings. These interactions occur across the boundary.

1.2.1 *Closed-system or control-mass*

A system where there is no matter transfer across the boundary is called a *closed-system* or a *control-mass*. Note, however, that energy interactions may occur across the boundary.

1.2.2 *Open-system or control-volume*

For an *open-system* or *control-volume*, matter may cross the system boundary. In addition, there may also be energy interactions across the boundary, which is usually stationary.

1.2.3 *Isolated system*

An *isolated-system* has no matter transfer or energy interactions across the boundary. Several different thermodynamic systems are considered in worked examples 1.1 and 1.2.

1.3 Properties of a System

A *property* of a system is a characteristic of the system that is usually measurable or could be calculated using measured quantities. Typical measurable properties of a system are pressure, temperature and volume. There are several other properties like internal energy, enthalpy and entropy, to be defined later, that could be calculated using the values of measured properties. It is important to distinguish between two general classes of thermodynamic properties: *intensive* properties and *extensive* properties.

1.3.1 *Intensive properties*

An *intensive property* is independent of the mass of a system. Pressure and temperature are examples of intensive properties. It is clear that the magnitude of an intensive property may vary spatially within a system, especially under dynamic conditions. We then speak of intensive property gradients within a system. For example, the pressure of a gas contained in a vertical cylinder varies due to the action of gravity.

1.3.2 *Extensive properties*

The value of an *extensive property* depends on the quantity of matter in the system. Mass and volume are examples of extensive properties. The value of an extensive property of a composite system may be obtained by adding the values of the property of the constituent sub-systems.

1.4 State of a System

The state of a system is characterized or described by the values of its properties. The minimum number of independent properties that are needed to uniquely identify the state depends on the complexity of the system. For *simple thermodynamic systems* two independent properties are sufficient to completely describe its state, for example temperature and pressure.

1.5 Some Basic Properties of Systems

In the following sections we review a few familiar thermodynamic properties that we have encountered in basic physics courses. These are pressure, temperature and density.

1.5.1 *Pressure*

Pressure is defined as the normal force per unit area exerted by a system on the boundary. In order to define the pressure at a location inside the system we consider the force, δF on one side of a small imaginary surface of area δA at the location. The pressure is given by, $P = \delta F/\delta A$. The pressure distribution in a system in an equilibrium state is assumed to be uniform. However, we need to remember that gravity will cause an increase in pressure from the top of the system to its bottom even under equilibrium conditions. In many applications this pressure variation due to gravity can be neglected.

Many techniques are available for the measurement of pressure. The more common pressure-measuring instruments include liquid manometers, Bourdon-tube pressure gages and pressure transducers. These instruments, in general, measure the gage pressure, which is the difference in pressure between a fluid and the surrounding atmosphere. The absolute pressure of the fluid is obtained from the relation

$$P_{absolute} = P_{gage} + P_{atmospheric}$$

If the absolute pressure is lower than that of the atmosphere it is called *vacuum* pressure.

A simple U-tube manometer consists of a vertical U-shaped tube with a liquid occupying its lower section and free to move in the two vertical legs of the tube as shown in Fig. 1.1. When the open ends of the legs are exposed to two different pressures, the difference in height of the liquid columns, H is related to the pressure difference, $(P_1 - P_2)$. *By Pascal's*

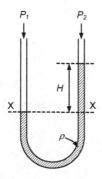

Fig. 1.1 U-tube manometer

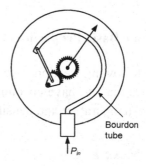

Fig. 1.2 Bourdon tube gage

principle the pressure is the same at every point on a horizontal plane through the *same continuous* liquid at *equilibrium*.

Therefore at the section x-x, the pressure in the two liquid columns of the U-tube are equal. Hence

$$P_1 = P_2 + H\rho g \tag{1.1}$$

where ρ is the density of the liquid and g is the acceleration due to gravity.

A *Bourdon-tube pressure gage* consists of a hollow tube of oval cross section, bent to a nearly circular shape as shown in Fig. 1.2. One end of the tube is open and is rigidly connected to the system in which the pressure is to be measured. The other end of the tube is closed and free to move. When the pressure inside the tube is larger than the ambient pressure acting on the outside of the tube, the tube tends to uncoil causing the free end of the tube to move. The reverse is true when the pressure on the inside is lower than the pressure acting on the outside. This displacement of the free end is amplified with the use of a linkage, a toothed-sector and a gear wheel and, finally translated to the movement of a needle over a dial. The movement of the needle is calibrated to read the gage pressure.

Pressure transducers use the change of a material property such as the electrical resistance with the strain caused by the applied pressure to measure the latter. There are a number of such transducers that are used for specialized applications. A series of interesting practical situations involving the calculation and measurement of pressure are presented in worked examples 1.7 to 1.16.

1.5.2 *Temperature*

Physiological sensations of hot and cold help us make qualitative statements on temperature levels. However, to measure temperature in an objective manner, we rely on material properties that are known to be affected by changes in temperature. Examples of such properties are the volume and pressure of a gas, the volume of a liquid, and the electrical resistance of a wire. Some of the more common temperature

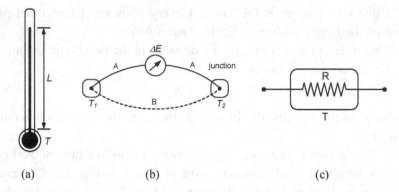

Fig. 1.3(a) Liquid-in-glass thermometer, (b) Thermocouple, (c) Resistance thermometer

measuring devices are: liquid-in-glass thermometers, thermocouples, resistance thermometers or resistance temperature detectors (RTD) and thermistors.

The *liquid-in-glass thermometer*, consists of a narrow tube attached to a bulb containing a liquid such as mercury or alcohol as shown in Fig. 1.3(a). The bulb is placed in the system in which the temperature is to be measured.

As the bulb comes to equilibrium with the system the liquid column gets longer due to the expansion of the liquid in the bulb. The length of the liquid in the tube is an indication of the temperature level of the system.

The working of the *thermocouple* is based on the Seebeck effect. In its simple form, a thermocouple consists of two wires of dissimilar metals made into a loop by joining them at the ends to form two junctions that are maintained at different temperatures (see Fig. 1.3(b)). The voltage generated in the loop is called the *thermoelectric effect* and it is dependent on the temperature difference of the junctions. If one of the junction temperatures is known, then the temperature of the other junction is related directly to the voltage measured in the loop.

The *resistance thermometer or RTD* is usually made of platinum which could be in the form of a wire or a film (see Fig. 1.3(c)). The resistance thermometer forms a component of an electrical circuit with a power source. The variation in the electrical resistance of the thermometer is related to the temperature being measured. The thermistor is basically a resistance thermometer made of a ceramic or polymer.

1.5.3 *Temperature scales*

Four temperature scales are encountered frequently in engineering practice. These are called the Celsius (t_c), Kelvin (T_K), Fahrenheit (t_F) and Rankine (T_R) temperature scales. We shall briefly discuss these scales using the mercury-in-glass thermometer, shown in Fig. 1.3(a), as our measuring system.

For the Celsius scale, the level of the mercury column in the capillary tube when the bulb is placed in ice water (ice point) is designated as 0°C. The corresponding level of mercury when the bulb is immersed in boiling water (steam point) is designated as 100°C. The length of the tube between the two levels is divided into 100 equal divisions and each division represents a change of temperature of a degree Celsius. In the Fahrenheit scale the ice point and the steam point are designated as 32 and 212 degrees respectively.

Since the above temperature scales depend on the properties of mercury difficulty arises when temperatures below –39°C, the freezing point of mercury, are measured. This situation may be partially remedied by using a different thermometric fluid which, however, may have different expansion properties from mercury. Therefore there is a need to search for a more universal calibration method. Fortunately, the behavior of gases contained in the rigid vessel under low pressures offers such a method.

Shown in Fig. 1.4(a) is a simplified schematic diagram of a constant-volume gas thermometer. The bulb B containing a dilute gas is immersed in the medium whose temperature is to be determined. At each measurement, the mercury column in leg A of the U-tube is adjusted to bring the top of the mercury to the fixed mark, O in the tube. This is achieved by moving the leg A1 of the U-tube vertically using the flexible connecting tube A2. Consequently, at each measurement the volume of gas in the bulb, and the tube up to O is constant. The pressure indicted by the height *H* of the mercury in leg A1 gives the gage pressure of the gas.

The pressures measured with the bulb B immersed in ice water and steam are plotted as Y and X in Fig. 1.4(b) for a particular mass of the gas. A line drawn through the points X and Y provides a calibration line for determining intermediate temperatures from measured gas pressures. An interesting feature of the gas thermometer is the intercept on the temperature axis of the line XY when extrapolated to zero pressure as seen in Fig. 1.4(b). If the bulb is filled with a different gas and the measurements are repeated, the new calibration line X_1Y_1 has the same intercept as before. When the ice point and the steam point are designated by 0°C and 100°C respectively, the constant intercept on the temperature axis corresponds to a value of –273.15°C, irrespective of the type of *dilute gas* in the bulb.

Therefore the intercept seems to offer a *natural fixed point* for the calibration of thermometers. Based on the above experimental observation, an absolute temperature scale has been defined by designating the intercept temperature as zero. Such a temperature scale is called the Kelvin (T_K) or the absolute scale. The second fixed point of the Kelvin scale, designated by 273.16 K, is the *triple point* of water when ice, water and vapor exist in equilibrium at a pressure of 0.6112 kPa. The Kelvin scale is sometimes called the *ideal-gas* temperature scale. The absolute temperature scale corresponding to the Fahrenheit (t_F) scale is called the Rankine (T_R) scale.

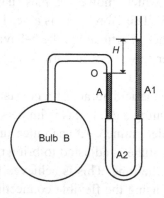

Fig. 1.4(a) Gas thermometer

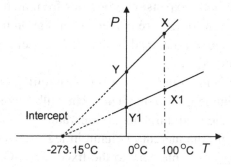

Fig. 1.4(b) Ideal gas temperature scale

The different temperature scales have the following relations:

$$T_K = 273.15 + t_c = 5T_R/9 \qquad (1.2)$$

$$T_R = 459.67 + t_F = 9T_K/5 \qquad (1.3)$$

1.5.4 *Density and specific volume*

Density is usually defined as the mass of a substance divided by its volume. If the volume and mass considered are relatively large, the density obtained from the above definition would be the *average density*. When we deal with a medium like the atmosphere where the density varies continuously with altitude, the following definition is more appropriate:

$$\rho = \lim_{\delta V \to \delta V'} \left(\frac{\delta M}{\delta V} \right) \qquad (1.4)$$

where ρ is the density, δM is the mass of a small volume δV of the substance. The limiting volume $\delta V'$ is the smallest volume which ensures that the medium is a *continuum*. In other words, if the volume considered in applying Eq. (1.4) is too small the mass within the volume could fluctuate over time because of the random movement of the molecules in and out of the volume. This is particularly relevant in dealing with rarefied gases. The specific volume of a substance is the inverse of the density: $v = 1/\rho$. Worked examples 1.19 and 1.20 deal with numerical calculations involving density and specific volume.

1.6 Macroscopic and Microscopic View Points

Thermodynamic systems can be characterized and analyzed using directly measurable properties like the pressure, temperature and volume. Such an approach is called a *macroscopic view point* of the system. Alternatively, a system may be described in terms of the characteristics of its microscopic constituents like molecules and atoms. The latter is called a *microscopic view point*. The analysis of engineering systems is carried out more readily using the macroscopic view point and is therefore used widely. The microscopic view point, however, enables us to gain a deeper understanding of the processes and interactions that occur in physical systems.

1.7   Thermodynamic Equilibrium

When a system is in an *equilibrium state*, all its intensive properties are spatially uniform within the system. For *mechanical* equilibrium there should be no pressure gradients within the system and the forces acting at the boundary of the system must be in balance. For *thermal* equilibrium the temperature gradients within the system are zero. *Chemical* equilibrium requires that the chemical composition within the system is uniform.

We should note that when a real system interacts with its surroundings, property gradients could occur within the system resulting in a *non-equilibrium* state. Under such conditions it is not possible to identify the state of the system using a set of unique properties that are *uniform* within the system.

1.7.1   *Quasi-equilibrium and non-equilibrium processes*

A system is said to undergo a *process* when its state changes due to the change of one or more of its properties. When a system undergoes a series of processes and returns to the initial state, the system is said to have executed a *cycle*, or a *cyclic process*.

We illustrate some important details of a thermodynamic process by considering as our system a gas confined in a cylinder by a frictionless piston as depicted in Fig. 1.5(a). This is a useful idealized experimental arrangement which we shall refer to frequently in this book.

In the initial state, the gas is in equilibrium with its pressure P determined by the mass of the piston M_p, the mass of the weights placed on the piston M_w, and the outside ambient pressure P_a acting on the piston. The mechanical equilibrium of the piston is governed by the equation of force balance:

$$PA = M_p g + M_w g + P_a A \qquad (1.5)$$

where A is the cross-sectional area of the piston and g is the acceleration due to gravity.

The pressure of the gas is indicated by the two pressure gages fitted to the cylinder at two locations. We choose the volume and the pressure of the gas as the two *independent* properties that define the state of the system. The gas is to be subjected to a process in which its pressure is to be reduced by removing all the weights on the piston. Due to the resulting imbalance of the forces acting on the piston, the gas will expand until a new equilibrium state is established. The pressure P_f of the gas is then given by the new equation of force balance:

$$P_f A = M_p g + P_a A \qquad (1.6)$$

We shall consider two practical ways to carry out the above process. In the first process, we simply remove *all* the weights on the piston simultaneously. This clearly would result in a large imbalance of the forces acting on the piston. The gas would undergo a very rapid expansion, with the piston accelerating through the cylinder. Following a few oscillations the piston would eventually come to rest with the gas attaining the final equilibrium state. Here we assume that the piston remains inside the cylinder during the entire process. As the gas expands rapidly the two pressure gages will read different values, indicating the non-uniformity of the pressure within the system during the process.

If we decide to plot a graph of the pressure versus the volume as in Fig. 1.5(b), we could only assign unique values to the initial and final equilibrium states of the gas. The intermediate values of the pressure are

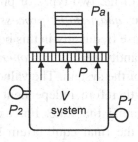

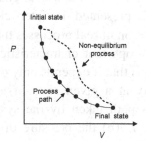

Fig. 1.5(a) Thermodynamic process   Fig. 1.5(b) Process path

not uniform and are also not readily measured. The above process is therefore called a *non-equilibrium* process because the system on its way from the initial state to the final state passes through a series of non-equilibrium states. We have indicated this process by the broken-line in Fig. 1.5(b) to remind us that the intermediate states have not been identified.

Now consider an alternative process in which the weights on the piston are removed one at a time, allowing sufficient time for the piston to attain a new equilibrium state following the removal of each small weight. The pressures indicated by the two gages will be much closer now because the change in pressure during each step is small. We are therefore able to plot the intermediate states of the gas in the pressure versus volume graph as shown in Fig. 1.5(b) where each of the points represents an intermediate state of the system.

Ideally, these points could be made as close as we desire by making the weight removed at each step correspondingly smaller. The system now passes through a series of *equilibrium* states on its way from the initial to the final state, and the process is called a *quasi-equilibrium* or *quasi-static* process. For such a process we could draw a smooth curve though the intermediate states to define the *path of the process*. Moreover, it is often possible to describe the process path by a simple mathematical relation between the two properties, which is sometimes called *the process law*. In examples 1.5 and 1.6 we shall discuss two situations involving quasi-equilibrium and non-equilibrium processes.

It has to be noted that we have discussed only the essential details needed to distinguish between quasi-equilibrium and non-equilibrium processes. Some additional characteristics of the two types of processes will be presented in later chapters. The *quasi-static process* is an idealization of real processes that we shall use frequently in this book.

An important characteristic of any quantity that is a *property* of a system is that it depends only on the state of the system. The value of the property at a given state of a system is therefore independent of the process path taken by the system in attaining the state. For example, in Fig. 1.5(b) the pressure of the gas in the final equilibrium state is the same at the end of both processes. Therefore a *property* of a system is called a *path-independent* quantity. We shall use this criterion in worked example 1.4 to determine whether or not a given derived quantity is a property of a system.

1.8   Temperature Measurement and the Zeroth-Law

When using any type of thermometer we actually measure one of its properties that is known to vary with temperature. This property is a length in the case of a liquid-in-glass thermometer, a voltage for a thermocouple or an electrical resistance in the case of a resistance thermometer. In order to establish a unified, common scale for temperature we need to measure the various sensor-properties under the same conditions. This process is called *calibration* and it is an important aspect of temperature measurement in engineering.

A typical temperature calibration set-up consists of a liquid bath whose temperature is accurately controlled with minimal spatial variation. The thermometers to be calibrated are placed in close proximity to each other in the liquid bath as shown in Fig. 1.6(a). The temperature of the bath is increased in small increments and the output of the thermometers being calibrated are recorded after steady equilibrium conditions are established. One of the thermometers, usually called the *primary standard*, has to be the more reliable of the thermometers. For engineering purposes, this could be an accurate liquid-in-glass thermometer. Calibration graphs are obtained by plotting the property of the thermometer being calibrated against the temperature indicated by

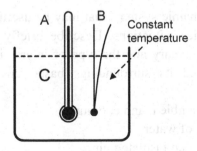

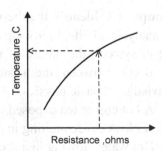

Fig. 1.6(a) Calibration of temperature sensors   Fig. 1.6(b) Calibration curve

the primary standard. For example in the calibration graph shown in Fig. 1.6(b), the output of the resistance thermometer in ohms is plotted against the temperature in °C indicated by the liquid-in-glass thermometer (Fig. 1.6(a)). The equality of the temperatures measured by the different thermometers when they are in equilibrium with the liquid bath is a consequence of the *zeroth law of thermodynamics*.

1.8.1 *The Zeroth-law of thermodynamics*

The zeroth law of thermodynamics which is based on experimental facts may be stated in the following form: Two systems A and B in thermal equilibrium with a third system C are in thermal equilibrium with each other. For instance, in the thermometer-calibration experiment, (see Fig. 1.6(a)) the output readings of the two temperature sensors A and B will become steady when they are individually in equilibrium with the liquid bath C. The zeroth law enables us to conclude that under these conditions the measured properties of the two sensors correspond to the same thermal state or temperature. Worked examples 1.17 and 1.18 illustrate the application of temperature calibration curves.

1.9 Worked Examples

In this section we present a series of worked examples to illustrate the engineering applications of the topics discussed in the preceding sections.

Example 1.1 Identify the thermodynamic systems that may be useful in the analysis of the following physical situations. Describe briefly the type of system, the nature of the boundary and the type of interactions that occur between the system and its surroundings based on the knowledge of basic physics.
 (i) A hot cup of tea exposed to the ambient and cooling.
 (ii) A block of ice melting in a tank of water.
 (iii) Hot water flowing in a section of an insulated pipe.

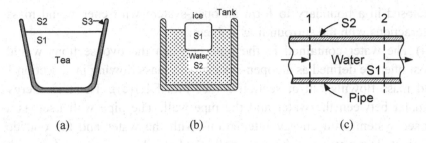

Fig. E1.1 Thermodynamic systems

Solution (i) The tea can be considered as a system (see Fig. E1.1(a)). The boundary consists of the inner surface of the cup and the exposed liquid surface. It is an open-system because there is mass transfer due to evaporation at the top surface. The mass of tea is therefore not constant. There is energy transfer across the entire boundary to the surroundings which means the system is non-isolated. The cup itself is a closed-system with rigid boundaries and a constant mass. There are energy interactions over its entire boundary and therefore the system is non-isolated. The tea and the cup can be combined to form a composite system by placing the boundary around the outer surface of the cup and the top surface of the tea.

(ii) A block of ice with a density of about 0.9 that of water will float in the water with part of the ice exposed to air (see Fig. E1.1(b)). The ice could be a system with its outer surface as the boundary. As the ice melts this boundary shrinks and the liquid formed mixes with the water in the tank. Therefore it is an open-system with a flexible boundary. The energy needed to melt the ice flows across the boundary making the system non-isolated.

The liquid water in the tank is an open-system which receives additional mass from the melting ice and transfers some mass at the top surface due to evaporation. It has energy interactions with the ice and the surroundings which includes the tank. The tank itself is a closed-system due to its fixed mass. However, it has energy interactions with the water and the outside ambient. The ice, water and the tank can be enclosed in a boundary to form an open-system with energy and mass interactions with the surrounding ambient.

(iii) The water contained in the pipe between the two sections would most often be defined as an open-system with mass flowing in at section 1 and mass flowing out at section 2 (see Fig. E1.1(c)). There is energy transfer between the water and the pipe wall. The pipe wall itself is a closed-system with energy interactions with the water and the outside ambient. The water and the pipe wall could also be an open system with energy transfers with the ambient.

Example 1.2 A cylinder fitted with a piston is connected to a rigid gas-tank through a valve as shown in Fig. E1.2. Initially, the valve is closed and the pressure in the tank is larger than in the cylinder. The valve is opened allowing gas to flow from the tank to the cylinder. The gas flow lifts the piston until it hits a lock at the top of the cylinder. The gas flow continues until the pressures in the tank and cylinder equalize. The outer surface of the entire set-up is thermally insulated.

 (i) Draw boundaries around the various components in Fig. E1.2 to show (a) closed systems, (b) open systems and (c) isolated systems.
 (ii) Identify the processes for which the categories in (i) apply.
(iii) Describe the interactions that occur between the systems and the surroundings.

Solution We consider two different processes P1 and P2 that occur over different time intervals. (i) P1-the process after the valve is opened till

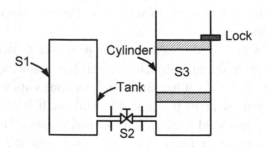

Fig. E1.2 Piston-cylinder set-up

the piston hits the lock at the top of the cylinder, and (ii) P2-the process from the time the piston hits the lock until the pressures in the tank and cylinder equalize.

Four system boundaries are identified. (i) S1-boundary surrounding the gas in the tank, (ii) S2-boundary surrounding the valve, (iii) S3-boundary surrounding the gas in the piston-cylinder arrangement, and (iv) S4-boundary surrounding the tank, the valve and piping and the cylinder-piston.

Initially, the systems S1, S2, S3 and S4 are all closed isolated systems because the mass within each system is constant and there are no energy interactions with the respective surroundings because of the thermal insulation.

During Process P1: S1 is an open-system with a rigid boundary and mass flow out. Due to the thermal insulation there is no energy interaction. System, S2, the valve, is an open-system with rigid boundaries and mass flow in and out of the system. There are no energy interactions. System, S3 is an open-system due to gas flow into the cylinder. It has a flexible boundary due to the movement of the piston. Moreover, the movement of the piston against the outside ambient constitutes an energy interaction.

The composite system S4, is a closed system because the total mass inside the system is constant. However, the movement of the piston makes the boundary of S4 flexible and results in an energy interaction between S4 and its surroundings.

During Process P2: S1 could still be an open-system owing to the possible oscillation of the gas following the sudden stoppage of the piston movement. S2 would also experience such changes of mass within it and therefore be an open-system. The same is true of system S3. However, S4 will have constant total mass and would be a closed-system. Due to the thermal insulation there would be no energy interactions at the boundaries of all 4 systems.

Example 1.3 Which of the characteristics of a mercury-vapor-filled light bulb listed below are properties? Which are not? Explain. The characteristics are: (i) mass of vapor, (ii) temperature and pressure of the

vapor, (iii) the number of hours of operation, and (iv) the total watt-hours
of energy consumed.

Solution (i) The mass of vapor within the fluorescent bulb is constant
and independent of the processes to which it is subjected. Therefore it is
a property. (ii) The pressure and temperature of the vapor are properties.
The operation of the light bulb would increase the temperature and
pressure. However, these same changes in pressure and temperature
could also be brought about by heating the bulb externally. The temperature and pressure once attained will be independent of the means by
which the change was brought about. Therefore these are properties of
the light bulb. (iii) The number of hours of operation is not a property of
the bulb because it depends on the operational history. (iv) The total
watt-hours of electrical energy consumed by the bulb depends on the
number of hours of operation and is therefore not a property.

Example 1.4 For a simple system, consisting of an ideal gas, the
specific volume, v and the absolute pressure, P are independent properties
as was discussed in Sec. 1.7.1. The dependent property T is defined by
the relationship, $Pv = RT$, where R is a constant. Four new quantities X,
Y, Z and W have been proposed as possible thermodynamic properties.
The changes in these quantities during an infinitesimal process are given
by the following expressions:

$$\text{(i)} \quad dX = Pdv + vdP, \qquad \text{(ii)} \quad dY = vdP - Pdv$$

$$\text{(iii)} \quad dZ = RdT + Pdv \quad \text{and} \quad \text{(iv)} \quad dW = v^2 dP + 2RTdv$$

Which of these suggested quantities are properties? Which are not?

Solution Consider a process that can be represented by the process law,
$P = F(v)$. The process-path is obtained by plotting the above relation on
a graph of P versus v. As was discussed in Sec. 1.7.1, a quantity whose
change is independent of the process-path and depends only on the initial
and final states is a property.

   We apply this criterion to each of the suggested quantities in turn
considering a change of state of the system from 1 to 2.

(i) Integrate the given relation, $dX = Pdv + vdP$ along the path $P = F(v)$ from state 1 to state 2. This gives

$$X_2 - X_1 = \int_1^2 [Pdv + vdP] = \int_1^2 d(Pv) = (Pv)_2 - (Pv)_1 .$$

The change in X is only dependent on the states 1 and 2 and not on the relation $P = F(v)$. Therefore X is a property of the gas.

(ii) Similarly, $\quad Y_2 - Y_1 = \int_1^2 [vdP - Pdv] = \int_1^2 d(Pv) - 2Pdv$

$$Y_2 - Y_1 = (Pv)_2 - (Pv)_1 - 2\int_1^2 F(v)dv$$

The evaluation of the second term on the right hand side in the above equation requires knowledge of the process-law or the process-path. Therefore the quantity Y is not a property because the magnitude of its change depends on the process path.

(iii) Differentiate the relation $Pv = RT$, and substitute for RdT to obtain

$$dZ = RdT + Pdv = Pdv + vdP + Pdv = d(Pv) + F(v)dv$$

Integrating the above equation

$$Z_2 - Z_1 = (Pv)_2 - (Pv)_1 + \int_1^2 F(v)dv$$

The magnitude of the change in Z depends on the process path due to the second term on the right hand side, and therefore Z is not a property.

(iv) Now

$$dW = v^2 dP + 2RTdv = v^2 dP + 2Pvdv = d(v^2 P)$$

Integrating the above equation we have

$$W_2 - W_1 = \int_1^2 d(v^2 P) = (v^2 P)_2 - (v^2 P)_1$$

The change in W is independent of the process path and depends only on the states 1 and 2. Therefore W is a property of the gas.

Example 1.5 Figure E1.5 shows a mass of gas trapped in a cylinder behind a piston of mass, $M = 5$ kg and cross sectional area, $A = 0.032$ m^2. Initially the pressure of the gas is $P_1 = 108$ kPa and a lock L_1 holds the piston in the position shown. The lock is removed and the gas expands pushing the piston until it hits the lock L_2. The final equilibrium pressure of the gas is $P_2 = 102$ kPa. The ambient pressure is $P_o = 100$ kPa.

(i) Calculate the force exerted by the locks in the initial and final equilibrium states.

(ii) Describe briefly the process undergone by the gas.

(iii) Describe an idealized process that could be used to effect the same change of pressure in the gas in a quasi-equilibrium manner.

Solution (i) The force balance for the initial equilibrium of the piston is [see Fig. E1.5(a)]

$$P_1 A = P_0 A + Mg + R_1$$

where R_1 is the force exerted by the lock on the piston. Substituting numerical values in the above equation

$$R_1 = (108 - 100) \times 10^3 \times 0.032 - 5 \times 9.8 = 256 - 49 = 207 \text{ N}$$

We have assumed that the acceleration due to gravity, $g = 9.8$ ms^{-2}. Similarly, for the final equilibrium state the force balance is

$$P_2 A = P_0 A + Mg + R_2$$

where R_2 is the force exerted by the lock. Hence

$$R_2 = (102 - 100) \times 10^3 \times 0.032 - 5 \times 9.8 = 64 - 49 = 15 \text{ N}$$

(ii) When the lock L_1 is removed, the piston undergoes an accelerated motion due to the lack of balance of forces. The air in the cylinder will experience rapid changes in pressure and volume. There will be sharp gradients in the intensive properties of the air which makes it a non-equilibrium process. Therefore it is not possible to assign spatially uniform single values for the intensive properties.

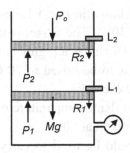

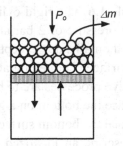

Fig. E1.5(a) Non-equilibrium process   Fig. E1.5(b) Quasi-equilibrium process

However, once the piston is brought to rest following the impact with lock, L_2 the properties of the air will progressively become spatially uniform and the system will attain a final equilibrium state. The properties of the initial and final states are uniform and could be assigned unique values. However, the determination of the property variations during the non-equilibrium process will require a complex analysis which is beyond the scope of this book.

(iii) We see that the force on the lock L_1 of 207 N is equivalent to placing a mass of about 21 kg (207/9.8) on the piston. In the final equilibrium state the force on lock L_2 of 15 N is equivalent to placing a mass of about 1.53 kg (15/9.8) on the piston. In the proposed idealized *quasi-equilibrium* process the lock L_1 is replaced with the additional mass of 21 kg on the piston thus maintaining the initial equilibrium state.

Suppose that the 21 kg is made up of 'small' masses placed in a container [Fig. E1.5(b)]. These small masses are now removed in steps until the final mass left on the piston is 1.53 kg, the force produced by lock L_2. These steps can be made as small as we like to cause the least imbalance of forces on the piston during the process. Consequently, the spatial variation of the pressure of the air can be made as small as we desire. The initial and final pressures of the air in the real process and the idealized process are the same. We would, however, find that the temperatures are different.

Example 1.6 The rigid cylindrical vessel shown in Fig. E1.6 contains a gas at a pressure of 2 bar and temperature 30°C. The vessel is thermally insulated on all sides except at the bottom which has a highly conducting metal surface. The temperature of the gas is to be raised to 70°C. Three alternative processes are to be compared. These are:

(i) Place the bottom on a gas heater with a flame temperature of 600°C.
(ii) Place the bottom surface in a hot water bath at 90°C.
(iii) Describe an idealized process that could be used to carry out the heating process in a quasi-equilibrium manner.

Solution (i) Heating the gas by applying a flame to the bottom surface of the cylinder is a non-equilibrium process in the extreme (Fig. E1.6(a)). The temperature gradient in the gas is very steep and there would be much turbulence. It is also difficult to control the final temperature attained by the gas.

(ii) Placing the bottom surface of the cylinder in a water bath would cause a temperature gradient in the gas due to the finite temperature difference between the water and the gas. This gradient, however, would be less steep than in (i) above.

(iii) The heating process could be carried out in a quasi-equilibrium manner by employing a series of heat sources with temperatures ranging from $(30 + \Delta T)$°C to $(70 + \Delta T)$°C where ΔT is a small temperature difference as shown in Fig. E1.6(b). For instance, if $\Delta T = 2$°C then the temperatures of the external heat sources will range from 32°C to 72°C in

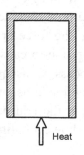

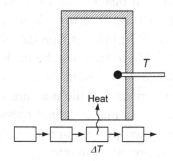

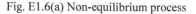

Fig. E1.6(a) Non-equilibrium process   Fig. E1.6(b) Quasi-static process

steps of 2°C. These heat sources are brought into communication with the bottom surface of the cylinder, one at a time, to allow heat flow due to the small temperature difference between the gas and heat source at any stage. The temperature gradient in the gas could be made as small as we desire by making ΔT small. The gas will undergo a quasi-equilibrium heating process that would allow us to assign unique values to the gas temperature, with the desired accuracy, at any stage during the process.

Example 1.7 A gas is trapped under a high pressure, P in a cylinder above a piston as shown in Fig. E1.7. The head of the piston has an ellipsoidal shape with a base radius of R. Obtain an expression for the axial force exerted by the gas on the piston. Work from first principles.

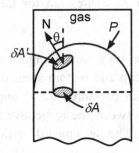

Fig. E1.7 Force on a cylinder head

Solution Consider the force on a small area $\delta A'$ on the surface of the piston head due to the pressure. Let the angle between the normal to this area and the vertical be θ. Then the vertical component of the force on the area is, $\delta F_v = P \delta A' \cos \theta$. But $\delta A' \cos \theta = \delta A$, is the projected area of $\delta A'$ on the horizontal plane. Therefore $\delta F_v = P \delta A$.

Integrating over the entire piston head we have, $F_v = PA$, where $A = \pi R^2$, is the total horizontal area, which is the area of cross section of the cylindrical section of the piston.

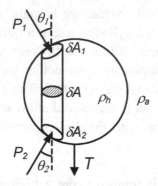

Fig. E1.8 Buoyancy force on balloon

Example 1.8 A balloon filled with helium gas is held down by a vertical rope as shown in Fig. E1.8. The volume of the balloon is V and its mass

when empty is M. The density of the helium and the air outside are ρ_h and ρ_a respectively. Obtain an expression for the tension in the rope. Work from first principles.

Solution Consider the forces on a small imaginary cylinder, of height z, within the balloon. Let the top and bottom areas of the cylinder be, δA_1 and δA_2 where the outside air pressures are P_1 and P_2 respectively. Let θ_1 and θ_2 be the angles between the respective surface normal at the two points and the vertical. The net upward vertical force on the small cylinder is

$$\delta F_v = P_2 \delta A_2 \cos \theta_2 - P_1 \delta A_1 \cos \theta_1 \qquad (E1.8.1)$$

But for the small areas, $\delta A_1 \cos \theta_1 = \delta A_2 \cos \theta_2 = \delta A$, the horizontal cross-sectional area of the cylinder. On substitution, the vertical force becomes

$$\delta F_v = P_2 \delta A - P_1 \delta A = (P_2 - P_1) \delta A \qquad (E1.8.2)$$

Considering the hydrostatic pressure variation in the air we have

$$P_2 - P_1 = z \rho_a g \qquad (E1.8.3)$$

where ρ_a is the density of air, g is the acceleration due to gravity and z is the vertical distance between the points 1 and 2.

Substituting in Eq. (E1.8.2) the force is

$$\delta F_v = z\rho_a g\delta A = \rho_a g\delta V \tag{E1.8.4}$$

where $\delta V = z\delta A$ is the volume of the small cylinder.

Integrating Eq. (E1.8.4), the resultant vertical force, which is usually called the *up-thrust*, becomes

$$F_v = \rho_a g V \tag{E1.8.5}$$

Consider the equilibrium of the balloon. The vertical force balance gives

$$T + Mg + V\rho_h g = F_v = V\rho_a g,$$

where T is the tension in the rope. Hence

$$T = (\rho_a - \rho_h)Vg - Mg$$

Note that the first term, $(\rho_a - \rho_h)Vg$ is called the *buoyancy force*.

Example 1.9 The variation of the density ρ with pressure P for an isothermal (constant temperature) atmosphere is given by

$$\rho(z) = \rho_0 P(z)/P_o$$

where the subscript o denotes the conditions on the earth's surface.

Obtain an expression for the variation of the pressure with height Z above the earth's surface.

Solution Consider the equilibrium of a small element of quiescent fluid of height δz and area δA at a height z above the earth's surface as shown in Fig. E1.9. Let the local pressure be P and the density ρ, both of which are functions of z. The vertical force balance on the fluid element gives

$$(P + \delta P)\delta A + \delta A \delta z\, \rho g = P\delta A$$

where g is the acceleration due to gravity.

Hence the governing differential equation for the pressure is

$$dP/dz = -\rho g \tag{E1.9.1}$$

The density variation with pressure is given as

$$\rho(z) = \rho_0 P(z)/P_o \tag{E1.9.2}$$

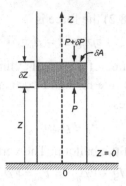

Fig. E1.9 Pressure variation in the atmosphere

Substituting for ρ in Eq. (E1.9.1) from Eq. (E1.9.2) we have

$$dP/dz = -g\rho_o P/P_o \qquad (E1.9.3)$$

The solution of the above differential equation is

$$\ln P = -(\rho_o g/P_o)z + c \qquad (E1.9.4)$$

The constant of integration c is found by applying the conditions on the earth's surface which states that at $z = 0$, $P = P_o$. Hence we obtain the pressure variation as

$$P/P_o = \exp(-\rho_o gz/P_o) \qquad (E1.9.5)$$

Example 1.10 Assuming the temperature of the atmosphere to vary with height above the earth's surface in a linear manner given by, $T(z) = (T_o - \alpha z)$ and the air density to be given by the expression, $\rho = P/RT$, obtain an expression for the variation of pressure with height in terms of the conditions at the earth's surface. The symbols have the same meaning as in worked example 1.9 above.

Solution Now the temperature and density variations are given as $T(z) = T_o - \alpha z$, where α is a constant, and $\rho = P/RT$. Eliminating T we obtain

$$\rho = \frac{P}{R(T_o - \alpha z)}$$

Substituting for ρ in the governing differential equation (E1.9.1), we have

$$\frac{dP}{dz} = -\frac{Pg}{R(T_o - \alpha z)}$$

Rearranging the above equation

$$\frac{dP}{P} = -\frac{gdz}{RT_o(1 - \alpha z/T_0)}$$

Integrate the above differential equation and substitute the pressure P_o at the earth's surface, $z = 0$ to determine the constant of integration. Hence we obtain the pressure variation with height as

$$P/P_o = (1 - \alpha z/T_o)^{(g/R\alpha)}$$

Example 1.11 A spherical balloon of radius 0.45 m and mass 0.11 kg is filled with helium. The balloon is to carry instruments weighing 0.07 kg to a height of 5.5 km above the earth's surface. The temperature of the atmosphere decreases with altitude at the rate of 0.0065 K.m⁻¹. The ambient temperature and pressure at sea level are 16°C and 101 kPa respectively. The gas constant, $R = 0.287$ kJ kg⁻¹ K⁻¹ and the acceleration due to gravity, $g = 9.8$ ms⁻². Assume that the volume of the balloon remains constant. Calculate (i) the temperature, pressure and density of the air at a height of 5.5 km, and (ii) the mass of helium in the balloon.

Solution The pressure variation with height z was obtained in worked example 1.10 as

$$P/P_o = (1 - \alpha z/T_o)^{(g/R\alpha)}$$

From the given data, $g = 9.8$ ms⁻², $\alpha = 0.0065$ K.m⁻¹, $z = 5500$ m, $T_o = 273 + 16 = 289$ K and $P_o = 101$ kPa.

The constant volume of the balloon is, $V = 4\pi r^3/3 = 0.382$ m³

Now the exponent in the above equation is given by

$$(g/R\alpha) = 9.8/(287 \times 0.0065) = 5.253$$

Substituting numerical values in the equation above we have

$$P/101 = (1 - 0.0065 \times 5500/289)^{5.253} = 0.499$$

Hence the pressure at a height of 5.5 km is

$$P = 101 \times 0.499 = 50.47 \text{ kPa}$$

The properties of air at 5.5 km are as follows:
Temperature,

$$T = T_o - \alpha z = (289 - 0.0065 \times 5500) = 253.25 \text{ K}$$

Density,

$$\rho = P/RT = 50.47 \times 10^3/(253.25 \times 287) = 0.694 \text{ kg m}^{-3}$$

(i) The equation of force balance for the balloon at 5.5 km is

$$(0.11 + 0.07)g = V(\rho - \rho_{He})g$$

The right hand side is the buoyancy force on the balloon which was derived in worked example 1.8, and the left hand side is the weight of the balloon and the instrument. Substituting numerical values in the above equation

$$(\rho - \rho_{He}) = 0.471$$

Hence
$$\rho_{He} = 0.223 \text{ kg m}^{-3}$$

The mass of helium in the balloon is

$$M_{He} = V\rho_{He} = 0.382 \times 0.223 = 0.085 \text{ kg}$$

Example 1.12 A rigid tank filled with oil has two cylindrical tubes with leak-proof pistons fitted in them as shown in Fig. E1.12. When a force of 300 N is applied to piston 1, whose area is 30 cm², calculate the maximum weight W that piston 2, of area 300 cm², can support in the position shown.

Solution Assume that the equilibrium pressure distribution in the tank is uniform. In terms of the forces and areas shown in Fig. E1.12, the uniform pressure is given by

$$P = F_1/A_1 = F_2/A_2$$

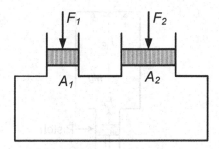

Fig. E1.12 Force on piston

Hence $$P = 300/30 = Wg/300$$

Assume that $g = 9.8$ ms^{-2}. Therefore $W = 306.12$ kg

Example 1.13 A cylindrical vessel of inner diameter 0.3 m, height 0.35 m and mass 40 kg has a smaller cylindrical tube attached to it axially. The tube is fitted with a leak-proof piston of diameter 0.025 m. The vessel is filled with oil of specific gravity 0.8 and supported vertically on the piston as shown in Fig. E1.13. Calculate the force exerted by the oil on the upper flat surface of the vessel.

Solution The mass of oil filling the cylinder is

$$M_o = (\pi/4)(0.3)^2 \times 0.35 \times 0.8 \times 1000 = 19.8 \text{ kg}$$

where the density of water is taken as 1000 kg m^{-3}. The vertical force on the piston due to the weight of the cylinder and the oil is

$$F_p = M_o g + M_c g = (19.8 + 40) \times 9.8 = 586 \text{ N}$$

where $g = 9.8$ ms^{-2}.

The area of the piston is

$$A_p = (\pi/4) \times 0.025^2 = 4.909 \times 10^{-4} \text{ m}^2$$

The oil pressure on the piston is

$$P_o = F_p/A_p = 586/4.908 \times 10^{-4} = 11.94 \times 10^5 \text{ N m}^{-2}$$

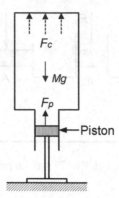

Fig. E1.13 Force on cylinder

Assume that the oil pressure in the cylinder is uniform. Then the force on the upper surface of the cylinder is:

$$F_c = P_o A_c = 11.94 \times 10^5 \times (\pi/4) \times 0.3^2 = 84.4 \text{ kN}$$

Example 1.14 Figure E1.14 shows a U-tube manometer used to measure the pressure difference in the two pipes A and B carrying oil of density 810 kgm^{-3}. The manometric liquid is mercury whose specific gravity is 13.6. The Bourdon-tube pressure gage fitted to A reads 1.5 bar and the local atmospheric pressure is 750 mm of mercury. The heights indicated in Fig. E1.14 are

$$h_1 = 400 \text{ mm}, \quad h_2 = 95 \text{ mm and } h_3 = 500 \text{ mm}.$$

Assume that the density of water is 1000 kgm^{-3}. Calculate (i) the reading of the Bourdon-tube pressure gage attached to B in bar and (ii) the absolute pressures in pipes A and B in kPa.

Solution The atmospheric pressure,

$$P_{atm} = h_a \rho_m g = (750/1000) \times (13.6 \times 1000) \times 9.8 = 99.96 \text{ kPa}$$

Units of pressure are: 1 bar $= 10^5$ N m$^{-2} = 10^5$ Pa $= 100$ kPa

Density of oil, $\rho_o = 0.81 \times 1000$ kg m^{-3}

Density of mercury, $\rho_m = 13.6 \times 1000$ kg m^{-3}

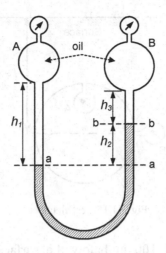

Fig. E1.14 Manometer

The pressures in the mercury columns at level a-a on the two vertical legs of the U-tube are the same. Hence

$$P_A + h_1\rho_o g = P_B + h_3\rho_o g + h_2\rho_m g$$

$$P_A + 3175.2 = P_B + 3969 + 12661.6$$

$$P_A - P_B = 13455.4 = 13.455 \text{ kPa} \qquad (\text{E1.14.1})$$

The Bourdon-tube gage reads the pressure relative to the surrounding ambient. It is therefore called a gage pressure.

Reading of Bourdon-tube gage A is, 1.5 bar = 150 kPa (gage)
From Eq. (E1.14.1) the reading of gage B is

$$P_B = 150 - 13.45 = 136.55 \text{ kPa (gage)}$$

The absolute pressures at A and B are

$$P_A(abs) = 150 + 99.96 = 249.96 \text{ kPa} = 2.5 \text{ bar}$$

$$P_B(abs) = 136.55 + 99.96 = 236.5 \text{ kPa} = 2.37 \text{ bar}$$

Example 1.15 A cylindrical test chamber of diameter 10 m is filled with atmospheric air at a pressure of 1.01 bar. It is sealed perfectly and

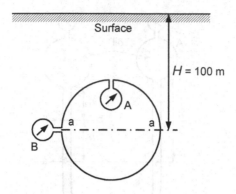

Fig. E1.15 Test chamber

lowered to a depth of 100 m below the surface of the ocean. Two Bourdon-tube pressure gages are installed in the chamber as shown in Fig. E1.15. There is also a mercury barometer in the chamber. The density of sea water is 1050 kgm⁻³. Determine (i) the readings of the two pressure gages in kPa, and (ii) the reading of the barometer.

Solution The pressure of the sea water at the central plane a-a (Fig. E1.15) of the cylinder is

$$P_{wb} = P_{atm} + H\rho_w g = (101 + 1029) = 1130 \ \text{kPa}$$

We have assumed that $g = 9.8 \ \text{ms}^{-2}$.

Because the vessel is sealed and rigid, the pressure inside is the same as that of the air originally introduced at the surface. The density of mercury is $\rho_m = 13.6 \times 10^3 \text{kg m}^{-3}$. Hence the reading of the mercury barometer is

$$H_b = P_{air} / g\rho_m$$

$$H_b = 101 \times 10^3 \times 10^3 / (13.6 \times 10^3 \times 9.8) = 757.8 \ \text{mm of Hg}$$

The Bourdon-tube of gage B is filled with air from the cylinder while the outside is exposed to sea water.

Therefore the reading of Bourdon-tube gage B is

$$P_B = (101 - 1130) \ \text{kPa} = -1029 \ \text{kPa}$$

Expressed as a vacuum pressure

$$P_B = (1029 \times 10^3)/(13.6 \times 10^3 \times 9.8) = 7.72 \text{ m}$$

of Hg vacuum relative to the surrounding water pressure.

The water pressure at the top of the cylinder where the gage A is attached is given by:

$$P_{wa} = 1130 - 5 \times 1050 \times 9.8 \times 10^{-3} = (1130 - 51.45) = 1078.55 \text{ kPa}$$

The Bourdon-tube of gage A has sea water inside the tube and is exposed to the air in the cylinder on the outside. Therefore the reading of gage B is

$$P_A = 1078.55 - 101 = 977.55 \text{ kPa}$$

Example 1.16 Two cylindrical vessels A and B, shown in Fig. E1.16, are placed inside a larger partially-evacuated cylindrical chamber C of mass 75 kg, inner diameter of 0.5 m and wall thickness 10 mm, to reduce the heat losses from the vessels. A mercury manometer whose ends are connected to the two vessels A and B indicates a difference in level of 54 mm. The mercury manometer attached to C shows a reading of 680 mm while the local ambient pressure is 750 mm mercury. The Bourdon-tube gage fitted to A reads 2.05 bar. Calculate (i) the absolute pressures in A, B and C in bar, (ii) the reading of the Bourdon-tube gage fitted to B, (iii) the force exerted by chamber C on the base, and (iv) the average pressure over the contact area between C and the base plate.

Solution The absolute pressure P_c in chamber C is given by

$$P_c + 680 = 750$$

$$P_c = 750 - 680 = 70 \text{ mm}$$

$$P_c = (70 \times 10^{-3} \times 13.6 \times 10^3 \times 9.8) = 9.33 \text{ kPa}$$

The ambient pressure outside C is

$$P_{atm} = 750 = (750 \times 10^{-3} \times 13.6 \times 10^3 \times 9.8) = 99.96 \text{ kPa}$$

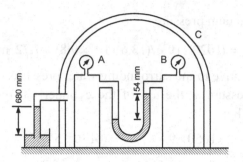

Fig. E1.16 Test chambers

The absolute pressures in A and B are related by

$$P_A = P_B + (54 \times 10^{-3} \times 13.6 \times 10^3 \times 9.8) \text{ kPa}$$

$$P_A - P_B = 7.197 \text{ kPa}$$

The reading of gage A is, $P_{Ag} = 205$ kPa

The Bourdon-tube gage A reads the gage pressure. Hence

$$P_A - P_C = 205 \text{ kPa}$$

The absolute pressure in A is

$$P_A = 205 + 9.33 = 214.33 \text{ kPa}$$

The absolute pressure in B is

$$P_B = 214.33 - 7.197 = 207.13 \text{ kPa}$$

The reading of gage B is given by

$$P_{Bg} = P_B - P_C = 207.13 - 9.33 = 197.8 \text{ kPa}$$

(i) The force balance on vessel C gives:

$$P_{atm}(\pi D_o^2/4) + M_c g = P_C(\pi D_i^2/4) + F_C$$

where the inner diameter, $D_i = 0.5$ m.

The outer diameter is given by

$$D_o = 0.5 + 2 \times 10/1000 = 0.52 \text{ m}$$

The mass of C is $M_C = 75$ kg

Substituting numerical values in the equation above we have

$$F_C = 99.96 \times \pi \times 0.52^2/4 - 75 \times 9.8 \times 10^{-3} - 9.33 \times \pi \times 0.5^2/4$$

Hence $\qquad\qquad\qquad F_C = 18.66 \text{ kN}$

(ii) The contact area between C and the base plate is

$$A_{Cb} = \pi \times (0.52^2 - 0.5^2)/4 = 0.016 \text{ m}^2$$

The average contact pressure is

$$R_C = F_C / A_{Cb} = 18.66/0.016 = 1164.9 \text{ kN m}^{-2}$$

Example 1.17 A platinum resistance thermometer is used as a primary standard to measure temperatures between 0 and 660°C. The measured electrical resistance is 10.950, 16.540 and 29.557 ohms at 0°C, 100°C and 444.6°C respectively. The calibration curve for the thermometer can be expressed in the form, $R = a_0 + a_1 t + a_2 t^2$, where R is the resistance in ohms, t is the temperature in °C. (i) Find the constants a_0, a_1 and a_2. (ii) Calculate the sensitivity of the thermometer, $(\delta R / \delta t)$ at 100°C.

Solution We substitute the three given calibration conditions in the assumed quadratic relationship

$$R = a_0 + a_1 t + a_2 t^2$$

to obtain the following three equations.

$$10.95 = a_0 + a_1 \times 0 + a_2 \times 0$$

$$16.54 = a_0 + a_1 \times 100 + a_2 \times 100^2$$

$$29.557 = a_0 + a_1 \times 444.6 + a_2 \times 444.6^2$$

(i) The simultaneous solution of the above equations gives:

$$a_0 = 10.95, \; a_1 = 59.9768 \times 10^{-3} \text{ and } a_2 = -40.7687 \times 10^{-6}$$

To estimate the sensitivity, we perform a first-order Taylor expansion of the calibration equation about the given value. Hence

$$\delta R = (dR/dt)\delta t$$

$$(dR/dt) = a_1 + 2a_2 t$$

$$(dR/dt) = 59.7968 \times 10^{-3} - 2 \times 40.7687 \times 10^{-6} \times 100 = 0.05182$$

(ii) The sensitivity is 0.05182 ohms per °C.

Example 1.18 A thermocouple is calibrated with the reference junction maintained at 0°C, the ice-point temperature, and the temperature of the test junction is varied from 30°C to 150°C. The voltmeter reading E (V) of the thermocouple and the temperature (°C) are correlated by fitting the quadratic equation: $E = a_1 t + a_2 t^2$. The constants are obtained as:

$$a_1 = 0.22 \times 10^{-3} \text{ V°C}^{-1} \text{ and } a_2 = -5.1 \times 10^{-7} \text{ V°C}^{-2}.$$

 (i) When the thermocouple is used to measure an unknown temperature, the voltmeter indicates 15.2 mV. Calculate the temperature in °C.
(ii) What is the sensitivity of the thermocouple, $(\delta E/\delta t)$ at 100°C.
(iii) If a temperature close 100°C is to be measured to an accuracy of 0.2°C, estimate the accuracy of the required voltmeter.

Solution The calibration equation is

$$E = a_1 t + a_2 t^2 = 0.22 \times 10^{-3} t - 5.1 \times 10^{-7} t^2$$

When the measured voltage is 15.2×10^{-3} V, we have

$$15.2 \times 10^{-3} = 0.22 \times 10^{-3} t - 5.1 \times 10^{-7} t^2$$

$$t^2 - 431.3725 t + 2.9804 \times 10^4 = 0$$

The solution of the above quadratic equation gives two positive roots:

$$t_1 = 86.4 \text{ °C and } t_2 = 344.98 \text{ °C}$$

Since the calibration was carried out in the range 30°C $\leq t \leq$ 130°C only the first value is physically acceptable. (i) Therefore the measured temperature is 86.39°C.

(ii) To estimate the sensitivity, we carry out a first-order Taylor expansion of the calibration equation about the given value to obtain

$$\delta E = (dE/dt)\delta t \qquad (E1.18.1)$$

$$(dE/dt) = a_1 + 2a_2 t$$

$$(dE/dt) = 0.22 \times 10^{-3} - 2 \times 5.1 \times 10^{-7} \times 100 = 1.18 \times 10^{-4}$$

The sensitivity at 100°C is 0.118 mV per °C.
(iii) The desired accuracy of the temperature is 0.2°C. We use Eq. (E1.18.1) to estimate the required accuracy of the voltmeter as, $\delta E = 0.2 \times 0.118 = 0.0236$ mV.

Example 1.19 A rigid vessel of volume 0.25 m³ contains a mixture of steam and water. The steam occupies 0.95 of the total volume. The densities of steam and water are 1.129 kg m⁻³ and 943 kg m⁻³ respectively. Calculate (i) the mass fractions of steam and water in the vessel, (ii) the mean density of the contents of the vessel, and (iii) the mean specific volume of the contents.

Solution Consider a volume of 0.25 m³ of mixture.
The mass of steam is given by

$$m_s = 0.25 \times 0.95 \times 1.129 = 0.2681 \text{ kg}$$

The mass of liquid water is

$$m_w = 0.25 \times 0.05 \times 943 = 11.7875 \text{ kg}$$

Therefore the mass fraction of steam is

$$x = m_s /(m_s + m_w) = 0.2681/(0.2681 + 11.7875) = 0.0222$$

The mass fraction of liquid water is

$$y = m_w /(m_s + m_w) = 11.7875/(0.2681 + 11.7875) = 0.9778$$

As expected, $\qquad\qquad x + y = 1$
(i) The mean density is given by

$$\rho_{mean} = (m_s + m_w)/V_{total} = (11.7875 + 0.2681)/0.25 = 48.22$$

The mean density is 48.22 kg m⁻³.

(ii) The mean specific volume is given by

$$v_{mean} = 1/\rho_{mean} = 1/48.22 = 0.0274$$

The mean specific volume is $0.0274 \ m^3 \ kg^{-1}$

Example 1.20 A block of ice is placed in a cylindrical tank of water. The diameter of the tank is 0.3 m. The density of ice and water, assumed constant, are 917 kg m^{-3} and 1000 kg m^{-3} respectively. When the ice melts completely the water level in the tank rises by 6.7 mm from the original level. Calculate the initial mass and volume of the block of ice.

Solution The total volume of water added to the tank by the melting ice is

$$V_w = \pi D^2 \delta h/4 = \pi \times 0.3^2 \times 6.7 \times 10^{-3}/4 = 4.7359 \times 10^{-4} \ m^3$$

The mass of water added is

$$M_w = 4.7359 \times 10^{-4} \times 1000 = 0.47359 \ kg$$

The mass of ice is equal to the mass of water added which is 0.47359 kg.
   The volume of ice is given by

$$V_{ice} = M_{ice}/\rho_{ice} = 0.47359/917 = 5.164 \times 10^{-4} m^3$$

Problems

P1.1 Identify the thermodynamic systems that may be useful in the analysis of the following physical situations. Draw the appropriate boundaries and describe the type of system, the nature of the boundary, and the type of interactions that occur between the system and its surroundings.
   (i) A hot block of metal cooling at the bottom of a tank of water.
   (ii) Current flowing through a section of an insulated electrical cable.
   (iii) Water filling a storage tank from its bottom.

P1.2 Which of the characteristics associated with the systems listed below are properties? Which are not? Explain.

(a) The system is the heating element of an electric iron. The characteristics are: (i) the mass, (ii) the electrical resistance, (iii) the temperature, (iv) the number of hours of operation and (v) the total watt-hours of electrical energy consumed.

(b) The system is the battery of an automobile. The characteristics are: (i) the mass, (ii) the voltage, (iii) the temperature, (iv) the internal resistance, (v) the number of hours of operation and (vi) the total watt-hours of energy delivered.

P1.3 A mass of gas at a temperature of 40°C and pressure of 2 bar is trapped in a cylinder behind a leak-proof piston. The temperature of the gas is to be increased to 65°C while its pressure is to be decreased to 1.25 bar. Describe how this change of state could be brought about in a quasi-equilibrium manner. You may make use of any external devices required and also make any assumptions about the walls of the cylinder.

P1.4 A gas is contained in a cylinder behind a frictionless piston of diameter 0.1 m and mass 25 kg. When an additional mass M is placed on the piston the gage pressure of the gas becomes 2.0 bar. The local barometric pressure is 775 mm of mercury. (a) Calculate (i) the mass of M and (ii) the absolute pressure of the gas in the cylinder. (b) The piston is held in this position with the aid of a lock on the outside while heat is supplied to the gas until its absolute pressure becomes 4 bar. Calculate the force on the lock in the final equilibrium state. [*Answers*: (a) (i) 135 kg, (ii) 3.034 bar, (b) 758 N]

P1.5 An artificial solar pond is a body of water with a density gradient created by adding salt to the water. Obtain an expression for the variation of pressure P with depth z assuming the density distribution to have the linear form, $\rho(z) = \rho_o + \alpha z$, where ρ_o is the density at the surface and α is a constant. The ambient pressure is P_o. [*Answer*: $P(z) = P_o + gz(\rho_o + \alpha z / 2)$]

P1.6 A U-tube manometer A is attached to two vertical tubes B and C as shown in Fig. P1.6. The cross-sectional areas of the tubes A, B and C are

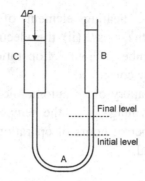

Fig. P1.6 Manometer

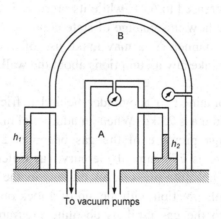

Fig. P1.7 Cylindrical chambers

80 mm^2, 600 mm^2 and 900 mm^2 respectively. The specific gravities of the liquids in B and C are 0.75 and 0.85 respectively. Initially, the interface between the two liquids is on the side of the U-tube connected to B. Due to an increase in the pressure applied to C the interface between liquids rises by 50 mm in the U-tube section A. Determine the increase in pressure in Pa. [*Answer*: 135 Pa]

P1.7 Figure P1.7 shows a cylindrical vessel A of inner diameter 300 mm and wall thickness 5 mm placed inside a cylindrical vessel B of inner diameter 750 mm and wall thickness 8 mm. The vessels are connected to

vacuum pumps as shown in the figure and are fitted with two Bourdon–
tube pressure gages and two mercury manometers.

The local barometric pressure is 750 mm of mercury. The readings of
the manometers are $h_1 = 280$ mm and $h_2 = 102$ mm. Calculate (i) the
absolute pressure in vessels A and B, (ii) the readings of the two
Bourdon tube gages in A and B, and (iii) the net vertical force on the
vessels A and B due to the pressure acting on their surfaces. [*Answers*:
(i) 49.1 kPa, 62.7 kPa, (ii) 13.6 kPa, 37.4 kPa, (iii) 1.26 kN, 18.4 kN]

P1.8 The air pressure at the foot of a mountain is 101 kPa while at the
top of the mountain the pressure and temperature are 75.7 kPa and 267 K
respectively. The temperature of the atmosphere decreases linearly with
altitude at the rate of 0.0065 $K.m^{-1}$. Calculate the height of the mountain.
[*Answer*: 2317 m]

P1.9 A rubber balloon when fully inflated has a spherical shape with a
radius of 0.3 m. The absolute pressure inside the balloon is 103 kN m^{-2}
and the local barometric ambient pressure is 750 mm of mercury. By
considering the equilibrium of a hemispherical portion of the balloon,
calculate the tension in the rubber membrane per unit circumferential
length. Neglect the effect of the thickness of the balloon. [*Answer*:
0.44 kN m^{-1}]

P.10 A new temperature scale called the *N*-scale is to be defined using
the ice point, at 0°C as 50°N and the steam point at 100°C as 500°N.
(i) Obtain the relationship between the N-scale and the Celsius scale of
temperature. (ii) If the M-scale is the absolute temperature corresponding
to the N-scale, calculate the absolute temperatures in M at the ice and
steam points. [*Answer*: (i) °N = 4.5 × °C + 50, (ii) M = 1179.18 + °N,
1229.2°M, 1679.2°M]

References

1. Cravalho, E.G. and J.L. Smith, Jr., *Engineering Thermodynamics*, Pitman, Boston,
   MA, 1981.

2. Jones, J.B. and G.A. Hawkins, *Engineering Thermodynamics*, John Wiley & Sons, Inc., New York, 1986.
3. Massey, B.S., *Mechanics of Fluids*, 5th edition, Van Nostrand Reinhold, U.K., 1983.
4. Streeter, V.L. and E.B. Wylie, *Fluid Mechanics*, 7th edition, McGraw-Hill, Inc., New York, 1978.
5. Zemansky, M., *Heat and Thermodynamics*, 5th edition, McGraw-Hill, Inc., New York, 1968.

Chapter 2

Properties of a Pure Substance

In many engineering systems, energy transfers are carried out using a working fluid. In the case of thermal power plants and refrigeration systems the working fluid undergoes changes in phase while it moves from one component to the next. Gas turbine plants on the other hand use working fluids that remain in the gaseous phase during its passage through the system.

In this chapter, we shall review the phase-change processes for a substance with a constant chemical composition, which is usually called a *pure substance*. Property data for a wide variety of pure substances are available in the form of tables and charts. For some working fluids, especially those that do not undergo phase change, it is possible to develop mathematical models to represent the property data analytically (see chapter 14).

2.1 The Pure Substance

A *pure substance* is homogeneous and has the same chemical composition throughout its volume. It may, however, be a mixture of different phases. For example, a mixture of steam and water where the vapor phase is in equilibrium with the liquid phase is a pure substance. A mixture of ice and liquid water is a pure substance in which the solid phase is in equilibrium with the liquid phase. In the absence of phase changes, mixtures of gases like air exhibit some of the characteristics of a pure substance. Some processes involving pure substances are discussed in worked example 2.1.

43

2.2 Phase Equilibrium in a Pure Substance

In this section we describe the important physical processes that occur when a pure substance undergoes phase change from a liquid to a vapor. These processes are best illustrated by referring to the following idealized experiment. Figure 2.1 depicts a fixed mass of liquid water in a cylinder behind a leak-proof piston on which weights are placed to subject the water to a desired pressure. If the weights are held fixed, mechanical equilibrium of the piston will ensure that the pressure of the water is unchanged, provided all processes are carried out in a quasi-equilibrium manner.

Heat is supplied to the water and changes in its temperature and specific volume are observed. In Fig. 2.2 we plot the variation of the temperature (T) with the specific volume (v) for different values of the constant pressure (P). During the first continuous process, from A to B, the volume of the liquid increases gently with temperature. At point B, the first vapor bubble appears in the liquid and the water is then said to be in a *saturated liquid state*. The temperature at B is the *saturation temperature* (T_s) or the *boiling point* corresponding to the prevailing pressure (P_s). The specific volume of liquid water is called the *saturated liquid specific volume* (v_f), the inverse of which is the *saturated liquid density*. Between states A and B, the liquid is in a *sub-cooled* or a *compressed-liquid* state because its temperature is below the saturation temperature corresponding to the pressure.

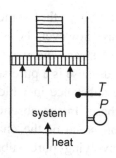
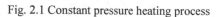

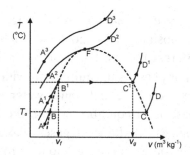

Fig. 2.1 Constant pressure heating process   Fig. 2.2 *T-v* diagram

As more heat is added, the vaporization process continues with a corresponding increase in the volume. The temperature, however, remains constant, as seen in the section BC of the curve, and the mixture of vapor and liquid that exist in equilibrium is called a *wet vapor*. The mass fraction of vapor in the mixture is defined as the *vapor quality* (x). At point C, where the last drop of liquid has just evaporated the vapor is in a *dry saturated* state. The specific volume of the vapor at C is the *saturated vapor specific volume* (v_g). We note that during the vaporization process, the saturation temperature is a function of the pressure and therefore remains constant like the latter. The pressure and temperature can no longer define the state of the system, as they are *not independent*. The quality of the vapor, however, is an independent property, and it forms the second of the two independent properties of the system.

If we now continue the heating process, both the volume and temperature of the vapor increase steadily as seen in the section CD of the graph. The vapor in this section is called a *superheated vapor* because it is at a temperature above the saturation temperature (boiling point) corresponding to the prevailing pressure. The specific volume is the *superheated vapor specific volume.*

If the above experiment is repeated with a larger weight on the piston and therefore a higher pressure of the water we would generate a graph $A^1B^1C^1D^1$, similar in all respects to ABCD, except, with a vaporization or *two-phase* section B^1C^1 that is somewhat shorter than BC. Further increase in pressure makes the vaporization section still shorter, until at a certain pressure known as the *critical pressure*, the usually discontinuous vaporization section becomes a point of inflection F with zero slope, as shown by the graph A^2FD^2 in Fig. 2.2. The temperature and specific volume at F, the *critical point,* are called the *critical temperature* and the *critical specific volume* respectively. The smooth curve drawn through the points B, B^1, F, C^1, C, enveloping the vaporization region, is known as the *saturation curve.*

If we repeat the experiment with a pressure greater than the critical pressure, the state of the water changes continuously without a sharp discontinuous section separating the sub-cooled liquid section from the superheated vapor section as the graph A^3D^3 shows. Water under these conditions is called a *supercritical fluid* which is neither a liquid nor a vapor as we usually define them.

Although we have used water as the pure substance in our discussions, the processes described and various terms defined would apply equally well to any pure substance such as nitrogen or oxygen or mercury. We tend to think of oxygen and nitrogen as gases because it is not often that we encounter nitrogen or oxygen in their liquid states, whereas with water we are familiar with many situations where it undergoes phase changes from solid to liquid to vapor. Similarly, we do not ordinarily think of mercury as a vapor.

2.3   Phase Diagrams

From the data generated in the experiment described in Sec. 2.2, we could now plot two other charts that would further clarify phase change processes. These are: (i) a graph of pressure (P) versus specific volume (v) at constant temperature (T) and (ii) a graph of temperature (T) against pressure (P) at constant specific volume (v).

The former chart (P-v) which is widely used in engineering analysis, is shown in Fig. 2.3. Consider the graph JKLM. At state J, the water is in a *compressed liquid* state because its pressure is larger than the saturation pressure corresponding to the prevailing temperature. As the pressure is decreased at constant temperature the volume of the liquid increases slowly until the state K is reached. At K the first vapor bubble appears in the liquid and vaporization commences. During the vaporization process KL, the pressure remains constant while the volume increases due to vapor generation. At point L the last drop of water has evaporated and the vapor is in a dry saturated state. When the pressure is decreased further, the volume increases rapidly with the vapor attaining a superheated state. The curve JKLM is known as an *isotherm* because the temperature is constant during the process.

If the process is repeated at a higher temperature we would generate a graph $J^1K^1L^1M^1$ where the vaporization section K^1L^1 is shorter than KL. The curve J^2FM^2 is the *critical isotherm* that has a point of inflection

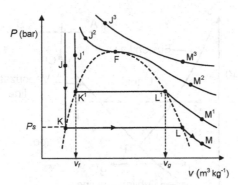

Fig. 2.3 *P-v* diagram

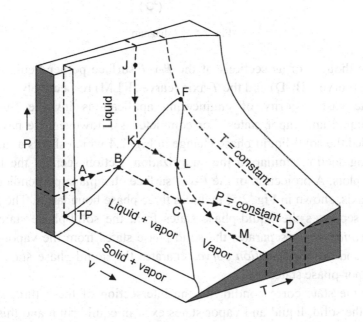

Fig. 2.4 *P-v-T* surface

with zero slope at the critical point F. For temperatures higher than the critical temperature, we obtain a continuous curve like J^3M^3 representing supercritical states of the fluid.

In Fig. 2.4 we show a three dimensional representation of the *P-v-T* data, which results in a surface. The *T-v* and *P-v* charts discussed above

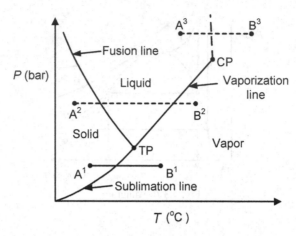

Fig. 2.5 *P-T* diagram

can be thought of as sections of the *P-v-T* surface perpendicular to the *P*-axis (curve ABCD) and the *T*-axis (curve JKLM) respectively.

The vast majority of engineering applications involve fluids in their liquid and vapor states. For completeness, however, we have also included the solid-liquid phase change in Fig. 2.4 which shows a melting/freezing section similar to the vaporization section seen in the liquid-vapor plots. A projection of the *P-v-T* surface on a plane perpendicular to the *v*-axis, shown in Fig. 2.5, includes three phase boundaries. The *fusion curve* separates the liquid-phase states from the solid-phase states, the *sublimation curve* separates the solid-phase states from the vapor-phase states and the *vaporization curve* separates the liquid-phase states from the vapor-phase states.

At the state corresponding to the intersection of these three curves (TP), the solid, liquid and vapor states exist in equilibrium and this state is known as the *triple point.* The pressure and temperature of water at the triple-point are 0.6112 kPa and 0.01°C respectively.

Consider the constant-pressure phase change processes represented by the three lines parallel to the *T*-axis in Fig. 2.5. For the process A^1B^1 where the pressure is below the triple point pressure, the substance changes directly to a vapor from a solid. When the pressure is larger than the triple point pressure but below the critical pressure, (A^2B^2) the phase change occurs from a solid to a liquid and then to a vapor as the temperature is increased. For pressures greater than the critical pressure, as for process A^3B^3, the change in properties is continuous without a distinction between liquid and vapor.

2.4 Independent Properties

We now focus our attention on a special type of pure substance called a *simple compressible substance*, which is a useful model for engineering analysis of real systems. When kinetic, gravity, surface, magnetic and electrical effects are absent the state of a pure substance can be defined by two *independent properties*. Such a substance is called a *simple compressible pure substance*. The *P-v-T* data in whichever form it is represented is termed the *equation of state* of the pure substance.

From our discussion thus far on water we note that for sub-cooled liquids, any two properties of the set *P-v-T* would be sufficient to define the state. In the case of a wet vapor the independent properties include the quality (*x*) and the saturation pressure or temperature. For superheated vapors, any two of three properties *P*, *v* and *T* would be adequate. This independent property requirement for simple compressible pure substances is sometimes called the *two-property rule*.

2.5 Tables and Charts of Property Data: Equation of State

Engineering analysis and computations require property data for a large number of working fluids. Typically these fluids include water, refrigerants and numerous gases. Although analytical forms for *P-v-T* variations would be very convenient to use, it is often not possible to develop accurate analytical expressions for all situations. Therefore we tend to rely on tabulated data and charts similar to those shown in Figs. 2.2 and 2.3. Extracts of two typical property tables for saturated water vapor and superheated water vapor are given in Tables 2.1 and 2.2. respectively.

Table 2.1 Properties of saturated water

T(°C)	P_s (kPa)	$v_f \times 10^3$ (m³ kg⁻¹)	$v_g \times 10^3$ (m³ kg⁻¹)
0.01	0.6112	1.002	206.1
5	0.8719	1.001	147.1
10	1.227	1.003	106.4

Table 2.2 Properties of superheated water vapor

P(kPa) [T_s(°C)]	T(°C)	100	150	200
50 [81.3]	v(m³ kg⁻¹)	3.42	3.89	4.356
75 [91.8]	v(m³ kg⁻¹)	2.271	2.588	2.901
100 [99.6]	v(m³ kg⁻¹)	1.696	1.937	2.173

In Table 2.1 the first column is the temperature while the second column gives the saturation pressure corresponding to that temperature. Columns three and four give the specific volume of saturated liquid, v_f and saturated vapor v_g respectively (also see Figs. 2.2 and 2.3). The mean specific volume, v_{sp} of a mixture of saturated liquid of mass, m_l and saturated vapor of mass, m_g is defined by:

$$(m_l + m_g)v_{sp} = m_g v_g + m_l v_f$$

$$v_{sp} = \frac{m_g v_g}{(m_g + m_l)} + \frac{m_l v_f}{(m_g + m_l)} = xv_g + (1-x)v_f$$

where the *quality*, x is given by

$$x = m_g /(m_l + m_g)$$

The first column of Table 2.2 lists the pressure and the saturation temperature for that pressure within square brackets. In the first row the superheated temperatures of interest are given. As can be seen these temperatures are larger than the saturation temperature for that pressure. The specific volume of the vapor is tabulated in the rest of the table. We will be using tabulated data similar to the above in the worked examples that are to follow in this chapter.

In the above tables we have only given the specific volume as the dependent property. However, when we discuss the first law of thermodynamics in chapter 4, we will introduce two new properties called internal energy and enthalpy. Following the discussion of the second law of thermodynamics in chapters 6 and 7, we add entropy to our list of dependent properties. The above tables will have additional columns to accommodate these new properties as seen in the book of tables entitled *Thermodynamic and Transport Properties of Fluids* edited by G.F.C. Rogers and Y.R. Mayhew [4]. A series of situations involving phase-change processes are presented in worked examples 2.2 to 2.14.

2.6 Ideal-Gas Equation of State

Under appropriate conditions the behavior of vapors is well approximated by an algebraic equation called the *ideal or perfect gas* equation of state. For a mass m of gas the ideal gas equation of state is

$$PV = mRT \qquad (2.1)$$

where R is the gas constant for the particular gas, T is the temperature in degree Kelvin, P is the pressure and V is the volume.

The molar-form of the ideal gas equation of state is an alternative representation which is sometimes more convenient to apply. We define a *mole* of gas as the amount of gas numerically equal to its molecular weight. For example, a gram-mole of helium would have a mass of 4.003 gm which is equal to the molecular weight of helium expressed in gm. Similarly, one *kmol* of a gas has a mass numerically equal to its molecular mass expressed in kg. The molar-form of the equation of state is obtained by introducing the molecular weight, M in Eq. (2.1).

$$PV = (m/M)(MRT) = n\overline{R}T \qquad (2.2)$$

where the number of moles of the gas is, $n = m/M$ and the universal gas constant is, $\overline{R} = MR = 8.3145 \times 10^3$ J·kmol^{-1}K^{-1}.

When the molecular mass of the gas M, is expressed in kg·kmol^{-1} the units of n are kmols. Typical units for the other variables of the ideal gas equation of state expressed by Eqs. (2.1) and (2.2) are as follows:

$$P = [\text{N m}^{-2}], \quad V = [\text{m}^3], \quad m = [\text{kg}], \quad T = [\text{K}],$$

$$R = [\text{J·kg}^{-1}\text{K}^{-1}] \quad \text{and} \quad \overline{R} = [\text{J·kmol}^{-1}\text{K}^{-1}]$$

The ideal gas model tends to represent the behavior of real gases with better accuracy (i) as pressure decreases, (ii) as temperature increases, and (iii) as molecular mass decreases. Worked examples 2.15 to 2.18 deal with processes involving ideal gases.

2.7 Microscopic View Point

The ideal gas equation of state and the kinetic theory of gases, which treats the gas molecules as tiny spheres, enable us to develop a microscopic interpretation of temperature. Consider a gas contained in a rigid vessel. The pressure of the gas is defined as the normal force on unit area of the wall of the container due to the change of momentum of the gas molecules colliding with the wall. Now we do not intend to include the details of this derivation here; instead, we quote below the expression for the pressure, P in terms of the properties of the molecules of the gas.

$$P = m'n'\overline{C}^2/3 = 2n'\left(m'\overline{C}^2/2\right)/3 \qquad (2.3)$$

where m' is the mass of a molecule, n' is the number of molecules per unit volume, $\overline{C}^2$ is the average of the squared-velocity of the molecules. Substituting for P from Eq. (2.3) in the molar-form of the ideal gas equation given by Eq. (2.2) we have

$$T = PV/n\overline{R} = 2Vn'\left(m'\overline{C}^2/2\right)\left(M/3m\overline{R}\right) \qquad (2.4)$$

Now $n'V$ is the number of gas molecules in the volume V, which when multiplied by m' gives the total mass m of the gas. Simplifying Eq. (2.4) we obtain

$$T = M\overline{C}^2/3\overline{R} \qquad (2.5)$$

This relationship shows that the absolute temperature of a gas is a measure of the average-square-speed of the constituent molecules.

A numerical application of the above equation is given in worked example 2.19.

2.8 Gas Laws

There are two gas laws called Boyle's law and Charles' law that we encounter in basic physics courses. Boyle's law states that 'the volume of a gas varies inversely as the pressure provided the mass and temperature of the gas are maintained constant'. It is easy to show that Boyle's law is a special case of the ideal gas equation. Now from the ideal gas equation, $PV = mRT$. For a constant mass of gas at a constant temperature the right hand side of the above equation is a constant. Hence it follows that, $P \propto 1/V$, which is Boyle's law.

Charles' law states that 'when the pressure and mass of a gas are maintained constant, the volume varies directly as the temperature'. Rearranging the ideal gas equation of state, $V/T = mR/P$. For the conditions stated in Charles' law, the right hand side of the above equation is constant. Therefore $V \propto T$.

2.9 Van der Waals Equation of State

Van der Waals introduced two modifications to the ideal gas equation with a view to accounting for (i) the finite size of the molecules by replacing the molar specific volume v by $(v\text{-}b)$ where b is an approximate representation of the volume of the molecules, and (ii) the attractive forces between the molecules by subtracting from the ideal gas pressure a quantity, a/v^2. If the actual pressure is P then the equivalent ideal gas pressure becomes, $(P + a/v^2)$.

The Van der Waals equation of state for a kmole of gas can be written in the form

$$(P + a/v^2)(v - b) = \overline{R}T .$$ (2.6)

This equation of state is particularly useful in predicting the properties at the critical state. Two applications of the Van der Waals equation of state are presented in worked examples 2.20 and 2.21.

2.10 Worked Examples

In the following worked examples the required property data for water and steam are obtained from the book of tables entitled *Thermodynamic and Transport Properties of Fluids* edited by G.F.C. Rogers and Y.R. Mayhew [4].

Example 2.1 Listed below are systems composed of different substances and phases. For the processes given, which of the systems are composed of pure substances?
  (i) A vessel contains a mixture of liquid water and steam. A cooling process produces more liquid water.
 (ii) A vessel has two compartments, one containing nitrogen, the other helium separated by a partition. The separating partition is removed and the gases mix.
(iii) A vessel contains moist air, which is a mixture of air and water vapor. The vessel is cooled and some water vapor condenses to water.
(iv) A cube of ice floats in a tank of liquid water. Heat absorption results in the melting of the ice to liquid water.

Solution (i) The chemical composition of the system composed of liquid water and steam is unchanged during the phase change process. Therefore system consists of a pure substance.
(ii) The system consists of the two gases, nitrogen and helium which are initially separated by the partition. Consequently, the chemical composition of the system is not uniform and it does not constitute a pure substance initially. However, after the partition is removed, and the gases mix completely to attain an equilibrium state, the composition can be assumed uniform for all practical purposes. Then the system can be treated as a pure substance because it exhibits some of the characteristics of a pure substance.
(iii) Initially the moist air can be assumed well mixed with a uniform chemical composition which makes the contents a pure substance. When condensation occurs due to cooling, only the water vapor will be converted to liquid water. The chemical composition of the moist air in

the vessel is then different from the pure condensed water. Therefore the system does not contain a pure substance in the final state.

(iv) The ice and the water constitute our system. Melting of the ice is a phase change process which maintains the chemical composition of the system unchanged. The system therefore constitutes a pure substance.

Example 2.2 A tank of volume 0.25 m³ contains 98 percent saturated water vapor by volume and 2 percent saturated liquid water at 30°C. The liquid and vapor are agitated and well mixed. Calculate (i) the mass fractions of the vapor and liquid, (ii) the quality of the mixture, (iii) the mean specific volume of the mixture, and (iv) the mean density of the mixture.

Solution Since the total volume, V_t = 0.25 m³, the vapor and liquid volumes are given by

$$V_g = 0.98 \times 0.25 = 0.245 \, \text{m}^3$$

and

$$V_l = 0.2 \times 0.25 = 0.005 \, \text{m}^3.$$

The saturated vapor and liquid specific volumes at 30°C are

$$v_g = 32.93 \, \text{m}^3 \, \text{kg}^{-1} \quad \text{and} \quad v_f = 0.10044 \times 10^{-2} \, \text{m}^3 \, \text{kg}^{-1}$$

respectively from the steam tables in [4].

The mass of vapor and liquid are given by

$$m_g = V_g / v_g = 0.245 / 32.93 = 7.44 \times 10^{-3} \, \text{kg}$$

$$m_l = V_l / v_f = 0.005 / 0.10044 \times 10^{-2} = 4.9781 \, \text{kg}$$

(i) The vapor mass fraction is

$$m_g / (m_g + m_l) = 1.4923 \times 10^{-3}$$

The liquid mass fraction is

$$m_l / (m_g + m_l) = 0.9985$$

(ii) The vapor quality is equal to the vapor mass fraction.

Table E2.3 Conditions of water

State	Pressure bar	Temperature °C	Quality x	Specific volume (m³ kg⁻¹)
1	26.0	226.0	?	0.05
2	?	300.0	0.8	?
3	15.0	325.0	?	?
4	?	275.0	?	0.20
5	128.0	?	0.6	?

Hence $x = 1.4923 \times 10^{-3}$

(iii) The mean specific volume of the mixture is given by

$$v_{sp} = xv_g + (1-x)v_f$$

Substituting the numerical values we have

$$v_{sp} = 1.4923 \times 10^{-3} \times 32.93 + (1 - 1.4923 \times 10^{-3}) \times 0.10044 \times 10^{-2}$$

$$v_{sp} = 5.014 \times 10^{-2} \text{ m}^3 \text{ kg}^{-1}$$

(iv) The mean density is given by

$$\rho_m = 1/v_{sp} = 1/5.014 \times 10^{-2} = 19.94 \text{ kg m}^{-3}$$

Example 2.3 The Table E2.3 above gives the properties of water under five different sets of conditions. Use the table of thermodynamic properties [4] to obtain the missing properties for each state.

Solution (i) State 1: At pressure, $P = 26$ bar the saturation temperature, $T_s = 226°C$ and the vapor specific volume, $v_g = 0.07689$ m³ kg⁻¹ [4]. Since the given specific volume of 0.05 m³ kg⁻¹ is less than v_g, the vapor is wet. At $P = 26$ bar, the liquid specific volume, $v_f = 0.1198 \times 10^{-2}$ m³ kg⁻¹ [4]. Note that this value is obtained by linear interpolation. We illustrate the procedure in the table below by extracting the relevant values from the data table in [4].

P (bar)	23.2	26	27.98
v_f (m³ kg⁻¹)	0.119×10^{-2}	?	0.1209×10^{-2}

If the quality is x, then

$$v_{sp} = xv_g + (1-x)v_f$$

Substituting the numerical values,

$$0.05 = 0.07689 \times x + 0.1198 \times 10^{-2} \times (1-x)$$

Hence $\qquad\qquad\qquad x = 0.6447$

(ii) State 2: Since the quality is given as $x = 0.8$, the steam is a wet vapor. The saturation pressure and vapor specific volume at $T_s = 300°C$ are obtained using data from [4] given below:

T_s (°C)	299.2	300	303.3
P (bar)	85	?	90
v_g (m³ kg⁻¹)	0.02192	?	0.02048

Linear interpolation gives $P = 85.98$ bar. From the data in [4]

$$v_g = 0.02164 \ \text{m}^3 \, \text{kg}^{-1} \text{ and, } v_f = 0.1404 \times 10^{-2} \, \text{m}^3 \, \text{kg}^{-1}.$$

The specific volume is

$$v_{sp} = 0.8 \times 0.02164 + 0.2 \times 0.1404 \times 10^{-2} = 0.01759 \ \text{m}^3 \, \text{kg}^{-1}$$

(iii) State 3: At a pressure of 15 bar the saturation temperature is $T_s = 198.8°C$ [4]. Since the given temperature of 325°C is greater than the saturation temperature of 198.8°C, the steam is superheated.
Superheated steam data from [4] at $P = 15$ bar are used to obtain v by linear interpolation

T (°C)	300	325	350
v (m³ kg⁻¹)	0.1697	?	0.1865

Hence the specific volume is

$$v = 0.1781 \ \text{m}^3 \, \text{kg}^{-1}$$

The quality, x is defined only for wet steam and therefore has no meaning for superheated steam.

(iv) State 4: We need to first determine whether the steam is superheated or wet by finding v_g at 275°C which is 0.032 m³ kg⁻¹ [4]. Since the given value of $v = 0.2$ m^3 kg^{-1} is greater than v_g, the steam is superheated. The procedure to find P is outlined below. Since data for 275°C is not tabulated, we first generate this column by linear interpolation. The values of pressure chosen are those for which the v-values are in the neighborhood of 0.2 m^3 kg^{-1} as shown below

T (°C)	250	275	300
P (bar)			
10	0.2328	?	0.2580
15	0.1520	?	0.1697

The v-values at $T = 275$ °C are given below:

v (m^3 kg^{-1})	0.1609	0.2	0.2454
P (bar)	15	?	10

A second linear interpolation gives the pressure, P at $v = 0.2$ m^3 kg^{-1} as 12.686 bar.

It should be noted that this somewhat tedious double-interpolation become necessary due mainly to the limitations of the intervals of data tabulation. With closely tabulated data we should be able to perform the computations with less effort.

(v) State 5: Since the quality $x = 0.6$, the steam is a wet vapor. The data from the saturated steam table [4] are used to obtain the temperature and specific volume.

P (bar)	125	128	130
T_s (°C)	327.8	?	330.8
v_g (m^3 kg^{-1})	0.01349	?	0.01278

At $P = 128$ bar, linear interpolation gives:

$$T_s = 329.6°C \quad \text{and} \quad v_g = 0.013064 \text{ m}^3 \text{ kg}^{-1}$$

From the data in [4], $\quad v_f = 0.1582 \times 10^{-2}$ m^3 kg^{-1}

Hence the required specific volume is

$$v_{sp} = 0.6 \times 0.013064 + 0.4 \times 0.1582 \times 10^{-2} = 0.00847 \text{ m}^3 \text{ kg}^{-1}$$

Example 2.4 A cylinder contains 0.2 kg of water at a pressure of 0.75 bar and temperature 50°C behind a leak-proof piston. Heat is added to the water while maintaining a constant pressure (*isobaric process*). Find (i) the initial volume of water, (ii) the temperature at which boiling commences, (iii) the quality of the contents when the volume is 0.35 m³, (iv) the volume when the steam in the cylinder is a dry saturated vapor, and (v) the volume of the steam when the temperature is 135°C. Sketch the heating process on a *T-V* diagram and a *P-V* diagram for water.

Solution (i) The saturation temperature at 0.75 bar is 91.8°C. Therefore at 50°C the water in the cylinder is in a *sub-cooled liquid* state. For sub-cooled water the effect of pressure on specific volume is quite small as seen from the data in [4]. A reasonable approximation is to use the saturated liquid specific volume at the given *temperature* while ignoring the effect of the higher pressure. Therefore the liquid specific volume at 50°C is

$$v_{l1} = 0.1012 \times 10^{-2} \text{ m}^3 \text{ kg}^{-1}.$$

(ii) Boiling commences at the saturation temperature corresponding to $P = 0.75$ bar, which is 91.8°C.

(iii) When the volume, $V_2 = 0.35$ m³, the specific volume is

$$v_2 = V_2/m = 0.35/0.2 = 1.75 \text{ m}^3 \text{ kg}^{-1}$$

At $P = 0.75$ bar, the saturated specific volumes are

$$v_{g2} = 2.217 \text{ m}^3 \text{ kg}^{-1} \text{ and } v_{f2} = 0.10374 \times 10^{-2} \text{ m}^3 \text{ kg}^{-1}$$

The specific volume is $v_2 = x_2 v_{g2} + (1 - x_2) v_{f2}$

Substituting numerical values,

$$1.75 = x_2 \times 2.217 + (1 - x_2) \times 0.10374 \times 10^{-2}$$

Hence $x_2 = 0.789$

(iv) The volume when the steam is dry saturated is

$$V_3 = m v_{g2} = 0.2 \times 2.217 = 0.4434 \text{ m}^3$$

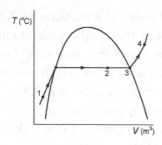

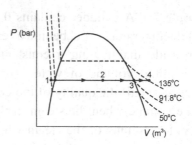

Fig. E2.4(a) *T-V* diagram   Fig. E2.4(b) *P-V* diagram

(v) When $T_4 = 135°C$ and $P_4 = 0.75$ bar, the steam is superheated because the given temperature is higher than the saturation temperature of 91.8°C. Therefore we use data from [4] for $P = 0.75$ bar to obtain the specific volume at $T_4 = 135°C$. Linear interpolation is carried out between 100°C and 150°C. This gives

$$v_4 = 2.4929 \text{ m}^3 \text{kg}^{-1}$$

Hence $V_4 = mv_4 = 0.2 \times 2.4929 = 0.4985 \text{ m}^3$

The *T-V* and *P-V* diagrams are shown in Figs. E2.4(a) and (b) where the numbers correspond to the various values calculated above.

Example 2.5 A constant mass of superheated steam at a pressure of 200 kPa, temperature of 165°C, and volume 0.04 m³ undergoes a cooling process at constant pressure (*isobaric process*). Find (i) the mass of steam, (ii) the temperature at which condensation of water begins, (iii) the quality of the steam when the volume is 0.026 m³, and (iv) the volume when the temperature is 90°C. Sketch the process on a *T-V* diagram and a *P-V* diagram for water.

Solution (i) At the initial state 1, when $T_1 = 165°C$ and $P = 200$ kPa (2 bar) the steam is superheated because the temperature is higher than the saturation temperature of 120.2°C, as seen from the data in [4].
We use the superheated steam data in [4] at $P = 2$ bar, and interpolate between 150°C and 200°C to obtain the specific volume as

$$v_1 = 0.9964 \text{ m}^3 \text{ kg}^{-1}$$

The mass of steam is

$$m = V_1/v_1 = 0.04/0.9964 = 0.04014 \text{ kg}.$$

(ii) Condensation of steam begins at the saturation temperature corresponding to 2 bar which is 120.2°C.

(iii) The quality of steam at $V_3 = 0.026$ m^3 is obtained from the equation

$$V_3/m = v_3 = x_3 v_g + (1 - x_3) v_f$$

At a pressure of 2 bar,

$$v_g = 0.8856 \text{ m}^3 \text{ kg}^{-1} \text{ and } v_f = 0.1060 \times 10^{-2} \text{ m}^3 \text{ kg}^{-1}$$

Substituting these values we have

$$0.026/0.04014 = x_3 \times 0.8856 + (1 - x_3) \times 0.1060 \times 10^{-2}$$

Hence $\qquad\qquad\qquad x_3 = 0.731$

(iv) At $P = 2$ bar and $T_4 = 90$°C, the water is a compressed liquid because its temperature is below the saturation temperature of 120.2°C.

The specific volume of the sub-cooled liquid is well approximated by the specific volume of saturated liquid at 90°C which is 0.1036 × 10^{-2} m^3 kg^{-1}. Therefore

$$V_4 = mv_f = 0.04014 \times 0.1036 \times 10^{-2} = 4.158 \times 10^{-5} \text{ m}^3$$

The *T-V* and *P-V* diagrams are shown in Figs. E2.5(a) and (b) where the numbers correspond to the various values calculated above.

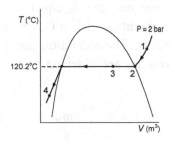

Fig. E2.5(a) *T-V* diagram

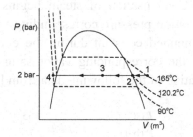

Fig. E2.5(b) *P-V* diagram

Example 2.6 A fixed mass of steam is subjected to a constant temperature (*isothermal process*) compression process. Initially, the pressure, volume and temperature of the steam are 1.7 bar, 170°C and 0.08 m^3 respectively. Find (i) the mass of steam, (ii) the pressure at which condensation of water just begins, (iii) the quality of steam when the volume is 0.01 m^3, and (iv) the volume when the pressure is 10 bar. Sketch the process on the *P-V* and *T-V* diagrams for water.

Solution (i) In the initial state 1 the steam is superheated because its temperature, $T_1 = 170\,°C$ is higher than the saturation temperature of 115.2°C, corresponding to the prevailing pressure of $P_1 = 1.7$ bar. We use the superheated steam data from [4] to perform a double-interpolation to obtain the specific volume of superheated steam at 170°C and 1.7 bar.

The pertinent data are as follows:

T (°C)	150	170	200
P (bar)			
1.5	1.285	? (1.3496)	1.445
1.7		??	
2.0	0.9602	? (1.0085)	1.081

This gives $v_1 = 1.2137 \ \text{m}^3\,\text{kg}^{-1}$.

Therefore the mass of steam is

$$m = V_1/v_1 = 0.08/1.2137 = 0.0659 \ \text{kg}$$

(ii) Condensation of steam begins when the pressure is equal to the saturation pressure corresponding to the temperature of 170°C, which is maintained constant during the entire process. The saturation pressure and the corresponding vapor saturation volume are obtained by inter-polation of the following data from [4]

T (°C)	165	170.0	170.4
P (bar)	7	?	8
v_g (m^3 kg^{-1})	0.2728	?	0.2403

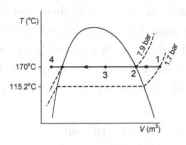

Fig. E2.6(a) *T-V* diagram   Fig. E2.6(b) *P-V* diagram

Hence $P_2 = 7.925$ bar and $v_g = 0.2427$ m^3 kg^{-1}

Also $v_f = 0.1114 \times 10^{-2}$ m^3 kg^{-1}.

Therefore condensation begins at a pressure of 7.925 bar.

(iii) Now when the volume is 0.01 m^3, the specific volume is

$$v_3 = V_3/m = 0.01/0.0659 = x_3 \times 0.2427 + (1 - x_3) \times 0.1114 \times 10^{-2}$$

Hence $x_3 = 0.623$

(iv) When the pressure is 10 bar, the water is in a compressed liquid state because its pressure is larger than that corresponding to the prevailing temperature of 170°C, which is 7.925 bar. We calculate the volume by approximating the specific volume with the saturation liquid specific volume at 170°C. Therefore

$$V_4 = mv_4 = 0.0659 \times 0.1114 \times 10^{-2} = 7.341 \times 10^{-5} \text{ m}^3$$

The *T-V* and *P-V* diagrams are shown in Figs. E2.6(a) and (b), where the numbers correspond to the various values calculated above.

Example 2.7 A quantity of water of mass 0.05 kg undergoes an *isothermal* expansion process from an initial state of 2 bar and 90°C. Find (i) the initial volume of water, (ii) the pressure at which boiling just begins, (iii) the quality of steam when the volume is 0.04 m^3, and (iv) the volume when the pressure has reached 0.1 bar. Sketch the *P-V* and the *T-V* diagrams for the process.

Solution (i) The initial pressure of water of 2 bar is higher than the saturation pressure corresponding to the prevailing temperature of 90°C. Therefore the water is in a compressed liquid state. We ignore the effect of pressure and assume the specific volume to be equal to that of saturated liquid at 90°C. From the data in [4], $v_f = 0.1036 \times 10^{-2}$ m³ kg⁻¹. Therefore the initial liquid volume is

$$V_1 = mv_1 = 0.05 \times 0.1036 \times 10^{-2} = 5.18 \times 10^{-5} \text{ m}^3$$

(ii) Boiling begins at the saturation pressure corresponding to 90°C which is 0.7011 bar [4].

(iii) The specific volume,

$$v_3 = V_3/m = 0.04/0.05 = 0.8 \text{ m}^3 \text{ kg}^{-1}$$

Now at 90°C, the saturated vapor and liquid specific volumes are

$$v_g = 2.361 \text{ m}^3 \text{ kg}^{-1} \text{ and } v_f = 0.1036 \times 10^{-2} \text{ m}^3 \text{ kg}^{-1}$$

$$v_3 = x_3 v_g + (1 - x_3) v_f$$

Substituting numerical values in the above equation

$$0.8 = x_3 \times 2.361 + (1 - x_3) \times 0.1036 \times 10^{-2}$$

Hence $x_3 = 0.3385$

(iv) When the pressure is 0.1 bar the steam is superheated because the temperature of 90°C is larger than the saturation temperature corresponding to 0.1 bar which is 45.8°C. Using the following superheated steam data from [4] we obtain the specific volume, v by interpolation

T (°C)	50	90	100
v (m³ kg⁻¹)	14.87	?	17.2

The required specific volume is 16.734 m³ kg⁻¹. Therefore the volume of steam is,

$$V_4 = mv_4 = 0.05 \times 16.734 = 0.8367 \text{ m}^3$$

The *T-V* and *P-V* diagrams are shown in Figs. E2.7(a) and (b) where the numbers correspond to the various values calculated above.

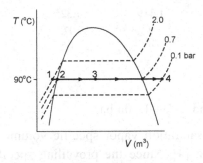

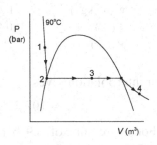

Fig. E2.7(a) *T-V* diagram   Fig. E2.7(b) *P-V* diagram

Example 2.8 A rigid vessel of volume 0.1 m³ contains 0.0758 kg of steam at an initial pressure of 1.5 bar. The steam is subjected to a *constant volume* cooling process. Find (i) the initial temperature of steam, (ii) the temperature and pressure at which condensation of water just begins, and (iii) the quality and temperature when the pressure is 0.9 bar. Sketch the process on a *P-V* and *T-V* diagram for water.

Solution (i) The specific volume is given by

$$v_1 = V/m = 0.1/0.072 = 1.319 \text{ m}^3 \text{ kg}^{-1}.$$

From the data in [4], at a pressure of 1.5 bar, the vapor specific volume is $v_g = 1.159$ m³ kg⁻¹. Since $v_1 > v_g$, the steam is superheated. We extract data from [4] for superheated steam to determine the temperature by using the specific volume at state 1.

v (m³ kg⁻¹)	1.286	1.319	1.445
T (°C)	150	?	200

This gives the required temperature as, $T_1 = 160.3$°C.
(ii) Since mass and volume are constant, the specific volume is also constant. When condensation begins the specific volume v_1 is equal to the saturated vapor specific volume at the prevailing pressure and temperature. We obtain the saturation pressure and temperature at the prevailing specific volume of 1.319 m³ kg⁻¹ using data from [4].

v_g (m³ kg⁻¹)	1.236	1.319	1.325
T (°C)	109.3	?	107.1
P (bar)	1.4	?	1.3

Linear interpolation gives

$$T_{s2} = 107.25°C \quad \text{and} \quad P_2 = 1.306 \text{ bar.}$$

(iii) For a pressure of 0.9 bar, the saturation vapor specific volume is $v_g = 1.869$ m³ kg⁻¹ from the data in [4]. Since the prevailing specific volume is less than the saturated vapor specific volume, the steam is a wet vapor. For pressure, $P_3 = 0.9$ bar, $T_3 = 96.7°C$,

$$v_g = 1.869 \text{ m}^3 \text{ kg}^{-1} \quad \text{and} \quad v_f = 0.1041 \times 10^{-2} \text{ m}^3 \text{ kg}^{-1}.$$

The specific volume is given by

$$v_3 = x_3 v_g + (1 - x_3) v_f$$

Substituting the numerical values we have

$$1.319 = x_3 \times 1.869 + (1 - x_3) \times 0.1041 \times 10^{-2}$$

Hence the quality, $x_3 = 0.705$.

The *T-V* and *P-V* diagrams are shown in Figs. E2.8(a) and (b) where the numbers correspond to the various values calculated above.

Example 2.9 A constant mass of steam of quality 0.4 and pressure 0.5 bar is heated at a *constant volume* of 0.3 m³. Find (i) the mass of steam, (ii) the temperature and pressure at which the steam becomes dry

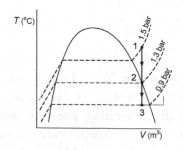

Fig. E2.8(a) *T-V* diagram

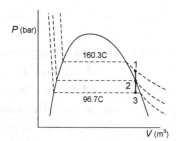

Fig. E2.8(b) *P-V* diagram

saturated, and (iii) the temperature of steam when the pressure is 1.5 bar. Sketch P-V and T-V diagrams for the process.

Solution (i) The vapor and liquid specific volumes at 0.5 bar pressure are obtained from the data in [4] as

$$v_g = 3.239 \text{ m}^3 \text{ kg}^{-1} \text{ and } v_f = 0.1029 \times 10^{-2} \text{ m}^3 \text{ kg}^{-1}$$

The specific volume at state 1 is

$$v_1 = x_1 v_g + (1 - x_1) v_f$$

Substituting numerical values we have

$$v_1 = 0.4 \times 3.239 + (1 - 0.4) \times 0.1029 \times 10^{-2} = 1.2962 \text{ m}^3 \text{ kg}^{-1}$$

The mass of steam is

$$m = V_1 / v_1 = 0.3/1.2962 = 0.2314 \text{ kg}$$

(ii) When the steam becomes dry saturated the specific volume v_1 is equal to the saturated vapor specific volume at the prevailing pressure and temperature.
Therefore

$$v_{g2} = v_1 = 1.2962 \text{ m}^3 \text{ kg}^{-1}$$

Using the data in [4] we obtain the saturation pressure and temperature.

v_g (m^3 kg^{-1})	1.325	1.2962	1.236
P (bar)	1.3	?	1.4
T (°C)	107.1	?	109.3

Interpolation gives

$$T_2 = 107.81°C \text{ and } P_2 = 1.332 \text{ bar}$$

(iii) When the pressure is 1.5 bar the saturated vapor specific volume is 1.159 m^3 kg^{-1}. The prevailing specific volume of steam of 1.2962 m^3 kg^{-1} is higher than the above value and therefore the steam is superheated. We obtain the temperature from the superheated steam data given below [4].

v_g (m^3 kg^{-1})	1.286	1.2962	1.445
T (°C)	150	?	200

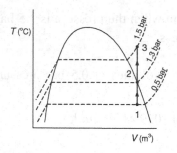 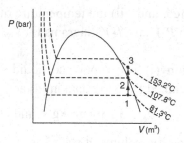

Fig. E2.9(a) T-V diagram   Fig. E2.9(b) P-V diagram

Interpolation gives $T_3 = 153.2\,°C$

The T-V and P-V diagrams are shown in Figs. E2.9(a) and (b) where the numbers correspond to the various values calculated above.

Example 2.10 A rigid vessel contains a mixture of liquid water and steam of total mass of 0.01 kg at a temperature of 80°C. When the mixture is heated at *constant volume* the state passes through the *critical point* of water. Calculate (i) the volume of the vessel, (ii) the initial quality of steam, and (iii) the initial mass of steam and water in the vessel. Sketch the process on the P-V and T-V diagrams for water.

Solution (i) Since the mass of steam is fixed and the vessel is rigid, the specific volume is constant during the heating process. At the critical point, which is the last data entry in the saturated steam table [4], the vapor specific volume is 0.00317 m³ kg⁻¹.
The volume of the vessel is,

$$V_1 = mv_1 = 0.01 \times 0.00317 = 3.17 \times 10^{-5} \ m^3.$$

(ii) At 80°C, the vapor and liquid specific volumes are obtained from the data in [4] as

$$v_g = 3.408 \ m^3 \ kg^{-1} \quad and \quad v_f = 0.1029 \times 10^{-2} \ m^3 \ kg^{-1}$$

The specific volume of steam at state 1 is,

$$v_1 = x_1 v_g + (1 - x_1) v_f$$

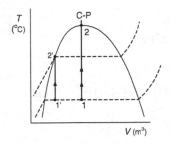

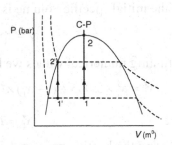

Fig. E2.10(a) *T-V* diagram   Fig. E2.10(b) *P-V* diagram

Substituting the numerical values

$$v_1 = x_1 \times 3.408 + (1 - x_1) \times 0.1029 \times 10^{-2} = 0.00317 \text{ m}^3 \text{ kg}^{-1}$$

Therefore the initial quality is $x_1 = 6.284 \times 10^{-4}$

(iii) The mass of vapor and liquid at state 1 are

$$m_{v1} = mx_1 = 0.01 \times 6.284 \times 10^{-4} = 6.284 \times 10^{-6} \text{ kg}$$

$$m_{l1} = m(1 - x_1) = 0.01 \times (1 - 6.284 \times 10^{-4}) = 9.9937 \times 10^{-3} \text{ kg}$$

The *T-V* and *P-V* diagrams are shown in Figs. 2.10(a) and (b) where the process is indicated by the line 1-2.

Example 2.11 A quantity of wet steam of mass 0.01 kg and temperature 70°C is contained in a rigid vessel of volume 1.4×10^{-5} m³. The steam undergoes a *constant volume* heating process. Calculate (i) the initial quality of the steam, and (ii) the pressure and temperature at which the steam becomes saturated liquid water. Sketch the *P-V* and *T-V* diagrams for the process.

Solution (i) The initial specific volume is,

$$v_1' = V_1'/m = 1.4 \times 10^{-5}/0.01 = 1.4 \times 10^{-3} \text{ m}^3 \text{ kg}^{-1}.$$

At a temperature of 70°C, the saturation vapor and liquid specific volumes are given by

$$v_g = 5.045 \text{ m}^3 \text{ kg}^{-1} \quad \text{and} \quad v_f = 0.1023 \times 10^{-2} \text{ m}^3 \text{ kg}^{-1}$$

Now the initial specific volume is,

$$v_1' = x_1' v_g + (1 - x_1') v_f$$

Substituting numerical values we have

$$v_1' = x_1' \times 5.045 + (1 - x_1') \times 0.1023 \times 10^{-2} = 1.4 \times 10^{-3} \text{ m}^3 \text{ kg}^{-1}$$

Hence $x_1' = 7.474 \times 10^{-5}$

(ii) During the heating process the specific volume remains constant. The steam turns completely to water when the saturated liquid specific volume is 1.4×10^{-3} m^3 kg^{-1}. The temperature and pressure are obtained from the data listed below from [4].

$v_f \times 10^2$ (m^3 kg^{-1})	0.1366	0.14	0.1404
P_2' (bar)	74.45	?	85.92
T_2' (°C)	290	?	300

Hence by linear interpolation, $T_2' = 298.9°C$ and $P_2' = 84.7$ bar.
The *T-V* and *P-V* diagrams are shown in Figs. E2.10(a) and (b) where numbers correspond to the various values calculated above.

Example 2.12 A rigid cylindrical vessel of inner diameter 1.2 m has a mixture of water and steam in equilibrium at a temperature of 200°C. It has a glass window through which the water level is visible. Some water is withdrawn through a valve fitted at the bottom and as a result the water level drops by 0.2 m. Calculate the mass of water withdrawn.

Solution Consider the cylindrical vessel shown in Fig. E2.12. Let the volume of the vessel between the initial and final liquid levels be, $\Delta V = A \delta L$, where A is the area of cross-section of the cylinder and δL is the change in the water level. This volume is initially occupied by saturated liquid. Finally, the same volume is filled with saturated vapor at the same temperature (assuming that there are no liquid drops). The conditions and the phases of water in the rest of the vessel are unaffected by the process.

From the law of conservation of mass it follows that the mass withdrawn from the cylinder is

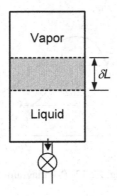

Fig. E2.12 Cylinder with steam and water

$$\Delta m = \Delta V / v_f - \Delta V / v_g$$

At 200°C, the data from [4] gives the saturated specific volumes as

$$v_g = 0.1273 \text{ m}^3 \text{ kg}^{-1} \quad \text{and} \quad v_f = 0.1157 \times 10^{-2} \text{ m}^3 \text{ kg}^{-1}$$

Now $$\Delta V = \pi \times 1.2^2 \times 0.2/4 = 0.226 \text{ m}^3$$

Hence $$\Delta m = 0.226/0.1157 \times 10^{-2} - 0.226/0.1273 = 193.6 \text{ kg}$$

Example 2.13 A quantity of wet steam at a pressure of 2 bar has a mass of 0.12 kg and a quality of 0.9. The steam undergoes a *polytropic process* where the pressure and volume vary according to the relation $PV^{1.2} = C$, where C is a constant. The final pressure of the steam is 1.5 bar. Calculate the final volume, temperature, and quality of the steam. Sketch the process on a *P-V* diagram for water.

Solution At a pressure of 2 bar, the saturation vapor and liquid specific volumes are obtained from [4] as

$$v_g = 0.8856 \text{ m}^3 \text{ kg}^{-1} \quad \text{and} \quad v_f = 0.1060 \times 10^{-2} \text{ m}^3 \text{ kg}^{-1}$$

The specific volume in the initial state 1 is

$$v_1 = x_1 v_g + (1 - x_1)v_f$$

Substituting numerical values

Engineering Thermodynamics

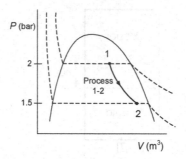

Fig. E2.13 *P-V* diagram

$$v_1 = 0.9 \times 0.8856 + (1 - 0.9) \times 0.106 \times 10^{-2} = 0.7971 \text{ m}^3 \text{ kg}^{-1}$$

The initial volume is, $V_1 = mv_1 = 0.12 \times 0.7971 = 0.0956 \text{ m}^3$.

Applying the polytropic process law we have

$$P_1 V_1^{1.2} = P_2 V_2^{1.2}$$

where P_2 and V_2 are the final pressure and volume respectively.

Substituting numerical values we have

$$2 \times (0.0956)^{1.2} = 1.5 (V_2)^{1.2} .$$

Hence $V_2 = 0.1215 \text{ m}^3$.

The specific volume in the final state is

$$v_2 = V_2 / m = 0.1215 / 0.12 = 1.0125 \text{ m}^3 \text{ kg}^{-1}$$

The saturation vapor specific volume at 1.5 bar is, $v_{g2} = 1.159 \text{ m}^3 \text{ kg}^{-1}$.
Since $v_2 < v_{g2}$, the steam is a wet vapor at state 2.
The saturation temperature at 1.5 bar is 111.4°C.
The specific volume in the final state is

$$v_2 = x_2 \times 1.159 + (1 - x_2) \times 0.1053 \times 10^{-2} = 1.0125 \text{ m}^3 \text{ kg}^{-1}$$

Hence the final quality, $x_2 = 0.8734$.

Example 2.14 Wet steam of mass 1.05 kg is contained in a cylinder behind a piston. Initially the temperature and quality are 110°C and 0.9 respectively. The movement of the piston is resisted by a linear spring with a spring constant of 10 kN m⁻¹. In the initial state the spring just touches the piston but exerts no force on it. The area of cross-section of the piston is 0.5 m². Heat is added to the steam and the piston begins to rise. (i) Obtain a relationship between the volume and pressure of the steam, and (ii) indicate the process on a *P-V* diagram for water.

Solution (i) The spring-loaded piston-cylinder apparatus is shown in Fig. E2.14(a). Since the spring force is initially zero, the equilibrium equation for the piston gives the initial pressure, P_1 as

$$P_1 A = P_{atm} A + Mg$$

$$P_1 = P_{atm} + Mg/A \qquad (E2.14.1)$$

where P_{atm} is the atmospheric pressure acting on the outside of the piston, A is the cross-sectional area of the piston, M is the mass of the piston and g is the acceleration due to gravity.

Assume that the heating process is carried out in a quasi-equilibrium manner. When the piston has moved a distance z the spring force is Kz where K is the spring force per unit compression. The equilibrium equation for the piston gives

$$PA = P_{atm} A + Mg + Kz$$

$$P = P_{atm} + Mg/A + Kz/A = P_1 + Kz/A \qquad (E2.14.2)$$

The change in volume of the steam is related to the piston displacement by the equation

$$V - V_1 = Az \qquad (E2.14.3)$$

Substituting for z from Eq. (E2.14.3) in Eq. (E2.14.2) we have

$$P = P_1 + K(V - V_1)/A^2 \qquad (E2.14.4)$$

Substituting the specific volume, v and the mass of steam, m in the above equation

$$P = P_1 + Km(v - v_1)/A^2 \qquad (E2.14.5)$$

This is the *P-V* relation for the expansion process which has the linear form, $P = a + bv$, where a and b are constants.

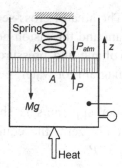

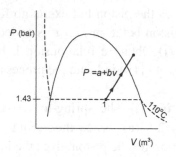

Fig. E2.14 (a) Piston-cylinder system   Fig. E2.14(b) *P-V* diagram

(ii) The initial pressure is the saturation pressure corresponding to 110°C which is 1.433 bar [4] and the specific volume of the wet steam is

$$v_1 = x_1 \times 1.2103 + (1 - x_1) \times 0.1052 \times 10^{-2} = 1.0894 \text{ m}^3 \text{ kg}^{-1}$$

where the vapor and liquid specific volumes have been obtained from [4].

Substituting the numerical values in Eq. (E2.14.5) we have

$$P = 1.433 \times 10^5 + 10 \times 10^3 \times 1.05 \times (v - 1.0894) / (0.5)^2$$

$$P = 0.9754 \times 10^5 + 0.42 \times 10^5 v$$

where the pressure P is in Nm^{-2} and the specific volume v is in m^3 kg^{-1}. The pressure-volume relationship is linear as shown in Fig. E2.14(b).

Example 2.15 An ideal gas of mass 1.02 kg and molecular mass, M = 29 kg·kmol^{-1}, is contained in a cylinder behind a piston. In the initial state, P_1 = 1.2 bar, T_1 = 32°C. The gas is compressed in a quasi-equilibrium process according to the polytropic law, $PV^{1.3}$ = constant, to a final pressure of 2.1 bar. The universal gas constant, $\overline{R}$ = 8.3145 kJ kmol^{-1} K^{-1}. Calculate

(i)   the number of kmoles of gas,
(ii) the volume and density in the initial state, and
(iii) the volume, temperature and density in the final state.

Solution (i) Number of kmoles of gas is,

$$n = m / M = 1.02 / 29 = 0.0352 \text{ kmol}$$

(ii) Applying the molar-form of the ideal gas equation to state 1

$$P_1 V_1 = n \overline{R} T_1$$

Substituting the properties of state 1, we have

$$1.2 \times 10^5 V_1 = 0.0352 \times 8.3145 \times 10^3 \times (32 + 273)$$

The initial volume is, $V_1 = 0.7439 \text{ m}^3$
The initial density is,

$$\rho_1 = m / V_1 = 1.02 / 0.7439 = 1.371 \text{ kg m}^{-3}$$

(iii) Applying the polytropic process law

$$P_1 V_1^{1.3} = P_2 V_2^{1.3}$$

Substituting the properties we have

$$1.2 \times (0.7438)^{1.3} = 2.1 \times V_2^{1.3}$$

Hence $\qquad\qquad V_2 = 0.4836 \text{ m}^3$

The density in the final state is

$$\rho_2 = m / V_2 = 1.02 / 0.4836 = 2.109 \text{ kg m}^{-3}$$

Applying the molar-form of the ideal gas equation to state 2, we obtain

$$P_2 V_2 = n \overline{R} T_2$$

Substituting the properties at state 2 in the above equation

$$2.1 \times 10^5 \times 0.4836 = 0.0352 \times 8.3145 \times 10^3 T_2$$

Hence the final temperature is, $T_2 = 347.2 \text{ K}$ which is 74.2°C.

Example 2.16 One kg of air is trapped in a cylinder behind a friction-less piston. The initial temperature and density of the air are 50°C and 1.08 kg m^{-3} respectively. The air is compressed until the volume is 1/6 of the initial volume and the pressure is 7 bar.
(a) Calculate (i) the initial volume and pressure of the air, and (ii) the final temperature of the air.
(b) If the piston is held fixed and the temperature of the air is reduced to 50°C by cooling, find the final pressure of the air.

Indicate these processes on a P-V diagram.
Assume that the molecular mass of air is 29 kg·kmol^{-1}.

Solution (a) (i) The initial volume of the air is

$$V_1 = m/\rho_1 = 1.0/1.08 = 0.926 \text{ m}^3$$

The gas constant R for air is related to the universal gas constant by

$$R = \bar{R}/M = 8.3145/29 = 0.2867 \text{ kJ kmol}^{-1}\text{K}^{-1}$$

Dividing the mass-form of the ideal gas equation, $P_1V_1 = mRT_1$, by the volume we obtain the density-form of the ideal gas equation. Hence

$$P_1 = mRT_1/V_1 = \rho_1RT_1$$

Substituting numerical values

$$P_1 = 1.08 \times 0.2867 \times 10^3 \times (273 + 50) = 1.0 \times 10^5 \text{ Nm}^{-2} = 1 \text{ bar}$$

For state 2, $V_2 = V_1/6 = 0.926/6 = 0.1543 \text{ m}^3$

Applying the ideal gas equation to state 2

$$P_2V_2 = mRT_2$$

Substituting numerical values in the above equation

$$7 \times 10^5 \times 0.1543 = 1.0 \times 0.2867 \times 10^3 T_2$$

Hence $T_2 = 376.77 \text{ K.}$

(b) Applying the ideal gas equation to state 3

$$P_3V_3 = mRT_3$$

Now for a constant mass of the ideal gas we have

$$mR = P_2V_2/T_2 = P_3V_3/T_3$$

The cooling process from state 2 to 3 is a constant volume process for which $V_2 = V_3$

Hence from the above equation it follows that

$$P_2/T_2 = P_3/T_3$$

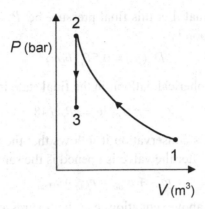

Fig. E2.16 P-V diagram

Substituting numerical values in the above equation we obtain

$$P_3 = 7 \times (273 + 50)/(1.0 \times 376.77) = 6.0 \text{ bar.}$$

The P-V diagram of the two processes is shown in Fig. E2.16.

Example 2.17 A rigid vessel A of volume 0.1 m³ is connected to a spherical elastic balloon B through an initially-closed valve as shown in Fig. E2.17. Initially, both A and B contain air at the ambient temperature of 25°C. The pressures in A and B are 300 kPa and 100 kPa respectively. The initial diameter of the balloon is 0.5 m. The valve is now opened, and remains open until the system attains the final equilibrium state with a uniform temperature of 25°C. It may be assumed that the pressure inside the balloon is directly proportional to its diameter. Determine the final pressure, the final volume and the mass of air in the balloon. Assume that air is an ideal gas.

Solution It is given that the pressure in the balloon is directly proportional to the diameter. Therefore let $D_b(m) = CP_b(bar)$ where C is a constant. We substitute the initial conditions given to find the constant C as 0.5 m per bar.

In the final equilibrium state the air temperatures in A and B are equal to 25°C. Since the valve remains open, the final equilibrium pressures in

A and B are also equal. Let this final pressure be P_f bar. Then the final diameter of the balloon is

$$D_f(m) = 0.5P_f(bar)$$

The volume of the spherical balloon in the final state is

$$V_{bf} = \pi D_f^{\ 3}/6 = \pi P_f^{\ 3}/48$$

From the law of mass conservation it follows that the total mass of air in A and B before and after the valve is opened is the same. Therefore

$$m_{Ab} + m_{Bb} = m_{Aa} + m_{Ba}$$

Each term of the above equation can be expressed in term of the pressure, volume and temperature using the mass-form of the ideal gas equation. Thus we obtain

$$\frac{P_{Ab}V_{Ab}}{RT_{Ab}} + \frac{P_{Bb}V_{Bb}}{RT_{Bb}} = \frac{P_{Aa}V_{Aa}}{RT_{Aa}} + \frac{P_{Ba}V_{Ba}}{RT_{Ba}} \qquad (E2.17.1)$$

The following are the pertinent numerical values and conditions:

$$P_{Ab} = 3 \text{ bar}, \ P_{Bb} = 1 \text{ bar}, \ V_{Ab} = V_{Aa} = 0.1 \text{ m}^3, \ P_{Aa} = P_{Ba} = P_f$$

$$V_{bf} = \pi P_f^{\ 3}/48, \ T_{Ab} = T_{Aa} = T_{Bb} = T_{Ba} = (273+25) = 298\,\text{K}$$

Substituting numerical values in Eq. (E2.17.1) we have

$$3 \times 0.1 + \pi(0.5)^3/6 = P_f \times 0.1 + P_f \times \pi P_f^{\ 3}/48$$

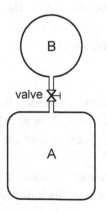

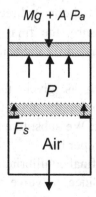

Fig. E2.17 Filling of balloon   Fig. E2.18 Piston-cylinder set-up

The above equation can be simplified to the form:

$$P_f^{\,4} + 1.528 P_f - 5.5837 = 0$$

The solution of the above equation obtained by trial-and-error is

$$P_f = 1.367 \text{ bar}$$

(ii) The final volume of the balloon is

$$V_{Ba} = \pi P_f^{\,3}/48 = \pi (1.367)^3/48 = 0.167 \text{ m}^3$$

(iii) Applying the ideal gas equation we obtain the final mass of air in the balloon as

$$m_{Ba} = P_f V_{Ba}/RT_{Ba}$$

$$m_{Ba} = 1.367 \times 10^5 \times 0.167/(0.2867 \times 10^3 \times 298) = 0.267 \text{ kg.}$$

Example 2.18 Figure E2.18 shows a vertical cylinder of cross sectional area 0.25 m², fitted with a leak-proof piston, containing 0.3 kg of air. Initially, the volume is 0.5 m³ and the temperature 500°C. The air is cooled and the piston descends until it hits two stops on the inside of the cylinder and the volume of air is then 0.25 m³. The cooling process is continued until the temperature of the air becomes 20°C. Calculate
(i) the initial pressure of the air,
(ii) the temperature of the air when the piston hits the stops, and
(iii) the pressure of the air and the force on the stops, when the temperature is 20°C.
Assume that air is an ideal gas with a gas constant of 0.287 kJ kg⁻¹ K⁻¹.

Solution (i) Applying the ideal gas equation to the initial state 1 we obtain the initial pressure as

$$P_1 = mRT_1/V_1 = 0.3 \times 0.287 \times (273 + 500)/0.5 = 1.33 \text{ bar}$$

(ii) We assume that the processes are carried in a quasi-static manner. During the first phase of the cooling process until the piston hits the stops, (state 1 to state 2) the force balance on the piston gives (see Fig. E2.18)

$$Mg + P_a A_p = P_1 A_p \qquad\qquad (E2.18.1)$$

Therefore the pressure of the air inside is constant. Hence

$$P_2 = P_1 = 1.33 \text{ bar.}$$

Applying the ideal gas equation to the two states 1 and 2 we have

$$mR = P_1V_1/T_1 = P_2V_2/T_2$$

Hence $$V_1/T_1 = V_2/T_2$$

Substituting numerical values in the above equation

$$T_2 = 773 \times 0.25/0.5 = 386.5 \text{ K}$$

(iii) After the piston hits the stops, the air undergoes a constant volume $(V_3 = V_2)$ cooling process from state 2 to state 3.

We apply the ideal gas equation to the final state 3 to obtain the pressure

$$P_3 = mRT_3/V_3 = 0.3 \times 0.287 \times 293/0.25 = 1.0091 \text{ bar}$$

As the air is cooled, the pressure inside is lowered and the stops provide the required upward force to balance the constant downward force due to the external pressure on the piston and the weight of the piston.

Force balance on the piston in the final state 3 gives

$$Mg + P_aA_p = F_{stop} + P_3A_p \qquad\qquad \text{(E2.18.2)}$$

Hence from Eqs. (E2.18.1) and (E2.18.2), the force on the stops is

$$F_{stop} = 0.25 \times (1.33 - 1.009) \times 10^5 = 8.025 \text{ kN}$$

Example 2.19 Under a standard atmosphere of 1.01 bar pressure and 273 K temperature, 22.4 m^3 of an ideal gas contains 6.02×10^{23} molecules. For these conditions, find the following quantities for carbon dioxide, CO_2 gas (i) mass of a molecule, (ii) number of molecules per unit volume, and (iii) the root-mean-square speed of the molecules.

Solution Under standard atmospheric conditions of 1.01 bar and 273 K temperature a kmole of ideal gas contains 6.02×10^{23} molecules, which is called *Avogadro's number*.

(i) The mass of a kmol of CO_2 is 44 kg. Therefore the mass of a molecule is,

$$m' = 44/(6.02 \times 10^{23}) = 7.309 \times 10^{-23} \text{ kg}$$

(ii) A kmole of an ideal gas under standard atmospheric conditions occupies 22.4m³. The number of molecules per unit volume is

$$n' = 6.02 \times 10^{23}/22.4 = 2.688 \times 10^{22}$$

(iii) The expression for the pressure from the simple kinetic theory [Eq. (2.3)] is

$$P = m'n'\overline{C}^2/3 = 7.309 \times 10^{-23} \times 2.688 \times 10^{22} \overline{C}^2/3 = 1.01 \times 10^5$$

Hence $$\overline{C}^2 = 0.1542 \times 10^6$$

Therefore the root-mean-square-speed is, $\left(\overline{C}^2\right)^{1/2} = 392.72 \text{ ms}^{-1}$.

Example 2.20 The temperature and pressure of water at its critical point are 647.15 K and 221.2 bar respectively. Use this data to determine the constants a and b for water in van der Waals equation of state, which in the molar-form can be written as

$$(P + a/v^2)(v - b) = \overline{R}T$$

where v is the volume of a mole of substance and $\overline{R}$ is the universal gas constant.

Solution In Sec. 1.2 we found that for an isothermal process, the P-V curve for a pure substance, has a point of inflection at the critical point. We now use this fact to determine the critical point properties using the van der Waals equation of state. The equation is first rearranged as follows:

$$P = \overline{R}T/(v - b) - a/v^2$$

At the point of inflection of the critical isotherm, $(T = T_c)$,

$$\frac{\partial P}{\partial v} = 0 \text{ and } \frac{\partial^2 P}{\partial v^2} = 0$$

Applying these conditions to the van der Waals equation of state we obtain the following two relations for the critical point properties:

$$\frac{2a}{v_c^3} = \frac{\overline{R}T_c}{(v_c - b)^2} \quad \text{and} \quad \frac{3a}{v_c^4} = \frac{\overline{R}T_c}{(v_c - b)^3}$$

From the above relations we have

$$b = v_c/3 \quad \text{and} \quad a = 9\overline{R}v_cT_c/8$$

Substituting the above relations for a and b in van der Waals equation of state

$$v_c = 3\overline{R}T_c/(8P_c)$$

Hence $$b = v_c/3 = \overline{R}T_c/8P_c$$

and $$a = 9\overline{R}T_cv_c/8 = 27\overline{R}^2T_c^2/64P_c$$

The critical temperature and pressure for water are:

$$T_c = (374.15 + 273) = 647.15 \text{ K} \quad \text{and} \quad P_c = 221.2 \times 10^5 \text{ N m}^{-2}$$

Substituting these values in the relationships for a and b we have

$$a = \frac{27 \times (8.3145)^2 \times 10^6 \times (647.15)^2}{64 \times 221.2 \times 10^5} = 5.52 \times 10^5 \text{ N m}^4 \text{ kmol}^{-2}$$

$$b = \frac{8.3145 \times 10^3 \times 647.15}{8 \times 221.2 \times 10^5} = 0.0304 \text{ m}^3 \text{ kmol}^{-1}$$

Example 2.21 Calculate the temperature of steam when the pressure is 120 bar and the specific volume is 0.025 m³ kg⁻¹ using (i) the ideal gas equation, and (ii) van der Waals equation of state.

Solution The molar mass of water is M = 18 kJ kmol⁻¹. Therefore the molar specific volume of steam is, $0.025 \times 18 = 0.45$ m³ kmol⁻¹.
(i) Using the ideal gas equation we have

$$T = Pv/\overline{R} = 120 \times 10^5 \times 0.45/(8.3145 \times 10^3) = 649.47 \text{ K}$$

(ii) For water the values of a and b in van der Waals' equation were calculated in worked example 2.20 as

$$a = 5.52 \times 10^5 \text{ N m}^4 \text{ kmol}^{-2} \quad \text{and} \quad b = 0.0304 \text{ m}^3 \text{ kmol}^{-1}$$

Substituting for a, b, P and v in van der Waals' equation

$$T = (P + a/v^2)(v - b)(1/\overline{R})$$

$$T = (120 \times 10^5 + 5.52 \times 10^5/0.45^2)(0.45 - 0.0304)(1/8314.5)$$

The temperature is, $T = 743$ K

The absolute temperature of steam predicted by the ideal gas model is about 12% lower.

Problems

P2.1 The table below gives the properties of water under five different sets of conditions. Use the table of thermodynamic properties [4] to obtain the missing properties for each state.

State	Pressure bar	Temperature °C	Quality x	Specific volume m³ kg⁻¹
1	1.5	?	0.7	?
2	2.7	130	?	?
3	26	400	?	?
4	?	150	?	0.30
5	?	280	0.6	?

P2.2 A fixed quantity of steam of mass 0.013 kg is heated at a constant pressure of 3 bar. The initial volume is 6.0×10^{-4} m³. Calculate (i) the initial quality and temperature of the steam, (ii) the quality when the volume is 7.2×10^{-4} m³, and (iii) the temperature when the volume is 9.5×10^{-3} m³. Indicate the heating process on a *P-V* and *T-V* diagram for water. [*Answers*: (i) 0.07447, 133.5°C, (ii) 0.0898, (iii) 208.8°C]

P2.3 The initial pressure, temperature and volume of a fixed quantity of steam are 5.5 bar, 350°C and 0.1 m³ respectively. The steam is compressed at constant temperature. Calculate (i) the mass of steam, (ii) the pressure at which condensation of steam just begins, and (iii) the quality of steam when the volume is 1.4×10^{-3} m³. Indicate the process on a *P-V* and *T-V* diagram for water. [*Answers*: (i) 0.1915 kg, (ii) 165.4 bar, (iii) 0.788]

P2.4 A rigid vessel of volume 0.02 m³ contains 0.054 kg of steam at an initial pressure of 4 bar. Heat is supplied to the steam from an external source. Calculate (i) the initial quality of the steam, (ii) the temperature and quality of the steam when the pressure becomes 4.5 bar, (iii) the pressure of the steam when it is just dry saturated, and (iv) the temperature when the pressure reaches 7 bar. Indicate the process on a *P-V* and *T-V* diagram for water. [*Answers*: (i) 0.8, (ii) 147.9°C, 0.894, (iii) 5.07 bar, (iv) 298.4°C]

P2.5 A fixed quantity of steam of mass 0.12 kg has an initial temperature and pressure of 375°C and 5.5 bar respectively. The steam undergoes a polytropic process according to the law $PV^{1.25}$ = constant. Calculate (i) the initial volume of the steam, and (ii) the volume and temperature when the pressure is 3 bar. Indicate the process on a *P-V* diagram for water. [*Answers*: (i) 0.064 m³, (ii) 0.104 m³, 294.7°C]

P2.6 A fixed quantity of air of mass 0.1 kg at 40°C and 100 kPa is compressed to a pressure of 1 MPa according to the law $PV^{1.4}$ = constant. It is then cooled to its initial temperature in a constant pressure process. Calculate (i) the initial volume of air, (ii) the volume and temperature at the end of compression, and (iii) volume at the end of the cooling process. Sketch the processes on a *P-V* diagram. [*Answers*: (i) 0.0898 m³, (ii) 0.0173 m³, 330°C, (iii) 0.00898 m³]

P2.7 Figure P2.7 shows a closed cylinder with a movable, frictionless piston which prevents any heat flow between the two spaces A and B. The cylinder contains a total of 1.2 kg of air at a uniform temperature of 30°C, and air occupies a total volume of 1.0 m³. The mass of air in space B is 0.6 kg. Heat is supplied to the space A, in a quasi-equilibrium manner until its final temperature is 200°C. During this process the air in space B is compressed by the piston according to the process law, $PV^{1.4}$ = constant. Calculate the final volume, pressure and temperature of the air in space B. [*Answers*: 0.4055 m³, 137 kPa, 49.6°C]

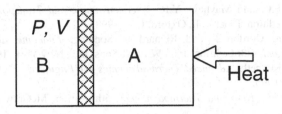

Fig. P2.7 Heating process in cylinder

P2.8 A rigid vessel contains one kmol of Butane (C_4H_6) gas under standard atmospheric conditions of 1.01 bar and 273 K. Calculate (i) number of molecules per unit volume, (ii) the mass of a molecule and (iii) the root-mean-square speed of the molecules. [*Answers*: (i) 2.67×10^{22} m^{-3}, (ii) 8.97×10^{-23} kg, (iii) 355 ms^{-1}]

P2.9 The critical pressure and temperature of nitrogen are 33.9 bar and 126.2 K respectively. Calculate (a) the values of a and b in the van der Waals equation of state for nitrogen, and (b) the pressure of nitrogen when the temperature is 100 K and the specific volume is 0.12 m^3 kg^{-1} using (i) the ideal gas equation, and (ii) the van der Waals equation. [*Answers*: (a) 137 kPa (m^3 kmol^{-1})^2, 0.0387 m^3 kmol^{-1} (b) (i) 247 kPa, (ii) 238 kPa]

P2.10 Calculate the specific volume of carbon dioxide at 30 bar and 300 K using (i) the ideal gas equation, and (ii) the van der Waals equation of state. For carbon dioxide the constants a and b in the van der Waals equation are 3.643 bar (m^3 kmol^{-1})^2 and 0.0427 m^3 kmol^{-1} respectively. [*Answers*: (i) 0.0189 m^3 kg^{-1}, (ii) 0.0162 m^3 kg^{-1}]

References

1. Cravalho, E.G. and J.L. Smith, Jr., *Engineering Thermodynamics*, Pitman, Boston, MA, 1981.
2. Jones, J.B. and G.A. Hawkins, *Engineering Thermodynamics*, John Wiley & Sons, Inc., New York, 1986.
3. Reynolds, William C. and Henry C. Perkins, *Engineering Thermodynamics*, 2nd edition, McGraw-Hill, Inc., New York, 1977.

4. Rogers, G.F.C. and Mayhew, Y.R., *Thermodynamic and Transport Properties of Fluids*, 5th edition, Blackwell, Oxford, U.K. 1998
5. Van Wylen, Gordon J. and Richard E. Sonntag, *Fundamentals of Classical Thermodynamics*, 3rd edition, John Wiley & Sons, Inc., New York, 1985.
6. Wark, Jr., Kenneth, *Advanced Thermodynamics for Engineers*, McGraw-Hill, Inc., New York, 1995.
7. Zemansky, M., *Heat and Thermodynamics*, 5th edition, McGraw-Hill, Inc., New York, 1968.

Chapter 3

Work and Heat Interactions

In this chapter we shall discuss the different types of work interactions that a thermodynamic system might experience. Although work is a familiar concept in mechanics, it needs to be generalized in a manner that fits the broad framework of thermodynamic systems and processes. We will consider work interactions in mechanical, elastic, electrical and magnetic systems. Heat interaction will be defined by using the concepts of temperature and equilibrium.

3.1 Concept of Work in Mechanics

Before we consider the application of the concept of work to thermodynamic systems, it is useful to recall some familiar details of its role in mechanics. Consider a simple mechanical system consisting of a body of mass M subjected to a force F as shown in Fig. 3.1.

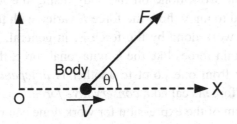

Fig. 3.1 Work done on a body

Let the displacement of the body in an inertial frame of reference be
x at time t. Application of Newton's second law in the x-direction gives
the equation of motion of the body as

$$F \cos\theta = M(dV/dt) \tag{3.1}$$

where V is the velocity in the x-direction and θ is the angle between the
force and the x-direction. Now the acceleration can be expressed as

$$(dV/dt) = V(dV/dx)$$

Therefore $\qquad\qquad F \cos\theta = MV(dV/dx) \tag{3.2}$

We integrate Eq. (3.2) from an initial state of the body defined by,
(x_1, V_1), the initial position and velocity respectively, to the corresponding final state (x_2, V_2). This gives

$$\int_{x_1}^{x_2} F \cos\theta dx = MV_2^2/2 - MV_1^2/2 \tag{3.3}$$

The left hand side of the above equation is the total work done by the
force on the body while the right hand side is the increase in kinetic
energy of the body.

It is interesting to note that the kinetic energy obtained by the
integration of Eq. (3.2) depends only the final and initial states of
the body and not on the manner in which the velocity changes
during the motion. Recalling our discussion in chapter 1 on the attributes
of properties, we conclude that the kinetic energy is a property of
the body.

To evaluate the work done on the body, using the left hand side of
Eq. (3.2), we need to know how the force F varies with the displacement
x. Therefore, the work done by the force is, in general, path-dependent.
However, for certain forces like the gravitational force, the work done in
moving the body from one point to another is *independent* of the path
followed. Such a force is called a *conservative force*.

The vector form of the expression for work done can be written as

$$W = \int_1^2 \hat{F}.d\hat{r}$$

where F and $\hat{r}$ are the force and position vectors respectively. The work, W however, is a scalar quantity. Note that the common units of work may be expressed as: $[W]$ = Nm (Newton-meter) = J (Joule).

3.2 Work Interactions in Thermodynamics

The interactions of a thermodynamic system with its surroundings occur at the system boundary. These interactions are usually the result of pressure forces, electrical currents, surface tension forces and others. For a concise formulation of the laws of thermodynamics we categorize all boundary interactions under two broad headings. These are called *work interactions* and *heat interactions*.

3.2.1 *Criterion for a work interaction*

It is possible to develop a criterion to determine unambiguously the nature of the interaction at a system boundary. A given boundary interaction is a *work interaction* if its sole effect on the surroundings could be resolved into the raising or lowering of a weight. In other words we should be able conceive of a device which when coupled to the given system (i) raises or lowers a weight in the surroundings using the same interaction, and (ii) leaves no other *permanent* effect on the surroundings. The concept is best illustrated using the piston-cylinder arrangement that has a gas, and an electrical heating element connected to a battery, as shown in Fig. 3.2(a).

Consider the gas as our system. As the gas expands due to heating, the piston will move up pushing the outside ambient air. There are two different boundary interaction that could be clearly identified. These are (i) the work done by the gas in raising the piston with the weight while pushing the outside air and (ii) the flow of electrical energy across the sections of the wire where they cross the system boundary.

The movement of the piston and weights placed on it satisfies our criterion of raising a weight and is therefore a work interaction. In fact, pushing the ambient air is equivalent to raising a weight of

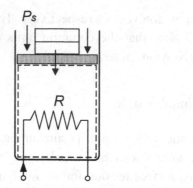 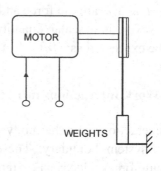

Fig. 3.2(a) Piston-cylinder arrangement   Fig. 3.2(b) Electrical work device

magnitude $P_s A$, where P_s is the ambient pressure and A is the area of the piston. We assume that the compression of the ambient air causes no other change in the surroundings due to its large volume.

The electricity flowing through the wires into the heater can be connected to an electric motor to raise a weight as shown in Fig. 3.2(b).

If we assume that the motor has a very high efficiency and negligible losses, then the flow of electricity across the boundary is also a work interaction according to the criterion stated above.

3.3   Work Done at a Moving Boundary

Consider the piston-cylinder arrangement shown in Fig. 3.3 which is subjected to a quasi-static expansion process. Recalling our discussion in chapter 1 this means that there is an external force on the piston that balances the force due to the gas pressure during the entire process. The gas expands due to small controlled reductions of the external force. Such an expansion is also called a *fully-resisted* expansion.

Assume that the pressure on the inner surface of the piston changes from P to ($P + \delta P$) when the piston moves through a small distance δz. The work done on the piston by the gas is equal to the product of the average force on the piston and its displacement. Therefore

$$\delta W = [P + (P + \delta P)]A\delta z/2 \tag{3.4}$$

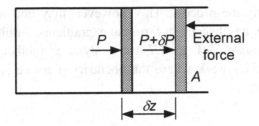

Fig. 3.3 Work done at a system boundary

where A is the area of the piston. Now the quantity, $A\ \delta z$ is equal to δV, the change in volume of the gas. Neglecting the second-order quantity $\delta P\ \delta z$, in Eq. (3.4), the work done on the piston by the gas can be written as

$$\delta W = P\ \delta V$$

In a differential form the expression for work becomes

$$\delta W = P\ dV \qquad (3.5)$$

The notation 'δ' is used to denote the infinitesimal quantity of work, instead of the usual 'd', to remind us that work is not a property but a path-dependent quantity.

In other words, the total work done depends on how the pressure varies with the volume during the process. When this pressure-volume relationship is known, the total work done could be obtained by integrating Eq. (3.5) from the initial state i to the final state f.

Thus
$$W = \int_{i}^{f} P\,dV \qquad (3.6)$$

A few important conditions and limitations should be noted when the above expression is used to calculate the work done by the gas. Clearly, the value of the pressure to be used in the expression has to be the pressure at the piston surface. For the quasi-equilibrium process considered here, the pressure at the piston surface is equal to the pressure of the gas in the cylinder because of the negligible pressure gradients within the gas during a *quasi-equilibrium* process. For *non-equilibrium* processes, the above expression could be used to calculate the work done provided we know the exact pressure variation at the surface of the piston during the process. This, however, may not be equal to the pressure of the gas because of pressure gradients within the system. In worked examples 3.1 to 3.7 we analyze a number of situations involving boundary work where the boundary pressure is a function of the displacement.

3.3.1 *Pressure-volume diagram*

Consider the gas in a piston-cylinder apparatus that undergoes a process in which the external force on the piston differs from the internal pressure force by only a small finite amount. For such a process, the pressure gradient within the cylinder is small enough for a single value to be assigned to the pressure at every stage of the process. The process path or the *P-V* diagram for this quasi-equilibrium process is shown in Fig. 3.4.

The expression for the work done by the gas, given by Eq. (3.6), can be interpreted as the area under the *P-V* diagram of the process. When the process is an expansion, work is done *by* the gas and the corresponding area under the curve I is positive. For a compression process, represented by curve II in Fig. 3.4, work is done *on* the gas by the external force and the area under the curve is negative.

A combined process, where the gas expands from state A to state B according to the *P-V* curve I and then returns to the original state following a compression process according to curve II, constitutes a *cycle*.

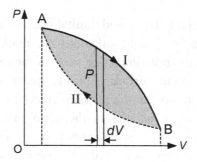

Fig. 3.4 Pressure-volume diagram

The net work done *by* the gas during the cycle is the algebraic sum of the areas under the two curves, which is equal to the area enclosed by the two curves. A cyclic process involving three processes is analyzed in worked example 3.14.

3.3.2 *Path dependence of work done*

Figure 3.5 shows the *P-V* diagrams for three possible processes by which the state of a gas changes from an initial state a to a final state b. The process a-c-b consists of a constant pressure process a-c (*P* = constant) followed by a constant volume process (*V* = constant) c-b. The curve a-d-b is a polytropic process with a process law, PV^n = constant. The third process consists of a constant volume process a-e followed by a constant pressure process e-b. Although the end states of the three processes are the same, the work done during each process is different because the areas under the three curves are different. This confirms that the work is not a property of the system because it is dependent on the path followed by the process. In other words, the evaluation of work, by integrating Eq. (3.6), requires knowledge of the dependence of *P* on *V* during the process or the process path.

In worked examples 3.10 and 3.11 we derive expressions for the work done in a polytropic expansion and an isothermal expansion respectively. These expressions are applied to two problems in worked examples 3.12 and 3.13.

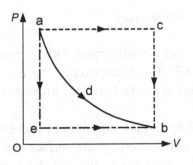
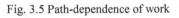

Fig. 3.5 Path-dependence of work

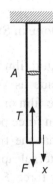

Fig. 3.6 Work done on a rod

3.4 Work Done in Extending a Solid Rod

Consider the work done during the extension of a solid rod shown in
Fig. 3.6. We assume that the rod is extended in a quasi-equilibrium
manner where the tension in the rod is balanced by the applied external
force F. The force F is increased in small controlled steps allowing time
for the system to attain equilibrium after each step. Let the normal stress
in the x-direction, induced in the rod due to the external pulling force,
be σ_x. The work done due to a small extension, dx is

$$\delta W = Fdx = \sigma_x A dx \qquad (3.7)$$

where A is the area of the rod.

The strain in the rod due to the extension is

$$d\varepsilon_x = dx/l \qquad (3.8)$$

where l is the length of the rod.

From Eqs. (3.7) and (3.8) we have,

$$\delta W = \sigma_x A l d\varepsilon_x = V \sigma_x d\varepsilon_x$$

where Al is equal to the volume, V of the rod.

The total work done is

$$W = V \int \sigma_x d\varepsilon_x \qquad (3.9)$$

A numerical problem on the stretching of a rod is considered in worked
example 3.8.

3.5 Work Done in Stretching a Liquid Surface

Consider a liquid film formed in a rectangular wire frame with a movable
rod on one side as shown in Fig. 3.7. The force exerted by the *two*
surfaces of the film on the movable rod of length L, is $2\tau L$, where τ is the
surface tension force per unit length.

Assume that the film is extended by a quasi-equilibrium process in
which the external applied force, F is balanced by the surface tension
force, T. The force F is increased in small steps allowing the film to
attain an equilibrium state at each step.

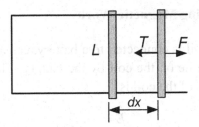

Fig. 3.7 Work done in stretching a surface film

The work done on the film due to an infinitesimal displacement *dx* is

$$\delta W = F\,dx = 2\tau L\,dx = \tau\,dA \qquad (3.10)$$

where $2L\,dx$ is equal to the total change of area dA of the film.

The total work done is

$$W = \int \tau\,dA$$

The above expression is used in worked example 3.9 to solve a numerical problem.

3.6 Systems Involving Electrical Work

Work is done in electrical systems when a charge *dq* is lifted through a potential difference *V*. Consider the charging of a capacitor using an external source. At a certain time during the process, the voltage between the plates will be *V*.

When an incremental charge *dq* is lifted through *V*, the work done *dW* is given by,

$$\delta W = V\,dq$$

The total work done in charging the capacitor is

$$W = \int V\,dq \qquad (3.11)$$

Similar expression could be used to obtain the work done in charging a reversible cell using an external source of electricity.

3.7   Systems Involving Magnetic Work

When an induction coil is connected to a battery, causing current to flow in the coil, work is done on the coil by the battery. The induced voltage between the terminals of the coil is

$$E = -L\frac{di}{dt} \tag{3.12}$$

where L is the self inductance of the coil and i is the current.

The rate of work input in sending the current through the potential difference is

$$\frac{\delta W}{dt} = -iE = iL\frac{di}{dt} \tag{3.13}$$

3.8   Heat Interactions

A heat interaction is an energy transfer that occurs between a system and its surroundings because of the difference in temperature between them. Consider the gas contained in a copper vessel as shown in Fig. 3.8. The gas is our system with a boundary coinciding with the inner surface of the vessel. If the initial temperature of the wall of the vessel is higher than the temperature of the gas, heat will flow from the wall to the gas. This boundary interaction is called a *heat interaction*.

If left for sometime this heat interaction will cease when the temperatures of the wall and the gas have become equal and thermal

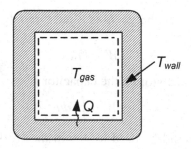

Fig. 3.8 Heat interaction between wall and gas

equilibrium is established. From a microscopic view point, the energy from the vibrating atoms at the surface of the wall transfer energy to molecules of gas due to random collisions with the latter. This is a disorganized form of energy transfer and is therefore not observed as work from a macroscopic view. As in the case of work, a system or body does not possess heat and therefore heat is not a property. The evaluation of the magnitude of the heat interaction during a process requires knowledge of the path of the process.

A system boundary that prevents heat interactions across it is called an *adiabatic boundary or wall.* Very low thermal conductivity insulation materials and the hollow evacuated space in a vacuum flask are approximations to adiabatic walls.

3.9 Comparison of Heat and Work

From our discussions thus far we notice that there are many similarities between heat and work interactions. These are summarized below.

(i) Heat and work are both energy in transit.
(ii) Systems or bodies do not possess heat or work.
(iii) Heat and work are not properties of a system.
(iv) The evaluation of the magnitude of the heat and work interactions between a system and its surroundings during a process requires knowledge of the process path.

Despite these similarities, there are some important differences between heat and work that will become evident when we introduce the second law of thermodynamics in chapter 6.

3.10 Worked Examples

Example 3.1 A sealed hollow cylinder of cross sectional area A and height H_o is pushed to the bottom of a large tank of water as shown in Fig. E3.1. Assuming that the height H of the water does not change due to its large volume, obtain an expression for the total work done on the cylinder by the external force when the process is carried out in a quasi-static manner. The ambient pressure is P_o.

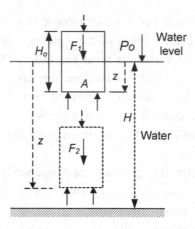

Fig. E3.1 Forces on an immersed cylinder

Solution Consider the vertical force balance on the cylinder when its lower surface is at a height z ($z < H_o$) below the water surface. The net upward force is

$$F_1 = A(P_o + z\rho_w g) - AP_o = Az\rho_w g$$

where ρ_w is the density of water, g is the acceleration due to gravity, and A is the cross sectional area of the cylinder. The total work done by the external force to fully immerse the cylinder until its top surface is at the water level is

$$W_1 = \int_0^{H_0} F_1 dz = \int_0^{H_o} Az\rho_w g dz = A\rho_w g(H_o^2/2)$$

Now consider the vertical forces on the cylinder when it is fully immersed, with its bottom surface at a height z below the water level. The net vertical force is equal to the difference in the pressure forces on the bottom and top surfaces of the cylinder. Hence the net upward force is

$$F_2 = Az\rho_w g - A(z - H_o)\rho_w g = AH_o\rho_w g$$

The work done in pushing the cylinder to the bottom of the tank is

$$W_2 = \int_{H_0}^{H} F_2 dz = \int_{H_o}^{H} AH_o\rho_w g dz = A\rho_w g H_o(H - H_o)$$

Therefore the total work done to immerse the cylinder to the bottom of the tank is

$$W = W_1 + W_2 = A\rho_w g H_o^2 / 2 + A\rho_w g H_o (H - H_o)$$

$$W = A\rho_w g H_o (H - H_o / 2)$$

Example 3.2 A rectangular tank, containing water, has a movable side wall which is held in equilibrium by an external force as shown in Fig. E3.2. The height, length and breadth of the volume V occupied by the water are Z, Y and b respectively. The force T is adjusted so that wall moves horizontally in the y-direction in a quasi-static manner. The ambient pressure is P_o.
(a) Obtain expressions for (i) the force of the water on the movable wall and (ii) the work done by the water when the wall is moved from position y_1 to y_2.
(b) Suggest a method to carry out the process in a quasi-static manner.

Solution (a) (i) The horizontal force on the movable wall of the tank is

$$F = \int_0^Z b(P_o + z\rho_w g)dz = bP_o Z + b\rho_w g(Z^2/2) \qquad \text{(E3.2.1)}$$

where Z is the height of the water in the tank.
(ii) The work done when the wall is moved a distance dy in a quasi-static manner is

$$\delta W = F dy \qquad \text{(E3.2.2)}$$

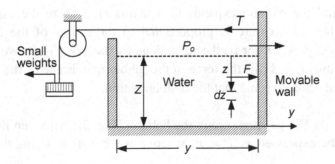

Fig. E3.2 Work done on movable-wall

Now the volume of water in the tank is constant. Therefore

$$V = bZy \qquad (E3.2.3)$$

Substituting for F and Z from Eq. (E3.2.1) and Eq. (E3.2.3) in Eq. (E3.2.2) we have

$$\delta W = b(P_o V / by)dy + b\rho_w g(V^2/2b^2 y^2)dy \qquad (E3.2.4)$$

The total work done in moving the wall from horizontal position y_1 to y_2 is obtained by integrating the above equation. Hence

$$W = P_o V \int_{y_1}^{y_2} (dy/y) + (\rho_w g V^2/2b)\int_{y_1}^{y_2} dy/y^2$$

$$W = P_o V \ln(y_2/y_1) + (\rho_w g V^2/2b)(1/y_1 - 1/y_2)$$

(b) Figure E3.2 shows a possible arrangement for moving the wall in a quasi-static manner subject to the water pressure. The weights attached to the rope are removed in small steps to allow the force of the water on the wall to exceed the tension in the rope holding the wall by a small value. As the wall moves out, the weights being removed from the end of the rope are placed at higher elevations. Therefore the work done by the water is stored as the potential energy of the weights removed as seen in the Fig. E3.2. We assume ideal conditions where the wall moves in a frictionless manner.

Example 3.3 An elastic balloon has a spherical shape of radius r_1 and contains gas at an initial pressure of P_1. The surrounding ambient pressure is P_o. The gas in the balloon receives heat in a quasi-static manner and the balloon expands to a radius r_1. During the expansion process the gas pressure is proportional to the radius of the balloon. Obtain an expression for (i) the work done by the gas, (ii) the work done on the ambient, and (iii) the force in the rubber membrane of the balloon in the final state. Assume that the balloon is thin.

Solution (i) The gas pressure in the balloon [Fig. 3.3(a)] when its radius is r may be expressed as $P = cr$, where c is a constant. Using the given

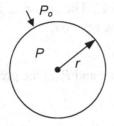

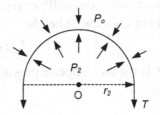

Fig. E3.3(a) Heated balloon   Fig. 3.3(b) Forces on balloon

initial conditions, $c = P_1/r_1$. The radial force due to the gas pressure is, $F = 4\pi r^2 P$. The work done by the gas is

$$\delta W_g = F.dr = 4\pi r^2 P dr = 4\pi r^2 c r dr = 4\pi c r^3 dr$$

The total work done by the gas is

$$W_g = 4\pi c \int_{r_1}^{r_2} r^3 dr = \pi P_1 \left(r_2^4 - r_1^4\right)/r_1$$

(ii) The work done in pushing the surrounding ambient at constant pressure is

$$W_o = \int_{r_1}^{r_2} P_o 4\pi r^2 dr = 4\pi P_o (r_2^3 - r_1^3)/3$$

(iii) Consider the force balance on one half of the balloon as shown in Fig. E3.3(b)

$$2\pi r_2 T + P_o \pi r_2^2 = P_2 \pi r_2^2$$

where T is the force per unit circumferential length in the thin membrane of the balloon. Therefore

$$T = r_2 (P_2 - P_o)/2 = r_2 \left(P_1 r_2 / r_1 - P_o\right)/2$$

Example 3.4 A spherical balloon, whose volume at sea level is V_0, rises slowly in the atmosphere up to a height of H. Assuming that the volume of the balloon varies inversely as the surrounding atmospheric pressure, obtain an expression for the work done by the skin of the balloon on the atmosphere when the balloon rises to a height H. The pressure, P of the atmosphere varies with height z according to the relation,

$$P = P_o \exp(-\rho_o gz/P_o)$$

where P is the atmospheric pressure at a height z and P_o is the pressure at sea level.

Solution We are given that the volume of the balloon is related to the surrounding atmospheric pressure by the expression

$$V = c/P \tag{E3.4.1}$$

where c is constant.

From the given conditions at sea level, $c = P_o V_o$

The work done on the atmosphere when the volume of the balloon changes by dV is,

$$\delta W = PdV \tag{E3.4.2}$$

Differentiating Eq. (E3.4.1) and substituting for c

$$dV = -cdP/P^2 = -P_o V_o dP/P^2 \tag{E3.4.3}$$

Substituting for dV from Eq. (E3.4.3) in Eq. (E3.4.2)

$$\delta W = PdV = -P_o V_o dP/P \tag{E3.4.4}$$

The total work done when the balloon moves to an elevation H where the surrounding atmospheric pressure is P is obtained by integrating Eq. (E3.4.4).

$$W = -P_o V_o \int_{P_o}^{P} dP/P = -P_o V_o \ln(P/P_o) \tag{E3.4.5}$$

Now the pressure of the atmosphere varies with height according to the expression

$$P = P_o \exp(-\rho_o gz/P_o) \tag{E3.4.6}$$

Recall that we derived the above expression earlier in worked example 1.9.

The pressure P at an elevation H is obtained by rearranging Eq. (E3.4.6) as

$$\ln(P/P_o) = -\rho_o gH/P_o \tag{E3.4.7}$$

Substituting from Eq. (E3.4.7) in Eq. (E3.4.5) the work done is

$$W = H\rho_o g V_o$$

Example 3.5 Figure E3.5 shows a U-tube with two vertical legs of cylindrical cross section containing oil of density ρ_o. The leg A, with a diameter D_1, is sealed at the top end and has gas trapped in the space above the oil. The oil surface in the other leg B, of diameter D_2, is exposed to the ambient pressure of P_o. Initially, the gas has a volume of V_1 and the level of oil in the two legs is the same.

The gas receives heat in a quasi-static manner from an external source. The expanding gas in A pushes some of the oil into leg B.
(i) Obtain the *P-V* relation of the gas. (ii) Obtain an expression for the work done by the gas when its volume changes from V_1 to V_2.

Solution (i) Consider the situation shown in Fig. E3.5 where the oil level in leg A has moved a distance x below the initial level. Let the rise in oil level in leg B be y. Equating the volume of oil displaced in the two legs we have

$$\pi D_1^2 x / 4 = \pi D_2^2 y / 4 \qquad (E3.5.1)$$

Now the volume of gas in tube A is

$$V = V_1 + \pi D_1^2 x / 4 \qquad (E3.5.2)$$

where V_1 is the initial volume of gas in A.

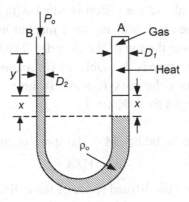

Fig. E3.5 Gas trapped in U-tube

The pressure of the gas in A is

$$P = P_o + (x+y)\rho_o g \qquad (E3.5.3)$$

Substituting from Eq. (E3.5.1) in Eq. (E3.5.3)

$$P = P_o + \rho_o g\left(1 + D_1^2/D_2^2\right)x \qquad (E3.5.4)$$

Substituting in Eq. (E3.5.4) for x from Eq. (E.5.2) we obtain the linear P-V relation as

$$P - P_o = \left(4\rho_o g/\pi\right)\left(1/D_1^2 + 1/D_2^2\right)(V - V_1) \qquad (E3.5.5)$$

(ii) The work done by the gas is given by

$$W = \int_{V_1}^{V_2} P dV \qquad (E3.5.6)$$

Substituting for P from Eq. (E3.5.5) in Eq. (E3.5.6)

$$W = \int_{V_1}^{V_2} P_o dV + \left(4\rho_o g/\pi\right)\int_{V_1}^{V_2}\left(1/D_1^2 + 1/D_2^2\right)(V - V_1)dV \qquad (E3.5.7)$$

Integrating Eq. (E3.5.7) we obtain the work done by the gas as

$$W = P_o(V_2 - V_1) + (2\rho_o g/\pi)\left(1/D_1^2 + 1/D_2^2\right)(V_2 - V_1)^2$$

Example 3.6 The piston-cylinder arrangement, shown in Fig. E3.6, contains air at a pressure of P_1 and volume V_1. The height of the cylinder is L and its cross sectional area is A. The thickness of the piston is l. The space in the cylinder above the piston is filled with sand of density ρ_o. Heat is supplied to the air in a quasi-static manner and the piston moves up spilling the sand over the top of the cylinder. The mass of the piston is M and the ambient pressure is P_o. Obtain expressions for (i) the pressure P of the gas when its volume is V, and (ii) the work done by the gas when its volume changes from V_1 to V_2.

Solution (i) Now the initial height of the space occupied by air is

$$x_1 = V_1/A$$

The initial height of the sand-filled space of the cylinder is

$$y_1 = L - l - x_1$$

Consider the situation when the inner surface of the piston has moved to a height x above the bottom of the cylinder. Then

$$x = V / A$$

The height of the sand filled volume is

$$y = L - l - x$$

The force balance on the piston gives

$$PA = Mg + P_o A + y A \rho_o g$$

$$PA = Mg + P_o A + A \rho_o g (L - l - V / A) \tag{E3.6.1}$$

The force balance on the piston in the initial state gives

$$P_1 A = Mg + P_o A + y_1 A \rho_o g$$

$$P_1 A = Mg + P_o A + A \rho_o g (L - l - V_1 / A) \tag{E3.6.2}$$

Subtracting Eq. (E3.6.2) from Eq. (E3.6.1) we obtain the linear P-V relation as

$$(P - P_1) = -\rho_o g (V - V_1) / A \tag{E3.6.3}$$

(ii) The work done by the gas is

$$W = \int_{V_1}^{V_2} P dV \tag{E3.6.4}$$

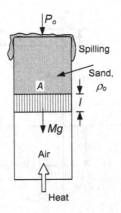

Fig. E3.6 Piston-cylinder set-up loaded with sand

Substituting for P from Eq. (E3.6.3) in Eq. (E3.6.4)

$$W = \int_{V_1}^{V_2} P_1 dV - (\rho_o g / A) \int_{V_1}^{V_2} (V - V_1) dV \qquad \text{(E3.6.5)}$$

Integrating Eq. (E3.6.5) we obtain the work done by the gas as

$$W = P_1(V_2 - V_1) - \rho_o g(V_2 - V_1)^2 / 2A$$

Example 3.7 A vertical cylinder of inner diameter D and length L contains ambient air at pressure P_o. It is slowly forced into a large tank of water as shown in Fig. E3.7. The air trapped in the cylinder is compressed by the water entering the cylinder under constant temperature conditions. Assume that air is an ideal gas. Obtain expressions for (i) the pressure of the air when the height of the cylinder above the water is H, and (ii) the work done in compressing the air from the initial pressure P_o to a final pressure P_f.

Solution (i) Let the height of the cylinder above the water surface be H and the depth of the water level inside the cylinder below the water surface in the tank be x as shown in Fig. E3.7. The pressure of the air is

$$P = P_o + x\rho_w g \qquad \text{(E3.7.1)}$$

where ρ_w is the density of water.

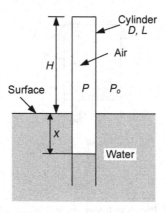

Fig. E3.7 Immersed-cylinder with trapped air

The volume of air is

$$V = A(x + H) \qquad (E3.7.2)$$

Substituting for x from Eq. (E3.7.2) in Eq. (E3.7.1)

$$P = P_o + \rho_w g(V/A - H) \qquad (E3.7.3)$$

Applying the ideal gas equation of state to the initial and final states of the air we have

$$mRT = P_o V_o = P_o AL = PV \qquad (E3.7.4)$$

where T is the constant temperature.

Substituting for V from Eq. (E3.7.4) in Eq. (E3.7.3) we have

$$P = P_o + \rho_w g(P_o AL/P - H) \qquad (E3.7.5)$$

$$P^2 - (P_o - H\rho_w g)P - \rho_w g P_o AL = 0 \qquad (E3.7.6)$$

The positive root of the quadratic equation Eq. (E3.7.6) gives the air pressure P when the height of the cylinder outside the water surface is H.
(ii) The work done by the air is

$$W = \int_{V_1}^{V_2} PdV \qquad (E3.7.7)$$

From Eq. (E3.7.4)

$$V = ALP_o/P \qquad (E3.7.8)$$

Differentiating Eq. (E3.7.8)

$$dV = -ALP_o dP/P^2 \qquad (E3.7.9)$$

Substituting for dV from Eq. (E3.7.9) in Eq. (E3.7.7)

$$W = -\int_{P_i}^{P_f} (ALP_o/P)dP = -ALP_o \ln\left(P_f/P_o\right)$$

Note that the final pressure P_f of the air is obtained by solving Eq. (E3.7.6) for a given value of H.

Example 3.8 A vertical steel rod of diameter 2 mm and length 0.5 m is stretched in a quasi-static manner by attaching weights to the bottom end in small steps from 100 kg to 200 kg. Assume that the temperature of the rod remains constant at 20°C. Calculate the work done on the rod. Young's modulus of steel at 20°C is 1.95×10^{11} Nm^{-2}.

Solution When the extension of the rod is x let the tension in the rod be T. The normal stress is, $\sigma_x = T/A$ and the normal strain is, $\varepsilon_x = x/L$, where L is length of the rod and A is its cross-sectional area.

The quantities L and A are both practically constant due to the small extension of the rod.

The Young's modulus, Y is defined as

$$Y = \sigma_x / \varepsilon_x = TL / Ax$$

Hence $T = AYx/L = kx$

where the constant $k = AY/L$

The work done in extending a rod is given by Eq. (3.9) as

$$W = V \int \sigma_x d\varepsilon_x \qquad\qquad (E3.8.1)$$

From the above equation, the total work done when the extension of the rod changes from x_1 to x_2 is

$$W = AL \int_{x_1}^{x_2} (T/A)(dx/L) = \int_{x_1}^{x_2} (AYx/L)dx$$

$$W = \int_{x_1}^{x_2} kxdx = k(x_2^2 - x_1^2)/2 \qquad\qquad (E3.8.2)$$

Let the tension in the rod for the two extensions be T_1 and T_2 respectively. Then

$$x_1 = T_1 L/AY \quad\text{and}\quad x_2 = T_2 L/AY \qquad\qquad (E3.8.3)$$

Substituting in Eq. (E3.8.2) from Eq. (E3.8.3) we have

$$W = L(T_2^2 - T_1^2)/2AY \qquad\qquad (E3.8.4)$$

Now the area of the rod is, $A = \pi \times \left(2 \times 10^{-3}\right)^2 / 4 = 3.14 \times 10^{-6}$ m^2

The initial tension, $\qquad T_1 = 100g = 981\,\text{N}$

and the final tension, $\qquad T_2 = 200g = 1962\,\text{N}$

where the acceleration due to gravity, g is taken as $g = 9.81\ \text{ms}^{-2}$

Substituting the numerical values in Eq. (E3.8.4), the work done is

$$W = 0.5 \times (1962^2 - 981^2)/(2 \times 3.14 \times 10^{-6} \times 1.95 \times 10^{11}) = 1.18\ \text{J}$$

Example 3.9 A thin liquid film is formed on a rectangular wire frame of length 0.04 m and breadth 0.03 m. One of the shorter sides of the frame is moved a distance of 0.01 m thus stretching the film. Assuming that the temperature of the film remains constant at 65°C during the process, calculate the work done on the film. The surface tension, σ of the film varies with temperature, t (°C) according to the relation

$$\sigma(Nm^{-1}) = 0.075(1 - t/374)^{1.2}$$

Solution Using the given relation, the surface tension at 65°C is calculated as 0.0596 Nm^{-1}.

The force exerted by the two-sided film on the movable side of length b of the rectangle is, $F = 2b\sigma$.

Since the temperature of the film is maintained constant, the surface tension and therefore the force, F remains constant during the extension process.

The work done in stretching the film with two sides from a length l_1 to l_2 is

$$W = \int_{l_1}^{l_2} 2b\sigma\,dx = 2b\sigma(l_2 - l_1)$$

Substituting numerical values in the above equation we have

$$W = 2 \times 0.04 \times 0.0596 \times 0.01 = 4.77 \times 10^{-5}\ \text{J}$$

Example 3.10 (a) A gas of mass m undergoes a quasi-static *polytropic* process, with a process law given by $PV^n = \text{constant}$. The pressure, volume, and temperature in the initial and final states are P_1, V_1, T_1 and

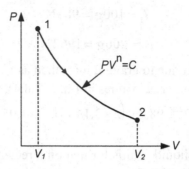

Fig E3.10 Work done in a polytropic process

P_2, V_2, T_2 respectively. (i) Obtain an expression for the work done by the gas during the process.

(ii) If the gas is an *ideal gas*, show that the work done is,

$$W_{12} = mR(T_1 - T_2)/(n-1)$$

Solution (i) The process path for the polytropic process is shown in Fig. E3.10. Since the process is carried out in a quasi-static manner, the work done is given by

$$W_{12} = \int_{V_1}^{V_2} PdV \qquad\qquad (E3.10.1)$$

From the polytropic relation we have, $P = C/V^n$, where C is a constant. Substituting for P in Eq. (E3.10.1), the work done becomes

$$W_{12} = \int_{V_1}^{V_2} (C/V^n)dV = (CV_2^{1-n} - CV_1^{1-n})/(1-n) \qquad (E3.10.2)$$

From the polytropic relation we have

$$CV_1^{-n} = P_1 \qquad \text{and} \qquad CV_2^{-n} = P_2$$

Substituting the above relations in Eq. (E3.10.2) the work done is

$$W_{12} = (P_1V_1 - P_2V_2)/(n-1) \qquad\qquad (E3.10.3)$$

(ii) Now for an *ideal gas* undergoing a polytropic process we obtain two additional relations by applying the equation of state,

$$P_1 V_1 = mRT_1 \quad \text{and} \quad P_2 V_2 = mRT_2 \quad \text{(E3.10.4)}$$

Substituting from Eq. (E3.10.4) in Eq. (E3.10.3), the work done by an *ideal gas* undergoing a polytropic process is

$$W_{12} = mR(T_1 - T_2)/(n-1)$$

Example 3.11 (i) An *ideal gas* of mass m undergoes a quasi-static *isothermal* process at temperature T. The initial and final pressure, and volume are P_1, V_1 and P_2, V_2 respectively. Show that the work done by the gas is,

$$W = mR \ln(P_1/P_2) = mR \ln(V_2/V_1).$$

(ii) Can these expressions be used to calculate the work done when wet steam expands under constant temperature conditions?

Solution (i) Figure E3.11 shows the expansion process of the ideal gas on a *P-V* diagram. Since the process is carried out in a quasi-static manner the work done by the gas is

$$W = \int_{V_1}^{V_2} P dV \quad \text{(E3.11.1)}$$

For an ideal gas, the equation of state gives the following relation for P:

$$P = mRT/V \quad \text{(E3.11.2)}$$

where R is the gas constant for the gas.

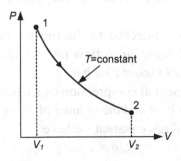

Fig. E3.11 Work done by an ideal gas in an isothermal process

Substituting for P from Eq. (E3.11.2) in Eq. (E3.11.1), and noting that T is constant, we obtain the work done as

$$W = \int_{V_1}^{V_2} PdV = \int_{V_1}^{V_2} mRTdV/V = mRT \ln(V_2/V_1) \qquad (E3.11.3)$$

Since the temperature T is constant during the *isothermal* process, the equation of state gives the following relation between initial and final volumes and pressures

$$V_2/V_1 = P_1/P_2 \qquad (E3.11.4)$$

Substituting from Eq. (E3.11.4) in Eq. (E3.11.3) the work done by the gas is

$$W = mR \ln(V_2/V_1) = mR \ln(P_1/P_2)$$

(ii) When wet steam undergoes an expansion under constant temperature, the above relation should *not* be used to calculate the work done because wet steam is *not an ideal gas*.

Example 3.12 A bicycle tire has air at a pressure of 3.8 bar. It is to be inflated using a reciprocating pump (Fig. E3.12a) with a stroke of 0.25 m. Initially the cylinder of the pump contains air at a pressure of 1.0 bar. Calculate the length of the stroke that will be swept before the air begins to enter the tire, (i) if the process is isothermal, and (ii) if the process follows the polytropic law $PV^{1.4}$ = constant. Which process requires more work? Assume that air is an ideal gas.

Solution The pump is attached to the tire through a non-return valve which will open and allow air to flow into the tire when the pressure of the air in the pump just exceeds 3.8 bar.

Consider the isothermal compression of air from 1 bar to 3.8 bar. Let the area of the piston be A and the volume of air be V_1.

Applying the ideal gas equation we have

$$mRT = P_1V_1 = P_2V_2$$

where 1 and 2 represent the initial and final states respectively. Let l_2 be the length of the cylinder occupied by air.

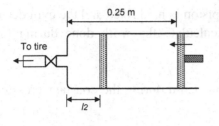
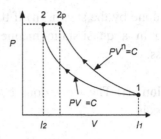

Fig. E3.12(a) Filling a tire with air   Fig. E3.12(b) *P-V* diagram

$$Al_2 \times 3.8 = 0.25A \times 1. \quad \text{Hence} \quad l_2 = 0.0658\,\text{m}$$

The distance moved by the piston is $(0.25 - 0.0658) = 0.1842$ m
The work done *by* the air during isothermal compression

$$W_1 = mRT \ln(V_2 / V_1) = P_1 V_1 \ln(0.0658A / 0.25A)$$

$$W_1 = -P_1 V_1 \times 1.3348$$

(i) For the polytropic process, $P_1 V_1^{1.4} = P_2 V_{2p}^{1.4}$

Therefore $\qquad V_{2p} / V_1 = (1/3.8)^{1/1.4} = 0.3854$

The length of the cylinder occupied by the air is

$$l'_{2p} = 0.25 \times 0.3854 = 0.0964\,\text{m}$$

The distance moved by the piston is $(0.25 - 0.0964) = 0.1536$ m
Work done in the polytropic compression is

$$W_2 = (P_1 V_1 - P_2 V_{2p}) / (n - 1)$$

$$W_2 = P_1 V_1 (1 - 0.3854 \times 3.8) / 0.4 = -1.16 P_1 V_1$$

Therefore $\qquad W_1 / W_2 = 1.3348 / 1.16 = 1.15$

The isothermal process requires about 15% more work. The *P-V* diagrams for the two processes are depicted in Fig. E3.12(b).

Example 3.13 A quantity of wet steam of mass 0.1 kg is trapped in a cylinder behind a leak-proof piston. The initial pressure and quality of the steam are 3 bar and 0.75 respectively. Heat is supplied to the steam at constant pressure until the volume becomes 0.1 m³. (i) Calculate the work done by the steam. (ii) If the piston is held fixed and the cylinder is cooled in a quasi-static manner, calculate the work done during the process.

Solution (i) The work done by the steam during the constant pressure process is

$$W_{1-2} = m\int_{v_1}^{v_2} Pdv = mP(v_2 - v_1)$$

At state 1, the steam is wet. Therefore $v_1 = x_1 v_{g1} + (1 - x_1)v_{f1}$

From tabulated data in [4] for saturated steam at 3 bar,

$$v_{g1} = 0.6057 \text{ m}^3 \text{ kg}^{-1} \quad \text{and} \quad v_{f1} = 0.1073 \times 10^{-2} \text{ m}^3 \text{ kg}^{-1}$$

The specific volumes are

$$v_1 = 0.75 \times 0.6057 + (1 - 0.75) \times 0.1073 \times 10^{-2} = 0.4545 \text{ m}^3 \text{ kg}^{-1}$$

and

$$v_2 = V_2/m = 0.1/0.1 = 1.0 \text{ m}^3 \text{ kg}^{-1}$$

$$W_{1-2} = mP(v_2 - v_1) = 0.1 \times 3 \times 10^5 \times (1.0 - 0.4545) = 16.36 \text{ kJ}.$$

(ii) When the piston is help fixed, the volume is constant. Therefore the boundary work done is

$$W_{2-3} = \int_{V_2}^{V_3} PdV = 0 \quad \text{because,} \quad dV = 0$$

Example 3.14 A quantity of air of mass 0.5 kg undergoes the *cyclic process* shown in Fig. E3.14. The pressure and temperature of state 1 are 1 bar and 30°C respectively. The air is compressed to a pressure of 8 bar during the *isothermal process*, 1-2. The air then expands at constant pressure to state 3. This is followed by the *polytropic process* 3-1 with a process-law, $PV^{1.4} = C$, which completes the cycle. Assume that air is an ideal gas with a molecular mass of 29 kg kmol^{-1}. The universal gas constant is $\bar{R} = 8.3145$ kJ kmol^{-1} K^{-1}.

Calculate (i) the pressure, volume, and temperature at states 1, 2 and 3, (ii) the work done by the air during each process, and (iii) the net work done by the air during the cyclic process 1-2-3-1.

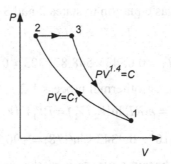

Fig. E3.14 Cyclic process

Solution The three processes undergone by the air are shown in Fig. E3.14 where the numbers indicate the end states of the processes.

The following data are given: $P_1 = 1$ bar, $T_1 = 273 + 30 = 303$ K, $P_2 = 8$ bar. Since the process 1-2 is isothermal, $T_2 = T_1 = 303$ K.

Using the ideal gas equation of state for state 1

$$P_1 V_1 = mRT_1$$

Hence $\quad V_1 = 0.5 \times (8.3145 \times 10^3 / 29) \times 303 / 10^5 = 0.4343\, \text{m}^3$

Since the process 1-2 is isothermal, it follows for the ideal gas equation that

$$P_1 V_1 = P_2 V_2$$

Therefore $\quad V_2 = 0.4343 \times (1/8) = 0.0543\ \text{m}^3$

Eliminating V between the ideal gas equation of state and the polytropic law we obtain the following useful relation between T and P:

$$T(1/P)^{(n-1)/n} = \text{constant}$$

Applying the above equation to states 1 and 3 we have

$$T_3 = T_1 (P_3 / P_1)^{(n-1)/n} \quad \text{where } n = 1.4$$

Substituting the given numerical data in the above equation

$$T_3 = 303 \times (8)^{0.4/1.4} = 548.87\ \text{K}$$

Because 2-3 is a constant pressure process, $P_3 = P_2 = 8$ bar.

Applying the ideal gas equation to states 2 and 3 we have

$$V_2/T_2 = V_3/T_3$$

$$V_3 = V_2 T_3/T_2 = 0.0543 \times 548.87/303 = 0.0983 \text{ m}^3.$$

The work done during the isothermal process 1-2 is

$$W_{1-2} = mRT_1 \ln(V_2/V_1) = P_1 V_1 \ln(V_2/V_1)$$

$$W_{1-2} = 10^5 \times 0.4343 \times \ln(1/8) = -90.32 \text{ kJ}$$

The work is negative because we derived the general formula for an expansion process in which work is done *by* the ideal gas. In the compression process from 1 to 2 work is done *on* the gas.

The work done during the constant pressure process 2-3 is

$$W_{2-3} = \int_{V_2}^{V_3} P dV = P \int_{V_2}^{V_3} dV = P_2(V_3 - V_2)$$

$$W_{2-3} = 8 \times 10^5 \times (0.0983 - 0.0543) = 35.24 \text{ kJ}$$

The work done during the polytropic process 3-2 is calculated using the formula derived in worked example 3.10 for an ideal gas.

$$W_{3-1} = mR(T_3 - T_1)/(n-1)$$

$$W_{3-1} = 0.5 \times 8.3145 \times 10^3 \times (548.86 - 303)/(29 \times 0.4) = 88.11 \text{ kJ}$$

The work is done *by* the gas during this expansion process and therefore as expected the work is positive.

The net work done is the *algebraic sum* of the work done in the three process

$$W_{net} = W_{1-2} + W_{2-3} + W_{3-1}$$

$$W_{net} = -90.32 + 35.24 + 88.11 = 33.03 \text{ kJ}$$

Note that the *net* work is the area of the *P-V* diagram enclosed by the three curves representing the individual processes.

Example 3.15 The spring-loaded piston cylinder arrangement shown in Fig. E3.15 contains 0.05 m³ of a gas at a pressure of 1.05 bar. The ambient pressure is 1.0 bar. The movement of the piston is restrained by

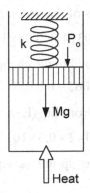

Fig. E3.15 Spring-loaded piston-cylinder set-up

a linear spring which initially touches the piston, but exerts no force on it. The diameter of the piston is 0.2 m. Heat is supplied to the gas in a quasi-static manner until volume and pressure are 0.065 m³ and 1.65 bar respectively. Calculate (i) the mass of the piston, (ii) the spring constant, (iii) the strain energy stored in the spring, (iv) the change in potential energy of the piston, (v) the work done in pushing the atmosphere, and (vi) the work done by the gas.

Solution The area of the piston is

$$A = \pi D^2 / 4 = \pi \times 0.2^2 / 4 = 0.03142 \ \text{m}^2$$

The initial and final volumes are

$$V_1 = 0.05 \ \text{m}^3 \quad \text{and} \quad V_2 = 0.065 \ \text{m}^3$$

The initial and final gas pressures are, $P_1 = 1.05$ bar and $P_2 = 1.65$ bar. The ambient pressure is $P_o = 1.0$ bar. The force balance on the piston in the initial state when the spring is uncompressed gives

$$P_1 A = Mg + P_o A \tag{E3.15.1}$$

where M is the mass of the piston and g is the acceleration due to gravity.

In the final state, the force balance equation is

$$P_2 A = Mg + P_o A + kl \tag{E3.15.2}$$

where l is the piston displacement and k is the spring constant.

(i) From Eq. (E3.15.1) the mass of the piston M is obtained as

$$M = A(P_1 - P_o)/g = 0.03142 \times 0.05 \times 10^5 / 9.81 = 16.01 \text{ kg.}$$

The piston displacement is related to the change in gas volume. Hence

$$l = (V_2 - V_1)/A = (0.065 - 0.05)/0.03142 = 0.477 \text{ m}$$

(ii) Subtracting Eq. (E3.15.1) from Eq. (E3.15.2) we have

$$k = A(P_2 - P_1)/l = 0.03142 \times 0.6 \times 10^5 / 0.477 = 3.95 \text{k Nm}^{-1}$$

(iii) The strain energy of the spring is equal to the work done in compressing it.

Therefore
$$E_{spr} = \int_0^l kx dx = kl^2/2$$

$$E_{spr} = 3.95 \times 10^3 \times 0.477^2/2 = 449.4 \text{ J}$$

(iv) The change in potential energy of the piston is

$$E_p = Mgl = 16.01 \times 9.81 \times 0.477 = 74.92 \text{ J}$$

(v) The work done in pushing the surrounding ambient is

$$W_a = \int_{V_1}^{V_2} P_o dV = P_o(V_2 - V_1)$$

$$W_a = 10^5 \times (0.065 - 0.05) = 1500 \text{ J}$$

(vi) The work done by the gas is equal to the sum of the strain energy of the spring, the increase in potential energy of the piston, and the work done in pushing the surrounding ambient. Therefore

$$W_g = E_{spr} + E_p + W_o = 449.6 + 74.91 + 1500 = 2024.3 \text{ J}$$

Alternatively, we could obtain the work done by the gas by first deriving the P-V relation for the process as we did in worked example 2.14, and then using the expression for boundary work,

$$W_g = \int_{V_1}^{V_2} P dV$$

Example 3.16 An initially flat balloon is filled with helium by connecting it to a tank of helium at a pressure of 10 bar and temperature 30°C. Until the balloon takes its final spherical shape, with a diameter of 6 m, the pressure within the balloon is approximately equal to the ambient pressure of 1.01 bar.

During the filling process, which occurs at constant temperature, the membrane of the balloon is not stretched. At the end of the process, the pressures in the tank and the balloon are equal to the ambient pressure. Calculate (i) the work done on the atmosphere during the filling process and (ii) the volume of the tank. Neglect the initial volume of the balloon.

Solution The volume of the fully-inflated balloon is

$$V_{b2} = \pi D^3 / 6 = \pi \times 6^3 / 6 = 113.1 \text{ m}^3$$

(i) During the filling process the expanding membrane of the balloon pushes the surrounding ambient air whose pressure is assumed constant. Therefore the work done on the ambient is

$$W = \int_0^{V_{b2}} P_o dV = P_o V_{b2} = 1.01 \times 10^5 \times 113.1 = 11.42 \text{ MJ}$$

P_o is the constant ambient pressure.
(ii) The total mass of helium in the tank and the balloon in the initial and final states is the same because of mass conservation. Assuming that helium is an ideal gas

$$m_{he} = P_1 V_t / RT = V_t P_o / RT + V_{b2} P_o / RT$$

where V_t is the volume of the tank and P_1 the initial pressure of the helium in the tank. It is given that the temperature T remains constant at 30°C. From the above equation, the volume of the tank is

$$V_t = P_o V_{b2} / (P_1 - P_o) = 1.01 \times 113.1 / (10 - 1.01) = 12.71 \text{ m}^3$$

Example 3.17 The cylinder shown in Fig. E3.17 is divided into two halves by a *thin* piston *of negligible mass.* Initially, the left half of the cylinder contains air at pressure 2 bar and temperature 25°C. The right half is evacuated and the piston is held in position with a locking pin.

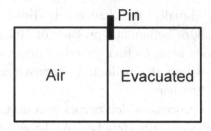

Fig. E3.17 Free expansion of a gas

The pin is now removed and the air expands pushing the piston. When the air attains its final equilibrium state filling the whole cylinder the temperature is 25°C. Calculate (i) the work done by the air and (ii) final pressure of the air.

Solution (i) When the locking pin is removed the air pressure acting on the piston will push it outwards. Because the space in front of the advancing piston is evacuated it does not encounter any resistance. Moreover, because the piston has negligible mass it does not gain any kinetic energy. Therefore the work done by the air is *zero*. However, if the mass of the piston was significant, then the work done by the air will be equal to the kinetic energy gained by the piston.

The air itself undergoes a non-equilibrium process with a time dependent pressure distribution within it. The complete dynamic analysis of this *unresisted expansion* process is beyond the scope of this book.

(ii) Applying the ideal gas equation of state to the initial and final equilibrium states 1 and 2, we have

$$m = P_{a1}V_1 / RT_1 = P_{a2}V_2 / RT_2$$

From the given data, $T_1 = T_2 = 298 \, \text{K}$

Hence $\qquad\qquad P_{a2} = P_{a1}V_1 / V_2 = 2 \times 1 / 2 = 1 \,$ bar.

Example 3.18 A fixed quantity of wet steam of mass 0.3 kg undergoes a polytropic expansion according to the relation $PV^{1.2} = C$, a constant. The initial pressure is 4.5 bar and the final pressure of 2.5 bar. The initial quality is 0.8. Calculate the final quality of the steam and the work done by the steam.

Solution The vapor and liquid specific volumes for the initial state 1, at 4.5 bar are [4]

$$v_{g1} = 0.4139 \text{ m}^3 \text{ kg}^{-1} \quad \text{and} \quad v_{f1} = 0.1088 \times 10^{-2} \text{ m}^3 \text{ kg}^{-1}$$

The initial specific volume is

$$v_1 = x_1 v_{g1} + (1 - x_1) v_{f1}$$

$$v_1 = 0.8 \times 0.4139 + 0.2 \times 0.1088 \times 10^{-2} = 0.3313 \text{ m}^3 \text{ kg}^{-1}$$

Therefore the initial volume is

$$V_1 = m v_1 = 0.3 \times 0.3313 = 0.099 \text{ m}^3$$

Applying the polytropic relation to states 1 and 2 we have

$$P_1 V_1^{1.2} = P_2 V_2^{1.2}$$

$$V_2 = 0.099 \times (4.5 / 2.5)^{1/1.2} = 0.1616 \text{ m}^3$$

The specific volume in the final state 2 is,

$$v_2 = 0.1616 / 0.3 = 0.538 \text{ m}^3 \text{ kg}^{-1}$$

In the final state 2 the vapor and liquid specific volumes are

$$v_{g2} = 0.7186 \text{ m}^3 \text{ kg}^{-1} \quad \text{and} \quad v_{f2} = 0.1068 \times 10^{-2} \text{ m}^3 \text{ kg}^{-1}$$

$$v_2 = 0.538 = x_2 v_{g2} + (1 - x_2) v_{f2}$$

$$0.538 = x_2 \times 0.7168 + (1 - x_2) \times 0.1068 \times 10^{-2}$$

Hence the quality in the final state is, $x_2 = 0.749$

(ii) The work done in the polytropic expansion is given by the expression

$$W = (P_1 V_1 - P_2 V_2) / (n - 1)$$

$$W = (4.5 \times 10^5 \times 0.099 - 2.5 \times 10^5 \times 0.1616) / 0.2 = 20.75 \text{ kJ}.$$

Example 3.19 The piston-cylinder arrangement shown in Fig. E3.19 is connected to a tank of volume 0.5 m³ containing argon at a pressure of 3 bar and temperature 30°C through a valve, which is initially closed. The piston rests at the bottom of the cylinder. The mass of the piston and the ambient pressure are such that a pressure of 1.8 bar has to act on the inside of the piston to *just* lift it. The valve is now opened, and remains open until the system attains the final equilibrium state with a uniform temperature of 30°C. Calculate the work done by the argon during the process.

Solution Figure E3.19 shows the initial and final positions of the piston in the cylinder. When the valve between the tank and the cylinder is opened gas from the tank will rush into the cylinder. This is a non-equilibrium process because of the steep pressure gradient between the tank and cylinder. The piston will gain kinetic energy and potential energy. Eventually the system will reach a final equilibrium state after some fluctuation of the various properties.

It is clear that for this non-equilibrium process we are not able to apply Eq. (3.6) to calculate the work done because the pressure on the inner surface of the piston is not known during the process.

Alternatively, we could consider the mechanical effect on the surrounding ambient due to the displacement of the piston. The work done by the gas is used to increase the potential energy of the piston, (E_p) and to do work against the ambient air (W_o). The gain in potential energy of the piston is

$$E_p = Mgz$$

where z is the height of the piston in the final equilibrium state above the initial position.

The work done in pushing the surrounding ambient air is

$$W_o = \int_0^{V_c} P_o dV = P_o V_c$$

where P_o is the constant ambient pressure and V_c is the volume of argon in the cylinder in the final equilibrium state.

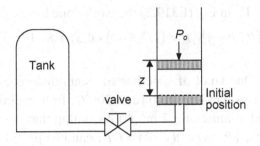

Fig. E3.19 Piston-cylinder set-up

Now $$z = V_c / A$$

where A is the area of the piston.

The force balance on the piston when it is just lifted by the gas is

$$P_i A = Mg + P_o A \qquad \text{(E3.19.1)}$$

The force balance equation for the piston in the final equilibrium state is

$$P_f A = Mg + P_o A$$

From the above equations we note that the final equilibrium pressure of the argon in the cylinder is equal to P_i. Moreover, because the valve remains open the pressure in the tank is also P_i.

Now the work done by the gas is

$$W_p = E_p + W_o = MgV_c / A + P_o V_c$$

Substituting from Eq. (E3.19.1) we have

$$W_p = V_c (Mg / A + P_o) = V_c P_i \qquad \text{(E3.19.2)}$$

By mass conservation, the total mass of argon in the initial and final states is the same. Applying the ideal gas equation to the initial and final equilibrium states

$$m = P_t V_t / RT = P_i (V_t + V_c) / RT$$

where V_t is the volume of the tank and P_t the initial pressure in the tank.

Hence $$V_c = \left(P_t / P_i - 1 \right) V_t$$

Substituting for V_c in Eq. (E319.2) the work done by the gas is

$$W_p = (P_t / P_i - 1)V_t P_i = (3/1.8 - 1) \times 0.5 \times 1.8 \times 10^5 = 60 \text{ kJ}$$

Example 3.20 One kmol of superheated steam undergoes a quasi-static expansion at a constant temperature of 200°C, from an initial volume of 3.2 m³ to a final volume of 5.2 m³. (i) Assuming that the van der Waals equation of state, $(P + a/v^2)(v - b) = \overline{R}T$, can be applied to superheated steam, calculate the work done by the steam. (ii) If superheated steam is treated as an ideal gas what is the work done?

Solution In worked example 2.20 we computed the values of the constants a and b in van Waals equation for steam as a = 5.52 × 10⁵ Nm⁴ kmol⁻² and b = 0.0304 m³ kmol⁻¹.

Since the expansion process is quasi-static we have

$$W = \int_{V_1}^{V_2} P dV$$

Rearranging the van der Waals equation we obtain P as

$$P = \overline{R}T/(v - b) - a/v^2$$

where v is the volume per kmol.

Therefore the work done in the constant temperature expansion is

$$W = \int_{v_1}^{v_2} \left(\overline{R}T/(v - b) - a/v^2 \right) dv$$

$$W = \overline{R}T \ln\left[(v_2 - b)/(v_1 - b) \right] + a\left(1/v_2 - 1/v_1 \right)$$

Substituting numerical values in the above equation

$$W = 8.3145 \times 10^3 \times 473 \times \ln\left(\frac{5.2 - 0.0304}{3.2 - 0.0304} \right)$$

$$+ 5.52 \times 10^5 \times \left(\frac{1}{5.2} - \frac{1}{3.2} \right)$$

$$W = 1923.86 - 66.346 = 1857.5 \text{ kJ}$$

(i) Using the expression obtained in worked example 3.11 for the work done by an ideal gas during an isothermal expansion we have

$$W_i = \overline{R}T \ln(V_2/V_1) = 8.3145 \times 10^3 \times 473 \times \ln(5.2/3.2) \text{ kJ}$$

$$W_i = 1909.38 \text{ kJ}$$

The prediction of the work done using the ideal gas equation is about 2.8% higher.

Problems

P3.1 A hollow conical vessel of height h and base radius r is pushed down vertically to the bottom of large tank of water in a quasi-static manner using an external force. The base of the cone is on top. The density of water is ρ_w and the height of water is H. Neglecting the weight of the vessel, show that the work done by the force is

$$W = r^2 h^2 g \rho_w \pi [H/3h - 1/4]$$

P3.2 A gas-filled balloon has a volume of V_0 at sea level where the ambient pressure is P_0. The balloon rises slowly in the atmosphere to a height H above sea level. The variation of atmospheric pressure P with height z is given by

$$P(z) = P_o \exp(-\rho_o g z / P_o),$$

where g is the acceleration due to gravity. The volume V of the balloon varies linearly with height according to the relation $V(z) = V_0 + \alpha z$, where α is a constant. Show that the total work done by the skin of the balloon on the atmosphere is given by

$$W = \alpha P_o{}^2 [1 - \exp(-\rho_o H g / P_o)]/\rho_o g$$

P3.3 A rigid spherical object of mass M and radius r_o is lifted from the bottom of a large body of water of depth H. The density of the water varies with depth in a linear manner given by the relation, $\rho_w(z) = a + bz$, where z is the depth of water below the surface and a and b are constants. The ambient pressure at the water surface is P_o.

Show that the total work done, W in lifting the object until the top surface just reaches the water surface is given by

$$W = g(H - 2r_o)(M - V_o a - V_o bH/2)$$

where V_O is the volume of the object. State all assumptions clearly.

P3.4 0.2 kg of air undergoes a cyclic process. The pressure and temperature at state 1 are 1.2 bar and 10°C respectively. The air receives heat at constant volume until the pressure is 1.6 bar. An isothermal expansion brings the air to a state where the pressure is 1.2 bar. The cycle is completed by a constant pressure compression of the air to state 1. Calculate (i) the work done in the three processes, and (ii) the net work done in the cyclic process. Draw the P-V and T-V diagrams for the cycle. Assume that air is an ideal gas. [*Answers*: $W_{12} = 0$, $W_{23} = 6.23$ kJ, $W_{31} = -5.4$ kJ, $W_{net} = 0.83$ kJ]

P3.5 A fixed quantity of superheated steam of mass 0.01 kg undergoes an *isothermal* expansion at a temperature of 150°C, from a pressure of 3 bar to 0.5 bar. (i) Plot the P-V diagram for this process by extracting the relevant data from the thermodynamic property tables [4]. (ii) Using a graphical or numerical method obtain the work done by the steam. (iii) Calculate the work done by assuming that steam is an ideal gas with a molecular mass of 18 kg kmol^{-1}. [*Answers*: (ii) 3.58 kJ, 3.50 kJ]

P3.6 A spring-loaded piston cylinder set-up contains 0.05 m^3 of a gas at a pressure of 1.1 bar. The ambient pressure is 1.0 bar. The movement of the piston is restrained by a linear spring which initially touches the piston but exerts no force on it. The diameter of the piston is 0.2 m. Heat is supplied to the gas in a quasi-static manner until volume and pressure are 0.07 m^3 and 1.65 bar respectively. (i) Obtain the P-V relation for the gas. (ii) Calculate the work done by the gas. (iii) Plot the P-V diagram for the process. [*Answers*: (i) $P(kPa) = 2754V - 27.7$, (ii) 2.75 kJ]

P3.7 A fixed quantity of air of mass 0.055 kg undergoes a cycle consisting of three processes. At state 1, the volume and pressure are

0.05 m³ and 1 bar. The air is compressed to a pressure of 1.8 bar in a polytropic process for which $PV^{1.3} = $ constant. A constant volume process 2-3 and a constant pressure process 3-1 bring the air back to its original state to complete the cycle. Calculate (i) the work done by the air in each process, and (ii) the net work done in the cycle. Draw *P-V* and *T-V* diagrams for the cycle. Assume that air is an ideal gas. [*Answers*: (i) $W_{12} = -2.423$ kJ, $W_{23} = 0$, $W_{31} = 1.82$ kJ, (ii) $W_{net} = -0.603$ kJ]

P3.8 A fixed quantity of wet steam of mass 0.05 kg is contained in a cylinder behind a piston. The cross sectional area and mass of the piston are 0.005 m² and 100 kg respectively. The ambient pressure is 1.01 bar. Initially the piston is held in position with a pin. The initial pressure and quality of steam are 3.5 bar and 0.2 respectively. The pin is now removed and the steam undergoes a non-equilibrium expansion process. In the final equilibrium state the steam has a quality of 0.6. (i) Calculate the work done by the steam. (ii) Draw the *P-V* and *T-V* diagrams for the processes. [*Answer*: (i) 3.94 kJ]

P3.9 A metal wire of diameter 0.5 mm and length 1.2 m is stretched in a quasi-static process at a constant temperature from an initial tension of 15 N to a final tension of 90 N. Calculate the work done in stretching the wire. The Young's modulus of the material of the wire at the temperature is 2.3×10^{11} Nm⁻². [*Answer*: 0.105 J]

P3.10 One kmol of a gas undergoes a constant temperature compression from an initial volume V_1 to a final volume V_2. (i) If the equation of state of the gas is

$$PV = \bar{R}T(1 + \alpha / V)$$

where α is a constant, obtain an expression for the work done on the gas. (ii) If the gas is assumed to be an ideal gas will the predicted work done be more or less? [*Answers*: (i) $W = \bar{R}T \ln(V_1/V_2) + \bar{R}T\alpha(1/V_2 - 1/V_1)$, (ii) $W = \bar{R}T \ln(V_1/V_2)$]

References

1. Cravalho, E.G. and J.L. Smith, Jr., *Engineering Thermodynamics*, Pitman, Boston, MA, 1981.
2. Jones, J.B. and G.A. Hawkins, *Engineering Thermodynamics*, John Wiley & Sons, Inc., New York, 1986.
3. Reynolds, William C. and Henry C. Perkins, *Engineering Thermodynamics*, 2nd edition, McGraw-Hill, Inc., New York, 1977.
4. Rogers, G.F.C. and Mayhew, Y.R., *Thermodynamic and Transport Properties of Fluids*, 5th edition Blackwell, Oxford, U.K. 1998.
5. Van Wylen, Gordon J. and Richard E. Sonntag, *Fundamentals of Classical Thermodynamics*, 3rd edition, John Wiley & Sons, Inc., New York, 1985.
6. Zemansky, M., *Heat and Thermodynamics*, 5th edition, McGraw-Hill, Inc., New York, 1968.

Chapter 4

The First Law of Thermodynamics

In the preceding chapter we categorized all energy interactions between a system and its surroundings either as work or heat. This was a useful starting-point for the introduction of the *First Law of Thermodynamics*, which is essentially the law of conservation of energy formulated in a unified manner to include all energy forms. The first law relates work and heat interactions that occur at the boundary of a system, to the energy possessed or stored by a system, which we shall call the *internal energy* of the system. The emphasis in this chapter will be the application of the first law to a closed system undergoing steady processes. In the next chapter we shall consider open systems and transient or unsteady processes.

4.1   First Law for a Cyclic Process

The first law of thermodynamics can be formulated in several different ways. We will adopt the approach, known as Poincare's formulation, because it makes use of the concepts that we are already familiar with from our work in the earlier chapters.

The first law states that when a *closed system* undergoes a *cyclic process*, the *net heat interaction* is equal to the *net work interaction*. This statement of the law can be further elaborated by referring to the cyclic process consisting of 'n' individual processes shown graphically in Fig. 4.1. In our sign convention, we shall take as *positive* quantities the work done *by* the system W, and the heat flow *into* the system Q. It should be noted that this choice of the signs for work and heat is arbitrary. In Fig. 4.1 we have indicated with arrows the heat flow into the

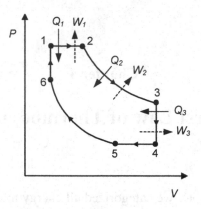

Fig. 4.1 Cyclic process

system and the work flow out of the system for the different processes that constitute the cycle. The first law for a cyclic process now translates to the following mathematical form:

$$Q_1 + Q_2 + \cdots + Q_i + \cdots + Q_n = W_1 + W_2 + \cdots + W_i + \cdots + W_n \qquad (4.1)$$

Expressed in the compact form Eq. (4.1) reads

$$\sum_{i=1}^{i=n} Q_i = \sum_{i=1}^{i=n} W_i \qquad (4.2)$$

In the cyclic-integral notation Eq. (4.2) can be written in the form

$$\oint \delta Q = \oint \delta W \qquad (4.3)$$

where δQ and δW are the heat input and work output during an infinitesimal change of state of the system. We recall that both heat and work are path-dependent quantities and are therefore not properties of the system. The symbol 'δ', instead of the more common 'd', is used to differentiate the changes of path-dependent quantities from those of properties.

Recall that we demonstrated the path-dependence of work done in chapter 3 and considered particular cases of path independent quantities in worked example 1.4. The first law is applied to cyclic processes in worked examples 4.2, 4.11 and 4.12.

4.2 First Law for a Change of State

Not all processes of importance in engineering thermodynamics are cyclic processes. There are numerous situations where we need to apply the first law to a single process which results in a change of state of the system. We shall now derive the form of the first law for a change of state using the cyclic-form of the law that we have already stated in Sec. 4.1.

Consider the cyclic process that consists of two processes, depicted in Fig. 4.2. The system undergoes a process A resulting in a change of state from 1 to 2, both of which are equilibrium states. The process B returns the system back to state 1 to complete a cycle. Therefore 1A2B1 constitutes a cyclic process to which we apply the first law to obtain the equation

$$Q_{12A} + Q_{21B} = W_{12A} + W_{21B} \tag{4.4}$$

We now select an alternative process C to return the state of the system from 2 to 1. Since 1A2C1 is a cyclic process we can apply the first law to obtain the equation

$$Q_{12A} + Q_{21C} = W_{12A} + W_{21C} \tag{4.5}$$

Subtracting Eq. (4.5) from Eq. (4.4) and rearranging we have

$$Q_{21B} - W_{21B} = Q_{21C} - W_{21C} \tag{4.6}$$

Equation (4.5) shows that the difference between the heat interaction and the work interaction for processes B and C is the same. Since the

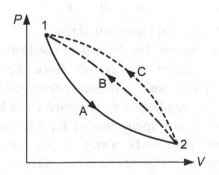

Fig. 4.2 Change of state

process C was chosen arbitrarily, we note that the quantity $(Q\text{-}W)$ is *independent* of the process path. Recalling our definition of a property, we can conclude that $(Q\text{-}W)$ is a property of the system. This new property is called the *internal energy* of the system and is denoted by the symbol E.

In summary, the first law for a single process states that although Q and W are path-dependent quantities, the difference $(Q\text{-}W)$ is *path-independent*, and is therefore a property of the system. It depends only on the state of the system. However, as for the physical nature of this new property the first law has little to add.

Equation (4.6) can be expressed in terms of the end-states 1 and 2 of a single process in the form

$$Q_{1-2} - W_{1-2} = \text{Increase in internal energy}$$

$$Q_{1-2} = (E_2 - E_1) + W_{1-2} \qquad (4.7)$$

where E_2 and E_1 are the internal energy of the system in states 2 and 1 respectively. Q_{1-2} is the heat *input*, and W_{1-2} is the work *output or done by* the system during the process.

The differential form of Eq. (4.7) for an infinitesimal change of state is

$$\delta Q = dE + \delta W \qquad (4.8)$$

where dE is the *increase* in the internal energy.

The rate-form of Eq. (4.8) is

$$\frac{\delta Q}{dt} = \frac{dE}{dt} + \frac{\delta W}{dt} \qquad (4.9)$$

where dt is the infinitesimal increment of time.

In Sec. 3.1 we showed that for a pure mechanical system, like the motion of a body subjected to an external force, the work done is equal to the increase in kinetic energy of the system. We hope to generalize this finding to more complicated systems involving both work and heat interactions. Although the application of Eq. (4.7) appears to be quite straightforward, we need to be aware of the following prerequisites before we input the various quantities in the equation.

(i)   The system, the boundary and the surroundings have to be clearly defined.

(ii) The interactions between the system and its surroundings that occur at the boundary have to be identified and categorized as work or heat.

(iii) All internal energy forms of the system that are likely to be affected by the process and the boundary interactions should be included in the term E.

The procedure is best illustrated by considering two practical examples of what are called *uncoupled-systems* and *coupled-systems*. The former category of systems is common in basic physics while the latter types are of particular importance in thermodynamics.

4.2.1 *An uncoupled-system*

Figure 4.3 shows a bucket of hot molten metal being hoisted vertically as may happen in a foundry. The bucket and its contents, of fixed mass M, is our system. Its temperature T may be assumed spatially uniform. The centre of mass, G of the bucket is at an elevation z above ground level and the vertical velocity of G is V. The hoisting cable attached to the top outer surface of the system has a tension F. The system interacts with the surrounding ambient which is at a uniform temperature T_o.

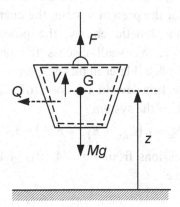

Fig. 4.3 An uncoupled system

Consider the vertical movement dz of the system that happens in a time dt. The work done on system by the external force can be expressed as

$$\delta W = -Fdz \qquad (4.10)$$

where the negative sign signifies that work is done *on* the system. From our knowledge of basic mechanics (also see Sec. 3.1)

$$E_{KE} + E_{PE} = MV^2/2 + Mgz \qquad (4.11)$$

where g is the acceleration due to gravity. E_{KE} and E_{PE} are the kinetic energy and the gravitational potential energy respectively. By virtue of its mass and temperature the system has thermal energy which we shall express as

$$E_{TH} = Mu(T) \qquad (4.12)$$

At this point we shall not elaborate on the physical nature of thermal energy per unit mass, $u(T)$ except to state that it depends on the temperature of the system. Since the system is at a higher temperature than the surroundings, there will be a heat flow out of the system, $-\delta Q$.

In order to apply the first law to the system we need to list under the category, internal energy, E all the different energy forms that the bucket and its contents may possess.

If from our experience, some of these energy forms are not affected by the interactions with the surroundings, then we may choose to ignore these energy forms. For the present system, the energy forms that appear to be pertinent are the kinetic energy, the potential energy and the thermal internal energy. We shall discuss the physical nature of the internal energy in more detail later in this chapter.

We now apply the first law in the form of Eq. (4.8) to the infinitesimal change of state of the system

$$\delta Q = d(E_{KE} + E_{PE} + E_{TH}) + \delta W \qquad (4.13)$$

Substituting the expressions from Eqs. (4.10), (4.11) and (4.12) in the above equation we have

$$-\delta Q = d(MV^2/2 + Mgz + Mu(T)) - Fdz \qquad (4.14)$$

Although we have combined all the boundary interactions and the internal energy forms in a single equation, Eq. (4.14), our experience tells us that the kinetic and potential energy are affected only by the work done by the tension of the cable and not by the heat loss to the surroundings. Similarly, the thermal internal energy of the system is not affected by the work done by the tension but depends on the heat losses. In view of these observations we can decompose Eq. (4.14) into two *uncoupled* equations as follows.

$$d\left(MV^2/2 + Mgz\right) = Fdz \tag{4.15}$$

and $$-\delta Q = d(Mu) \tag{4.16}$$

The main point that emerges from this example is that for *uncoupled-systems* each form of boundary interaction only affects one form of internal energy. In the present case the work interaction affects only the kinetic and potential energies as it happens in most mechanical systems. The heat interaction on the other hand, affects only the thermal internal energy, again a familiar situation in heat transfer. The example also helped to cast some light on the physical nature of the new property called internal energy which arose from our formulation of the first law. Two uncoupled systems are considered in worked examples 4.5 and 4.6.

4.2.2 A coupled-system

Consider the piston-cylinder apparatus shown in Fig. 4.4, where a *light* piston of cross sectional area A is attached to a vertical shaft that pushes the piston with a force F. The ambient pressure is P_o. The gas in the cylinder is at a temperature T and pressure P, and it receives a quantity of heat δQ from an external source.

The piston is at a level z from the top of the cylinder at time t. As a result of the two boundary interactions the piston moves a distance dz in a time dt. The gas in the cylinder is our system. The work done on the system by the force F and the ambient pressure is given by

$$\delta W = (Fdz + P_o Adz)$$

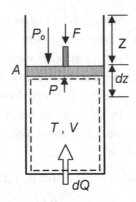

Fig. 4.4 A coupled system

Since the piston has negligible mass, the force balance on the piston gives

$$PA = F + P_o A$$

Hence $$\delta W = P\,A dz \qquad (4.17)$$

The internal energy of the gas is

$$E_{TE} = mu(T,P) \qquad (4.18)$$

where m is the mass and u is the thermal internal energy per unit mass.

Applying the first law in the form of Eq. (4.8) we have

$$\delta Q = dE_{TH} + \delta W \qquad (4.19)$$

Substituting in Eq. (4.19) from Eqs. (4.17) and (4.18)

$$mdu(T,P) = \delta Q - P(T,V)A dz \qquad (4.20)$$

We observe from Eq. (4.20) that the internal energy of the system is affected by both the heat interaction and the work interaction. Therefore the system is called a *coupled-system*. Such systems have wide engineering applications, especially in thermal power systems.

4.3 Internal Energy – A Thermodynamic Property

Our formulation of the first law predicted the existence of a property called the internal energy. In order to understand the physical nature of this new property we need to recall some of the common energy transformations from our earlier studies of basic physics. Two forms of energy that arise in the study of mechanics are the kinetic energy and the gravitational potential energy. Both these forms of energy are possessed by bodies or systems in motion and are therefore forms of internal energy (see Sec. 3.1). Kinetic and potential energy are associated with the system as a whole and are often called energy forms of the *bulk motion*.

Another common form of energy is the internal thermal energy associated with the temperature of a system. Application of the simple kinetic theory to a monatomic ideal gas in Sec. 2.7 revealed that the absolute temperature of the gas is a measure of the mean translational kinetic energy of the molecules. For more complicated polyatomic gases, the temperature is related to the number of different modes of molecular motion that include translation, rotation, vibration and others. The thermal internal energy, usually denoted by U, plays a very important role in thermodynamics and will be explored in greater detail in the subsequent sections of this chapter.

Several other forms of internal energy, although not frequently encountered in this book, need to be mentioned here. A compressed or extended spring possesses strain energy which is a form of internal potential energy. A fluid film like a soap bubble has surface energy which is also a form of internal energy. A system subjected to an electric or a magnetic field could possess stored energy by virtue of its interaction with these fields as it does when it is displaced in a gravitational field. There are also forms of internal energy that could undergo transformation to other forms of internal energy as it happens in combustion processes and nuclear fission reactions.

In the case of combustion, the chemical binding energy of the fuel, a form of internal energy, is first transformed into thermal internal energy of the products of the chemical reaction which manifests as an increase of temperature. The thermal internal energy in turn produces a heat interaction or a work interaction at the boundary of the combustion system depending on the application.

In the case of a fuel element in a nuclear reactor, the fission reaction converts the nuclear binding energy of the fuel first into thermal internal energy which then results in a heat interaction at the boundary between the fuel element and the coolant.

In summary, the internal energy of a system could, in general, include a number of different energy forms. These include, kinetic energy, potential energy, thermal internal energy, chemical binding energy, surface energy, nuclear binding energy and others. However, in most of the systems and applications of interest in this book we would be dealing mainly with only three forms of internal energy. These are the kinetic energy, the gravitational potential energy and the thermal internal energy. Expressed mathematically

$$E = \left(MV^2/2 + Mgz + U\right) \tag{4.21}$$

where M is the mass of the system, V is the velocity, g is the acceleration due to gravity, and z is the vertical height above a datum level where the potential energy is taken as zero. U is the thermal internal energy associated with microscopic-level motion of the molecules that constitute the system.

4.4 State Postulate

The state postulate is a statement on the number of independent properties required to specify the state of a system. For a system involving *only compressible work*, (PdV) *two intensive properties* are sufficient to completely specify an equilibrium state of the system. The above statement is sometimes called the *two-property rule*.

4.5 Internal Energy and Heat Capacities

Having discussed the nature of internal energy, E in general terms, we will now consider a few special situations where the nature of a substance allows us to develop methods to calculate and also to measure the internal energy. These include simple compressible substances and pure substances. It is to be noted that by internal energy we now mean the part of E that depends on the temperature and we shall denote it by the symbol U.

4.5.1 *Heat capacity at constant volume*

From the state postulate it follows that for a simple compressible system the thermal internal energy U may be expressed in terms of the temperature T and the volume V

$$U = U(T, V) \tag{4.22}$$

The heat capacity at constant volume, C_V is defined as

$$C_V = \left(\frac{\partial U}{\partial T} \right)_V \tag{4.23}$$

The heat capacity C_V is a property of the system because it is the slope of the graph of U versus T at constant V. We recall that U, T and V are properties of the system.

We now consider a *constant volume* process that would enable the measurement of C_V. The substance is contained within a rigid boundary and is heated in a quasi-static manner by supplying a small quantity of heat δQ which results in an increase of temperature of dT. Since boundary is rigid the work done δW is zero. The system as a whole is stationary, and therefore the change in bulk kinetic energy and potential energy are also zero.

Applying the first law in the form of Eq. (4.8) we have

$$\delta Q = dU \tag{4.24}$$

It follows from Eqs. (4.24) and (4.23) that

$$\left(\frac{\delta Q}{dT} \right)_{const.V} = \left(\frac{dU}{dT} \right)_{const.V} = \left(\frac{\partial U}{\partial T} \right)_V = C_V \tag{4.25}$$

Equation (4.25) shows that a constant volume heating process offers a possible method to determine C_V by measuring the heat input and the rise in temperature in the process. It should be noted that Eq. (4.25) which relates C_V to the heat input, δQ is only true for a *constant volume* process. However, from the general definition, given by Eq. (4.23), we see that C_V is a system property which is independent of the processes involved. It is clear that C_V is not defined for phase change processes like evaporation and condensation where the temperature of the system is constant.

4.5.2 *Enthalpy*

We now introduce a new property called *enthalpy* that is closely related to the thermal internal energy U. Enthalpy is most useful in the analysis of processes in open-systems which will be the subject of the next chapter. As a prelude, we use a constant pressure process to demonstrate the usefulness of enthalpy as a property.

Consider the piston-cylinder arrangement where the pressure, P of the gas within the cylinder is maintained constant by keeping the weights on the piston fixed. A small quantity of heat, δQ is supplied to the gas in a quasi-static manner resulting in an increase in volume dV. The work done by the gas is PdV. Applying the first law to this process we have

$$\delta Q = dU + PdV \qquad (4.26)$$

Since the *pressure is constant* during the process we could rewrite Eq. (4.26) as

$$\delta Q = d(U + PV) \qquad (4.27)$$

Now the quantity $(U + PV)$ is a property because U, P and V are all properties of the system. This new property is called *enthalpy* which is denoted by the symbol H.

Hence we could write Eq. (4.27) in the form

$$\delta Q = d(U + PV) = dH \qquad (4.28)$$

It should be emphasized that the change in enthalpy is equal to the magnitude of the heat interaction only for a *constant pressure* process. However, being a property, the enthalpy is a function of the state of the system and therefore independent of the processes undergone by the system.

4.5.3 *Heat capacity at constant pressure*

The system property called the heat capacity at constant pressure, C_P is related directly to the enthalpy of the system. Following the two-property rule we can express the enthalpy as a function of T and P.

$$H = H(T, P)$$

We define C_P as

$$C_P = \left(\frac{\partial H}{\partial T}\right)_P \qquad (4.29)$$

The heat capacity C_P is a property because it is the slope of the graph of H versus T at constant P. Note that the three quantities H, T and P are properties of the system.

A constant pressure process offers a method to determine C_P by measuring the heat input and the corresponding rise in temperature. Consider the piston-cylinder arrangement where heat is supplied to the system at constant pressure. Applying Eq. (4.28) to the constant pressure process we have

$$\delta Q = dH \qquad (4.30)$$

It follows from Eqs. (4.30) and (4.29) that

$$\left(\frac{\delta Q}{dT}\right)_{const.P} = \left(\frac{dH}{dT}\right)_{const.P} = \left(\frac{\partial H}{\partial T}\right)_P = C_P \qquad (4.31)$$

It should be noted that Eq. (4.31) which relates C_P to the heat input, δQ is only true for a *constant pressure* process.

However, from the general definition given Eq. (4.29) we see that C_P is a system property which is independent of the processes involved. It is clear that C_P is not defined for phase change processes where the temperature of the system is constant.

4.6 Properties of Ideal Gases

The ideal gas is a useful model for engineering analysis of thermodynamic systems.

In chapter 2 we introduced the equation of state of the ideal gas which is a relationship between the properties P, V and T. In this section we shall extend the ideal gas model to include its energy-related properties.

4.6.1 *Internal energy, enthalpy and heat capacities of an ideal gas*

The two-property rule for a simple compressible pure substance states that, in general, the thermal internal energy, U is a function of two other properties. However, using the first and second laws of thermodynamics and the equation of state it *can be proved* that the internal energy of an *ideal gas* is a function of *temperature only* (see chapter 14). Therefore

$$U = U(T) \tag{4.32}$$

Now the enthalpy of an ideal gas is given by

$$H = U(T) + PV \tag{4.33}$$

Substituting in Eq. (4.33) from the ideal gas equation of state, $PV = mRT$, we have

$$H = U(T) + mRT = H(T) \tag{4.34}$$

Equation (4.34) shows that the enthalpy of an *ideal gas* is also a function of *temperature only*.

Hence the heat capacities of an ideal gas can be expressed in the form

$$C_V = \left(\frac{\partial U}{\partial T} \right)_V = \left(\frac{dU}{dT} \right) \tag{4.35}$$

$$C_P = \left(\frac{\partial H}{\partial T} \right)_P = \left(\frac{dH}{dT} \right) \tag{4.36}$$

We are now able to write two simple relations between the internal energy, the enthalpy, and the heat capacities of an *ideal gas*. These follow directly from Eqs. (4.35) and (4.36) as

$$dU = C_V dT \tag{4.37}$$

and

$$dH = C_P dT \tag{4.38}$$

Differentiation of Eq. (4.34) gives

$$dH = dU + mRdT \tag{4.39}$$

Substituting from Eqs. (4.37) and (4.38) in the above equation we have

$$C_P dT = C_V dT + mRdT \tag{4.40}$$

Hence

$$C_P - C_V = mR \tag{4.41}$$

We define the heat capacities per unit mass or the *specific heat capacities* as

$$c_v = C_V / m \quad \text{and} \quad c_p = C_P / m$$

Equation (4.41) can be expressed in terms of the *specific heat capacities* as

$$c_p - c_v = R \tag{4.42}$$

The molar-from of Eq. (4.42) is

$$\bar{c}_p - \bar{c}_v = \bar{R} \tag{4.43}$$

where $\bar{c}_v$ and $\bar{c}_p$ are the specific heat capacities per mole of the ideal gas and $\bar{R}$ is the *universal gas constant*.

Summarized below are the typical units of the various energy-related properties introduced in this chapter:

$$[U] = \text{J}, \quad [H] = \text{J}, \quad [C_V] = \text{J K}^{-1}, \quad [C_P] = \text{J K}^{-1},$$

$$[c_v] = \text{J kg}^{-1} \text{K}^{-1}, \; [c_p] = \text{J kg}^{-1} \text{K}^{-1}, \quad [R] = \text{J kg}^{-1} \text{K}^{-1}$$

$$[\bar{c}_v] = \text{J kmol}^{-1} \text{K}^{-1}, \; [\bar{c}_p] = \text{J kmol}^{-1} \text{K}^{-1}, \; [\bar{R}] = \text{J kmol}^{-1} \text{K}^{-1}$$

4.6.2 *Heat capacities and kinetic theory*

In Sec. 2.7 we applied the kinetic theory to an ideal monatomic gas to obtain an expression for the temperature in terms of the molecular kinetic energy. We shall now extend this approach to relate the internal energy and the heat capacities of an ideal gas to the energy of the constituent molecules.

Consider the monatomic gas, the molecules of which are free to move in three mutually perpendicular directions. The translation in each direction constitutes a degree of freedom and therefore the monatomic gas has 3 degrees of freedom. The internal energy per mole of gas can be written as

$$\bar{u} = N_A \left(m' \bar{C}^2 / 2 \right) \tag{4.44}$$

where m' is the mass of a molecule, $\bar{C}$ is the average molecular speed and N_A is the number of molecules per mole, which is called Avogadro's number. But, $m'N_A = M$, the molar mass of the gas. Therefore from Eq. (4.44), we have

$$\bar{u} = M\left(\overline{C}^2/2\right) \tag{4.45}$$

In Sec. 2.7 we obtained the following equation for the temperature.

$$T = M\overline{C}^2/3\overline{R} \tag{4.46}$$

From Eqs. (4.45) and (4.46) it follows that

$$\bar{u} = 3\overline{R}T/2 \tag{4.47}$$

Now the enthalpy per mole of the gas is given by

$$\bar{h} = \bar{u} + \overline{R}T = 5\overline{R}T/2 \tag{4.48}$$

The molar specific heat capacities of the ideal gas are obtained as

$$\bar{c}_v = d\bar{u}/dT = 3\overline{R}/2 \tag{4.49}$$

$$\bar{c}_p = d\bar{h}/dT = 5\overline{R}/2 \tag{4.50}$$

In order to extend the above analysis to diatomic gases, we invoke the *principle of equipartition of energy* which states that the total energy of a molecule is shared equally between the different degrees of freedom of the molecule. Hence for a monatomic gas which has three translational degrees of freedom, it follows from Eq. (4.47) that the internal energy per degree of freedom is $\overline{R}T/2$. Moreover, from Eq. (4.49) the molar specific capacity at constant volume per degree of freedom becomes $\overline{R}/2$.

Now consider a diatomic molecule which can be modeled as two point masses connected by a spring. The centre of mass of the molecule has *three* translational degrees of freedom. In addition, the molecule can rotate about *two* mutually perpendicular directions. Being point masses, the rotation about the axis connecting the masses will acquire no rotational kinetic energy. Furthermore, the two masses could vibrate along the connecting axis as a spring-mass system with strain energy stored in the spring and vibrational kinetic energy stored in the two masses. The latter motion constitutes *two* degrees of freedom.

Therefore the diatomic molecule has 7 degrees of freedom. From the principle of equipartition of energy it follows that the diatomic molecule could have a molar specific capacity of up to $7\overline{R}/2$. The number of degrees of freedom that are excited and therefore participate in the internal energy storage depends on the temperature of the gas. At low temperatures only the three translational modes are excited, and $\bar{c}_v = 3\bar{R}/2$. At intermediate temperatures the two rotational modes are also excited and $\bar{c}_v = 5\bar{R}/2$. When temperature is very high all 7 modes participate in the energy sharing.

The variation of $\bar{c}_v$ with temperature predicted by the kinetic theory is depicted in Fig. 4.5. The gradual change seen between the values for the different degrees of freedom is because all the molecules of the gas do not undergo the transition from one level to the next at the same temperature. Table 4.1 gives the molar specific heat capacities of three gases obtained using the tabulated data from [4]. Notice that the values predicted by the kinetic theory, listed below Table 4.1, are in good agreement with the corresponding values in the table.

Table 4.1 Molar Specific heat capacities of three gases*

$T_s(K)$	Gas	CO	N_2	H_2
300	c_v	20.8	20.8	20.3
	$c_p = c_v + R$	29.1	29.1	28.6
600	c_v	22.1	21.8	20.7
	$c_p = c_v + R$	30.4	30.1	29.1

*The values of c_v [kJ kmol^{-1} K^{-1}] and $c_p = c_v + R$, predicted by the kinetic theory are: $3R/2 = 12.47$, $5R/2 = 20.78$, $7R/2 = 29.1$

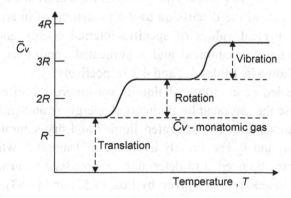

Fig. 4.5 Variation of $\bar{c}_v$ predicted by the Kinetic Theory

4.7   Temperature Dependence of Heat Capacity

Accounting for the temperature dependence of the specific heat capacities of ideal gases in engineering computations could be quite tedious. When the temperature change involved in a process is not too large, the specific heat capacities of the ideal gas may be assumed constant. From Fig. 4.5 we notice this to be true as long as the number of degrees of freedom of the molecules remains the same. This approximation is very common in engineering analysis and we shall use it in the worked examples in this chapter.

However, when the temperature range of interest is relatively wide, as in combustion processes, ignoring the temperature dependence of the specific heat capacities may lead to significant errors in the computed quantities. In such situations, the temperature dependence could be included in the analysis by fitting polynomials to the tabulated heat capacity data in the form

$$C_v = f(T) = a_o + a_1 T + a_2 T^2 + a_3 T^3 + \cdots \tag{4.51}$$

where a_1, a_2 and a_3 are constants. The temperature dependence of the specific heat capacity is considered in worked example 4.19.

4.8   Internal Energy and Enthalpy of a Pure Substance

In Sec. 2.5 we discussed the use of tabulated property data for a pure substance that undergoes phase change focusing mainly on P, v and T. We shall now extend the discussion to the tabulation of internal energy and enthalpy. Typical values of specific internal energy and specific enthalpy of water for saturated and superheated conditions, extracted from [4], are shown in Tables 4.2 and 4.3 respectively.

For sub-cooled or compressed liquids we ignore the effects of the pressure and use the saturated liquid internal energy u_f, and enthalpy h_f at the desired temperature. For saturated liquids and dry saturated vapors the values of u_g and h_g are directly available in Table 4.2. where linear interpolation may be needed to determine intermediate values. For wet vapors we use the expressions given by Eqs. (4.52) and (4.53), similar to

Table 4.2 Internal energy and enthalpy of saturated water and steam

P (bar)	T_s (°C)	u_f (kJ kg⁻¹)	u_g (kJ kg⁻¹)	h_f (kJ kg⁻¹)	h_g (kJ kg⁻¹)
0.05	32.9	138	2420	138	2561
10	179.9	762	2584	763	2778

Table 4.3 Internal energy and enthalpy of superheated water vapor

P (bar) [T_s(°C)]	T(°C)	300	350	400
5 [151.8]	u (kJ kg⁻¹)	2804	2883	2963
	h (kJ kg⁻¹)	3065	3168	3272
50 [263.9]	u (kJ kg⁻¹)	2700	2810	2907
	h (kJ kg⁻¹)	2927	3070	3196

that used for the specific volume, to compute the mean internal energy and enthalpy of the mixture.

$$u = xu_g + (1-x)u_f \tag{4.52}$$

$$h = xh_g + (1-x)h_f \tag{4.53}$$

where x is the vapor quality.

For superheated vapors the internal energy and enthalpy are obtained by interpolation if data at the required temperature and pressure are not directly tabulated (see Table 4.3).

In summary, a procedure similar to that used in chapter 2 to extract the specific volume from tabulated data can be adopted to obtain the internal and the enthalpy for pure substances. It is important to note that the ideal gas equation of state is *not* applicable to wet steam and superheated steam.

For solids, the internal energy U and the enthalpy H are equal because of their very low compressibility. Therefore the following relation may be used to describe their behavior

$$dU = dH = mcdT \tag{4.54}$$

where c is the specific heat capacity.

Far below their critical temperature many liquids have a very low compressibility, similar to solids. In these situations they may be treated incompressible, and as a first-order approximation, Eq. (4.54) may be used to determine the change in internal energy. The temperature dependence of the specific heat capacity may be included in engineering computations by fitting a polynomial, similar to Eq. (4.51), to the tabulated specific heat capacity data.

4.9 Worked Examples

Example 4.1 The heat input Q, the work output W, the initial and final internal energies U_1 and U_2 respectively, and the increase in internal energy ΔU for 5 processes of a closed system are tabulated below. Obtain the missing quantities.

Process	Q_{in} (kJ)	W_{out} (kJ)	U_1 (kJ)	U_2 (kJ)	ΔU (kJ)
a	12	6	81	?	?
b	25	−8	?	−13	?
c	−10	?	?	60	−18
d	10	−6	?	−10	?
e	?	25	58	?	16

Solution We shall use the first law for a change of state to determine the missing quantities in the table above. The two relevant equations are:

$$Q_{in} = U_2 - U_1 + W_{out} \tag{E4.1.1}$$

and

$$\Delta U = U_2 - U_1 \tag{E4.1.2}$$

(a) $12 = U_2 - 81 + 6$, $U_2 = 87$, $\Delta U = 6$

(b) $25 = -13 - U_1 - 8$, $U_1 = -46$, $\Delta U = 33$

(c) $-10 = -18 + W_{out}$, $W_{out} = 8$, $U_1 = 78$

(d) $10 = -10 - U_1 - 6$, $U_1 = -26$, $\Delta U = 16$

(e) $Q_{in} = 16 + 25$, $Q_{in} = 41$, $U_2 = 74$

Example 4.2 Figure E4.2 shows the *P-V* diagram for the change of state of a closed system from a to b. The internal energies at states a and 2 are 2 kJ and 42 kJ respectively. During the process a-1-b the heat input and the work output of the system are 82 kJ and 32 kJ respectively. When the same change of state is achieved through the process a-2-b, the work output of the system is 12 kJ. When the state of the system is changed from b to a through the process b-3-a, the work output is 22 kJ. Calculate (i) the heat input during the process a-2-b, (ii) the heat interaction during the process b-3-a, and (iii) the heat interactions during the processes a-2 and 2-b.

Solution The given data are:

$$U_a = 2, \qquad U_2 = 42, \qquad W_{a1b} = 32,$$
$$Q_{a1b} = 82, \qquad W_{a2b} = 12, \qquad W_{b3a} = 22$$

Applying the first law to a change of state

$$Q_{in} = U_2 - U_1 + W_{out}$$

The above equation is now applied to the various processes shown in Fig. E4.2

$$Q_{a1b} = U_b - U_a + W_{a1b} = 82 = U_b - U_a + 32$$

$$U_b - U_a = 50 \text{ kJ}$$

$$Q_{a2b} = U_b - U_a + W_{a2b} = 50 + 12 = 62 \text{ kJ}$$

$$Q_{b3a} = U_a - U_b + W_{b3a} = -50 + 22 = -28 \text{ kJ}$$

$$W_{a2b} = W_{a2} + W^*{}_{2b} = 12 = W_{a2} + 0, \qquad W_{a2} = 12 \text{ kJ}$$

$$W_{a1b} = W^*{}_{a1} + W_{1b} = 32 = W_{1b} + 0, \qquad W_{1b} = 32 \text{ kJ}$$

$$Q_{a2} = U_2 - U_a + W_{a2} = (42 - 2) + 12 = 52 \text{ kJ}$$

$$Q_{a2b} = Q_{a2} + Q_{2b} = 52 + Q_{2b} = 62, \qquad Q_{2b} = 10 \text{ kJ}$$

*Note that the work interaction in a quasi-static constant-volume process is zero.

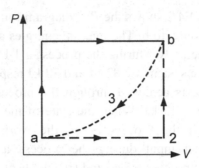

Fig. E4.2 *P-V* diagram

Example 4.3 In a heat treatment operation, a steel component of
mass 0.75 kg at an initial temperature of 200°C is cooled to 50°C by
placing it in a steel tank containing water at an initial temperature
of 25°C. The steel tank has a mass of 1.2 kg and is well-insulated.
The specific heat capacities of water and steel are 4.2 kJ kg⁻¹ K⁻¹
and 0.48 kJ kg⁻¹ K⁻¹ respectively. (a) Calculate the mass of water.
(b) Considering (i) the steel component, (ii) the tank, and (iii) the water
as closed systems, calculate the heat interaction, the work interaction,
and the change of internal energy.

Solution Consider the steel component and the tank of water as a closed
system. The subscripts *s*, *w* and *t* represent quantities related to the steel
component, the water and the tank respectively. (see Fig. E4.3). Choose
0°C as the reference state where the internal energy *U* is zero. Then the
initial internal energy of the system is

$$U_1 = (m_s c_s) \times 200 + (m_w c_w) \times 25 + (m_t c_t) \times 25$$

The final internal energy is

$$U_2 = (m_s c_s) \times 50 + (m_w c_w) \times 50 + (m_t c_t) \times 50$$

where *m* is the mass and *c* is the specific heat capacity at constant
volume. Note that for a solid and a liquid the heat capacities at constant
volume and constant pressure can be assumed equal because of their
relatively low compressibility.

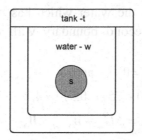

Fig. E4.3 Heat treatment operation

Applying the first law to the closed system we have

$$Q_{12} = U_2 - U_1 + W_{12} \tag{E4.3.1}$$

Since the tank is well-insulated, $Q_{12} = 0$. Assuming the tank to be rigid, $W_{12} = 0$. Substitute these conditions in Eq. (E4.3.1). Hence

$$U_2 = U_1$$

Substituting the given numerical data in the above equation

$$0.75 \times 0.48 \times (200 - 50) + 1.2 \times 0.48 \times (25 - 50)$$
$$= m_w \times 4.2 \times (50 - 25)$$

The mass of water, $m_w = 0.377$ kg.

We recall that in applying the first law to a closed system in the form of Eq. (E4.3.1), we follow the sign convention where the heat inflow, the work output, and the increase in internal energy are positive quantities.

Apply Eq. (E4.3.1) to the steel component which has a boundary with the water. Assume that the change in volume of the steel is negligible.

$$W_{s \to w} = 0$$
$$\Delta U_s = m_s c_s (50 - 200) = -0.75 \times 0.48 \times 150 = -54 \text{ kJ}$$
$$Q_{w \to s} = \Delta U_s + W_{s \to w} = -54 \text{ kJ}$$

Apply Eq. (E4.3.1) to the steel tank which has a boundary with the water. Neglecting any change in volume,

$$W_{t \to w} = 0$$
$$\Delta U_t = m_t c_t (50 - 25) = 1.2 \times 0.48 \times 25 = 14.4 \text{ kJ}$$
$$Q_{w \to t} = \Delta U_t + W_{t \to w} = 14.4 \text{ kJ}$$

Apply Eq. (E4.3.1) to the water which has one boundary with the steel component and a second boundary with the tank. Neglect any change in volume.

$$W_{w \to t} = 0 \qquad \text{and} \qquad W_{w \to s} = 0$$

$$\Delta U_w = m_w c_w (50 - 25) = 0.377 \times 4.2 \times 25 = 39.59 \text{ kJ}$$

$$Q_{s \to w} + Q_{t \to w} = \Delta U_w + W_{w \to t} + W_{w \to s} = 39.6 \text{ kJ}$$

Example 4.4 A well-insulated steel vessel of mass 1.2 kg contains 4 kg of water. A paddle wheel immersed in the water is driven by a motor that receives power from a storage battery as shown in Fig. E4.4. The heat capacity of the paddle wheel is 0.8 kJ K^{-1} and the thin non-conducting shaft connecting it to the motor has negligible heat capacity. The temperatures of the water, the wheel, and the vessel which may be assumed uniform, is found to be increasing at the rate of 0.012°C s^{-1} at a certain instant. The specific heat capacities of water and steel are 4.2 kJ kg^{-1} K^{-1} and 0.48 kJ kg^{-1} K^{-1} respectively.

Considering (i) the vessel, (ii) the paddle wheel and water, (iii) the motor, (iv) the battery and (v) the composite system as closed-systems, calculate the heat interaction, the work interaction and the rate of change of internal energy in Watts.

Solution The subscripts v, w, m, b, s and a denote quantities related to the vessel, the wheel and water, the motor, the battery, the composite system and the surroundings respectively.

To each system we apply the rate-form of the first law

$$\dot{Q} = \dot{U} + \dot{W} \qquad \text{(E4.4.1)}$$

where the 'dot' over a quantity represents differentiation, d/dt, t being time.

Consider the vessel as a closed system with an interacting boundary with the water and the wheel. The vessel is insulated on the outside. Applying the first law to the vessel we have

$$\dot{Q}_{w \to v} = \dot{U}_v + \dot{W}_{v \to w} \qquad \text{(E4.4.2)}$$

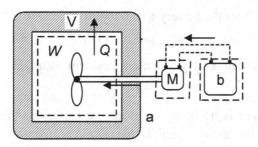

Fig. E4.4 Insulated tank with paddle wheel

The work interaction, $\dot{W}_{v\rightarrow w} = 0$. The rate of change of internal energy is

$$\dot{U}_v = m_v c_v (dT_w / dt) = 1.2 \times 0.48 \times 0.012 = 6.912 \text{ W}$$

Substituting in Eq. (E4.4.2)

$$\dot{Q}_{w\rightarrow v} = 6.912 \text{ W}$$

The water and the paddle wheel form a closed system, the boundary being the inside of the vessel and the shaft penetrating it. Applying the first law to the water and paddle wheel we have

$$\dot{Q}_{v\rightarrow w} = \dot{U}_w + \dot{W}_{w\rightarrow v} + \dot{W}_{w\rightarrow m} \qquad \text{(E4.4.3)}$$

Now $\qquad\qquad\qquad \dot{W}_{w\rightarrow v} = 0$

and $\qquad\qquad \dot{U}_w = m_w c_w (dT_w / dt) + m_{wh} c_{wh} (dT_{wh} / dt)$

$$\dot{U}_w = 4 \times 4.2 \times 0.012 + 0.8 \times 0.012 = 211.2 \text{ W}$$

Substituting in Eq. (E4.4.3)

$$-6.912 = 211.2 + 0 + \dot{W}_{w\rightarrow m}$$

$$\dot{W}_{w\rightarrow m} = -218.11 \text{ W}$$

Applying the first law to the motor as a closed system we have

$$\dot{Q}_{a\rightarrow m} = \dot{U}_m + \dot{W}_{m\rightarrow w} + \dot{W}_{m\rightarrow b}$$

Assume that the motor experiences no heat losses or internal energy changes. Therefore from the above equation

$$\dot{W}_{m\rightarrow w} + \dot{W}_{m\rightarrow b} = 0 = 218.11 + \dot{W}_{m\rightarrow b}$$

The work input from the battery to the motor is

$$\dot{W}_{b\rightarrow m} = 218.11 \text{ W}$$

Now consider the battery as a closed system. Applying the first law

$$\dot{Q}_{a\rightarrow b} = \dot{U}_b + \dot{W}_{b\rightarrow m}$$

Assume that there is no heat loss from the battery to the surroundings. Therefore from the above equation

$$\dot{U}_b + \dot{W}_{b\rightarrow m} = 0$$

$$\dot{U}_b = -\dot{W}_{b\rightarrow m} = -281.11 \text{ W}$$

Apply the first law to the composite system consisting of the vessel, the motor and the battery. The work interaction and the heat interaction with the surroundings are both zero. Therefore

$$\dot{Q}_{a\rightarrow s} = \dot{U}_s + \dot{W}_{s\rightarrow a} = \dot{U}_w + \dot{U}_v + \dot{U}_b + \dot{W}_{s\rightarrow a}$$

$$\dot{U}_s = 6.912 + 211.2 - 218.11 = 0$$

Example 4.5 A steel projectile of mass 0.003 kg moving at 250 ms^{-1} is fired into a stationary block of mass 5 kg of negligible thermal capacity. Initially the projectile and the block are at 20°C. Calculate (i) the speed of the block and the projectile soon after they start to move as a composite body and (ii) the temperature of the projectile at that stage. The heat capacity of steel is 0.48 kJ kg^{-1} K^{-1}.

Solution We apply the law of conservation of momentum to determine the speed after impact when the projectile and the block move as a composite body. Thus

$$m_p V_p = (m_p + m_b)V_{sf} \qquad \text{(E4.5.1)}$$

where V is the velocity, m is the mass and subscripts p and b denote quantities related to the projectile and the block respectively. Let V_{sf} be the velocity of the composite body after the impact.

Substituting numerical values in Eq. (E4.5.1)

$$V_{sf} = 3\times10^{-3} \times 250/(5+3\times10^{-3}) = 0.15 \text{ ms}^{-1}$$

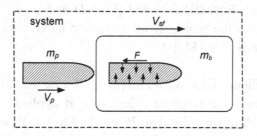

Fig. E4.5 Projectile in block

As the projectile plows into the block it does work to overcome the frictional force. This frictional work, which occurs at the interface between the projectile and the block results in a heat interaction. As a consequence the internal thermal energy of the projectile and the block increase. However, the heat interaction between the system and surroundings is negligibly small. Also there is no work transfer to the surroundings, because the frictional work is internal to the system consisting of the projectile and the block. The block has negligible thermal capacity and therefore it does not store any thermal internal energy.

We write the following expressions for the total internal energy of the projectile and the block before and after the impact. The thermal internal energy is taken to be zero at the reference state of 0°C.

$$E_i = m_p V_{pi}^2 / 2 + m_p c_p T_{pi} + m_b c_b T_{bi}$$

$$E_f = (m_p + m_b) V_{sf}^2 / 2 + m_p c_p T_{pf} + m_b c_b T_{bf}$$

Substituting numerical values in the two equations above we have

$$E_i = 0.003 \times (250)^2 / 2 + 0.003 \times 0.48 \times 10^3 \times 20 \qquad \text{(E4.5.2)}$$

$$E_f = 5.003 \times (0.15)^2 / 2 + 0.003 \times 0.48 \times 10^3 \times T_{pf} \qquad \text{(E4.5.3)}$$

Consider the projectile and the block as a closed system for which the work and heat interactions are both zero. Applying the first law

$$Q = E_f - E_i + W \qquad \text{(E4.5.4)}$$

$$E_f - E_i = 0 \qquad \text{(E4.5.5)}$$

Substituting from Eqs. (E4.5.2) and (E4.5.3) in Eq. (E5.4.5) we obtain the temperature of the projectile as it starts to move with the block as a composite body as $T_{pf} = 85.1°C$

Example 4.6 Figure E4.6 shows a braking-device used to stop a freely-rotating flywheel of a machine. The brake is applied by pressing the stationary disc A against the disc B attached to the flywheel. The heat exchange between A and B and the surroundings is negligible. At some instant when the brake is in operation the speed of the flywheel is 10 r s⁻¹, and the speed is decreasing at the rate of 0.5 r s⁻². The temperature of disc B is observed to be increasing at the rate of 0.02°C s⁻¹.

The effective moment of inertia of the flywheel is 5.0 kg m². The thermal capacities of the discs A and B are 0.3 kJ K⁻¹ and 0.75 kJ K⁻¹ respectively. Calculate (i) the rate of increase of the temperature of disc A, and (ii) the heat interactions for discs A and B.

Solution Let the subscripts w, a, b, s and o represent quantities related to the wheel, the disc A, the disc B, the composite system and the surroundings respectively. Applying the rate-form of the first law to the whole system we have

$$\dot{Q}_{s \to o} = \dot{E}_s + \dot{W}_{s \to o} \qquad (E4.6.1)$$

The total internal energy, $\dot{E}_s$ consists of the rotational kinetic energy of the flywheel and the thermal internal energies of the two discs A and B. The heat transfer and the work transfer from the composite system to the surroundings are both zero. Therefore

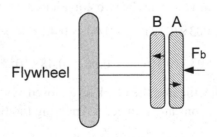

Fig. E4.6 Flywheel with disc brake

$$\dot{E}_s = \frac{d}{dt}\left[I\omega^2/2\right] + \dot{U}_a + \dot{U}_b = 0 \qquad (E4.6.2)$$

$$\dot{E}_s = \left[I\omega(d\omega/dt) + (mc)_a(dT_a/dt) + (mc)_b(dT_b/dt)\right] = 0$$

Substituting the given numerical values we have

$$\dot{E}_s = \left[-5\times10\times0.5\times10^{-3} + 0.3\times(dT_a/dt) + 0.75\times0.02\right] = 0$$

Hence $\qquad\qquad dT_a/dt = 0.033\ °Cs^{-1}.$

Now consider the disc A as a closed system. The boundary is drawn *just* inside A so that the rubbing interface between A and B is outside the boundary. Therefore the work interaction due to the frictional torque at the interface is outside this boundary. Applying the first law to this system we have

$$\dot{Q}_{i\to a} + \dot{Q}_{o\to a} = \dot{U}_a + \dot{W}_{a\to o} + \dot{W}_{a\to i} \qquad (E4.6.3)$$

where the subscript i denotes the interface. The heat interaction with the surroundings, $\dot{Q}_{o\to a} = 0$, the work interaction, $\dot{W}_{a\to o}$ with the surroundings and the work interaction, $\dot{W}_{i\to a}$ with the interface are zero.

Hence Eq. (E4.6.3) becomes

$$\dot{Q}_{i\to a} = \dot{U}_a = (mc)_a(dT_a/dt) = 0.3\times0.0330 = 0.0099\ \text{kW}$$

Therefore the heat flow rate into disc A is 0.0099 kW.

Applying Newton's second law to the system

$$-I(d\omega/dt) = \Gamma_f \qquad (E4.6.4)$$

where Γ_f is the frictional torque at the rubbing interface between A and B. Multiplying Eq. (E4.6.4) by ω we have

$$-I\omega\frac{d\omega}{dt} = \omega\Gamma_f \qquad (E4.6.5)$$

Equation (E4.6.5) can be written as

$$-\frac{d}{dt}\left(I\omega^2/2\right) = \Gamma_f(d\theta/dt) = \dot{W}_f \qquad (E4.6.6)$$

where θ is the angle of rotation, so that the angular speed, $\omega = d\theta/dt$.

The rate of frictional work done is $\dot{W}_f$. Equation (E4.6.6) shows that the rate of decrease of the rotational kinetic energy of the wheel is equal to the rate of work done by the frictional torque at any instant. Substituting numerical values in Eq. (E4.6.6)

$$\dot{W}_f = -I\omega(d\omega/dt) = 5\times10\times0.5\times10^{-3} = 0.025 \text{ kW}$$

Now apply the first law to the interface between discs A and B where the frictional work is being done.

$$\dot{Q}_{a\to i} + \dot{Q}_{b\to i} = \dot{U}_i + \dot{W}_f$$

The internal energy storage in the interface, U_i is negligible.

Substituting numerical values in the above equation we have

$$-0.0099 + \dot{Q}_{b\to i} = 0 - 0.025$$

Therefore the heat flow into disc B is, $\dot{Q}_{i\to b} = 0.0151$ kW

In summary, the rate of decrease in the rotational kinetic energy of the wheel is equal to the rate of frictional work done at the interface between A and B. The frictional work in turn is converted to two heat interactions with A and B at their interface. The thermal internal energy of A and B increase as a result of these heat flows.

Example 4.7 A well-insulated rigid vessel, divided by a partition, has 0.2 kmol of carbon dioxide at 30°C and 2 bar in one section and 3 kmol of hydrogen gas at 40°C and 4 bar in the other section. The partition is removed and the gases mix. Determine (i) the final equilibrium temperature and pressure of the gas mixture, and (ii) the volume of the vessel. The specific heat capacity, c_p of hydrogen and carbon dioxide are 14.35 J kg^{-1} K^{-1} and 0.84 J kg^{-1} K^{-1} respectively.

Solution Consider the vessel as a closed system. Applying the first law to the system we have

$$Q_{in} = U_f - U_i + W_{out}$$

Since the volume of the vessel is fixed, the boundary work done is zero. Being well-insulated, the heat interaction with the surroundings is also zero. Therefore the final internal energy, U_f is equal to the initial internal energy U_i.

$$m_h c_{vh} T_{fh} + m_c c_{vc} T_{fc} = m_h c_{vh} T_{ih} + m_c c_{vc} T_{ic} \qquad \text{(E4.7.1)}$$

where the subscripts h and c refer to hydrogen and carbon dioxide respectively.

Using the relation $c_p - c_v = \overline{R}/M$, M being the molar mass, we obtain the specific heat capacities for hydrogen and carbon dioxide as

$$c_{vh} = 14.35 - 8.3145/2 = 10.192 \ \text{J kg}^{-1} \ \text{K}^{-1}$$

$$c_{vc} = 0.84 - 8.3145/44 = 0.651 \ \text{J kg}^{-1} \ \text{K}^{-1}$$

The masses of hydrogen and carbon dioxide are:

$$m_h = n_h M_h = 3 \times 2 = 6 \ \text{kg}$$

$$m_c = n_c M_c = 0.2 \times 44 = 8.8 \ \text{kg}$$

In the final equilibrium state, the temperature, T_f of the gas mixture is uniform.

Hence $$T_{fh} = T_{fc} = T_f$$

Substituting numerical values in Eq. (E4.7.1) we have

$$T_f = \frac{8.8 \times 0.651 \times 300 + 6 \times 10.192 \times 313}{8.8 \times 0.651 + 10.192 \times 6} = 311.88 \,\text{K}$$

Applying the molar-form of the ideal gas equation to the initial state we obtain

$$V = V_1 + V_2 = n_h \overline{R} T_{ih} / P_{ih} + n_c \overline{R} T_{ic} / P_{ic}$$

$$V = 3 \times 8.3145 \times 313 / 400 + 0.2 \times 8.3145 \times 303 / 200 = 22.04 \,\text{m}^3$$

Applying the ideal gas equation to the final state

$$P_f = \frac{(n_h + n_c) \overline{R} T_f}{V} = \frac{3.2 \times 8.3145 \times 311.88}{22.04} = 376.5 \ \text{kPa}$$

Example 4.8 A rigid vessel A, containing 0.35 m³ of nitrogen at 65°C and 6 bar, is connected to a rigid vessel B, that is initially evacuated (see Fig. E4.8). The valve in the connecting pipe is initially closed. The valve is now opened, and nitrogen flows into B until the pressure and temperature in B become 1.5 bar and 25°C respectively. The valve is then closed and the final equilibrium pressure and temperature in A are 4 bar and 50°C respectively. The specific heat capacity, c_p of nitrogen is 1.041 kJ kg⁻¹ K⁻¹. Calculate (i) the volume of B and (ii) the total heat

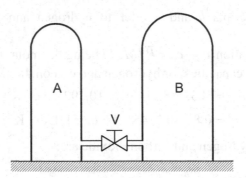

Fig. E4.8 Gas flow between two vessels

interaction between the nitrogen and the surroundings. Assume that the volumes of the pipe and valve are negligibly small.

Solution The gas constant for nitrogen is

$$R = \overline{R}/M = 8.3145/28 = 0.297 \text{ kJ kg}^{-1} \text{ K}^{-1}$$

The initial and final mass of nitrogen in vessel A and the final mass of nitrogen in vessel B are obtained by applying the ideal gas equation in the form, $m = PV/RT$. The volumes of the pipe and valve are negligible.

Initial mass in A is

$$m_{Ai} = \frac{0.35 \times 6 \times 10^5}{0.297 \times 10^3 \times 338} = 2.09 \text{ kg}$$

Similarly, the final mass in A is

$$m_{Af} = \frac{0.35 \times 4 \times 10^5}{0.297 \times 10^3 \times 323} = 1.459 \text{ kg}$$

The mass of nitrogen transferred to B is

$$m_{Bf} = m_{Ai} - m_{Af} = 2.09 - 1.459 = 0.631 \text{ kg}$$

The volume of B is obtained by applying the ideal gas equation to the nitrogen in B

$$V_B = \frac{0.631 \times 0.297 \times 10^3 \times 298}{1.5 \times 10^5} = 0.372 \text{ m}^3$$

Consider a closed system consisting of the vessels A and B, the pipe and the valve. The work done by this system is zero because the boundary is fixed. Applying the first law we obtain

$$Q_{in} = U_f - U_i = (m_{Af}c_v T_{Af} + m_{Bf}c_v T_{Bf}) - m_{Ai}c_v T_{Ai}$$

Now $\qquad c_v = c_p - R = 1.041 - 0.297 = 0.744 \text{ kJ kg}^{-1} \text{ K}^{-1}$

Hence $\qquad Q_{in} = (1.459 \times 0.744 \times 323 + 0.631 \times 0.744 \times 298)$

$$- 2.09 \times 0.744 \times 338 = -35.06 \text{ kW}$$

The rate of heat flow out of the system is 35.06 kW.

Example 4.9 A well-insulated rigid vessel has two compartments that are separated by a membrane. One section of the vessel contains 0.9 kg of water at 7 bar and 72°C. The other section is evacuated. The membrane ruptures and the water fills the entire volume of the vessel. The final equilibrium pressure is 0.14 bar. Calculate (i) the quality and temperature of the steam in the final state and (ii) the volume of the vessel.

Solution (i) Consider the water in the compartment as a closed system. Applying the first law for the change of state from the initial (*i*) to the final (*f*) we have

$$Q_{in} = U_f - U_i + W_{out} \qquad \text{(E4.9.1)}$$

When the membrane separating the two compartments ruptures the water undergoes a free expansion without any resistance. In other words, the volume of water increases to fill the entire vessel but it encounters no resistance in doing so. Therefore the work done in such a process, usually called a *'fully-unresisted'* expansion, is zero. Since the vessel is well-insulated the heat exchange with the surroundings is also zero. Hence from Eq. (E4.9.1) we have

$$U_f = U_i \qquad \text{(E4.9.2)}$$

In the initial state the water at 72°C is a sub-cooled liquid because its temperature is below the saturation temperature at 7 bar, which is 165°C. For the sub-cooled liquid we ignore the effect of pressure and obtain the internal energy from the steam tables in [4] by selecting the saturated liquid internal energy at 72°C. Therefore $u_i = 302$ kJ kg^{-1}. In the final state at 0.14 bar we *assume* that the vessel contains wet steam with a quality x_f. Hence

$$u_{fs} = x_f u_g + (1 - x_f) u_f \qquad\qquad \text{(E4.9.3)}$$

where $u_f = 220$ kJ kg^{-1} and $u_g = 2446$ kJ kg^{-1}

are respectively the saturated liquid and vapor internal energies at 0.14 bar.

Substituting in Eq. (E4.9.2) from Eq. (E4.9.3) we have

$$m_s[x_f u_g + (1 - x_f) u_f] = U_i = m_s \times 302$$

where m_s is the mass of water. Substituting numerical values

$$[x_f \times 2446 + (1 - x_f) \times 220] = 302 \qquad\qquad \text{(E4.9.4)}$$

Hence $x_f = 0.0368$

The final temperature is the saturated temperature at 0.14 bar which from the steam tables [4] is 52.6°C. Note that if the quality, x_f given by Eq. (E4.9.4), was greater than one, then we conclude that the steam is superheated after the expansion and obtain the final temperature from the superheated steam tables.

(ii) The final specific volume is given by

$$v_{fs} = x_f v_g + (1 - x_f) v_f \qquad\qquad \text{(E4.9.5)}$$

The saturated liquid and vapor specific volumes at 0.14 bar are obtained from the steam tables in [4] as

$$v_f = 0.1013 \times 10^{-2} \text{ m}^3 \text{ kg}^{-1} \quad \text{and} \quad v_g = 10.69 \text{ m}^3 \text{ kg}^{-1}$$

respectively. Substituting these values in Eq. (E4.9.5) we have

$$v_{fs} = 0.0368 \times 10.69 + (1 - 0.0368) \times 0.1013 \times 10^{-2} \text{ m}^3 \text{ kg}^{-1}$$

$$v_{fs} = 0.3948 \text{ m}^3 \text{ kg}^{-1}$$

In the final state the steam occupies the entire vessel. Therefore volume of the vessel is given by

$$V_v = m_s v_{fs} = 0.9 \times 0.3948 = 0.3553 \text{ m}^3$$

Example 4.10 An ideal gas undergoes a *quasi-static, adiabatic* expansion from an initial state (P_1, V_1, T_1) to a final state (P_2, V_2, T_2).
(a) Obtain in differential-form the P-V relation for the process in terms of the specific heat capacities c_p and c_v of the ideal gas.
(b) If the specific heat capacities are *constant*, obtain the P-V relation for the process.
(c) Obtain expressions for the work done by the gas, the heat input and the change in internal energy of the gas.

Solution (a) Consider a closed system consisting of the ideal gas of mass m. The specific heat capacities of an ideal gas are functions of temperature only.

Apply the first law to an infinitesimal change of state when the volume, pressure, and temperature are V, P and T respectively.

$$\delta Q = dU + \delta W \qquad (E4.10.1)$$

Since the process is adiabatic the heat transfer is zero. The process is quasi-static and therefore the boundary work is given by PdV. For the ideal gas the increase in internal energy is $mc_v dT$. Substituting these expressions in Eq. (E4.10.1) we obtain

$$0 = mc_v dT + PdV \qquad (E4.10.2)$$

The equation of state of the ideal gas gives

$$PV = mRT \qquad (E4.10.3)$$

where R is the gas constant of the ideal gas.

We now manipulate Eqs. (E4.10.1), (E4.10.2) and (E4.10.3) to obtain the differential form of the P-V relation for the process.
Differentiating Eq. (E4.10.3)

$$PdV + VdP = mRdT \qquad (E4.10.4)$$

Substituting for dT from Eq. (E4.10.4) in Eq. (E4.10.2) we have

$$0 = (c_v / R)(PdV + VdP) + PdV \qquad (E4.10.5)$$

For an ideal gas $R = c_p - c_v$

Substituting for R in Eq. (E4.10.5) and simplifying the resulting equation

$$(c_p / c_v) P dV + V dP = 0 \qquad\qquad (E4.10.6)$$

$$\gamma dV / V + dP / P = 0 \qquad\qquad (E4.10.7)$$

where $\gamma = c_p / c_v$.

In general, for an ideal gas γ is a function of T because v_p and c_p are functions of temperature. If this functional dependence is known, then in principle, Eq. (E4.10.7) could be integrated to obtain the P-V relation for the quasi-static, adiabatic process.

(b) Now consider a special model of the ideal gas for which c_p and c_v are constants. Then, $\gamma = c_p / c_v$ is also a constant, which allows easy integration of Eq. (E4.10.7) to obtain

$$\gamma \ln(V) + \ln(P) = \ln(C) \qquad\qquad (E4.10.8)$$

where $\ln(C)$ is the constant of integration. Equation (E4.10.8) can be rearranged to the form

$$PV^\gamma = C \qquad\qquad (E4.10.9)$$

which is the required P-V relation for a quasi-static adiabatic process of an ideal gas.

(c) In example 3.10 we obtained an expression for the work done in a quasi-static, polytropic process, $[\, PV^n = C \,]$ as

$$W = (P_1 V_1 - P_2 V_2)/(n-1) \qquad\qquad (E4.10.10)$$

Comparing Eq. (E4.10.9) with the polytropic law we find that for a quasi-static adiabatic process, $n = \gamma$. Hence using Eq. (E4.10.10) we obtain the work done in a quasi-static adiabatic process as:

$$W = (P_1 V_1 - P_2 V_2)/(\gamma - 1) \qquad\qquad (E4.10.11)$$

Now for the adiabatic process the heat interaction is zero.

Applying the first law the increase in internal energy is

$$U_2 - U_1 = -W = -(P_1 V_1 - P_2 V_2)/(\gamma - 1) \qquad (E4.10.12)$$

Substituting from the ideal gas equation in Eq. (E4.10.12) we have

$$U_2 - U_1 = -mR(T_1 - T_2)/(\gamma - 1)$$

$$U_2 - U_1 = -m(c_p - c_v)(T_1 - T_2)/(c_p/c_v - 1)$$
$$U_2 - U_1 = mc_v(T_2 - T_1)$$

Notice that we could have written the above equation directly for the increase in internal energy of an ideal gas.

Example 4.11 0.2 kg of air initially at 0.6 bar and 20°C undergoes the cycle consisting of the three quasi-static processes shown in Fig. E4.11. The air is compressed isothermally to a pressure of 1.0 bar during the process 1-2. From 2 to 3 the air is heated in a constant volume process until the pressure is 2 bar. The adiabatic expansion from 3 to 1 returns the air to the initial state, completing the cycle. Calculate (i) the work done, the change in internal energy and the heat transfer during the processes 1-2, 2-3 and 3-1 and (ii) the net work done during the cycle. Assume that air is an ideal gas with constant specific heat capacities, $c_v = 0.718$ kJ kg^{-1} K^{-1} and $c_p = 1.005$ kJ kg^{-1} K^{-1}.

Solution We assume that air is an ideal gas with constant specific capacities.

For air, $\quad R = c_p - c_v = 1.005 - 0.718 = 0.287$ kJ kg^{-1} K^{-1}

and $\qquad\qquad \gamma = c_p/c_v = 1.005/0.718 = 1.4$

Consider the isothermal compression from state 1 to 2, shown in Fig. E4.11. The work done by an ideal gas during an isothermal process was obtained in worked example 3.11 as

$$W = mRT \ln(P_1/P_2)$$

Substituting numerical values in the above equation we have

$$W_{12} = 0.2 \times 0.287 \times 293 \times \ln(0.6/1.0) = -8.59 \text{ kJ}$$

Now for the isothermal process 1-2, $T_2 = T_1$. Hence the change in internal energy of the ideal gas is

$$U_2 - U_1 = mc_v(T_2 - T_1) = 0$$

Applying the first law to the process 1-2 we have

$$Q_{12} = U_2 - U_1 + W_{12} = 0 - 8.591 = -8.591 \text{ kJ}$$

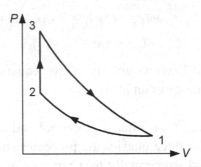

Fig. E4.11 A cyclic process

Apply the ideal gas equation to the states 2 and 3

$$mR = P_2V_2/T_2 = P_3V_3/T_3$$

Since $V_2 = V_3$ for the constant volume process 2-3,

$$P_2/T_2 = P_3/T_3$$
$$1/298 = 2/T_3$$

Hence $T_3 = 596\,\text{K}$

The work done in the constant volume process, 2-3, $W_{23} = 0$. The change in internal energy is

$$U_3 - U_2 = mc_v(T_3 - T_2)$$
$$U_3 - U_2 = 0.2 \times 0.718 \times (596 - 298) = 42.79\ \text{kJ}$$

Applying the first law to process 2-3

$$Q_{23} = U_3 - U_2 + W_{23} = 42.79 + 0 = 42.79\ \text{kJ}$$

Consider the quasi-static, adiabatic process 3-1. The work done is obtained by using the following expression derived in worked example 4.10

$$W_{31} = (P_3V_3 - P_1V_1)/(\gamma - 1) = mR(T_3 - T_1)/(\gamma - 1)$$
$$W_{31} = 0.2 \times 0.287 \times (596 - 298)/(1.4 - 1) = 42.76\ \text{kJ}$$

The change in internal energy is

$$U_1 - U_3 = mc_v(T_1 - T_3)$$
$$U_1 - U_3 = 0.2 \times 0.718 \times (298 - 596) = -42.79\ \text{kJ}$$

The heat transfer $\qquad\qquad Q_{31} = 0$

The net work done in the cycle is given by

$$W_{net} = W_{12} + W_{23} + W_{31} = -8.59 + 0 + 42.76 = 34.2 \text{ kJ.}$$

Example 4.12 A fixed quantity of steam of mass 0.4 kg at 200°C and 0.5 bar undergoes a cyclic process consisting of three quasi-static processes. The steam is compressed *isothermally* in a quasi-static process until the pressure is 4 bar. The steam then undergoes a constant volume process to 0.5 bar. This is followed by a constant pressure expansion to the initial state of 200°C. Calculate (i) the work done, the change in internal energy, and heat transfer for each process, (ii) the net work done in the cycle, and (iii) the net heat input.

Solution Shown in Fig. E4.12 is the cycle executed by the steam. Consider the isothermal compression from 1-2. Since steam is *not an ideal gas* we must not use the expressions derived in worked example 3.11 for ideal gases.

We will, therefore, extract the *P-V* data from the steam tables in [4] and estimate the work done numerically or graphically. The values of specific volume of superheated steam at different pressures for a constant temperature of 200°C are tabulated below. The work done during the quasi-static process 1-2 may be obtained using the expression

$$W_{12} = \int_{V_1}^{V_2} P\,dV$$

The integration is carried out numerically using the trapezoidal-rule by dividing area under the *P-V* curve to a number of trapeziums. The work represented by the area of a typical trapezium is

$$\Delta W_i = (P_i + P_{i+1})(V_{i+1} - V_i)/2 \qquad (i = 1,6)$$

The resulting numerical values are given in Table E4.12.

The total work done per kg of steam is

$$\sum_{i=1}^{i=6} \Delta W_i = 464.4 \text{ kJ kg}^{-1}$$

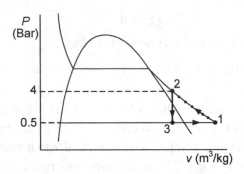

Fig. E4.12 Cyclic process for steam

Table E4.12 Data for isothermal process

I	1	2	3	4	5	6	7
$P_i \times 10^{-5}$ (Pa)	4.0	3.0	2.0	1.5	1.0	0.75	0.5
V_i (m³ kg⁻¹)	0.5345	0.7166	1.081	1.445	2.173	2.901	4.356
$\Delta W_i \times 10^{-5}$ (J kg⁻¹)	0.637	0.911	0.637	0.91	0.637	0.9093	-

The total work done is

$$W_{12} = -0.4 \times 464.2 = -185.68 \text{ kJ}$$

The work done is negative because in the compression process work is done on the steam. The internal energies of the superheated steam at states 1 and 2 are obtained from the data tables in [4] as:

$$u_1 = 2660 \text{ kJ kg}^{-1} \quad \text{and} \quad u_2 = 2648 \text{ kJ kg}^{-1}$$

The increase in internal energy for the process 1-2 is

$$U_2 - U_1 = m_s(u_2 - u_1) = 0.4 \times (2648 - 2660) = -4.8 \text{ kJ}$$

Applying the first law to process 1-2

$$Q_{12} = U_2 - U_1 + W_{12} = -4.8 - 185.68 = -190.48 \text{ kJ}$$

Consider the constant volume process 2-3. The quality at state 3 is found by equating the specific volumes at 2 and 3. Hence

$$0.5345 = x_3 \times 3.239 + (1 - x_3) \times 0.1029 \times 10^{-2}$$

$$x_3 = 0.1648$$

The internal energy at 3 is given by

$$u_3 = x_3 u_{g3} + (1 - x_3) u_{f3}$$

$$u_3 = 0.1648 \times 2483 + (1 - 0.1648) \times 340 = 693.06 \ \text{kJ kg}^{-1}$$

The change in internal energy for process 2-3 is

$$U_3 - U_2 = m_s(u_3 - u_2)$$

$$U_3 - U_2 = 0.4 \times (693.06 - 2648) = -781.98 \ \text{kJ}.$$

For the constant volume process 2-3, the work done, $W_{23} = 0$

Applying the first law to process 2-3

$$Q_{23} = U_3 - U_2 + W_{23} = -781.98 \ \text{kJ}$$

Consider the constant pressure process 3-1. The increase in internal energy is

$$U_1 - U_3 = m_s(u_1 - u_3)$$

$$U_1 - U_3 = 0.4 \times (2660 - 693.06) = 786.78 \ \text{kJ}$$

The work done during the constant pressure process 3-1 is

$$W_{31} = m_s P_1(v_1 - v_3)$$

$$W_{31} = 0.4 \times 0.5 \times 10^5 \times (4.356 - 0.5345) = 76.43 \ \text{kJ}$$

Applying the first law to the process 3-1 we have

$$Q_{31} = U_1 - U_3 + W_{31} = 786.78 + 76.43 = 863.21 \text{kJ}$$

The net work output for the cycle is

$$W_{net} = W_{12} + W_{23} + W_{31}$$

$$W_{net} = -185.68 + 0 + 76.43 = -109.25 \text{kJ}$$

The net heat input for the cycle is

$$Q_{net} = Q_{12} + Q_{23} + Q_{31}$$

$$Q_{net} = -190.48 - 781.98 + 863.21 = -109.25 \ \text{kJ}$$

As expected from the first law, for the cyclic process the net work output is equal to the net heat input.

Example 4.13 An ideal gas of mass 0.8 kg is trapped in a well-insulated cylinder behind a frictionless insulated piston as shown in Fig. E4.13. The piston-cylinder set-up has a mass of 2.0 kg and a specific heat capacity of 0.8 kJ kg^{-1} K^{-1}. The gas is compressed in a quasi-static manner from an initial state of 16 bar and 80°C to a final state where the volume is 40 percent of the initial volume. There is thermal equilibrium between the gas and the piston-cylinder set-up during the process due of the high thermal conductivity of the latter. The constant specific heat capacities of the gas are, $c_v = 0.72$ kJ kg^{-1} K^{-1} and $c_p = 1.0$ kJ kg^{-1} K^{-1}.
(a) Obtain an expression for the P-V relation for the process.
(b) Calculate (i) the final temperature of the system, (ii) the work done on the gas and the change in internal energy, (iii) the change in internal energy of the piston-cylinder set-up, and (iv) the heat interaction between the gas and the piston-cylinder set-up.

Solution Let m be the mass and c the specific heat capacity at constant volume. The subscripts c and g denote quantities related to the piston-cylinder set up and the gas respectively.

Consider a quasi-static process involving an infinitesimal change of state of the system consisting of the gas and the piston-cylinder set-up. The work done by the gas is given by

$$\delta W = PdV \qquad \text{(E4.13.1)}$$

Since the piston-cylinder set-up and the gas are in thermal equilibrium their temperatures are equal. The total change in internal energy is

$$dU = m_g c_g dT + m_c c_c dT \qquad \text{(E4.13.2)}$$

Since the system is well-insulated,

$$\delta Q = 0 \qquad \text{(E4.13.3)}$$

Applying the first law to the closed system we have

$$\delta Q = dU + \delta W \qquad \text{(E4.13.4)}$$

Substituting in Eq. (E4.13.4) from Eqs. (E4.13.1) to (E4.13.3) we obtain

$$(m_g c_g + m_c c_c)dT + PdV = 0 \qquad \text{(E4.13.5)}$$

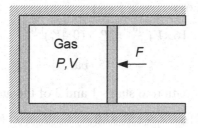

Fig. E4.13 Piston-cylinder set-up

Applying the ideal gas equation

$$PV = m_g R_g T \qquad \text{(E4.13.6)}$$

Differentiating Eq. (E4.13.6) we obtain

$$PdV + VdP = m_g R_g dT \qquad \text{(E4.13.7)}$$

Substituting for dT in Eq. (E4.13.5) from Eq. (E4.13.7)

$$(m_g c_g + m_c c_c)(VdP + PdV) + m_g R_g PdV = 0$$

$$(m_g c_g + m_c c_c + m_g R_g)PdV + (m_g c_g + m_c c_c)VdP = 0$$

$$\left(\frac{m_g c_g + m_g R_g + m_c c_c}{m_g c_g + m_c c_c} \right)\left(\frac{dV}{V} \right) + \frac{dP}{P} = 0 \qquad \text{(E4.13.8)}$$

Upon integration Eq. (E4.13.8) gives the P-V relation as

$$PV^\lambda = C \qquad \text{(E4.13.9)}$$

where the exponent λ is given by:

$$\lambda = \left(\frac{m_g c_g + m_g R_g + m_c c_c}{m_g c_g + m_c c_c} \right)$$

and C is the constant of integration.

(b) For the ideal gas

$$R_g = c_p - c_v = 1.0 - 0.72 = 0.28 \text{ kJ kg}^{-1} \text{ K}^{-1}$$

Substituting numerical values in the expression above for λ we obtain

$$\lambda = \left(\frac{0.8 \times 0.72 + 0.8 \times 0.28 + 2.0 \times 0.8}{0.8 \times 0.72 + 2.0 \times 0.8} \right) = 1.1029$$

Applying Eq. (E4.13.9) to states 1 and 2 we have

$$16 \times V_1^{1.1029} = P_2 \times (0.4 V_1)^{1.1029}$$

Hence $P_2 = 43.95$ bar.

Applying ideal gas equation to states 1 and 2 of the gas

$$mR_g = 16 \times V_1 / (273 + 80) = 43.95 \times (0.4 V_1) / T_2$$

(i) Hence $T_2 = 387.86$ K

(ii) The work done by an ideal gas, during a polytropic process, $PV^n = C$ was derived in worked example 3.10. Comparing the above process-law with Eq. (E4.13.9), $n = \lambda$. Hence

$$W_{12} = m_g R_g (T_1 - T_2) / (\lambda - 1)$$

$$W_{12} = 0.8 \times 0.28 \times (353 - 387.87) / (1.1029 - 1) = -75.91 \text{ kJ}$$

The increase in internal energy of the ideal gas is

$$U_{g2} - U_{g1} = m_g c_g (T_2 - T_1)$$

$$U_{g2} - U_{g1} = 0.8 \times 0.72 \times (387.87 - 353) = 20.085 \text{ kJ}$$

(iii) The increase in internal energy of the piston-cylinder set up is

$$U_{c2} - U_{c1} = m_c c_c (T_2 - T_1)$$

$$U_{c2} - U_{c1} = 2.0 \times 0.8 \times (387.87 - 353) = 55.8 \text{ kJ}$$

(iv) Applying the first law to the piston-cylinder set up

$$Q_{g \to c} = U_{c2} - U_{c1} + W_c$$

The boundary work done by the piston-cylinder set-up is zero because its volume is constant. Hence

$$Q_{g \to c} = U_{c2} - U_{c1} = 55.8 \text{ kJ}$$

Applying the first law to the gas we have

$$Q_{c \to g} = U_{g2} - U_{g1} + W_{12}$$

$$Q_{c \to g} = 20.085 - 75.91 = -55.8 \text{ kJ}$$

Example 4.14 A cylinder of total volume 1.2 m^3 is divided into two compartments A and B of equal volume by a thin, frictionless, adiabatic piston as shown in Fig. E4.14. Initially, compartment A contains air and B helium, both at 30°C and 100 kPa. The cylinder is well-insulated except for one face of A which is a diathermal wall. Heat is supplied to A in a quasi-static manner until a final equilibrium state, where the pressure is 220 kPa, is attained. Calculate (i) the final temperature of the helium and the air, (ii) the work done by the air, and (iii) the heat input to the air. Assume that helium and air are ideal gases with the following specific heat capacities. For helium $c_v = 3.12$ kJ kg^{-1} K^{-1}, $c_p = 5.21$ kJ kg^{-1} K^{-1} and for air $c_v = 0.72$ kJ kg^{-1} K^{-1} and $c_p = 1.00$ kJ kg^{-1} K^{-1}.

Solution Let subscripts i and f denote the initial and final states of the system. From the given data we have $V_{Ai} = V_{Bi} = 0.6$ m^3, and $T_{Ai} = T_{Bi} = 303$ K.

The force balance on the frictionless piston in the initial and final equilibrium states gives:

$$P_{Ai} = P_{Bi} = 100 \text{ kPa} \quad \text{and} \quad P_{Af} = P_{Bf} = 220 \text{ kPa}.$$

The helium gas in B, which is treated as an ideal gas with constant specific heat capacities, undergoes a quasi-static adiabatic process. In worked example 4.10, we derived the *P-V* relation for such a process as

$$PV^{\gamma} = C \tag{E4.14.1}$$

For helium, $\gamma = c_p / c_v = 5.21/3.12 = 1.67$

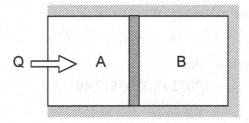

Fig. E4.14 Piston-cylinder arrangement

Applying Eq. (E4.14.1) to the initial and final states we have

$$100 \times (0.6)^{1.67} = 220 \times (V_{Bf})^{1.67}$$

Hence $V_{Bf} = 0.374 \text{ m}^3$

Now $V_{Af} = 1.2 - 0.374 = 0.826 \text{ m}^3$

Applying the ideal gas equation to the initial and final equilibrium states of the gases we obtain

$$mR = P_i V_i / T_i = P_f V_f / T_f$$

Substituting the pertinent numerical data for helium and air in the above equation

$$T_{Bf} = 220 \times 0.374 \times 303 / (100 \times 0.6) = 415.51 \text{ K}$$

$$T_{Af} = 220 \times 0.826 \times 303 / (100 \times 0.6) = 917.69 \text{ K}$$

Work done by the helium gas in B during the quasi-static adiabatic process is

$$W_B = (P_{Bi} V_{Bi} - P_{Bf} V_{Bf}) / (\gamma - 1)$$

$$W_B = (100 \times 0.6 - 220 \times 0.374) / (1.67 - 1) = -33.25 \text{ kJ}$$

The work done by the air in A is used to compress the helium gas in B. Therefore

$$W_A = -W_B = 33.25 \text{ kJ}$$

Applying the ideal gas equation, the mass of air is

$$m_A = 100 \times 0.6 / (0.28 \times 303) = 0.7072 \text{ kg}$$

The increase in internal energy of the air is

$$U_{Af} - U_{Ai} = m_A c_{vA} (T_{Af} - T_{Ai})$$

Substituting numerical values

$$U_{Af} - U_{Ai} = 0.7072 \times 0.72 \times (917.69 - 303) = 313 \text{ kJ}.$$

Applying the first law to the air in A we have

$$Q_A = U_{Af} - U_{Ai} + W_A = 313 + 33.25 = 346.25 \text{ kJ}.$$

Example 4.15 A vertical cylinder of cross-sectional area 0.1 m² is connected to a rigid pressure vessel of volume 0.4 m³ through a valve as shown in Fig. E4.15. Initially, the vessel contains air at 30°C and 6 bar and the cylinder is filled with air at 18°C and 1 bar. The piston is at a height of 0.15 m. The valve is now opened and air flows from the pressure vessel to the cylinder. The load on the piston is such that when the air pressure in the cylinder reaches 3.5 bar, the load just begins to move up. The load is moved a distance of 1 m while the air pressure in the cylinder remains constant at 3.5 bar. The load then hits a rigid stop but air from the vessel continues to flow into the cylinder until the pressure in the system is uniform and the temperature of the air is 18°C. Calculate (i) the final pressure in the system, and (ii) the heat transfer between the system and the surroundings. Assume that air is an ideal gas with $c_v = 0.718$ kJ kg⁻¹ K⁻¹ and $R = 0.287$ kJ kg⁻¹ K⁻¹. Neglect the volume of the piping and the valve.

Solution We use the subscripts v and c to denote quantities related to the pressure vessel and the cylinder respectively. Let i represent the initial state and f the final state.

Applying the ideal gas equation to the initial states of the pressure vessel and the cylinder respectively, we obtain

$$m_{ci} = 0.1 \times 0.15 \times 100/(0.287 \times 291) = 0.01796 \text{ kg}$$

$$m_{vi} = 0.4 \times 600/(0.287 \times 303) = 2.76 \text{ kg}$$

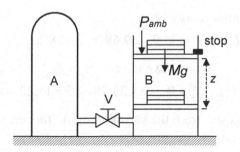

Fig. E4.15 Tank connected to piston-cylinder set-up

The total mass of air is

$$m = (0.01796 + 2.76) = 2.778 \text{ kg}$$

The final volume of air is

$$V_f = 0.4 + (1.0 + 0.15) \times 0.1 = 0.515 \text{ m}^3$$

Applying the ideal gas equation to the final equilibrium state we have

$$P_f V_f = mRT_f$$

$$P_f = 2.778 \times 0.287 \times 291 / 0.515 = 450.5 \text{ kPa}$$

The work done by the air in the system is used to lift the piston with the load Mg, and to push the ambient air. Therefore

$$W = Mg(z_f - z_i) + P_{amb}(V_f - V_i)$$

where z is the distance moved by the piston. If the area of the piston is A, then

$$W = (Mg/A + P_{amb})(V_f - V_i)$$

But $\qquad\qquad (Mg/A + P_{amb}) = P = 3.5$ bar,

the pressure required to just lift the piston. Hence

$$W = P(V_f - V_i) = 3.5 \times 100 \times (0.515 - 0.4 - 0.015) = 35 \text{ kJ}$$

Assume that the reference state for internal energy is $0°C$.
Then the final and initial internal energies are

$$U_f = 2.778 \times 0.718 \times 18 = 35.9 \text{ kJ}$$

$$U_i = (2.76 \times 0.718 \times 30) + (0.01796 \times 0.718 \times 18) = 59.68 \text{ kJ}$$

The increase in internal energy is

$$U_f - U_i = 35.9 - 59.68 = -23.78 \text{ kJ}$$

Applying the first law to the air we have

$$Q_{in} = U_f - U_i + W_{out} = -23.78 + 35 = 11.22 \text{ kJ}$$

which is the heat transfer from the surroundings to the air.

Example 4.16 A cylinder of diameter 0.15 m contains air behind a piston held in position by a rigid stop as shown in Fig. E4.16. The initial

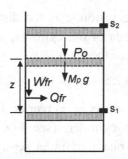

Fig. E4.16 Piston-cylinder set-up with friction

volume, pressure and temperature of the air are 5×10^{-3} m³, 8 bar, and 25°C respectively. The mass of the piston is 250 kg and the ambient pressure is 1 bar. The stop is removed and the piston moves up until it hits a second rigid stop when the volume of the air is 7.5×10^{-3} m³. The frictional force between the piston and the cylinder is *not negligible*. The cylinder and the piston are well-insulated and their heat capacities are negligible. Calculate the pressure and temperature in the final equilibrium state.

Solution The area of the piston is

$$A = \pi D^2/4 = \pi \times (0.15)^2/4 = 0.01767 \, \text{m}^2$$

Apply the ideal gas equation to obtain the mass of air as

$$m = 8 \times 10^2 \times 5 \times 10^{-3}/(0.287 \times 298) = 0.04677 \text{ kg}$$

When the stop is removed, the air does work to lift the piston, to push the surrounding air and to overcome the frictional force. The piston may also gain kinetic energy. After the piston hits the second stop and comes to rest the kinetic energy is converted to thermal energy due to the deformation of the piston and the stop. As we do not have sufficient information to estimate the latter effect using a simple model we assume that the kinetic energy gained is negligible. Therefore the work done by the air is

$$W_a = M_p g z + P_o(V_f - V_i) + W_{fr}$$

where M_p is the mass of the piston, z is the distance moved, P_o is the ambient pressure, A is the area of the cylinder, and W_{fr} is the frictional work. The change in volume of the air is

$$(V_f - V_i) = Az = (7.5 - 5.0) \times 10^{-3} = 2.5 \times 10^{-3} \ \text{m}^3$$

Now the frictional work done is converted to heat at the interface between the piston and the cylinder, (see Fig. E4.16). This heat, in general, is transferred to the piston-cylinder set-up and the air. Since the piston and cylinder have negligible thermal capacity, the internal energy stored in them can be neglected. Therefore all the heat generated by friction, Q_{fr} is transferred to the air in the cylinder. Hence

$$Q_{i \to a} = Q_{fr} = W_{fr} \qquad \text{(E4.16.1)}$$

Applying the first law to the air we have

$$Q_{fr} = (U_2 - U_1) + M_p gz + P_o A(V_f - V_i) + W_{fr} \qquad \text{(E4.16.2)}$$

Substituting from Eq. (E4.16.1) in Eq. (E4.16.2)

$$0 = U_2 - U_1 + M_p gz + P_o(V_f - V_i)$$

$$m_a c_v (T_2 - T_1) + M_p g(V_f - V_i)/A + P_o(V_f - V_i) = 0$$

Substituting the numerical values we obtain

$$m_a c_v (T_2 - T_1) + (250 \times 9.8 \times 10^{-3}/0.01767 + 100) \times 2.5 \times 10^{-3} = 0$$

$$0.04677 \times 0.718 \times (T_2 - 298) + 0.596 = 0$$

Hence $T_2 = 280.28$ K

Note that if the heat capacity of the cylinder is not negligible, then a fraction of the heat generated by friction at the interface will be absorbed by the cylinder as internal energy. The rest of the heat will flow into the air. The analysis of such a scenario is beyond the scope of this book.

Example 4.17 A spring-loaded-piston-cylinder arrangement shown in Fig. E4.17 contains 3.5×10^{-4} m^3 of air at 30°C. The mass of the piston is 10 kg and the ambient pressure is 100 kPa. Initially, the linear spring exerts a force of 500 N on the piston, and it is compressed by 0.015 m.

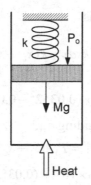

Fig. E4.17 Spring-loaded piston cylinder set-up

The diameter of the piston is 0.09 m. Heat is supplied to the air in a quasi-static manner until the pressure of the air is 300 kPa. Calculate (i) the spring constant, (ii) the work done on the spring, (iii) the work done on the surrounding air, (iv) potential energy gained by the piston, (v) the work done by the air, (vi) the final temperature of the air, (vii) the change in internal energy of the air, and (viii) the heat transfer to the air. Assume that air is an ideal gas with $c_v = 0.718$ kJ kg^{-1} K^{-1} and $R = 0.287$ kJ kg^{-1} K^{-1}.

Solution (i) The spring constant is
$$k = 500/0.015 = 33.33 \text{ kN m}^{-1}$$

Area of the piston is
$$A = \pi \times 0.09^2/4 = 6.36 \times 10^{-3} \text{ m}^2$$

Let the compression of the spring in the final state be x_1 m.

Force balance on the piston for the final equilibrium state gives
$$P_f A = Mg + P_o A + kx_1$$
$$300 \times 6.36 \times 10^{-3} = 10 \times 9.81 \times 10^{-3}$$
$$+ 100 \times 6.36 \times 10^{-3} + 33.33 \times x_1$$

Hence
$$x_1 = 0.0352 \text{ m}$$

(ii) Work done on the spring is

$$W_s = \int_{x_0}^{x_1} kx\,dx = k(x_1^2 - x_0^2)/2$$

$$W_s = 33.33 \times 10^3 \times (0.0352^2 - 0.015^2)/2 = 16.9 \text{ J}$$

(iii) Work done on the surrounding air,

$$W_o = P_0 A(x_1 - x_0)$$

$$W_o = 100 \times 10^3 \times 6.36 \times 10^{-3} \times (0.0352 - 0.015) = 12.85 \text{ J}$$

(iv) Potential energy gained by the piston,

$$W_p = Mg(x_1 - x_0) = 10 \times 9.81 \times (0.0352 - 0.015) = 1.98 \text{ J}$$

(v) Work done by the air is given by

$$W_a = W_s + W_o + W_p = 16.92 + 12.85 + 1.98 = 31.75 \text{ J}$$

(vi) The initial pressure is,

$$P_i = P_o + Mg/A + kx_o/A$$

$$P_i = 100 + (10 \times 9.81 \times 10^{-3} + 33.33 \times 0.015)/(6.36 \times 10^{-3})$$

Hence $P_i = 194.04$ kPa

Applying the ideal gas equation we obtain the mass of air as

$$m = P_1 V_1 / RT_1$$

$$m = 194.04 \times 3.5 \times 10^{-4}/(0.287 \times 303) = 0.781 \times 10^{-3} \text{ kg}$$

The final volume of air is

$$V_f = V_i + A(x_1 - x_0)$$

$$V_f = 3.5 \times 10^{-4} + 6.36 \times 10^{-3} \times (0.0352 - 0.015) = 0.478 \times 10^{-3} \text{ m}^3$$

Applying the ideal gas equation to the final state

$$T_f = P_f V_f / mR$$

$$T_f = 300 \times 0.478 \times 10^{-3}/(0.781 \times 10^{-3} \times 0.287) = 639.76 \text{ K}$$

(vii) The change in internal energy of the air is

$$U_f - U_i = mc_v(T_f - T_i)$$

$$U_f - U_i = 0.781 \times 10^{-3} \times 0.718 \times (639.76 - 303) = 188.84 \text{ J}$$

(viii) Applying the first law to the air we obtain the heat input as

$$Q_a = U_f - U_i + W_a = 188.84 + 31.75 = 220.6 \text{ J}$$

Example 4.18 A well-insulated piston-cylinder device has 0.5 kg of air at 30°C and 3 bar. Installed inside the cylinder is a small heating element of negligible thermal capacity as shown in Fig. E4.18. A steady current of 6 A is passed through the heating element for 30s. The air expands slowly at constant pressure until the volume increases by 40 percent. Assume that air is an ideal gas, and $c_v = 0.718$ kJ kg^{-1} K^{-1} and $R = 0.287$ kJ kg^{-1} K^{-1}. Calculate (i) the work done by the air, (ii) the change in internal energy of the air, and (iii) the resistance of the heating element.

Solution From the ideal gas equation we obtain the initial volume of the air as

$$V_i = 0.5 \times 0.287 \times 303 / (3 \times 100) = 0.1449 \text{ m}^3$$

The final volume is

$$V_f = 1.4V_i = 1.4 \times 0.1449 = 0.2029 \text{ m}^3$$

Work done by the air in the constant pressure process is

$$W_a = P(V_f - V_i) = 300 \times (0.2029 - 0.1449) = 17.4 \text{ kJ}$$

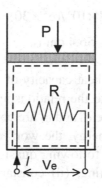

Fig. E4.18 Piston-cylinder set-up with electrical heater

Applying the ideal gas equation to the initial and final states

$$mR = PV_f / T_f = PV_i / T_i$$

Hence $T_f = 1.4 T_i = 1.4 \times 303 = 424.2$ K

Change in internal energy of the air is

$$U_f - U_i = mc_v(T_f - T_i)$$

$$U_f - U_i = 0.5 \times 0.718 \times (424.2 - 303) = 43.51 \, \text{kJ}$$

We apply the first law to the closed system consisting of the air and the heater. In chapter 3, we showed that electrical energy supplied to the heating element is a work interaction. Since the cylinder is well-insulated, the heat interaction between the system and the surroundings is zero.

Applying the first law we have

$$0 = U_f - U_i + W_a + W_{el}$$

$$W_{el} = -43.51 - 17.4 = -60.91 \text{ kJ}$$

The total electrical work supplied to the system is

$$W_{el} = V_e I \tau = I^2 R \tau$$

where V_e is the voltage across the heater, R is the resistance of the heater, I is the current and τ is time.

Substituting the given numerical values in the above equation

$$R = 60.91 \times 10^3 / (6^2 \times 30) = 56.4 \Omega$$

The resistance of the heater is 56.4 ohms.

Example 4.19 The specific heat capacity of an ideal gas varies linearly with temperature according to the relationship, $c_v = c_0 + c_1 T$ in the temperature range T_1 to T_2, where c_0 and c_1 are constants. Obtain expressions for the change in internal energy, the work done, and the heat transfer when the gas undergoes the following quasi-static processes from T_1 to T_2 (i) constant volume heating, (ii) constant pressure expansion, and (iii) adiabatic process.

Solution (i) Consider a quasi-static constant volume process. The work done is zero. The change in internal energy is given by

$$U_2 - U_1 = \int_{T_1}^{T_2} dU = m\int_{T_1}^{T_2} c_v dT = m\int_{T_1}^{T_2} (c_o + c_1 T)dT$$

Integrating the above equation we have

$$U_2 - U_1 = mc_o(T_2 - T_1) + mc_1(T_2^2 - T_1^2)/2$$

$$U_2 - U_1 = m[c_0 + c_1(T_1 + T_2)/2](T_2 - T_1) = mc_m(T_2 - T_1)$$

where $\qquad\qquad c_m = c_0 + c_1(T_1 + T_2)/2$,

is the specific heat capacity evaluated at the arithmetic mean temperature of the process, $\qquad\qquad T_m = (T_1 + T_2)/2$

The heat transfer is obtained by applying the first law as

$$Q_{in} = U_2 - U_1 = mc_m(T_2 - T_1)$$

(ii) Consider a constant pressure process. The change in internal energy is the same as in (i) above because the internal energy is a property and therefore independent of the process. For an ideal gas, the change on internal energy depends only on the initial and final temperatures.

For a constant pressure process, the work done by the ideal gas is

$$W = P(V_2 - V_1) = mR(T_2 - T_1)$$

Applying the first law, the heat input is

$$Q_{in} = U_2 - U_1 + W = mc_m(T_2 - T_1) + mR(T_2 - T_1)$$

$$Q_{in} = m(c_m + R)(T_2 - T_1)$$

(iii) For an adiabatic process the heat transfer is zero. The change in internal energy is the same as in (i) above because it is a property and is therefore path- independent. Applying the first law

$$Q_{in} = (U_2 - U_1) + W_{out}$$

$$W_{out} = -(U_2 - U_1) = mc_m(T_1 - T_2)$$

Example 4.20 In a heat transfer experiment a copper sphere of diameter 0.06 m is cooled in a stream of air at a constant temperature of 20°C as shown in Fig. E4.20. The initial temperature of the sphere is 60°C. The

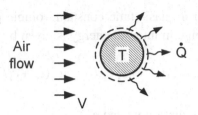

Fig. E4.20 Cooling of a copper sphere

rate of heat flow (kW) from the outer surface of the sphere to the air is given by

$$\delta Q / dt = 0.5 A (T - 20),$$

where $A(\text{m}^2)$ is the surface area of the sphere and $T(^\circ\text{C})$ is the temperature of the sphere, which may be assumed uniform because of the high thermal conductivity of copper. The density and specific heat capacity of copper are 8900 kg m^{-3} and 0.4 kJ kg^{-1} K^{-1} respectively.

(a) Obtain in differential-form an equation for the time variation of the temperature of the sphere. (b) Calculate the temperature of the sphere 90 s after it is placed in the air stream. (c) What is the final steady temperature of the sphere?

Solution Consider the copper sphere as a closed system. We apply the rate-form of the first law to the system

$$\delta Q / dt = dU / dt + \delta W / dt \qquad (\text{E4.20.1})$$

The boundary work done is zero because the volume of the sphere is constant. The heat transfer rate is given as

$$\delta Q / dt = -0.5 A (T - 20) \qquad (\text{E4.20.2})$$

The internal energy of the system is given by

$$U = mcT \qquad (\text{E4.20.3})$$

where m is the mass and c the specific heat capacity of copper. The reference state for the internal energy is at 0°C. Note that for a solid like copper, the c_v and c_p are assumed equal because of the low compressibility.

Substituting in Eq. (E4.20.1) from Eqs. (E4.20.2) and (E4.20.3) we have

$$-0.5A(T-20) = d(mcT)/dt \qquad \text{(E4.20.4)}$$

The area and mass of a sphere are given by

$$A = 4\pi r^2 \quad \text{and} \quad m = 4\pi r^3 \rho/3 \qquad \text{(E4.20.5)}$$

where r is the radius of the sphere and ρ is the density of copper.

Substituting from Eq. (E4.20.5) in Eq. (E4.20.4) we obtain

$$dT/dt = -0.5 \times 3 \times (T-20)/(rc\rho)$$

Substituting the given numerical values in the above equation

$$dT/dt = -0.01404 \times (T-20) \qquad \text{(E4.20.6)}$$

Equation (E4.20.6) is a first-order differential equation which can be written as

$$dT/(T-20) = -0.01404 dt \qquad \text{(E4.20.7)}$$

Integrating both sides of Eq. (E4.20.7) and substituting the initial temperature at time $t = 0$ as 60°C we have

$$(T-20)/40 = \exp(-0.01404t) \qquad \text{(E4.20.8)}$$

Hence the temperature after 90 s is given by the equation

$$(T_1 - 20)/40 = \exp(-0.01404 \times 90) = 0.2826$$

Thus $T_1 = 31.3°C$

From Eq. (E4.20.8), the steady-state temperature as $t \to \infty$ is 20°C.

Problems

P4.1 A well-insulated tank, of negligible thermal capacity contains 15 kg of liquid at 25°C. A paddle-wheel of mass 1.5 kg is placed within the liquid, and rotated by a motor which receives a steady energy input of 0.5 kW from a battery. The specific heat capacities of the liquid and the wheel are 1.2 kJ kg^{-1} K^{-1} and 0.85 kJ kg^{-1} K^{-1} respectively.

The motor is switched on for 300s, and then switched off. (a) Calculate the final temperature of the liquid and wheel assuming that it is uniform. (b) Treating (i) the liquid and wheel, (ii) the battery, and (iii) the composite system as closed systems, calculate the work transfer, the heat transfer, and the change in internal energy. State all assumptions clearly. [*Answers*: (a) 32.8°C, (b) (i) $W_{in} = \Delta U_{liq} = 150\,\text{kJ}$, (ii) $-\Delta U_{bat} = W_{out} = 150$ kJ, (iii) $\Delta U_{system} = 0$ kJ]

P4.2 A strip of metal of mass 0.05 kg and specific heat capacity 0.48 kJkg^{-1}K^{-1} is pressed against the surface of a grinding wheel of radius 0.35 m with a radial force of 0.045 kN. The coefficient of friction between the surfaces of the wheel and the metal is 0.55. The wheel is rotating at a steady speed of 8 r s^{-1}.
(a) Calculate temperature rise in the metal strip in 20 s. Assume that there is no heat loss to the surroundings and that the grinding wheel has a low thermal conductivity.
(b) Treating the metal strip as a closed system calculate the heat interaction, the work interaction, and the change in internal energy. State all assumptions clearly. [*Answers*: (a) 57.75°C, (b) At the sliding surface $\dot{W} = \dot{Q}$, for the strip, $\dot{Q} = \dot{U}$]

P4.3 A quantity of wet steam of mass 0.15 kg is trapped in a cylinder behind the leak-proof piston. The initial pressure and quality of the steam are 3 bar and 0.7 respectively. Heat is supplied to the steam at constant pressure until the volume becomes 0.1 m^3. (i) Calculate the work done, the change in internal energy, and the heat transfer. (ii) If the piston is held fixed, and the cylinder is cooled in a quasi-static manner until condensation of steam just begins, calculate the work done, change in internal energy, and the heat transfer during the process.
[*Answers*: (i) 10.9 kJ, 98.1 kJ, 109 kJ, (ii) 0 kJ, −9.47 kJ, −9.47 kJ]

P4.4 A fixed quantity of air of mass 0.4 kg undergoes a cycle consisting of three processes. The pressure and temperature of state 1 are 1.0 bar and 30°C respectively. During the process 1-2 the air is compressed isothermally to a pressure of 8 bar. The air then expands at constant pressure to state 3. The final process 3-1 is a *polytropic process*, with the condition, $PV^{1.3} = $ *constant.* Assume that air is an ideal gas with $c_v = 0.718\,\text{kJ kg}^{-1}\,\text{K}^{-1}$ and $c_p = 1.005\,\text{kJ kg}^{-1}\,\text{K}^{-1}$. Calculate (i) the work done, the change in internal energy, and the heat transfer during the three

processes, 1-2, 2-3 and 3-1, (ii) the ratio of the net work done to the total heat supplied to the air, and (iii) the efficiency of the cycle.
[Answers: (i) $W_{12} = -72.33$ kJ, $U_{12} = 0$, $Q_{12} = -72.33$ kJ, $W_{23} = 21.42$ kJ, $U_{23} = 53.58$ kJ, $Q_{23} = 75.0$ kJ, $U_{31} = -53.58$ kJ, $W_{31} = 71.39$ kJ, $Q_{31} = 17.8$ kJ, (ii) $W_{net} = 20.48$ kJ, $Q_{in} = 92.8$ kJ, (iii) $\eta = W_{net}/Q_{in} = 22\%$]

P4.5 A fixed quantity of wet steam of mass 0.4 kg undergoes a polytropic expansion according to the relation, $PV^{1.2} = constant$, from an initial pressure of 4.5 bar and quality 0.8 to a final pressure of 2.5 bar. Calculate the work done, the change in internal energy, and the heat transfer during the polytropic process. [Answers: $W_{12} = 27.87$ kJ, $\Delta U_{12} = -52.2$ kJ, $Q_{12} = -24.33$ kJ]

P4.6 A rigid tank of volume 0.25 m^3 contains argon at a pressure of 4 bar and temperature 30°C. The tank is connected to a piston-cylinder apparatus, through a valve, which is initially closed, and the piston rests at the bottom of the cylinder. The mass of the piston and the ambient pressure are such that a pressure of 2 bar has to act on the inside of the piston to *just* lift it. The valve is now opened, and remains open until the system attains the final equilibrium state with a uniform temperature of 30°C. Calculate (i) the work done by the argon during the process, (ii) the change in internal energy of the argon, and (iii) the heat transfer to the system. Assume that argon is an ideal gas with $c_v = 0.312$ kJ kg^{-1} K^{-1} and $c_p = 0.52$ kJ kg^{-1} K^{-1}. [Answers: (i) $W_{12} = 50$ kJ, (ii) $\Delta U_{12} = 0$, (iii) $Q_{12} = 50$ kJ]

P4.7 A well-insulated rigid tank is divided into two compartments 1 and 2 by a non-conducting membrane. Initially, the compartment 1 contains nitrogen at 100 kPa and 300 K while compartment 2 has nitrogen at 300 kPa and 400 K. The volume of compartment 1 is three times the volume of compartment 2. The membrane is now ruptured, and the gases in the two compartments mix to attain a final equilibrium state. Calculate the final pressure and temperature of the nitrogen. [Answers: 150 kPa, 342.9 K]

188 Engineering Thermodynamics

P4.8 A thin electrical heating element of resistance 100 ohms and thermal capacity 0.05 kJK⁻¹ is located inside a piston-cylinder set-up. The well-insulated cylinder contains 0.2 kg of air with a uniform initial temperature of 20°C. A current of 2 A is passed through the heater for 20 s and then switched off. The air undergoes a constant pressure process, and the system attains a final equilibrium state. Calculate (i) the final equilibrium temperature and (ii) the work done by the air. [*Answers*: (i) 51.87°C, 1.83 kJ]

P4.9 The cylinder, shown in Fig. P4.9, is divided into two chambers A and B by the *thin* spring-loaded piston which is free to move inside the cylinder in a frictionless manner. The chamber A contains 0.5×10^{-3} kg of helium gas while the chamber B, where the spring is located, is evacuated. The length of the cylinder is 1.8 m and its cross sectional area is 0.005 m². The linear spring has a spring constant of 2 kN m⁻¹. The initial pressure and temperature of the helium are 100 kPa and 30°C respectively. The gas is heated slowly until the temperature is 200°C. Calculate (i) the change in internal energy of the gas, (ii) the work done by the gas, and (iii) the heat transfer during the process. Assume that helium is an ideal gas with c_v = 3.116 kJ kg⁻¹ K⁻¹ and c_p = 5.193 kJ kg⁻¹ K⁻¹. [*Answers*: (i) 0.265 kJ, (ii) 0.0537 kJ, (iii) 0.3185 kJ]

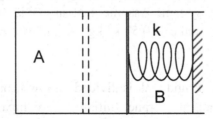

Fig. P4.9 Spring-loaded piston-cylinder set-up

P4.10 A stream of air at a constant temperature of 15°C passes over a 0.75 kW electrical heater. Before the current is switched on the heater temperature is 15°C. The rate of heat flow (kW) from the outer surface of the heater to the air can be expressed as

$$\delta Q/dt = 0.01 \times (T - 15),$$

where T (°C) is the temperature of the heater which may be assumed uniform. The heat capacity of the heater is 0.5 kJ K^{-1}. Calculate (i) the temperature of the heater 30 s after it is switched on, and (ii) the final steady temperature of the heater. [*Answers*: (i) 48.8°C, (ii) 90°C]

References

1. Cravalho, E.G. and J.L. Smith, Jr., *Engineering Thermodynamics*, Pitman, Boston, MA, 1981.
2. Jones, J.B. and G.A. Hawkins, *Engineering Thermodynamics*, John Wiley & Sons, Inc., New York, 1986.
3. Reynolds, William C. and Henry C. Perkins, *Engineering Thermodynamics*, 2nd edition, McGraw-Hill, Inc., New York, 1977.
4. Rogers, G.F.C. and Mayhew, Y.R., *Thermodynamic and Transport Properties of Fluids*, 5th edition, Blackwell, Oxford, U.K. 1998.
5. Tester, Jefferson W. and Michael Modell, *Thermodynamics and Its Applications*, 3rd edition, Prentice Hall PTR, New Jersey, 1996.
6. Van Wylen, Gordon J. and Richard E. Sonntag, *Fundamentals of Classical Thermodynamics*, 3rd edition, John Wiley & Sons, Inc., New York, 1985.
7. Zemansky, M., *Heat and Thermodynamics*, 5th edition, McGraw-Hill, Inc., New York, 1968.

Chapter 5

First Law Analysis of Open Systems

In the last chapter we applied the first law of thermodynamics to closed systems where there was no mass exchange between the system and the surroundings. However, many engineering systems, especially those involving energy conversion devices, consist of components where working fluids flow in and out of them. In the present chapter we propose to extend the formulation of the first law to such open systems, which are also called control volumes.

5.1 Open Systems: An Example

Consider the temperature-controlled water bath, shown schematically in Fig. 5.1, as an example of an open system or a control volume. Cold

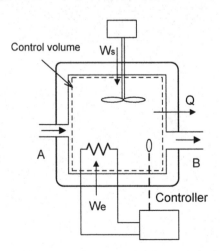

Fig. 5.1 Constant-temperature water bath

190

water enters the bath through the inlet pipe connected at A to the metal tank, and leaves at B through an exit pipe. An electrical resistance heater located inside the tank supplies heat to the water while the motor-driven stirrer helps to maintain a uniform water temperature in the bath. The tank is thermally insulated to minimize heat losses to the surroundings.

The boundary of the tank, indicated by a broken-line in Fig. 5.1, constitutes an *open system* or a *control volume* where mass exchange occurs at points A and B. The system experiences an input of shaft work through the stirrer. In addition, the electrical energy supplied to the heater enters the control volume as a work input. The water at A does work to push the water already in the pipe to enter the control volume. Similarly, the water leaving at B does work to push the water in front of it in the pipe. The latter forms of work are usually called *flow work*. The heat loss through the insulation to the surroundings constitutes a heat interaction.

There are several modes of operation of the water bath that need to be analyzed, say for design purposes. When the inlet and outlet flow rates, the mass of water in the bath, and the various properties like the water temperatures at the inlet, outlet and through out the bath are all constant, the system is said to be in a *steady state* mode of operation. However, when the inlet and outlet flow rates are unequal the level of water in the bath will vary with time. Moreover, when the electricity supply to the heating element is switched on with cold water in the bath, or switched off when the water is hot, the water temperature will vary with time. In order to maintain a constant supply-water temperature the controller has to periodically switch the heater on and off. These situations are called unsteady *or transient* modes of operation of the system, and are important from the point of view of design.

In the next section we shall develop the mathematical form of the first law which will enable us to analyze both steady-state and transient modes of operation of systems that fall into the category of *open systems* or *control volumes*.

5.2   General Form of the First Law for Control Volumes

The first law of thermodynamics will now be derived for a control volume. This will be done by revising the first law for a closed system

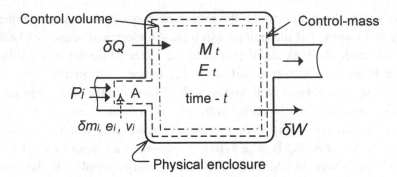

Fig. 5.2(a) Control-mass and control-volume boundaries at time t

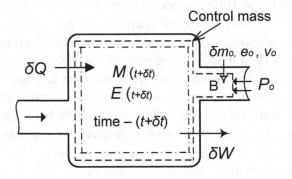

Fig. 5.2(b) Control-mass and control-volume boundaries at time $(t + \delta t)$

or control-mass, which was developed in Sec. 4.2. Consider the control volume depicted schematically in Fig. 5.2(a), and then again in (b).

Although indicated separately for clarity, the boundaries of the control volume, the control mass, and the physical enclosure are strictly coincident except for the parts in the ducts A and B. We focus our attention on the changes experienced by a fixed mass of matter contained within the boundary of the control mass shown in Fig. 5.2(a) at time t and in Fig, 5.2 (b) at time $(t + \delta t)$.

At time t, the boundary includes the mass M_t in the control volume and the mass δm_i in a small section of the inlet duct at A. During the small time interval δt, the boundary of the control mass deforms to the shape in Fig. 5.2.(b) where the mass in duct A has just entered the control volume while a mass δm_o has been pushed out of the control volume into a small section of the exit duct at B.

In our analysis, we consider a single inlet duct and a single exit duct which penetrate the control volume at locations A and B respectively. The use of single streams, in the derivation to follow, is not a limitation because it is possible, in a straightforward manner, to write the expressions for multiple streams.

We now apply the first law for the change of state of the fixed mass of matter within the *control mass* boundary. Hence

$$\delta Q_{in} = E_2 - E_1 + \delta W_o \qquad (5.1)$$

The meanings of the various terms in Eq. (5.1) are as follows:

E_1 is the internal energy of the of the control mass at time t, which consists of the internal energy $E_{c,t}$, of the mass in the control volume, and the internal energy δe_i of the mass in the inlet duct A. Therefore

$$E_1 = E_{c,t} + \delta e_i = E_{c,t} + \delta m_i e_i \qquad (5.2)$$

where e_i is the internal energy per unit mass of the matter in the inlet duct at A.

Similarly, the internal energy of the *control mass* at time $(t + \delta t)$ is

$$E_2 = E_{c,(t+\delta t)} + \delta e_o = E_{c,(t+\delta t)} + \delta m_o e_o \qquad (5.3)$$

where e_o is the internal energy per unit mass of the matter in the exit duct at B.

The work done by the system during the process is δW_o. It consists of two parts. The first part could include shaft work, like the work done by the stirrer and the electrical work supplied to the heater in the case of the water bath shown in Fig 5.1. We lump all these types of work in a separate term δW_c, and call it the *external work* interaction of the control volume. The second part of the work is associated with the deformation of the *control mass* boundary from the shape in Fig. 5.2(a) to the shape in Fig. 5.2(b). The identifiable changes of shape occur at the points A and B which are essentially constant pressure processes due to the small volume changes involved. Any work associated with boundary deformations elsewhere in the control volume will have been already included in δW_c. The net work done *by* the system during the boundary deformations at the inlet and exit ducts is

$$\delta W_b = P_o\delta V_o - P_i\delta V_i = P_o v_o\delta m_o - P_i v_i\delta m_i \tag{5.4}$$

where v_i and v_o are the specific volumes of the fluid at A and B respectively.

The total work interaction may be written as

$$\delta W_o = \delta W_c + P_o v_o\delta m_o - P_i v_i\delta m_i \tag{5.5}$$

Substituting in Eq. (5.1) from Eqs. (5.2), (5.3) and (5.5) we have

$$\delta Q_{in} = E_{c,(t+\delta t)} + e_o\delta m_o - E_{c,t}$$

$$- e_i\delta m_i + \delta W_c + P_o v_o\delta m_o - P_i v_i\delta m_i \tag{5.6}$$

Rearranging Eq. (5.6)

$$E_{c,(t+\delta t)} - E_{c,t} = (e_i + P_i v_i)\delta m_i - (e_o + P_o v_o)\delta m_o + \delta Q_{in} - \delta W_c \tag{5.7}$$

Equation (5.7) is cast in the rate-form by dividing by the time interval δt

$$\frac{E_{c,(t+\delta t)} - E_{c,t}}{\delta t} = (e_i + P_i v_i)\frac{\delta m_i}{\delta t} - (e_o + P_o v_o)\frac{\delta m_o}{\delta t} + \frac{\delta Q_{in}}{\delta t} - \frac{\delta W_c}{\delta t} \tag{5.8}$$

We now take the limit as δt approaches zero to obtain the differential-form of Eq. (5.8)

$$\frac{dE_{c,t}}{dt} = (e_i + P_i v_i)\frac{dm_i}{dt} - (e_o + P_o v_o)\frac{dm_o}{dt} + \frac{\delta Q_{in}}{dt} - \frac{\delta W_c}{dt} \tag{5.9}$$

If in applying Eq. (5.9), we restrict the types of energy of the fluid at the inlet and exit ports of the control volume to thermal internal energy, kinetic energy, and gravitational potential energy, then

$$e = u(T,P) + \frac{V^2}{2} + gz \tag{5.10}$$

where $u(T,P)$ is the specific internal energy associated with the temperature, $V^2/2$ and gz are respectively the kinetic and the potential energy per unit mass. Hence the term

$$e + Pv = u + \frac{V^2}{2} + gz + Pv = h + \frac{V^2}{2} + gz \tag{5.11}$$

where, $h = u + Pv$ is the enthalpy per unit mass of the fluid, which was defined earlier in Sec. 4.5.2. Using Eq. (5.4) we could interpret the term Pv that is included in the enthalpy term, h as the work done in forcing a unit mass of fluid into the control volume. Therefore this work term is sometimes called *flow-work*.

Substituting from Eq. (5.11) in Eq. (5.9), the rate-form of the first law becomes

$$\frac{dE_{c,t}}{dt} = (h_i + \frac{V_i^2}{2} + g_i z_i)\frac{dm_i}{dt} - (h_o + \frac{V_o^2}{2} + gz_o)\frac{dm_o}{dt} + \frac{\delta Q_{in}}{dt} - \frac{\delta W_c}{dt}$$

(5.12)

If we have more than one inlet and exit port, it is clear that we can generalize Eq. (5.12) by summing the first and second terms on the right hand side over all the inlet streams and exit streams respectively. This gives

$$\frac{dE_{c,t}}{dt} = \sum_{in}(h_i + \frac{V_i^2}{2} + gz_i)\dot{m}_i - \sum_{out}(h_o + \frac{V_o^2}{2} + gz_o)\dot{m}_o + \dot{Q}_{in} - \dot{W}_c$$

(5.13)

where the 'dot' over a quantity represents its derivative with respect to time, d/dt.

5.3 Mass Conservation Law for Control Volumes

Since the mass within the boundary of the *control mass*, shown in Figs. 5.2(a) and (b) is constant during the time interval δt, we have

$$M_{c,(t+\delta t)} + \delta m_o = M_{c,t} + \delta m_i \qquad (5.14)$$

Rearranging Eq. (5.14) and dividing by δt we obtain

$$\frac{M_{c,(t+\delta t)} - M_{c,t}}{\delta t} = \frac{\delta m_i}{\delta t} - \frac{\delta m_o}{\delta t} \qquad (5.15)$$

We now take the limit as δt approaches zero to obtain the differential-form of Eq. (5.15) as

$$\frac{dM}{dt} = \dot{m}_i - \dot{m}_o \qquad (5.16)$$

For situations involving multiple streams, we can generalize Eq. (5.16) as

$$\frac{dM}{dt} = \sum_{in} \dot{m}_i - \sum_{out} \dot{m}_o \qquad (5.17)$$

5.4 Steady-Flow Energy Equation (SFEE)

There are many engineering applications where the systems operate under *steady-state* conditions. In such systems, as was mentioned in Sec. 5.1, there is no time variation of the mass and energy of the matter within the control volume. Moreover, the properties of the fluids at the inlet and exit ducts remain unchanged. The general form of the first law, expressed mathematically by Eq. (5.13), then takes the following simplified form:

$$\sum_{in} (h_i + \frac{V_i^2}{2} + gz_i)\dot{m}_i + \dot{Q}_{in} = \sum_{out} (h_o + \frac{V_o^2}{2} + gz_o)\dot{m}_o + \dot{W}_c$$

$$(5.18)$$

The mass conservation equation can be written as:

$$\sum_{in} \dot{m}_i = \sum_{out} \dot{m}_o \qquad (5.19)$$

Equation (5.18), often called the *steady-flow-energy-equation* (SFEE), is widely used in the energy analysis of engineering systems.

For applications with one inlet and one exit stream, Eqs. (5.18) and (5.19) can be further simplified as:

$$(h_i + \frac{V_i^2}{2} + gz_i)\dot{m}_i + \dot{Q}_{in} = (h_o + \frac{V_o^2}{2} + gz_o)\dot{m}_o + \dot{W}_c \qquad (5.20)$$

$$\dot{m}_i = \dot{m}_o = \dot{m} \qquad (5.21)$$

From Eqs. (5.20) and (5.21) it follows that

$$(h_i + \frac{V_i^2}{2} + gz_i) + \frac{\dot{Q}_{in}}{\dot{m}} = (h_o + \frac{V_o^2}{2} + gz_o) + \frac{\dot{W}_c}{\dot{m}} \qquad (5.22)$$

It is interesting to note that the various terms of Eq. (5.22) represent energy, heat and work per unit mass of fluid flowing through the control volume. The typical units of the terms in Eq. (5.18) are kJ s^{-1} or kW,

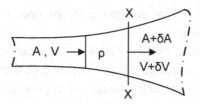

Fig. 5.3 Mass flow rate in a duct

while in Eq. (5.22) the units are kJ kg^{-1}. We shall make use of these different forms of the first law and the mass balance equation for control volumes in solving the worked examples to follow in this chapter. First we need to derive an additional result, as follows.

5.5 Fluid Mass Flow Rate in a Duct

Figure 5.3 depicts a duct of variable cross-sectional area through which a fluid is flowing. Consider the fluid occupying a small control volume of width, δx just downstream of a section x-x of the duct. Assume that the entire mass of fluid in this volume will flow across x-x during the time δt. Therefore the mass flow across the section is

$$\frac{\delta m}{\delta t} = \frac{(A + A + \delta A)\delta x \rho}{2\delta t} \qquad (5.23)$$

where ρ is the density of the fluid.

Neglecting the second-order quantity $\delta x \delta A$ we can express the mass flow rate as

$$\dot{m} = A\rho V \qquad (5.24)$$

The average velocity of the fluid is, $V = \delta x / \delta t$.

5.6 Some Steady-Flow Devices

Many devices commonly found in engineering systems, for the most part, operate under steady conditions. In this section we present a number of such devices mainly to illustrate the application of the steady-flow-energy equation (SFEE), and the mass conservation equation to the device. The primary function and principle of operation of the device will only be described briefly.

Although the emphasis in this section is on steady-state operation, it should be noted that transient modes of operation of the device that occur during start-up, shut-down, and change of inputs are of considerable practical importance. These aspects, however, are not addressed in this chapter.

5.6.1 *Nozzles and diffusers*

A nozzle is essentially a duct with a specially designed shape (Fig. 5.4), whose main function is to convert a part of the enthalpy of the fluid entering it to kinetic energy at the exit. Nozzles are used to produce high-speed jets of fluids for various applications, including the input to drive turbines.

Consider a control volume surrounding the nozzle with an inlet port i and an exit port e. Applying the mass conservation equation [Eq. (5.21)] to the control volume we have

$$\dot{m}_i = \dot{m}_e = \dot{m} \qquad\qquad (5.25)$$

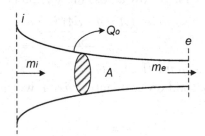

Fig. 5.4 A Nozzle

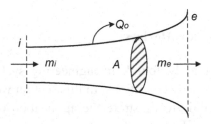

Fig. 5.5 A diffuser

Using Eq. (5.24), the above equation may be expressed in the form

$$\dot{m} = A_i\rho_i V_i = A_e\rho_e V_e \qquad (5.26)$$

where A is the area of cross-section of the port, ρ is the density, and V is the mean velocity of the fluid.

The work output of the nozzle is zero, and the difference in potential energy of the fluid at the inlet and outlet ports is negligible. However, the heat interaction between the fluid and surroundings, $\dot{Q}_o$ is not always zero. Applying the SFEE, [Eq. (5.20)] to the control volume we have

$$\dot{m}\left(h_i + V_i^2/2\right) = \dot{m}\left(h_e + V_e^2/2\right) + \dot{Q}_o \qquad (5.27)$$

The diffuser, shown in Fig. 5.5, is used to convert a part of the kinetic energy of the fluid at the inlet port i to enthalpy at the exit port e. In practice, diffusers are often used to reduce the speed of a fluid stream before it is discharged to the ambient. As for the nozzle, by applying the equation of mass conservation and the SFEE to a control volume surrounding the diffuser we could obtain two equations, similar to Eqs. (5.25) and (5.27), that are applicable to the diffuser. The application of the above analysis is illustrated in worked examples 5.1, 5.8 and 5.9.

5.6.2 *Turbines and compressors*

The primary function of a turbine, shown schematically in Fig. 5.6, is to convert a part of the enthalpy of the fluid entering it at the inlet i to a work output, $\dot{W}_o$ at the output shaft. A typical turbine consists of a large wheel mounted on a concentric shaft. A series of blades are fixed to the periphery of the wheel, and the fluid flows over these blades to produce a torque by virtue of its momentum change. The torque drives the external load connected to the shaft, thus delivering a work output.

Consider a control volume surrounding the turbine with an inlet port i and an exit port e. Applying the mass conservation equation [Eq. (5.21)] to the control volume we have

$$\dot{m}_i = \dot{m}_e = \dot{m} \qquad (5.28)$$

Using Eq. (5.24) the above equation may be expressed in the form

$$\dot{m} = A_i\rho_i V_i = A_e\rho_e V_e \qquad (5.29)$$

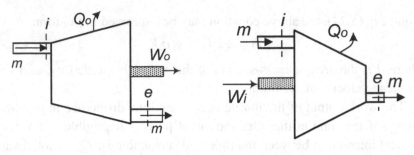

Fig. 5.6 Turbine   Fig. 5.7 Compressor

where A is the area of cross-section of the port, ρ is the density and V is the mean velocity of the fluid.

The difference in potential energy of the fluid at the inlet and outlet ports of the control volume is usually negligible. However, the heat interaction between the fluid and surroundings, $\dot{Q}_o$ is not necessarily zero. Applying the SFEE to the control volume we have

$$\dot{m}\left(h_i + V_i^2/2\right) = \dot{m}\left(h_e + V_e^2/2\right) + \dot{Q}_o + \dot{W}_o \qquad (5.30)$$

The main function of a compressor, shown schematically in Fig. 5.7, is to increase the pressure of the fluid entering it at the inlet port i. The fluid flows over a series of blades mounted on a wheel which is rotated by an external work input as depicted in Fig. 5.7.

Considering a control volume surrounding the compressor, and applying the mass conservation equation and the SFEE we could obtain three equations similar to Eqs. (5.28) to (5.30) to analyze the compressor. The application of these equations is illustrated in worked examples 5.2, 5.3 and 5.11.

5.6.3 *Mixing chambers and heat exchangers*

The mixing chamber shown in Fig. 5.8 is commonly used in the process industry to mix two or more inlet streams to produce a single exit stream. Consider a control volume surrounding the mixing chamber with two inlet ports 1 and 2 and an exit port 3. Applying the mass conservation equation [Eq. (5.19)] we obtain

$$\dot{m}_1 + \dot{m}_2 = \dot{m}_3$$

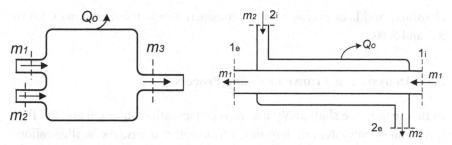

Fig. 5.8 Mixing chamber   Fig. 5.9 Heat exchanger

The kinetic and potential energy of the fluid streams are usually negligible compared to the enthalpies. Applying the SFEE [(Eq.5.18)] to the control volume we have

$$\dot{m}_1 h_1 + \dot{m}_2 h_2 = \dot{m}_3 h_3$$

The heat exchanger, depicted in Fig. 5.9, has two streams of fluid that flow in separate passages and therefore do not mix as in a mixing chamber.

The main function of the heat exchanger is to transfer heat from a high temperature fluid to a low temperature fluid. Heat exchangers have a wide range of engineering applications including various types of heaters and coolers.

Consider a control volume surrounding the heat exchanger (Fig. 5.9) with two inlet ports $1i$ and $2i$ and exit ports $1e$ and $2e$. The mass flow rates of the two streams are $\dot{m}_1$ and $\dot{m}_2$ respectively. Since the fluid streams do not mix, these mass flow rates are the same at the respective inlet and exit ports. The kinetic and potential energy of the streams are usually negligible compared to the enthalpy. Furthermore, the work interaction of the control volume is zero. Applying the SFEE [Eq. (5.18)] we have

$$\dot{m}_1 h_{1i} + \dot{m}_2 h_{2i} = \dot{m}_1 h_{1e} + \dot{m}_2 h_{2e} + \dot{Q}_o \qquad (5.31)$$

Condensers and evaporators are two special types of heat exchangers where one of the fluid streams undergoes phase change during its passage through the heat exchanger. These heat exchangers are used widely in steam power plants and refrigeration systems. Mixing chambers and heat exchangers are considered in worked examples 5.4 to 5.7 and 5.10.

5.7 Analysis of a Transient Filling Process

In this section we shall apply the mass conservation equation and the first law to an open system undergoing a *transient process*. As an illustration, we consider the arrangement for filling a pressure vessel with a fluid from a high pressure source of the fluid as depicted in Fig. 5.10.

Initially, the vessel contains a given mass of fluid at a given temperature and pressure. It is connected to a pipe carrying fluid at a high pressure through a valve, which is initially closed. The valve is then opened, and fluid is allowed to flow into the vessel until the pressure reaches a specified value. The valve is then closed.

Consider a control volume, with a single inlet port, i enclosing the vessel. Since there is no exit port the conditions of the fluid in the vessel during the filling process will be unsteady or time dependent. The conditions of the fluid in the pipe, however, are assumed constant. Let the mass and internal energy of the fluid in the vessel at time t be M and E respectively. The instantaneous rate of fluid flow through the valve is $\dot{m}_1$.

Applying the unsteady mass balance equation [Eq. (5.16)] we obtain

$$\frac{dM}{dt} = \dot{m}_1 \tag{5.32}$$

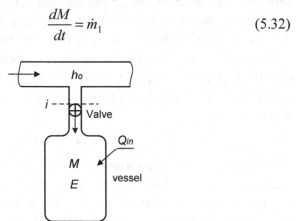

Fig. 5.10 Filling process

The kinetic energy and the potential energy of the fluid are assumed to be negligible compared to the internal energy. Furthermore, the work transfer between the control volume and the surroundings is zero.

Applying the unsteady energy equation, [Eq. (5.13)] subject to the above assumptions we have

$$\frac{dE}{dt} = h_0 \dot{m}_1 + \dot{Q}_{in} \qquad (5.33)$$

where h_0 is the constant specific enthalpy of the fluid in the pipe and $\dot{Q}_{in}$ is the instantaneous heat flow rate to the control volume from the surroundings.

From Eqs. (5.32) and (5.33) we obtain

$$\frac{dE}{dt} = h_0 \frac{dM}{dt} + \dot{Q}_{in} \qquad (5.34)$$

Let the subscripts i and f denote the quantities related to the initial and final states of the fluid in the vessel. Integrating Eq. (5.32),

$$M_f - M_i = \int_i^f \dot{m}_1 dt = \int_i^f dm_1 = \Delta m_1 \qquad (5.35)$$

where Δm_1 is the total mass of fluid that entered the vessel during the process.

Integrating Eq. (5.34) between the initial and final states,

$$E_f - E_i = h_0 (M_f - M_i) + \int_i^f \dot{Q}_{in} dt$$

$$E_f - E_i = h_0 (M_f - M_i) + \Delta Q_{in} \qquad (5.36)$$

where ΔQ_{in} is the total heat flow into the control volume during the process. Equation (5.36) can be expressed in terms of the specific internal energy and the mass. Thus we have

$$M_f u_f - M_i u_i = h_0 (M_f - M_i) + \Delta Q_{in} \qquad (5.37)$$

We note that Eq. (5.37) is independent of the nature of the fluid and it could therefore be applied to situations involving both a gas like air or a pure substance like steam. The forgoing analysis is applied in worked examples 5.12 to 5.15 to filling problems involving both air and steam.

Discharge of a high-pressure fluid from a vessel is another transient flow situation that is of considerable practical importance. These discharging processes can be analyzed by applying the unsteady energy equation and the mass conservation equation. Some practical applications are illustrated in worked examples 5.16 to 5.20.

5.8 Worked Examples

Example 5.1 Air flows through a horizontal converging nozzle whose area of cross-section varies from 3×10^{-3} m² at the entrance to 2×10^{-3} m² at the exit. The pressure, temperature, and air speed at the entrance are 100 kPa, 27°C and 60 ms^{-1} respectively. At the exit of the nozzle the pressure is 80 kPa and the temperature is 50°C. Assume that air is an ideal gas with $c_v = 0.716$ kJ kg^{-1} K^{-1} and $R = 0.287$ kJ kg^{-1} K^{-1}. Calculate (i) the mass flow rate of air, (ii) the air speed at the exit, and (iii) the rate of heat exchange between the air and the surroundings.

Solution (i) The density of the air at the entrance section 1 is obtained from the ideal gas equation as

$$\rho_1 = m/V_1 = P_1/RT_1 = 100/(0.287 \times 300) = 1.161 \text{ kg m}^{-3}$$

The mass flow rate at the entrance section 1 is given by Eq. (5.24).

$$\dot{m} = A_1\rho_1 V_1 = 3 \times 10^{-3} \times 1.161 \times 60 = 0.209 \text{ kg s}^{-1}$$

(ii) The density of air at the exit section 2 is

$$\rho_2 = P_2/RT_2 = 80/(0.287 \times 323) = 0.863 \text{ kg m}^{-3}$$

The air flow through the nozzle is steady, and therefore the mass flow rate at the exit section 2 is also 0.209 kg s^{-1}. Using Eq. (5.24) we have

$$\dot{m} = 0.209 = A_2\rho_2 V_2$$

The air velocity at the exit section 2 is

$$V_2 = 0.209/(2 \times 10^{-3} \times 0.863) = 121.1 \text{ ms}^{-1}$$

(iii) Consider the control volume bounded by the nozzle wall with inlet and exit ports at sections 1 and 2 respectively. Apply the steady-flow-energy-equation (SFEE) to this control volume.

$$(h_1 + \frac{V_1^2}{2} + gz_1)\dot{m} + \dot{Q}_{in} = (h_2 + \frac{V_2^2}{2} + gz_2)\dot{m} + \dot{W}_c$$

The nozzle is horizontal and therefore there is no change in gravitational potential energy from inlet to exit. Also there is no work output from the control volume. Hence $\dot{W}_c = 0$. The specific enthalpy of the air, which is treated as an ideal gas, is given by $c_p T$. Applying these conditions, the above equation becomes:

$$\dot{Q}_{in} = -(c_p T_1 + \frac{V_1^2}{2})\dot{m} + (c_p T_2 + \frac{V_2^2}{2})\dot{m}$$

Substituting the numerical data we have

$$\dot{Q}_{in} = -(1.003 \times 300 + \frac{60^2}{2 \times 10^3}) \times 0.209$$

$$+ (1.003 \times 323 + \frac{121.12^2}{2 \times 10^3}) \times 0.209$$

The heat flow rate into the nozzle is

$$\dot{Q}_{in} = 1.157 + 4.821 = 5.978 \text{ kW}$$

Example 5.2 Atmospheric air at 100 kPa and 25°C undergoes an adiabatic, steady-flow process, to 400 kPa in an axial-flow compressor. At the discharge section of the compressor, where the area is 0.05 m², the air temperature and velocity are 200°C and 105 ms⁻¹ respectively. Assume that air is an ideal gas. Calculate (i) the density of air at the discharge section, (ii) the mass flow rate of air, and (iii) the power input to drive the compressor.

Solution Consider the control volume shown in Fig. E5.2. The conditions of the air at the entrance section *i-i* of the compressor are not known. Therefore we consider an imaginary boundary in the surrounding air, sufficiently far from the inlet, where the velocity of the air is negligibly small. We assume that the given inlet conditions of temperature and pressure apply at this boundary, indicated by 1 in the figure.

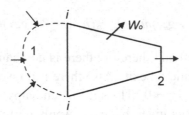

Fig. E5.2 Axial-flow compressor

(i) The density of the air at the exit section, 2 is obtained by applying the ideal gas equation.

$$\rho_2 = m/V_2 = P_2/RT_2 = 400/(0.287 \times 473) = 2.947 \text{ kg m}^{-3}$$

(ii) The mass flow rate at the exit section 2 is given by Eq. (5.24)

$$\dot{m} = A_2\rho_2V_2 = 0.05 \times 2.947 \times 105 = 15.47 \text{ kg s}^{-1}$$

(iii) Consider the control volume bounded by the imaginary inlet surface 1, the compressor walls and the exit section 2. The change in potential energy between the inlet and exit, and the velocity at the inlet section 1 are assumed negligible. Since the compressor is adiabatic the heat loss to the surroundings is zero. Applying the SFEE [Eq. (5.18)], subject to the above assumptions, we have

$$h_1\dot{m} = (h_2 + \frac{V_2^2}{2})\dot{m} + \dot{W}_c$$

Treating air as an ideal gas, the above equation may be written as

$$c_pT_1\dot{m} = (c_pT_2 + \frac{V_2^2}{2})\dot{m} + \dot{W}_c$$

Substituting the numerical data we obtain

$$\dot{W}_o = 1.003 \times 298 \times 15.47 - (1.003 \times 473 + \frac{105^2}{2 \times 10^3}) \times 15.47$$

$$\dot{W}_o = -2.715 - 0.085 = -2.8 \text{ MW}$$

The work input to the compressor is 2.8 MW.

Example 5.3 Superheated steam at 20 bar and 380°C enters a steam turbine with a velocity of 60 ms⁻¹. The pressure, quality and velocity of the steam at the discharge section of the turbine are 0.75 bar, 0.95 and 200 ms⁻¹ respectively. The entrance section of the turbine has an area of 0.025 m² and it is 3 m above the exit section. The rate of heat loss from the turbine casing to the surroundings is 8.5 kW. Calculate (i) the mass flow rate of steam, and (ii) the power output of the turbine.

Solution (i) Consider the steam flow at the entrance section 1 of the turbine, shown in Fig. E5.3. The steam is superheated at section 1 and the following properties are obtained from the steam tables [4] at 20 bar and 380°C.

$$v_1 = 0.1461 \text{ m}^3 \text{ kg}^{-1} \quad \text{and} \quad h_1 = 3204 \text{ kJ kg}^{-1}$$

The density at the entrance section is

$$\rho_1 = 1/v_1 = 1/0.1461 = 6.8446 \text{ kg m}^{-3}$$

The mass flow rate is obtained by applying Eq. (5.24) to section 1.

$$\dot{m} = A_1 \rho_1 V_1 = 0.025 \times 6.8446 \times 60 = 10.27 \text{ kg s}^{-1}$$

(ii) At the discharge section 2, the steam has a quality, $x_2 = 0.95$. The data for saturated steam at 0.75 bar is obtained from [4].
    The specific enthalpy at section 2 is

$$h_2 = x_2 h_g + (1 - x_2) h_f$$

$$h_2 = 0.95 \times 2662 + 0.05 \times 384 = 2548.1 \text{ kJ kg}^{-1}$$

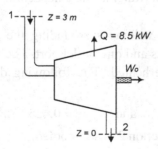

Fig. E5.3 Steam turbine

Consider the control volume bounded by the walls of the turbine, and the inlet and exit sections as shown in Fig. E5.3. Take the exit section 2 as the datum level ($z_2 = 0$) for the gravitational potential energy of the steam. Applying the SFEE we have

$$(h_1 + \frac{V_1^2}{2} + gz_1)\dot{m} + \dot{Q}_{in} = (h_2 + \frac{V_2^2}{2} + gz_2)\dot{m} + \dot{W}_o$$

Substituting the given numerical values in the above equation

$$\dot{W}_o = (3204 + \frac{60^2}{2 \times 10^3} + \frac{9.81 \times 3}{10^3}) \times 10.27$$

$$-8.5 - (2548 + \frac{200^2}{2 \times 10^3}) \times 10.27$$

$$\dot{W}_o = 6.7371(\Delta h) - 0.1869(\Delta KE) + 3.0 \times 10^{-4}(\Delta PE) - 0.0085(Q_{loss})$$

The work output is: $\dot{W}_o = 6.54$ MW

From the numerical values we observe that the change in kinetic energy and potential energy are much smaller than the change in enthalpy.

Example 5.4 The mixing chamber shown in Fig. E5.4 has a stream of water entering at 500 kPa and 15°C, and a stream of steam entering at 500 kPa and 200°C. Water flows at the rate of 0.6 kg s^{-1} through a pipe of cross sectional area 5×10^{-3} m^2 while the steam flows at 0.1 kg s^{-1} through a pipe of area 2.5×10^{-4} m^2. The condensed steam and water leave at 500 kPa through a pipe of large cross sectional area. Assuming that the chamber is well-insulated and operates under steady conditions, calculate the temperature of the water at the exit section.

Solution Consider a control volume surrounding the mixing chamber (Fig. E5.4) with two inlet ports and one outlet port. The steam entering at 500 kPa at section 1 is superheated. The following data are obtained from the steam tables [4].

$$h_1 = 2857 \text{ kJ kg}^{-1} \quad \text{and} \quad v_1 = 0.4252 \text{ m}^3 \text{ kg}^{-1}.$$

The water entering through section 2 is sub-cooled.

The properties are therefore obtained from the saturated steam tables [4] at 15°C, ignoring the effect of pressure.

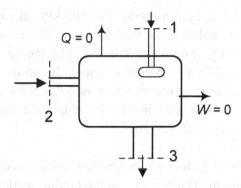

Fig. E5.4 Steam-water mixing chamber

$$h_2 = 62.9 \text{ kJ kg}^{-1} \quad \text{and} \quad v_2 = 0.1001 \times 10^{-2} \text{ m}^3 \text{ kg}^{-1}$$

The fluid velocities at 1 and 2 are calculated by applying Eq. (5.24).

$$V_1 = \dot{m}_1 / A_1 \rho_1 = \dot{m}_1 v_1 / A_1 = 0.1 \times 0.4252 / (2.5 \times 10^{-4}) = 170.08 \text{ ms}^{-1}$$

$$V_2 = \dot{m}_2 / A_2 \rho_2 = 0.6 \times 0.1001 \times 10^{-2} / (5 \times 10^{-3}) = 0.12 \text{ ms}^{-1}$$

The mass flow rate of the fluid stream at the exit section 3 is given by the mass balance equation [Eq. (5.19)] as:

$$\dot{m}_3 = \dot{m}_1 + \dot{m}_2 = 0.1 + 0.6 = 0.7 \text{ kg s}^{-1}$$

Due to the large cross-sectional area of the exit duct at 3, the velocity V_3 is negligible. Moreover, we neglect any changes in potential energy of the three streams. Since the chamber is well–insulated, $\dot{Q}_{in}$ is zero. The work output $\dot{W}_o$ is also zero. Applying the SFEE subject to the above conditions we have

$$(h_1 + \frac{V_1^{\,2}}{2})\dot{m}_1 + (h_2 + \frac{V_2^{\,2}}{2})\dot{m}_2 = (h_3 + \frac{V_3^{\,2}}{2})\dot{m}_3$$

Substituting the given numerical values in the above equation

$$(2857 + \frac{170.08^2}{2 \times 10^3}) \times 0.1 + (62.9 + \frac{0.1201^2}{2 \times 10^3}) \times 0.6 = h_3 \times 0.7$$

Therefore $\qquad\qquad h_3 = 464.12 \text{ kJ kg}^{-1}$

Now the stream 3 leaves at a pressure of 500 kPa, at which pressure the saturation liquid enthalpy is 640 kJ kg^{-1}. Since the calculated enthalpy of 464.1 kJ kg^{-1} of stream 3 is less than the above value we conclude that stream 3 is a sub-cooled liquid. Ignoring the effect of pressure, we seek the temperature at which the saturated liquid enthalpy is 464.1 kJ kg^{-1}. From the data in [4], we obtain the temperature of stream 3 by interpolation as $T_3 = 110.7\,°C$.

Example 5.5 Figure E5.5 shows a steady-flow mixing section of an air conditioning duct system. The pressure in the mixing section is uniform. The volume flow rate and the temperature of the air streams 1 and 2 are respectively 1.2 m^3 s^{-1}, 8°C and 2.4 m^3 s^{-1}, 38°C. Assume that the heat loss from the air to the surroundings is negligible and that air is an ideal gas with constant specific heat capacities. Calculate (i) volume flow rate, and (ii) the temperature of the air at the exit section 3.

Solution Consider the control volume shown in Fig. E5.5 which has two inlet streams at sections 1 and 2 and one exit stream at section 3. Mass balance for the control volume gives

$$\dot{m}_1 + \dot{m}_2 = \dot{m}_3 \qquad\qquad (E5.5.1)$$

Substituting for mass from the ideal gas equation in Eq. (E5.5.1)

$$\frac{P_1\dot{V}_1}{RT_1} + \frac{P_2\dot{V}_2}{RT_2} = \frac{P_3\dot{V}_3}{RT_3} \qquad\qquad (E5.5.2)$$

where $\dot{V}$ is the volume flow rate.

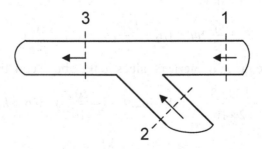

Fig. E5.5 Mixing section of duct system

Applying the SFEE to the control volume, neglecting the kinetic and potential energy of the streams, we have

$$h_1 \dot{m}_1 + h_2 \dot{m}_2 = h_3 \dot{m}_3 \tag{E5.5.3}$$

Substituting for mass from the ideal gas equation and the expressions for the enthalpy of an ideal gas in Eq. (E5.5.3) we obtain

$$\frac{c_p T_1 P_1 \dot{V}_1}{RT_1} + \frac{c_p T_2 P_2 \dot{V}_2}{RT_2} = \frac{c_p T_3 P_3 \dot{V}_3}{RT_3} \tag{E5.5.4}$$

Note that the reference state for enthalpy is taken as 0 K. Since it is given that the pressure in the mixing section is uniform, $P_1 = P_2 = P_3$. Hence Eqs. (E5.5.2) and (E5.5.4) can be simplified as

$$\frac{\dot{V}_1}{T_1} + \frac{\dot{V}_2}{T_2} = \frac{\dot{V}_3}{T_3} \tag{E5.5.5}$$

$$\dot{V}_1 + \dot{V}_2 = \dot{V}_3 \tag{E5.5.6}$$

Substituting the given numerical values in Eqs. (E5.5.6) and (E5.5.5)

$$1.2 + 2.4 = \dot{V}_3$$

$$\frac{1.2}{(273 + 8)} + \frac{2.4}{(273 + 38)} = \frac{3.6}{T_3}$$

From the above equation we obtain the temperature of stream 3 as 300.3 K or 27.3°C.

Example 5.6 The ejector pump shown in Fig. E5.6 uses a stream of superheated steam at 15 bar and 250°C entering at 2 to raise the pressure of a stream of saturated liquid water entering at 1 at 30°C from 100 kPa to 300 kPa. The mass flow rate of steam is 40% of the mass flow rate of water. Assume that the heat loss from the ejector to the surroundings and the kinetic and potential energy of the streams are negligible. Determine the condition of the stream at the exit 3 of the ejector.

Solution Consider the control volume with two inlet streams 1 and 2 and one exit stream 3 as shown in Fig. E5.6. The water at section 1 is sub-cooled while the steam at section 2 is superheated. We obtain the following data from [4]:

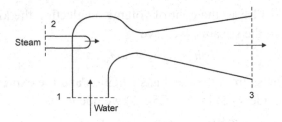

Fig. E5.6 Steam ejector-pump

$$h_1 = 125.7 \text{ kJ kg}^{-1} \quad \text{and} \quad h_2 = 2925 \text{ kJ kg}^{-1}$$

Let the mass flow rate of water at 1 be $\dot{m}_1$. Then $\dot{m}_2 = 0.4\dot{m}_1$.

Applying the mass balance equation under steady conditions

$$\dot{m}_3 = \dot{m}_1 + \dot{m}_2 = \dot{m}_1 + 0.4\dot{m}_1 = 1.4\dot{m}_1$$

Applying the SFEE to the control volume, neglecting the kinetic and potential energy of the streams, we have

$$\dot{m}_3 h_3 = \dot{m}_1 h_1 + \dot{m}_2 h_2 = 125.7\dot{m}_1 + 2925 \times 0.4\dot{m}_1 = 1295.7\dot{m}_1$$

From the two equations above, $h_3 = 925.5 \text{ kJ kg}^{-1}$.

Now the fluid leaving at section 3 has a pressure of 300 kPa. The saturation liquid and vapor enthalpies at this pressure are:

$$h_{f3} = 561 \text{ kJ kg}^{-1} \quad \text{and} \quad h_{g3} = 2725 \text{ kJ kg}^{-1}$$

Since $$h_{f3} < h_3 < h_{g3}$$

we conclude that the steam at 3 is wet. The quality is obtained from

$$x_3 h_{g3} + (1 - x_3)h_{f3} = h_3$$

Substituting the relevant numerical values in the above equation

$$2725x_3 + 561(1 - x_3) = 925.5$$

Therefore $x_3 = 0.168$, and from the data in [4], the saturation temperature at 3 bar is 133.5 °C.

Example 5.7 Figure E5.7 shows a heat exchanger for cooling oil using cold water. Oil flows through the inner tube at the rate of 0.65 kg s^{-1} while water flows through the annulus at the rate of 0.55 kg s^{-1}. The oil is to be cooled from 75°C to 50°C. The temperature of the entering water is

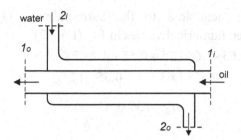

Fig. E5.7 Oil cooling heat exchanger

28°C. Oil and water may be treated as pure substances with constant specific heat capacities of 1.65 kJ kg^{-1} K^{-1} and 4.2 kJ kg^{-1} K^{-1} respectively.

Calculate the water temperature at the exit, (i) if there is no heat loss from the heat exchanger to the surroundings, and (ii) if heat is lost to the surroundings at the rate of 6 kW.

Solution Figure E5.7 shows the heat exchanger with two inlet streams and two outlet streams. Since the two fluid streams do not mix they have the same mass flow rates at the respective inlet and outlet ports under steady conditions. We denote quantities associated with oil and water by the subscripts 1 and 2 respectively.

Consider a control volume surrounding the heat exchanger. The inlet and outlet ports are indicated by i and o respectively.

Applying the SFEE to the control volume, neglecting the potential and kinetic energy of the fluids, we have

$$\dot{m}_1 h_{1i} + \dot{m}_2 h_{2i} + \dot{Q}_{in} = \dot{m}_1 h_{1o} + \dot{m}_2 h_{2o} \qquad (E5.7.1)$$

The water and oil are treated as pure substances with constant specific heat capacities. Therefore Eq. (E5.7.1) can be written as

$$\dot{m}_1 c_1 T_{1i} + \dot{m}_2 c_2 T_{2i} + \dot{Q}_{in} = \dot{m}_1 c_1 T_{1o} + \dot{m}_2 c_2 T_{2o} \qquad (E5.7.2)$$

When there is no heat loss from the heat exchanger, $\dot{Q}_{in} = 0$.

Substituting the given numerical values in Eq. (E5.7.2)

$$0.65 \times 1.65 \times 75 + 0.55 \times 4.2 \times 28 = 0.65 \times 1.65 \times 50 + 0.55 \times 4.2 T_{2o}$$

Hence $\qquad\qquad\qquad T_{2o} = 39.6\,°C$

(ii) When there is heat loss to the surroundings, $\dot{Q}_{in} = -6.0\,\text{kW}$. Substituting the given numerical values in Eq. (E5.7.2)

$$0.65 \times 1.65 \times 75 + 0.55 \times 4.2 \times 28 - 6.0$$
$$= 0.65 \times 1.65 \times 50 + 0.55 \times 4.2T_{2o}$$

Hence $T_{2o} = 37.0\,°\text{C}$

Example 5.8 A throttling calorimeter for measuring the quality of wet steam flowing in a pipe is shown in Fig. E5.8. A small quantity of steam is bled-off from a pipe carrying steam, through a throttle valve, into a well-insulated chamber, where the temperature and pressure are measured. The temperature and pressure of the steam in the pipe are 700 kPa and 165°C respectively, and the corresponding values in the chamber are 50 kPa and 90°C. (a) Calculate the quality of steam in the pipe. (b) Is there a limit to the steam quality that can be measured using the throttling calorimeter?

Solution (a) Figure E5.8 shows a control volume surrounding the throttling valve, with one inlet and one outlet stream denoted by 1 and 2 respectively. The kinetic energy and potential energy of the streams are negligible compared to their enthalpies. The valve and pipes are well insulated, and therefore the external heat loss can be neglected. Also, there is no external work transfer from the control volume. Any work done due to friction is transferred to the steam as heat.

Applying the SFEE to the control volume subject to the above conditions

$$\dot{m}_1 h_1 = \dot{m}_2 h_2 \qquad\qquad (E5.8.1)$$

Mass balance for the control volume gives

$$\dot{m}_1 = \dot{m}_2 \qquad\qquad (E5.8.2)$$

From Eqs. (E5.8.1) and (E5.8.2) we have

$$h_1 = h_2 \qquad\qquad (E5.8.3)$$

At section 2, $P_2 = 0.5$ bar and $T_2 = 90\,°\text{C}$. From the tabulated data in [4], the steam is superheated at 2. Hence $h_2 = 2662.67$ kJ kg^{-1}. Let the

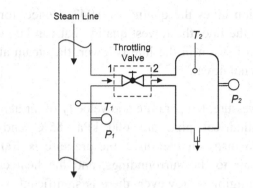

Fig. E5.8 Throttling calorimeter

quality at section 1 be x_1. The saturated liquid and vapor enthalpies at section 1, where the pressure is 700 kPa are obtained from [4] as

$$h_{f_1} = 697 \text{ kJ kg}^{-1} \quad \text{and} \quad h_{g1} = 2764 \text{ kJ kg}^{-1}$$

Equation (E5.8.3) may be written as

$$h_1 = h_{g1}x_1 + (1 - x_1)h_{f1} = h_2 \qquad \text{(E5.8.4)}$$

Substituting numerical values

$$2764 \times x_1 + 697 \times (1 - x_1) = 2662.67$$

Hence the quality of the steam in the line, $x_1 = 0.951$

(b) There is a lower limit to the quality x_1 of the steam in the pipe that can be measured by the throttling calorimeter. If x_1 is below this minimum value the steam leaving the throttling valve will be wet steam, whose pressure and temperature are not independent properties. Therefore for the calorimeter to work the steam at 2 must be superheated. For superheated steam the pressure and temperature are independent properties that enable the enthalpy to be determined.

As an example consider the present problem with dry saturated steam at 0.5 bar leaving section 2. Then

$$h_2 = h_{g2} = 2645 \text{ kJ kg}^{-1}$$

Substituting in Eq. (E5.8.4) we have

$$2764 \times x_1 + 697 \times (1 - x_1) = 2645 \text{ kJ kg}^{-1}$$

The above equation gives the quality as 0.942. Hence, for the same pressure of steam in the line, the lowest quality that can be measured is 0.942. If the quality of steam in the line is lower, the steam at 2 will be wet and its quality is unknown.

Example 5.9 The pressure, temperature and velocity of air at a section 1, of a horizontal cylindrical pipe are 600 kPa, 85°C and 90 m s^{-1} respectively. At a downstream section 2, the pressure is 400 kPa. The heat loss from the air to the surroundings, and the heat capacity of the pipe wall are negligible. However, there is significant wall friction in the pipe. (a) Assuming that air is an ideal gas, calculate the velocity and temperature at section 2. (b) When a fluid with a constant density of 600 kg m^{-3} and specific heat capacity 1.5 kJ kg^{-1} K^{-1} flows in the same pipe, the pressure drop between sections 1 and 2 due to friction is 0.1 m of water. Calculate the change in temperature of the fluid between sections 1 and 2.

Solution Consider the control volume bounded by the pipe wall, the inlet section 1 and the outlet section 2. Applying the equation of mass balance

$$A_1 V_1 \rho_1 = A_2 V_2 \rho_2 \tag{E5.9.1}$$

Using the ideal gas equation to express the density in terms of the pressure and temperature, Eq. (E5.9.1) can be written as

$$V_1 \frac{P_1}{RT_1} = V_2 \frac{P_2}{RT_2} \tag{E5.9.2}$$

Substituting the numerical values in Eq. (E5.9.2)

$$\frac{90 \times 600}{358} = \frac{400 V_2}{T_2}$$

$$\frac{V_2}{T_2} = 0.377 \tag{E5.9.3}$$

The heat transfer and work transfer from the control volume to the surroundings are zero. Since the pipe wall has negligible thermal capacity all the work done due to wall friction is transferred to the air as heat. There is no change in the potential energy of the air because the pipe is horizontal. Applying the SFEE, subject to the above conditions,

$$(c_p T_1 + \frac{V_1^2}{2})\dot{m} = (c_p T_2 + \frac{V_2^2}{2})\dot{m} \qquad (E5.9.4)$$

Substituting the given numerical values in the above equation

$$1.003 \times 358 + \frac{90^2}{2 \times 10^3} = 1.003 T_2 + \frac{V_2^2}{2 \times 10^3}$$

$$T_2 + 0.4985 \times 10^{-3} V_2^2 = 362.037 \qquad (E5.9.5)$$

Eliminating V_2 between Eqs. (E.5.9.3) and (E5.9.5) we obtain

$$T_2^2 + 14.114 \times 10^3 T_2 - 5.11 \times 10^6 = 0 \qquad (E5.9.6)$$

The positive root of Eq. (E5.9.6) gives, $T_2 = 353.2$ K
The velocity at section 2 is obtained from Eq. (E5.9.3) as

$$V_2 = 133.16 \text{ ms}^{-1}$$

(b) Since the density of the fluid and the cross-sectional area of the pipe are constant, we deduce from the mass balance equation that $V_1 = V_2$.

The heat transfer and work transfer between the control volume and the surroundings are zero. Applying the SFEE to the control volume we have

$$h_1 = h_2 \qquad (E5.9.7)$$

Treat the fluid as an incompressible pure substance with a constant specific heat capacity and density. Using the relation, $h = u + Pv$ we express Eq. (E5.9.7) in the form

$$(c_v T_1 + \frac{P_1}{\rho}) = (c_v T_2 + \frac{P_2}{\rho})$$

Hence $\qquad\qquad c_v(T_2 - T_1) = \dfrac{P_1 - P_2}{\rho} \qquad\qquad (E5.7.8)$

Now the pressure difference between sections 1 and 2 is given by

$$P_1 - P_2 = h\rho g = 0.1 \times 10^3 \times 9.81 = 0.98 \text{ kPa}$$

Substituting in Eq. (E5.7.8) we obtain

$$(T_2 - T_1) = \frac{0.98}{1.5 \times 600} = 1.09 \times 10^{-3} \text{ K}$$

The above equation shows that the drop in pressure due to friction, which in fluid mechanics is commonly called a 'pressure loss', results in a rise in temperature of the fluid. This is because the work done due to friction is converted to a heat flow into the fluid resulting in an increase in the internal energy of the fluid.

Example 5.10 Atmospheric air containing 0.023 kg of water vapor per kg of dry air enters an air conditioner with an enthalpy of 88 kJ per kg of dry air. The air leaving the air conditioner has 0.018 kg of water vapor per kg of dry air and its enthalpy is 70 kJ per kg of dry air. The dry air flow rate is 8 kg s^{-1}. The enthalpy of the condensate is 104 kJ per kg of liquid. Calculate (i) the mass flow rate of the condensate, and (ii) the rate of heat removal from the air.

Solution Atmospheric air is a mixture of dry air and moisture or water vapor that flows as a single fluid stream. However, for the purpose of our analysis we shall treat the water vapor and the dry air as two separate streams.

When air is cooled in the air conditioner, some of the water vapor is condensed to liquid water and leaves as condensate. However, the mass flow rate of *dry air* remains constant.

Consider the control volume, depicted in Fig. E5.10, with two inlet streams and three outlet streams. The dry air flow rate is the same at the inlet and outlet. Therefore the mass flow rates of the water vapor streams may be obtained by multiplying the given flow rates by the constant dry air flow rate.

(i) The mass balance equation for water gives

$$0.023 \times 8 = 0.018 \times 8 + \dot{m}_c$$

Hence the mass flow rate of condensate is,

$$\dot{m}_c = 0.04 \text{ kg s}^{-1}.$$

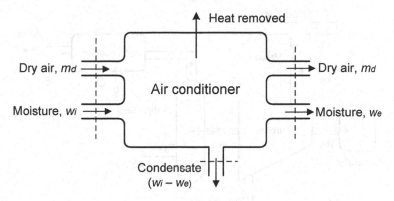

Fig. E5.10 Air conditioner

(ii) Assume that the kinetic and potential energy of the fluid streams are negligible. The work transfer from the control volume is also zero. Applying the SFEE we have

$$88 \times 8 + \dot{Q}_{in} = 70 \times 8 + 104 \times 0.04$$

Therefore the heat removal rate is,

$$\dot{Q}_{re} = -\dot{Q}_{in} = 139.8 \text{ kW}.$$

Example 5.11 Figure E5.11 shows a steam turbine that drives an air compressor and an electric generator. The turbine receives steam from a steam line in which the pressure and temperature are 40 bar and 425°C respectively. A throttle valve reduces the pressure to 35 bar before the steam enters the turbine. The pressure and quality of the steam leaving the turbine are 0.2 bar and 0.9 respectively. The steam flow rate is 0.15 kg s^{-1}.

Air enters the compressor at a pressure of 1 bar and 20°C and leaves the after-cooler at 100 bar and 40°C. The air flow rate is 0.02 kg s^{-1}. The power input to the compressor is 10.5 kW. The heat loss from the turbine to the surroundings is negligible. Calculate (i) the power output of the turbine, (ii) the power input to the electric generator, and (iii) the total heat flow from the air as it passes through the compressor and the after-cooler.

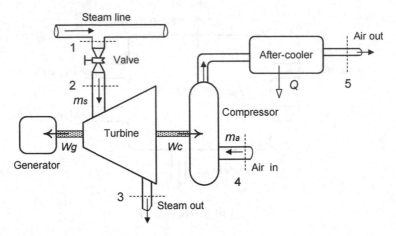

Fig. E5.11 Steam turbine-compressor-generator system

Solution The steam in the line is superheated and the enthalpy is obtained from the data in [4] as $h_1 = 3272$ kJ kg^{-1}. Consider a control volume surrounding the throttling valve. We assume that the kinetic and potential energy of the steam at the inlet and outlet and the heat and work transfers to the surroundings are negligible (also see worked example 5.8).

Applying the SFEE we have

$$h_2 = h_1 = 3272 \text{ kJ kg}^{-1}$$

The outlet stream of steam is wet with a quality of 0.9. The enthalpy is given by

$$h_3 = x_3 h_{g3} + (1 - x_3)h_{h3}$$

$$h_3 = 0.9 \times 2609 + 0.1 \times 251 = 2373.2 \text{ kJ kg}^{-1}$$

Consider a control volume surrounding the steam turbine. Apply the SFEE, neglecting the kinetic and potential energy of the steam and any heat losses to the surroundings.

$$\dot{m}_s h_2 = \dot{m}_s h_3 + \dot{W}_o$$

Substituting numerical values

$$0.15 \times 3272 = 0.15 \times 2373.27 + \dot{W}_o$$

Hence the work output of the turbine is, $\dot{W}_o = 134.8$ kW.

(ii) The work balance for the turbine, compressor, and generator gives

$$\dot{W}_o = \dot{W}_c + \dot{W}_g$$

Therefore work input to the electric generator is

$$\dot{W}_g = 134.8 - 10.5 = 124.3 \,\text{kW}$$

Consider a control volume surrounding the air compressor and the after-cooler. Assume that the kinetic energy and potential energy of the air are negligible. Applying the SFEE we have

$$h_4 \dot{m}_a + \dot{Q}_{in} = h_5 \dot{m}_a + \dot{W}_c$$

Treating air as an ideal gas the above equation may be written as

$$c_p T_4 \dot{m}_a + \dot{Q}_{in} = c_p T_5 \dot{m}_a + \dot{W}_c$$

Substituting numerical values we have

$$1.003 \times 20 \times 0.02 + \dot{Q}_{in} = 1.003 \times 40 \times 0.02 - 10.5$$

The heat input to the air is, $\dot{Q}_{in} = -10.1$ kW. Hence the heat transfer from the air to the surroundings is 10.1 kW.

Example 5.12 A well-insulated rigid vessel of volume 0.4 m³ contains air at 100 kPa and 28°C. The vessel is filled with additional air by connecting it to a pipe carrying air at 180°C and 800 kPa, through a valve. The valve is closed when the pressure in the vessel is 580 kPa. Assume that air is an ideal gas with $c_v = 0.716$ kJ kg⁻¹ K⁻¹ and $R = 0.287$ kJ kg⁻¹ K⁻¹. (a) Calculate (i) the final temperature of the air in the vessel, and (ii) the mass of air added to the vessel. (b) How will the answer to (a) change if the vessel was initially evacuated?

Solution Consider a control volume, with a single inlet port, enclosing the vessel. Since there is no outlet port the conditions of the air in the vessel during the filling process are time dependent. The conditions of the air in the pipe, however, are constant. Let the mass and internal energy of air in the vessel at time t be M and E respectively. The rate of air flow through the valve is $\dot{m}_1$.

Applying the unsteady mass balance equation [Eq. (5.16)]

$$\frac{dM}{dt} = \dot{m}_1 \qquad\qquad \text{(E5.12.1)}$$

The kinetic and potential energy of the air are assumed to be negligible compared to the internal energy. Moreover, the heat and work transfer between the control volume and the surroundings are zero. Applying the unsteady energy equation, [Eq. (5.13)] subject to the above conditions, we have

$$\frac{dE}{dt} = h_0 \dot{m}_1 \qquad\qquad \text{(E5.12.2)}$$

where h_0 is the constant specific enthalpy of the air in the pipe.

Let the subscripts i and f denote the quantities related to the initial and final state of the air in the vessel. Integrating Eq. (E5.12.1)

$$M_f - M_i = \int_i^f \dot{m}_1 dt = \int_i^f dm_1 = \Delta m_1 \qquad\qquad \text{(E5.12.3)}$$

where Δm_1 is the total mass of air that entered the vessel.

Integrating Eq. (E5.12.2) we have

$$E_f - E_i = \int_i^f h_0 \dot{m}_1 dt = h_0 \int_i^f dm_1 = h_0 \Delta m_1 \qquad\qquad \text{(E5.12.4)}$$

From Eqs. (E5.12.3) and (E5.12.4) it follows that

$$E_f - E_i = h_0(M_f - M_i) \qquad\qquad \text{(E5.12.5)}$$

We now assume that air is an ideal gas. Substituting for the mass from the ideal gas equation, and the relevant expressions for the internal energy and enthalpy in Eq. (E5.12.5) we have

$$\frac{P_f V_v (c_v T_f)}{R T_f} - \frac{P_i V_v (c_v T_i)}{R T_i} = c_p T_0 \left(\frac{P_f V_v}{R T_f} - \frac{P_i V_v}{R T_i} \right) \qquad \text{(E5.12.6)}$$

where V_v is the constant volume of the vessel, and T_0 the constant temperature of the air in the pipe. Simplifying Eq. (E5.12.6) we obtain

$$P_f - P_i = \gamma T_0 \left(\frac{P_f}{T_f} - \frac{P_i}{T_i} \right) \qquad \text{(E5.12.7)}$$

$\gamma = c_p / c_v$, is the ratio of the specific heat capacities.
(a) Substituting the given numerical values in Eq. (E5.12.7)

$$580 - 100 = 1.4 \times 453 \times \left(\frac{580}{T_f} - \frac{100}{301} \right)$$

Therefore the final temperature of the air in the vessel is, $T_f = 532.6 \, \text{K}$.
(ii) The mass of air added is given by Eq. (E5.12.3) as

$$\Delta m_1 = M_f - M_i = \left(\frac{P_f V_v}{R T_f} - \frac{P_i V_v}{R T_i} \right) \qquad \text{(E5.12.8)}$$

Substituting numerical values in the above equation

$$\Delta m_1 = \left(\frac{580 \times 0.4}{0.287 \times 532.6} - \frac{100 \times 0.4}{0.287 \times 301} \right) = 1.055 \, \text{kg.}$$

(b) If the vessel is initially evacuated then, $P_i = 0$. Substituting in Eq. (E5.12.7) we obtain the final temperature as

$$T_f = \gamma T_0 = 1.4 \times 453 = 634.2 \, \text{K}$$

The mass of air added is given by Eq. (E5.12.8) as

$$\Delta m_1 = \left(\frac{580 \times 0.4}{0.287 \times 634.2} \right) = 1.27 \, \text{kg}$$

Example 5.13 Steam is allowed to flow from a large reservoir maintained at a constant pressure of 3 MPa into a rigid insulated vessel of volume $4.0 \, \text{m}^3$ containing dry saturated steam at 1 MPa. When the pressure and temperature of the steam in the vessel are 2 MPa and 325°C the flow of steam is stopped. Calculate (i) the constant temperature of the steam in the reservoir, and (ii) mass of steam that flows into the vessel.

Solution Consider a control volume surrounding the vessel and the valve controlling the flow of steam, as shown in Fig. 5.10. We assume that the kinetic and potential energy of the steam are negligible, and that the work and heat transfer between the control volume and the surroundings are both zero. This unsteady filling process is similar in all respects to the situation considered is Sec. 5.7, and the problem in worked example 5.12.

Therefore the starting point of our solution could be Eq. (5.37) which is independent of the working fluid.

$$M_f u_f - M_i u_i = h_0(M_f - M_i) \qquad (E5.13.1)$$

We can express Eq. (E5.13.1) in terms of the specific volume as

$$\left(\frac{V_v}{v_f}\right)u_f - \left(\frac{V_v}{v_i}\right)u_i = h_0\left(\frac{V_v}{v_f} - \frac{V_v}{v_i}\right) \qquad (E5.13.2)$$

From the given data the steam in the vessel in the initial state at 10 bar is dry saturated with

$$v_i = 0.1944 \text{ m}^3 \text{kg}^{-1} \quad \text{and} \quad u_i = 2584 \text{ kJ kg}^{-1}.$$

In the final state the steam is superheated and the specific volume and internal energy are obtained from [4] as:

$$v_f = 0.1320 \text{ m}^3 \text{kg}^{-1} \quad \text{and} \quad u_f = 2817.5 \text{ kJ kg}^{-1}$$

Substituting these values in Eq (E5.13.2) we have

$$\left(\frac{4}{0.1320}\right) \times 2817.5 - \left(\frac{4}{0.1944}\right) \times 2584 = h_0\left(\frac{4}{0.1320} - \frac{4}{0.1944}\right)$$

Therefore the enthalpy of the steam in the reservoir,

$$h_0 = 3311.5 \text{ kJ kg}^{-1}$$

Since the pressure of the steam in the reservoir is 30 bar, we conclude from the tabulated data in [4], that the steam is superheated with a temperature of 436 °C.

(ii) The mass of steam that flows into the vessel is given by Eq. (5.35) as

$$\Delta m = \left(\frac{V_v}{v_f} - \frac{V_v}{v_i}\right)$$

$$\Delta m = \left(\frac{4}{0.1320} - \frac{4}{0.1944}\right) = 9.73 \text{ kg}$$

Example 5.14 A rigid tank of volume 0.7 m³ is connected to a large reservoir of air at 15 bar and 450 K through a valve. Initially, the valve is closed and the tank contains air at 1 bar and 300 K. The valve is now opened, and air is allowed to flow into the tank until its pressure and temperature are 10 bar and 400 K respectively. Assume that air is an ideal gas. Calculate (i) the heat transfer between the tank and the surroundings, and (ii) the mass of air that flows into the tank.

Solution This example is similar in many respects to the filling process described in worked example 5.12 except that there is heat transfer between the air and the surroundings in the present case. We refer to the schematic diagram in Fig. 5.10.

Neglecting the kinetic and potential energy of the air, the mass and energy balance equations may be written as

$$\frac{dM}{dt} = \dot{m}_1 \qquad (E5.14.1)$$

$$\frac{dE}{dt} = h_0 \dot{m}_1 + \dot{Q}_{in} \qquad (E5.14.2)$$

where $\dot{Q}_{in}$ is the instantaneous rate of heat flow into the control volume.
From Eqs. (E5.14.1) and (E5.14.2) we have

$$\frac{dE}{dt} = h_0 \frac{dM}{dt} + \dot{Q}_{in} \qquad (E5.14.3)$$

Integrating Eq. (E5.14.3) between the initial and final states

$$E_f - E_i - h_0(M_f - M_i) = \int_i^f \dot{Q}_{in} dt = \Delta Q_{in} \qquad (E5.14.4)$$

where ΔQ_{in} is the total heat received by the air from the initial state to the final state.

We now substitute for mass from the ideal gas equation and the relevant expressions for the internal energy and enthalpy in Eq. (E5.14.4).

$$\Delta Q_{in} = \left(\frac{P_f V_t}{RT_f}\right) c_v T_f - \left(\frac{P_i V_t}{RT_i}\right) c_v T_i - c_p T_0 \left(\frac{P_f V_t}{RT_f} - \frac{P_i V_t}{RT_i}\right)$$

$$\Delta Q_{in} = \left(\frac{P_f V_t}{R}\right)c_v - \left(\frac{P_i V_t}{R}\right)c_v - c_p T_0 \left(\frac{P_f V_t}{RT_f} - \frac{P_i V_t}{RT_i}\right)$$

Substituting the given numerical values in the above equation

$$\Delta Q_{in} = \frac{0.7 \times 0.716}{0.287}(1000 - 100) - \frac{1.003 \times 450 \times 0.7}{0.287}\left(\frac{1000}{400} - \frac{100}{300}\right)$$

Hence $\Delta Q_{in} = -813.5$ kJ. The heat transfer from the air to the surroundings is 813.5 kJ.

Example 5.15 A spring loaded piston-cylinder set-up is connected through a valve at the bottom to a high pressure air line carrying air at pressure P_g and temperature T_o (K). The linear spring has a spring constant k and the mass and area of the piston are M_p and A respectively. The external pressure on the piston is P_a. Initially, the spring is uncompressed and the piston sits at the bottom of the cylinder. Air is allowed to flow slowly into the cylinder until the pressure inside the cylinder becomes P_g.

Obtain expressions for (i) the final volume of the air in the cylinder, and (ii) the final temperature of the air in the cylinder. Neglect any heat exchange between the air and the surroundings.

Solution Consider a control volume enclosing the piston-cylinder set-up and the valve as shown in Fig. E5.15. The air pressure required to *just* lift the piston from the bottom is obtained from the force balance on the piston as:

$$P_i = P_a + \frac{M_p g}{A} \tag{E5.15.1}$$

Let the compression of the spring in the final equilibrium state be x. The force balance on the piston gives

$$AP_f = AP_a + Mg + kx = AP_g \tag{E5.15.2}$$

Hence
$$x = \frac{A}{k}(P_g - P_a) - \frac{M_p g}{k} \tag{E5.15.3}$$

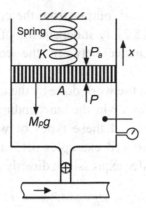

Fig. E5.15 Filling a spring-loaded piston-cylinder set-up

The final volume of the air in the cylinder is

$$V_f = Ax$$

Substituting for x from Eq. (E5.15.3), in the above equation, the final volume becomes

$$V_f = \left(\frac{A^2}{k}\right)(P_g - P_a) - \frac{M_p g A}{k} \qquad \text{(E.5.15.4)}$$

This example is similar to worked example 5.12 except that we now have work transfer between the control volume and the surroundings. We assume that the kinetic and potential energy of the air are negligible and that the heat transfer is zero. The mass and energy balance equations for the control volume may be written as:

$$\frac{dM}{dt} = \dot{m}_1 \qquad \text{(E5.15.5)}$$

$$h_0 \dot{m}_1 = \frac{dE}{dt} + \dot{W}_o \qquad \text{(E5.15.6)}$$

Eliminating $\dot{m}_1$ between Eqs. (E5.15.5) and (E5.15.6), and integrating the resulting equation from the initial to the final states we have

$$h_0(M_f - M_i) - (E_f - E_i) = \int_i^f \dot{W}_o dt = \Delta W_o \qquad \text{(E5.15.7)}$$

where ΔW_o is the total work output from the control volume. It should be noted that Eq. (E5.15.7) is independent of the type of fluid and the modes of work transfer between the control volume and the surroundings.

In the present problem the work done by the control volume is used to lift the piston, to push the air in the surroundings, and to compress the spring. We have encountered these types of work transfers in earlier worked examples (see worked examples 3.15 and 4.17). Therefore we shall write the work transfer expressions directly as follows:

$$\Delta W_o = Mgx + AP_a x + \frac{1}{2}kx^2 \qquad (E5.15.8)$$

Since the cylinder contains no air in the initial state, $M_i = 0$ and $E_i = 0$. Hence Eq. (E5.15.7) becomes

$$h_0 M_f - E_f = \Delta W_o \qquad (E5.15.9)$$

Substituting in Eq. (E5.15.9) from Eq. (E5.15.8)

$$h_0 M_f - E_f = Mgx + AP_a x + \frac{1}{2}kx^2 \qquad (E5.15.10)$$

where x is given by Eq. (5.15.3).

Assuming that air is an ideal gas with constant specific heat capacities, Eq. (E5.15.10) can written as:

$$c_p T_0 \left(\frac{P_g V_f}{RT_f} \right) - c_v \left(\frac{P_g V_f}{R} \right) = Mgx + AP_a x + \frac{1}{2}kx^2$$

The final air temperature T_f is obtained by substituting for V_f from Eq. (E5.15.4), and for x from Eq. (E5.15.3) in the above equation.

Example 5.16 A well-insulated rigid vessel contains 0.25 kg of air at 3 bar and 100°C. Some air is discharged to the surroundings from the vessel through a valve until the pressure inside the vessel is 1 bar. Assume that air is an ideal gas. Calculate (i) the final temperature of the air in the vessel, and (ii) the final mass of air in the vessel.

Solution Consider the control volume with a single exit port enclosing the vessel. The air in the vessel during the discharge process is assumed to be well-mixed so that the enthalpy of the air leaving is the same as the enthalpy of the air in the vessel.

Neglect the kinetic and potential energy of the air. The heat and work transfer between the control volume and the surroundings are zero.

Applying the conservation equations of mass and energy we obtain

$$\frac{dM}{dt} = -\dot{m}_1 \qquad \text{(E5.16.1)}$$

$$\frac{dE}{dt} = -\dot{m}_1 h \qquad \text{(E5.16.2)}$$

where M and E are respectively the mass and internal energy of the air in the vessel at time t. The instantaneous rate of discharge of air is $\dot{m}_1$.

Assume that air is an ideal gas. Eliminating $\dot{m}_1$ between Eqs. (E5.16.1) and (E5.16.2) and substituting the expressions for the internal energy and enthalpy we have

$$\frac{d(Mc_vT)}{dt} = c_pT\frac{dM}{dt}$$

$$\frac{d(MT)}{dt} = \gamma T\frac{dM}{dt} \qquad \text{(E5.16.3)}$$

where T is the temperature of the air, and γ is the ratio of the specific heat capacities. From Eq. (E5.16.3) we obtain

$$M\frac{dT}{dt} + T\frac{dM}{dt} = \gamma T\frac{dM}{dt}$$

$$M\frac{dT}{dt} = (\gamma - 1)T\frac{dM}{dt} \qquad \text{(E5.16.4)}$$

Integrating Eq. (E5.16.4) we have

$$\frac{T}{M^{(\gamma-1)}} = C \qquad \text{(E5.16.5)}$$

where C is the constant of integration. Substituting for mass from the ideal gas equation of state in Eq. (E5.16.5) we obtain

$$T = C\left(\frac{PV_o}{RT}\right)^{(\gamma-1)}$$ (E5.16.6)

where V_o is the volume of the vessel.

Applying Eq. (E5.16.6) to the initial and final states of the air in the vessel

$$\frac{T_f}{T_i} = \left(\frac{P_f}{P_i}\right)^{\frac{(\gamma-1)}{\gamma}} = \left(\frac{1}{3}\right)^{0.286} = 0.73$$

(i) Hence the final air temperature is

$$T_f = 0.73 \times 373 = 272.3 \ \text{K}$$

Applying Eq. (E5.16.5) to the initial and final states

$$\frac{T_f}{M_f^{(\gamma-1)}} = \frac{T_i}{M_i^{(\gamma-1)}}$$

Substituting numerical data in the above equation

$$\frac{272.3}{M_f^{(\gamma-1)}} = \frac{373}{0.25^{(\gamma-1)}}$$

(ii) Hence the final mass of air in the vessel, $M_f = 0.114$ kg

Example 5.17 A pressure vessel contains 2.6 kg of air at 2.5 bar and 35°C. Air is discharged from the vessel until the pressure is 1.4 bar. During the process the temperature of the air in the vessel is maintained at 35°C by supplying heat from the surroundings. Assume that air is an ideal gas with $c_v = 0.716$ kJ kg^{-1} K^{-1} and $R = 0.287$ kJ kg^{-1} K^{-1}. Calculate (i) the volume of the vessel, (ii) the mass of air left in the vessel, and (iii) the heat transfer from the surroundings.

Solution Applying the ideal gas equation to the initial state of air

$$P_i V_0 = M_i R T_i$$ (E5.17.1)

where V_o is volume of the vessel. Substituting numerical values in Eq. (E5.17.1)

$$2.5 \times 100 \times V_0 = 2.6 \times 0.287 \times 308$$

Therefore the volume of the vessel is, $V_0 = 0.919 \, \text{m}^3$.

Applying the ideal gas equation to the final state

$$P_f V_0 = M_f R T_f$$

Substituting numerical values in the above equation

$$1.4 \times 100 \times 0.919 = M_f \times 0.287 \times 308$$

Hence the final mass of air in the vessel is, $M_f = 1.46 \, \text{kg}$.

We assume that during the discharge process (i) the kinetic and potential energy of the air are negligible, (ii) the work transfer is zero, and (iii) the properties of the air in the vessel are uniform.

Applying the conservation equations of mass and energy (also see worked example 5.16) we have

$$\frac{dM}{dt} = -\dot{m}_1 \qquad \text{(E5.17.2)}$$

$$\dot{Q}_{in} = \frac{dE}{dt} + h\dot{m}_1 \qquad \text{(E5.17.3)}$$

Elimination $\dot{m}_1$ between Eqs. (E5.17.2) and (E5.17.3) and substituting the expressions for the internal energy and enthalpy of an ideal gas we obtain

$$\dot{Q}_{in} = \frac{d(Mc_v T)}{dt} - c_p T \frac{dM}{dt} \qquad \text{(E5.17.4)}$$

Integrating Eq. (E5.17.4) from the initial to the final states we have

$$\Delta Q_{in} = (c_v - c_p)T(M_f - M_i) = RT(M_i - M_f)$$

Substituting numerical values in the above equation

$$\Delta Q_{in} = 0.287 \times 308 \times (2.6 - 1.46) = 100.8 \, \text{kJ}$$

Example 5.18 A rigid vessel of volume $1.2 \, \text{m}^3$ contains $0.7 \, \text{m}^3$ of dry saturated vapor and $0.5 \, \text{m}^3$ of saturated water at a pressure of 5 bar. Heat is added to the water while the pressure inside is kept constant by releasing vapor through a pressure relief valve at the top. Calculate the total heat input required to evaporate half of the mass of liquid in the vessel.

Solution We assume that (i) the vapor and liquid in the vessel are in equilibrium during the heating process, and (ii) only saturated vapor leaves through the valve at the top of the vessel.

The specific volumes of the saturated vapor and liquid at 5 bar are obtained from the tabulated data in [4] as

$$v_g = 0.3748 \text{ m}^3 \text{ kg}^{-1} \quad \text{and} \quad v_f = 0.1091 \times 10^{-2} \text{ m}^3 \text{ kg}^{-1}$$

Let subscripts i and f denote conditions in the initial and final states.

The initial masses of vapor and liquid are

$$m_{vi} = 0.7 / v_g = 0.7 / 0.3748 = 1.868 \text{ kg}$$

$$m_{li} = 0.5 / v_f = 0.5 /(0.1091 \times 10^{-2}) = 458.29 \text{ kg}$$

The final mass of liquid

$$m_{lf} = 458.29 / 2 = 229.15 \text{ kg}$$

The final volume of vapor is

$$V_{vf} = 1.2 - m_{lf} v_f = 1.2 - 229.15 \times 0.1091 \times 10^{-2} = 0.95 \text{ m}^3$$

The final mass of vapor is

$$m_{vf} = 0.95 / v_g = 0.95 / 0.3748 = 2.535 \text{ kg}$$

Apply the conservation equations of mass and energy to the discharge process (also see worked example 5.16). Neglect the kinetic and potential energy of the steam. The work transfer is zero. Hence we have

$$\frac{dM}{dt} = -\dot{m}_1 \qquad\qquad (E5.18.1)$$

$$\dot{Q}_{in} = \frac{dE}{dt} + h_g \dot{m}_1 \qquad\qquad (E5.18.2)$$

where E and M are respectively the total internal energy and mass of the contents of the vessel. The instantaneous rate of discharge of vapor through the relief valve is $\dot{m}_1$. Eliminate $\dot{m}_1$ between Eqs. (E5.18.1) and (E5.18.2), and integrate the resulting equation from the initial to the final state to obtain

$$\Delta Q_{in} = \int_i^f \dot{Q}_{in} dt = (E_f - E_i) - h_g(M_f - M_i)$$

$$\Delta Q_{in} = (m_{vf}u_g + m_{lf}u_f) - (m_{vi}u_g + m_{li}u_f)$$

$$- h_g(m_{vf} + m_{lf} - m_{vi} - m_{li})$$

$$\Delta Q_{in} = (m_{vi} - m_{vf})(h_g - u_g) + (m_{li} - m_{lf})(h_g - u_f)$$

Substituting numerical values in the above equation

$$\Delta Q_{in} = -0.667 \times (2749 - 2562) + 229.14 \times (2749 - 639)$$

The total heat input is, $\Delta Q_{in} = 483.36 \times 10^3$ kJ

Example 5.19 A well-insulated tank of volume 0.15 m³ contains air at 1.5 bar and 20°C. An electric heating element placed inside the tank receives power at a constant rate of 150 W from an external source. The pressure inside is maintained constant at 1.5 bar by a regulating valve which allows air to leave the tank at a controlled rate. Assume that air is an ideal gas. Calculate the temperature of the air 3 minutes after the heater is switched on.

Solution Consider the control volume with one outlet port enclosing the tank and the valve. The electrical energy flowing to the heater across the control volume boundary constitutes a work input. There is no heat exchange between the control volume and the surroundings. Assume that the kinetic energy and potential energy of the air are negligible.

Applying the mass and energy conservation equations to the control volume we have

$$\frac{dM}{dt} = -\dot{m}_1 \qquad (E5.19.1)$$

$$h\dot{m}_1 + \frac{dE}{dt} - \dot{W}_{in} = 0 \qquad (E5.19.2)$$

where $\dot{W}_{in}$ is the rate of electrical work input. E and M are respectively the total internal energy and mass of the air in the tank. The instantaneous rate of discharge of air through the regulating valve is $\dot{m}_1$.

Eliminate $\dot{m}_1$ between Eqs. (E5.19.1) and (E5.19.2), and substitute the relevant expressions for the mass, the internal energy, and the enthalpy of an ideal gas to obtain

$$-c_pT\frac{d}{dt}\left(\frac{PV_o}{RT}\right)+\frac{d}{dt}\left(\frac{PV_oc_vT}{RT}\right)-\dot{W}_{in}=0$$

$$-c_pT\frac{d}{dt}\left(\frac{PV_o}{RT}\right)-\dot{W}_{in}=0$$

$$T\frac{d}{dt}\left(\frac{1}{T}\right)=-\left(\frac{\dot{W}_{in}R}{c_pPV_o}\right)=-\lambda \qquad (E5.19.3)$$

In Eq. (E5.19.3), λ is a constant because the pressure, P in the vessel is maintained constant by the regulating valve and the electrical work input, $\dot{W}_{in}$ is constant.

We now solve the first-order differential equation (E5.19.3) to obtain the time variation of the temperature. Equation (E5.19.3) may be written as

$$\frac{dT}{dt}=\lambda T$$

The solution of the above equation is

$$T(t)=T_0e^{\lambda t} \qquad (E5.19.4)$$

where T_0 is the air temperature at time $t = 0$.

Substituting numerical values in the Eq. (E5.19.3)

$$\lambda=\frac{150\times0.287}{1.003\times1.5\times10^5\times0.15}=1.908\times10^{-3}\,\text{s}^{-1}$$

The temperature after 3 minutes is given by Eq. (E5.19.4) as

$$T(180)=293\exp(1.908\times10^{-3}\times180)=413.1\,\text{K}$$

Example 5.20 Air flows from a source at P_s (Nm^{-2}) and T_0 (K) into a well-insulated rigid tank as shown in Fig. E5.20. During the filling process the pressure inside the tank is maintained at P_0 (Nm^{-2}) by discharging air through a valve from the tank in a controlled manner. The

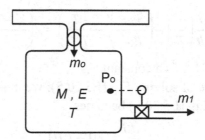

Fig. E5.20 Filling a tank at constant pressure

air flow rate into the tank, $\dot{m}_0$ (kg s^{-1}) is constant. Assume that air is an ideal gas. Obtain an expression for the variation of the air temperature in the tank with time.

Solution A control volume enclosing the tank with one inlet port and one exit port is shown in Fig. E5.20. The kinetic and potential energy of the air are assumed to be negligible. The heat and work transfer between the control volume and the surroundings are both zero. The conservation equations of mass and energy are written as follows:

$$\frac{dM}{dt} = \dot{m}_0 - \dot{m}_1 \qquad (E5.20.1)$$

$$\frac{dE}{dt} = \dot{m}_0 h_0 - h\dot{m}_1 \qquad (E5.20.2)$$

Eliminating $\dot{m}_1$ between Eqs. (E5.20.1) and (E5.20.2), and substituting expressions for the internal energy and enthalpy of the ideal gas we obtain

$$\frac{d(Mc_v T)}{dt} = \dot{m}_0 c_p T_0 - c_p T\dot{m}_0 + c_p T\left(\frac{dM}{dt}\right) \qquad (E5.20.3)$$

Substituting for mass, M from the ideal gas equation in Eq. (E5.20.3)

$$\frac{d}{dt}\left(\frac{P_0 V_t c_v T}{RT}\right) = \dot{m}_0 c_p T_0 - c_p T\dot{m}_0 + c_p T\frac{d}{dt}\left(\frac{P_0 V_t}{RT}\right) \qquad (E5.20.4)$$

Equation (E5.20.4) can be written in the simplified form:

$$\frac{d}{dt}\left(\frac{1}{T}\right)=\left(\frac{\dot{m}_o R}{P_o V_t}\right)\left(\frac{T-T_0}{T}\right)=\lambda\left(\frac{T-T_0}{T}\right) \qquad (E5.20.5)$$

where $\lambda = \dot{m}_o R / P_o V_t$, is a constant.

For the purpose of solving Eq. (E5.20.5), we make the transformation $\theta = 1/T$. Hence Eq. (E5.20.5) becomes

$$\frac{d\theta}{dt} = \lambda\left(1 - T_0\theta\right) \qquad (E5.20.6)$$

The solution of Eq. (E5.20.6) is

$$\frac{1-T_0\theta}{1-T_0\theta_1} = \exp\left(-\lambda T_0 t\right) \qquad (E5.20.7)$$

where $T_1 = 1/\theta_1$ is the temperature at time $t = 0$.

We now rewrite Eq. (E5.20.7) in a form that gives the time variation of the air temperature T more directly. Hence we have

$$\frac{T_0}{T(t)} = 1 + \left(\frac{T_0 - T_1}{T_1}\right)\exp\left(-\frac{\dot{m}_o R T_0 t}{P_o V_t}\right)$$

Example 5.21 A well-insulated rigid vessel contains dry saturated steam at 1.4 bar. The vessel is connected to a steam line carrying steam at 7 bar and 200°C through a valve, which is initially closed. The valve is opened, and steam is allowed to flow into the vessel until the pressure becomes 7 bar. Neglect heat losses from the vessel to the surroundings. Calculate the final temperature of the steam.

Solution We consider a control volume that surrounds the vessel, and the valve controlling the flow of steam, shown in Fig. 5.10. The kinetic and potential energy of the steam are assumed negligible, and the work and heat transfer between the control volume and the surroundings are zero.

Following the analysis of a filling process in Sec. 5.7, the starting point of our solution could be Eq. (5.37) which is

$$M_f u_f - M_i u_i = h_0(M_f - M_i) \qquad (E5.21.1)$$

We can express Eq. (E5.21.1) in terms of the specific volumes as

$$\left(\frac{V_v}{v_f}\right)u_f - \left(\frac{V_v}{v_i}\right)u_i = h_0\left(\frac{V_v}{v_f} - \frac{V_v}{v_i}\right) \tag{E5.21.2}$$

From the given data the steam in the vessel in the initial state is dry saturated. From the steam tables in [4] we obtain

$$u_i = 2517 \text{ kJ kg}^{-1} \quad \text{and} \quad v_i = 1.236 \text{ m}^3 \text{ kg}^{-1}$$

The steam in the supply line is superheated with $h_0 = 2846$ kJ kg^{-1}.

Substituting these numerical values in Eq. (E5.21.2) we obtain

$$u_f / v_f - (2517/1.236) = 2846 \times \left(1/v_f - 1/1.236\right)$$

The above equation gives the following linear relation between the internal energy and the specific volume in the final state:

$$u_f = 2846 - 266.18 v_f \tag{E5.21.3}$$

In the final state the pressure in the vessel is 7 bar. The superheated steam data from [4] gives an additional relationship between u_f and v_f at 7 bar as tabulated below.

Table E5.21 Conditions of steam at 7 bar

T (°C)	200	250	300	350	400
v_f (m^3 kg^{-1})	0.3001	0.3364	0.3714	0.4058	0.4397
u_f (kJ kg^{-1})	2636	2720	2800	2880	2961

We first plot a graph of u_f against v_f using the data tabulated above. Then on the same graph the linear relation given by Eq. (E5.21.3) is plotted. The point of intersection of the two curves gives the required final conditions as:

$$u_f = 2752 \text{ kJ kg}^{-1} \quad \text{and} \quad v_f = 0.35 \text{ m}^3 \text{ kg}^{-1}$$

The temperature of superheated steam for these conditions is obtained by linear interpolation as 270°C.

Problems

P5.1 Air enters a compressor at 1.05 bar and 32°C with negligible velocity. At the discharge section the air is at 5 bar and 250°C, and has a velocity of 145 ms^{-1}. The mass flow rate of air is 8.5 kg s^{-1}. The power input to the compressor is 2.2 MW. Calculate (i) the area of the discharge section, and (ii) the rate of heat transfer between the compressor and the surroundings. Assume that air is an ideal gas.
[*Answers*: (i) 0.0176 m^2, (ii) –252 kW]

P5.2 Gas enters a turbine at a steady rate of 1.5 kg s^{-1} with a pressure and temperature of 5 bar and 920 K respectively. The gas leaves at 1 bar and 700 K through an exit section of area 0.02 m^2. The velocity of the gas at inlet is negligible. The power output of the turbine is 280 kW. Assume that the gas is ideal with $c_v = 0.657$ kJ kg^{-1} K^{-1} and $R = 0.189$ kJ kg^{-1} K^{-1}. Calculate (i) the velocity of the gas at exit, and (ii) the heat transfer to the surroundings.
[*Answers*: (i) 99.23 ms^{-1}, (ii) 6.56 kW]

P5.3 Steam enters a converging nozzle at 27 bar and 300°C with negligible velocity. At the exit the pressure is 15 bar and the velocity is 520 ms^{-1}. The steam flow rate is 2.4 kg s^{-1}. Calculate (i) the state of the steam at exit, and (ii) the area of the nozzle at exit. Neglect any heat losses to the surroundings.
[*Answers*: (i) superheated steam at 350.5°C, (ii) 8.61×10^{-4} m^2]

P5.4 In a hydroelectric power system, water at 12°C enters the intake pipe at 100 kPa. At the exit of the discharge pipe the water has a temperature of 12.3°C and the pressure is 100 kPa. The difference in elevation between the intake and the exit is 270 m and the velocities of the water at these locations are negligible. The water flow rate is 200 kg s^{-1}. The specific heat capacity of water is 4.19 kJ kg^{-1} K^{-1}. Calculate the power output of the system. Why is there an increase in temperature of the water? Neglect any heat exchange with the surroundings.
[*Answer*: 1.67 MW]

P5.5 Water is heated in a heat exchanger by condensing steam that flows at the rate of 1.4 kg s^{-1}. The water flow rate is 5.2 kg s^{-1}, and the inlet and outlet temperatures are 15°C and 65°C respectively. Wet steam enters with a quality of 0.35 at a pressure of 100 kPa, and the condensed steam leaves as a sub-cooled liquid at a temperature of 90°C. The specific heat capacity of water is 4.2 kJ kg^{-1} K^{-1}. Calculate the heat transfer between the heat exchanger and the surroundings.
[*Answer*: −70.56 kW]

P5.6 A steam turbine receives steam at 4 MPa and 450°C with a mass flow rate of 20 kg s^{-1}. At the exit the steam is dry saturated at 0.5 MPa. The inlet and exit velocities of the steam are 250 ms^{-1} and 70 ms^{-1} respectively. The power output is 10 MW. Calculate (i) the areas of the inlet and exit sections of the turbine, and (ii) the heat transfer to the surroundings.
[*Answers*: (i) 6.4×10^{-3} m^2, 0.107m^2, (ii) 2196 kW]

P5.7 A rigid tank contains 2.5 kg of air at 2.5 bar and 315 K. Air is discharged from the tank until the pressure is 1.5 bar. During the process, the temperature inside the tank is maintained constant at 315 K by supplying heat in a controlled manner. Assume that air is an ideal gas. Calculate (i) the mass of air discharged, and (ii) the total heat supplied.
[*Answers*: (i) 1.0 kg, (ii) 90.4 kJ]

P5.8 A rigid vessel of volume 7 m^3 contains dry saturated steam at 1.4 bar. The vessel is connected to a steam line carrying steam at 7 bar and 200°C through a valve, which is initially closed. The valve is opened, and steam is allowed to flow into the vessel until the pressure becomes 7 bar. The final temperature of the steam in the vessel is 250°C. Calculate (i) mass of steam that enters the vessel, and (ii) the heat transfer between the vessel and the surroundings.
[*Answers*: (i) 15.15 kg, (ii) −758.6 kJ]

P5.9 A rigid vessel of volume 1.2 m³ contains 0.6 m³ of vapor and 0.6 m³ of liquid at 250.3°C, initially. Liquid is withdrawn slowly from the bottom of the vessel while the temperature inside is maintained constant by supplying heat from an external source. In the final state, the total mass of steam and water in the vessel is half the initial mass. Calculate the total heat transfer to the vessel.
[*Answer*: 10.5 MJ]

P5.10 A rigid tank of volume 0.55 m³ contains steam at 5 bar and 175°C. Steam is released from the vessel while the pressure inside is maintained constant by supplying heat from an external source. The final steam temperature is 200°C. Calculate (i) the mass of steam released, and (ii) the total heat supplied.
[*Answers*: (i) 0.082 kg, (ii) 71.5 kJ]

References

1. Cravalho, E.G. and J.L. Smith, Jr., *Engineering Thermodynamics*, Pitman, Boston, MA, 1981.
2. Jones, J.B. and G.A. Hawkins, *Engineering Thermodynamics*, John Wiley & Sons, Inc., New York, 1986.
3. Reynolds, William C, and Henry C. Perkins, *Engineering Thermodynamics*, 2nd edition, McGraw-Hill, Inc., New York, 1977.
4. Rogers, G.F.C. and Mayhew, Y.R., *Thermodynamic and Transport Properties of Fluids*, 5th edition. Blackwell, Oxford, U.K. 1998.
5. Van Wylen, Gordon J. and Richard E. Sonntag, *Fundamentals of Classical Thermodynamics*, 3rd edition, John Wiley & Sons, Inc., New York, 1985.
6. Wark, Jr., Kenneth, *Advanced Thermodynamics for Engineers*, McGraw-Hill, Inc., New York, 1995.

Chapter 6

The Second Law of Thermodynamics

In the last two chapters of this book we applied the first law of thermodynamics to closed and open systems considering both quasi-static and non-quasi-static processes. A question that arises naturally is whether all processes that satisfy the first law could be carried out in practice. In order to explore this question we need to introduce the second law of thermodynamics which teaches us that there are additional conditions for the feasibility of a process. In particular, the irreversibility of natural processes is an important consequence of the second law.

There are several statements of the second law, which at first appear to be unrelated. However, by treating any one of the different statements of the second law as a postulate, all the other statements could be deduced logically. The first law predicted the existence of a property, which was called the internal energy of a system. In a similar manner, we shall derive a new property called the entropy by applying the second law to a system. In developing the second law we propose to follow the historical route that involves cyclic processes and heat engines. From an engineering perspective, this approach is deemed to be more practical, and closely related to the subject matter covered in the preceding chapters.

6.1 The Heat Engine Cycle

One of the main functions of a majority of energy conversion systems is to generate motive power or work using stored energy from various fuels. The steam power plant, which operates on the vapor power cycle, is a common example of such an energy conversion system. The various

241

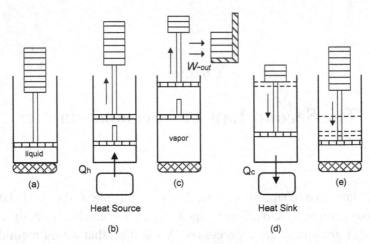

Fig. 6.1 A vapor-power cycle producing a net work output

processes that constitute a typical vapor power cycle could be carried out using the piston-cylinder arrangement shown in Fig. 6.1. Initially, (Fig. 6.1(a)) the well-insulated cylinder contains a compressed liquid with the weights on the piston generating the desired high pressure of the cycle. The bottom section of the insulation around the cylinder is now removed and a heat source is applied to the cylinder to heat the liquid (Fig. 6.1(b)). During this process, the liquid first reaches its saturation temperature and then undergoes phase change at constant pressure while the piston is pushed out.

After the evaporation process is complete, the bottom of the cylinder is insulated and the weights on the piston are removed in steps to expand the vapor (Fig. 6.1(c)). The potential energy gained by the weights constitutes the external work delivered by the system to the surroundings. When the weight on the piston is reduced to an amount that generates the required low pressure of the cycle, the bottom insulation is removed and a cold heat sink is applied to the bottom of the cylinder (Fig. 6.1(d)). This process is continued until all the vapor is condensed at a constant pressure.

The heat sink is now withdrawn and the bottom of the cylinder is again insulated. Weights are then placed on the piston in steps to increase the pressure of the liquid to the required high pressure, thereby completing the cycle (Fig. 6.1(e)).

At the beginning of each cycle of operation, a new set of weights is placed on the piston and these are moved to a higher elevation thereby delivering a net work output to the surroundings. The heat supplied to convert the sub-cooled liquid to a vapor constitutes the energy input to the cycle.

We used the piston-cylinder arrangement to illustrate the operation of a power cycle mainly to relate it to our work in the earlier chapters. However, actual vapor power cycles are more complicated in design and they make use of separate components to carry out the various processes of the cycle.

The essential sub-components of a vapor power cycle are depicted schematically in Fig. 6.2. Each component is a control volume through which the working fluid passes periodically. Let us follow a 'packet' of fluid starting from 1, the entry point to the boiler, where the fluid receives heat from an external heat source. The heat source could be the furnace atmosphere which is maintained at a high temperature by burning a fuel like oil or natural gas.

In the case of a nuclear power plant, the heat source is the reactor where the fission reaction generates the required thermal energy. The fluid enters the boiler at 1 as a sub-cooled liquid with a temperature below the saturation temperature corresponding to the pressure of the boiler. As the packet of fluid passes through the boiler its temperature

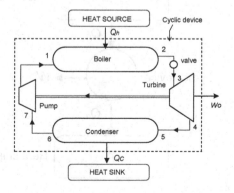

Fig. 6.2 Simple vapor-power cycle

first increases to the saturation value, and the fluid then undergoes phase change to emerge at the exit 2 of the boiler as a saturated vapor.

After passing through the ducting the fluid packet enters the turbine at point 3. During its passage through the turbine the fluid packet expands and leaves at 4, releasing part of its enthalpy as external shaft work of the turbine. We described the operation of a typical turbine in Sec. 5.6.2.

At 5, the fluid enters the condenser where it rejects heat at constant pressure to a cold heat sink. A typical heat sink is the atmosphere or a body of water such as a river. The fluid packet leaves the condenser at 6 as a saturated liquid and enters a feed-pump at 7 to be pumped back to the boiler at 1. The feed pump is operated by using a fraction of the work produced by the turbine.

We see that each fluid packet undergoes a cyclic process during its passage through the plant. The above description of the vapor power cycle is very brief in that it includes only the essential details needed to introduce the second law in the next section.

We notice that the various processes undergone by each 'packet' of fluid in the 'plant diagram' shown in Fig. 6.2 and the fixed mass of fluid in the 'piston-cylinder set-up' in Fig. 6.1 are essentially the same. By focusing our attention mainly on the heat and work interactions among the system, the heat source, the heat sink, and the surroundings we can draw the energy flow diagram of the heat engine as shown in Fig. 6.3.

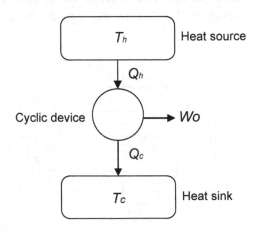

Fig. 6.3 Energy-flow diagram for heat engine

During each cycle of operation, a quantity of heat Q_h flows from the heat source at temperature T_h to the 'cyclic device' which delivers to the surroundings a *net* quantity of work W_o. A quantity of heat Q_c is rejected to the heat sink at temperature T_c during the cycle. This simplified diagram is a very useful abstraction of the more complicated systems depicted in Figs. 6.1 and 6.2. Therefore, in the discussions to follow in this chapter we shall use the energy flow diagram in Fig. 6.3 to represent a typical *heat engine cycle*.

6.1.1 *Efficiency of a heat engine cycle*

The efficiency η of the heat engine cycle is defined as the net work output per unit heat input. This can be expressed as:

$$\eta = \frac{W_o}{Q_h} \qquad (6.1)$$

Applying the first law to the cycle we have

$$Q_h - Q_c = W_o \qquad (6.2)$$

From Eqs. (6.1) and (6.2) we obtain

$$\eta = 1 - \frac{Q_c}{Q_h} \qquad (6.3)$$

It is seen that the efficiency of the cycle has an economic significance because in a typical power plant the heat source is maintained at the high temperature by burning fuel, which constitutes the main operating energy cost of the plant. The net work delivered to the surroundings, on the other hand, is the desired output of the plant. Therefore the maximization of the efficiency of the cycle should be the primary objective of power plant design.

6.2 The Reversed Heat Engine Cycle

In dealing with topics related to the second law we also encounter an important class of energy conversion systems called *reversed heat engine cycles*. In practical terms these are refrigerators or heat pumps as they are sometimes called. Whereas heat engines deliver work by absorbing heat from high temperature sources, reversed heat engines transfer heat from bodies at low temperatures to bodies at high temperatures. The latter process does not occur without the aid of an external energy input.

Illustrated in Fig. 6.4 is a typical reversed heat engine cycle which operates using a working fluid that undergoes phase change. It should be noted that this is an idealized cycle that differs somewhat from the ideal vapor-compression refrigeration cycle due mainly to practical reasons.

Let us now follow the passage of a 'packet' of fluid through the various sub-components of the plant shown in Fig. 6.4. The fluid packet enters the evaporator as a liquid or a wet-vapor at 1. As it passes through the evaporator the fluid absorbs heat from the cold body because its temperature is below that of the cold body. In a typical refrigerator the cold body is the refrigerated space.

At 2 the fluid is usually a saturated vapor, and at 3 it enters the compressor where its pressure is raised to the condenser pressure. This compression process requires an external work input. During the passage through the condenser from 5 to 6 the fluid rejects heat to a heat sink, to attain a liquid state at the exit 6. In order for this heat transfer to take place the working-fluid temperature in the condenser has to be higher than the temperature of the heat sink, which in the case of a practical refrigerator is the atmosphere. At 7 the fluid enters an expander which reduces its pressure to the evaporator pressure to complete the cycle.

The work produced during the expansion process could be made use of to reduce the external work input to the compressor. However, in

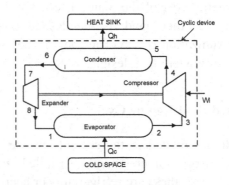

Fig. 6.4 A reversed heat engine cycle

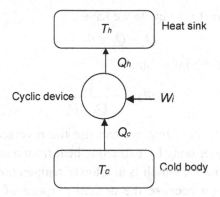

Fig. 6.5 Energy-flow diagram for a reversed heat engine cycle

practical vapor compression cycles, the expander is replaced with an expansion valve for practical reasons that we will discuss in a later chapter.

The processes of the reversed heat engine cycle could also be carried out using a piston-cylinder set-up similar to that shown in Fig 6.1. As for the heat engine, we can draw the energy flow diagram for the reversed heat engine to indicate the main energy interactions as shown in Fig. 6.5. We note that the directions of the heat and work interactions for the reversed heat engine are opposite to those in the heat engine cycle.

6.2.1 *Coefficient of performance of a reversed heat engine cycle*

The performance indicator for the reversed heat engine is called the *coefficient of performance* (COP). In defining the COP we need to identify the desired energy interaction, which is the main function of the reversed heat engine, and the input energy form. Two situations need to be distinguished for this purpose.

When the objective of the reversed engine is to extract heat from the cold body using the work input we call the device a *refrigerator*, the COP of which is defined as:

$$(COP)_{ref} = \frac{Q_c}{W_i} \qquad (6.4)$$

Applying the first law to the cycle we have

$$Q_h - Q_c = W_i \qquad (6.5)$$

From Eqs. (6.4) and (6.5) we obtain

$$(COP)_{ref} = \frac{Q_c}{Q_h - Q_c} \qquad (6.6)$$

It is interesting to note that we can use the reversed heat engine to transfer heat to a hotter body by extracting heat from a source such as the atmosphere or the ground which is at a lower temperature. Such a device is called a *heat pump* because the desired purpose of the device is to transfer heat to the hotter body. In recent years heat pumps have found wide spread application in the heating of homes and buildings. The COP of the heat pump is defined as

$$(COP)_{hp} = \frac{Q_h}{W_i} \qquad (6.7)$$

Applying the first law to the cycle we have

$$Q_h - Q_c = W_i \qquad (6.8)$$

From Eqs. (6.7) and (6.8) we obtain

$$(COP)_{hp} = \frac{Q_h}{Q_h - Q_c} \qquad (6.9)$$

From Eqs. (6.6) and (6.9) it follows that

$$(COP)_{hp} = 1 + (COP)_{ref} \qquad (6.10)$$

Having discussed the main features of heat engines and reversed heat engines we now pose the following questions concerning the design and operation of such devices:

(i) Is the condenser in the vapor power cycle shown in Fig. 6.2 really necessary? After all it absorbs a portion of the energy from the heat source which could otherwise have been used to produce more work.

(ii) Is there an upper limit to the efficiency of the heat engine cycle?

(iii) Are there processes and conditions that have an adverse effect on the performance of heat engine cycles?

(iv) In a reversed heat engine, is it really necessary to have a work input to transfer heat from a cold body to a hot body?

(v) Is there an upper limit to the COP of a reversed heat engine?

From a historical perspective, it was the search for the answers to these interesting and relevant questions that lead to the formulation of the *second law of thermodynamics*.

6.3 The Second Law of Thermodynamics

The second law of thermodynamics has been stated in many different forms, some of which appear at first to have no relation to the other forms. If we accept any one of the statements of the law as a postulate, then the other forms can be proved by using logical arguments. However, the statement of the second law that we start with cannot be derived from any other law of nature. Two of the well-known statements of the second law, called the *Kelvin-Planck statement*, and the *Clausius statement*, are directly related to heat engines and reversed heat engines respectively.

The Kelvin-Planck statement (K-P-S): It is impossible to construct a device that will operate in a cyclic manner and produce no effect other than produce work while exchanging heat with bodies at a single fixed temperature.

We can elaborate the practical relevance of this statement by referring to the heat engine cycle shown in Fig. 6.2. In this case, the components within the boundary indicated by the broken-lines, is the cyclic device. Heat flows to the device from the high temperature source and there is a net production of work. The Kelvin-Planck statement states that it is impossible for this heat engine cycle to operate if a fraction of the heat input is not rejected to the heat sink. In the absence of the heat sink, the heat flow from the heat source is equal to the work production by virtue of the first law. Therefore the efficiency of the heat engine is 100 percent according to Eq. (6.1). Thus far, all efforts to construct a cyclic device that converts all the heat received from a source at a single temperature *continuously* to work have failed. Such an engine is called a *perpetual motion machine of the second kind.*

The Clausius statement (C-S): It is impossible to construct a device that operates in a cyclic manner and produces no effect other than the transfer of heat from a colder body to a hotter body. This statement has a direct bearing on the operation of reversed heat engine cycles of the type shown in Fig. 6.4.

Consider the components within the broken-line boundary in Fig. 6.4 as the cyclic device. The Clausius statement states that it is impossible for this reversed heat engine cycle to operate if the external work input to the device is withdrawn. In the absence of the work input, heat entering from the cold body is equal to the heat transferred to the hot body by virtue of the first law. Moreover, the COP of the cycle is infinity according to Eq. (6.4). Therefore according to the Clausius statement, it impossible to construct a refrigerator or heat pump that operates in a cycle manner without an input of work. Also, the COP of the refrigerator always has a finite value.

6.3.1 *Equivalence of the Kelvin-Planck and Clausius statements*

We shall now prove that the K-P-S and C-S of the second law are equivalent. That is if a device which violates the C-S can be constructed then the device also violates the K-P-S, and vice versa. Shown in Fig. 6.6 is a cyclic device CS which transfers steadily an amount of heat Q_c from a cold body to a hot body with no work input. Such a device violates the C-S.

We now operate a heat engine HE between the hot body and the cold body, whose design is such that it receives an amount $(Q_c + W)$ of heat from the hot body, produces an amount W of work and rejects an amount Q_c of heat to the cold body. Consider the units within the boundary indicated by the broken-lines as a composite cyclic device. It receives a net amount of heat W from the hot body and coverts all of it to work thus violating the K-P-S.

To prove the converse consider the cyclic device KPS shown in Fig. 6.7. It receives an amount Q_h of heat from the hot body and converts all of it to work, violating the K-P-S. We now operate a heat pump HP between the two bodies using the work input from KPS. The heat pump extracts an amount Q_c of heat from the cold body and transfers an amount $(Q_c + W)$ of heat to the hot body. Consider the composite cyclic device within the broken-line boundary.

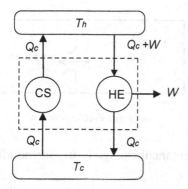
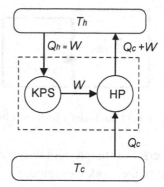

Fig. 6.6 Equivalence of CS and KPS   Fig. 6.7 Equivalence of KPS and CS

This device extracts an amount Q_c of heat from the cold body and transfers the same amount of heat on a *net* basis to the hot body, with no external work input, thereby violating the C-S.

Any device that violates the C-S will violate the K-P-S and therefore the second law of thermodynamics. Such a device is called a perpetual motion machine of the second kind or PMM2. Any device that produces an energy output larger than the energy input will violate the first law of thermodynamics. Such a device is called a perpetual motion machine of the first kind or a PMM1.

6.4 Reversible and Irreversible Processes

Reversible processes play an important role in deducing some important consequences of the second law. A *reversible process* is a process which after it has occurred can be reversed *without* leaving any change, either in the *system* or the *surroundings*. We now examine the reversibility of some natural processes in light of the above definition.

The C-S of the second law is a formalization of our everyday experience that heat flows unaided from a body at a higher temperature to a body at a lower temperature (Fig. 6.8). The reversal of this natural process, however, requires the use of a heat pump with an external work input, which has to be positive according to the C-S. Therefore the

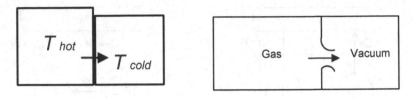

Fig. 6.8 Heat transfer irreversibility   Fig. 6.9 Free expansion

reversal of the process will leave a permanent change in the surroundings due to the work extracted.

There are numerous other processes that seem to always follow a preferred direction naturally, although the first law would allow these processes to occur both in the forward as well as the reverse directions. Let us consider three such processes to which we applied the first law earlier in chapter 4.

The well-insulated vessel shown in Fig. 6.9 is divided into two compartments by a diaphragm, with one compartment evacuated and the other containing a gas. If the diaphragm ruptures the gas will expand to fill the whole vessel.

To reverse this process, we need to compress the gas using a work input from the surroundings while transferring heat to a cold body to control the temperature of the gas. This way the gas could be returned to its initial state but there is a permanent change in the surroundings due to the work extracted and the heat transferred to the cold body. To restore the surrounding to its original state, the heat transferred to the cold body will have to be completely reconverted to work. The latter process, however, would violate the K-P-S.

The paddle wheel shown in Fig. 6.10 is immersed in water contained in an insulated tank. When the weight attached to the rope is released the rotation of the paddle wheel that follows will increase the temperature of the water due to frictional effects. To reverse this process we could, in principle, operate a cyclic heat engine with the water as the heat source and a cold body as the heat sink. The work output of the engine could be used to partially lift the weight. Since some heat has to be rejected to the heat sink, as required by the K-P-S, all the work originally transferred by the paddle wheel to the water cannot be extracted. Therefore, an

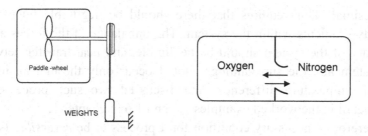

Fig. 6.10 Frictional heating   Fig. 6.11 Irreversible mixing of gases

additional work input will be required to lift the weight to the original height. Although the water and the weight have been returned to their initial states, permanent changes have occurred in the surroundings. These changes, however, cannot be reversed without violating the K-P-S.

In the third example we consider the two different gases contained in the two compartments of the vessel shown in Fig. 6.11. If the diaphragm separating the two sections ruptures, the gases will expand to fill whole vessel. The reversal of this process will require a complex gas separation process that will leave permanent changes in the surroundings.

We can summarize the common characteristics of the various processes described in the above examples as follows:

(i) The initial equilibrium state of the system is maintained by the constraints on the system. In the first and third examples (Figs. 6.9 and 6.11) they are the separating diaphragms, while the platform holding weight is the constraint in the second example (Fig. 6.10).

(ii) When the constraint is removed suddenly, the system passes through a series of non-equilibrium states involving rapid dynamic changes before the final equilibrium state is attained.

(iii) The system can only be returned to the initial state by energy exchanges with the surroundings that leave permanent changes in the latter if the second law is not to be violated.

The question that follows immediately is how do we carry out a process reversibly? We have discussed the answer to this question partially when we considered quasi-equilibrium processes in chapter 1. At any stage in a quasi-equilibrium process the constraints on the system are controlled in such a manner that deviation from equilibrium is infinitesimal. This requires that there should be negligible intensive - property gradients within the system. The unbalance of the forces at the boundary of the system should be negligible, and heat transfer between the system and the surroundings should occur only through an infinitesimal temperature difference. We discussed two such processes in some detail in the worked examples 1.5 and 1.6 in chapter 1.

Therefore a necessary condition for a process to be *reversible* is that it should be carried out in a *quasi-static* manner. The only additional condition for reversibility is the complete absence of solid or fluid *friction effects* during the process. Frictional effects when present convert some of the work input to a heat interaction. According to the second law this heat cannot be completely converted back to work using a heat engine because such a device would violate the K-P-S. Therefore the presence of frictional effects renders a process irreversible.

In summary, a process will be reversible when: (i) it is performed in a quasi-static manner, and (ii) it does *not* involve any frictional effects. Worked examples 6.1, 6.2 and 6.3 discuss the reversibility of several processes of practical importance.

6.4.1 *Types of irreversible processes*

Consider the vapor power cycle shown in Fig. 6.2. Assume that the processes undergone by the working fluid in the boiler occur in a quasi-static manner without frictional effects. However, let there be a finite temperature difference between the heat reservoir and the fluid in the boiler when heat is transferred to the latter. We call this situation an *external thermal irreversibility* because the irreversible process occurs outside the system, while the system itself undergoes a reversible change.

Similarly, the conversion of the work input of the paddle wheel to internal energy of the water due to frictional effects (Fig. 6.10) is an *external mechanical irreversibility*. In this case, the frictional irreversibility occurs at the boundary between the paddle wheel and the water, which is the system of interest. Here we assume that the temperature and pressure gradients in the water are negligible.

The gas expansion following the rupture of the diaphragm in Fig. 6.9 is an *internal mechanical irreversibility*. The irreversible process, which in this case, is the free expansion of the gas across a finite pressure difference, occurs inside the system. An *internal thermal irreversibility* is a heat flow process that takes place due to a finite temperature difference within a system as in Fig. 6.8. The mixing of the two gases shown in Fig. 6.11 is called a *chemical irreversibility*.

6.4.2   *Reversible heat engines and thermal reservoirs*

The reversible process is a useful idealization of real processes that enables us to analyze them conveniently because the properties of the system are spatially uniform at every stage. We now extend this concept to heat engine cycles by defining a *reversible heat engine cycle* as a cycle in which all the processes are reversible.

Therefore the cycle can be completely reversed leaving no permanent effects on the surroundings. The heat and work interactions of the reversed cycle are equal in magnitude but opposite in direction to the corresponding interactions in the direct cycle.

A *thermal or heat reservoir* is an idealized heat source or sink that enables us to carry out heat transfer processes in a reversible manner. It is defined as a body whose mass is such that the absorption or rejection of a quantity of heat of any magnitude will not result in an appreciable change in its temperature or any other thermal property. We define an *ideal heat engine cycle* as a reversible heat engine operating between a high temperature reservoir and a low temperature reservoir. This cycle is reversible both internally and externally.

6.5   Some Consequences of the Second Law

We deduced from the K-P-S of the second law that the thermal efficiency of any heat engine has to be less than 100 percent. A question that follows directly is: what is the maximum possible efficiency that a heat engine can have? The answer to this question was provided by a French engineer, Nicolas Leonard Sadi Carnot in 1824. He developed a reversible cycle, commonly known as the *Carnot cycle*, that could be carried out with a piston-cylinder arrangement using an ideal gas as the working fluid.

The expression for the efficiency of the Carnot cycle, however, is independent of the detailed design of the heat engine, and therefore it should be applicable to all reversible cycles.

We shall now prove the latter statement, commonly known as the *Carnot principle*, using the K-P-S of the second law. The Carnot principle states that:

(i) No heat engine can be more efficient than a *reversible* engine operating between the same high temperature heat reservoir and the low temperature heat reservoir.

(ii) The efficiency of all reversible heat engines operating between the same high and low temperature reservoirs is the same.

We shall prove the first statement by comparing the efficiency of any engine EX and a reversible engine ER by making use of the arrangement shown in Fig. 6.12. Here the reversible engine is reversed and operated as a reversible heat pump. The engine EX receives a quantity of heat Q_h

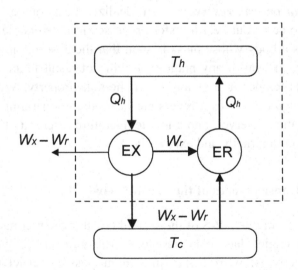

Fig. 6.12 The Carnot principle for heat engines

from the high temperature reservoir and produces a quantity of work W_x as its output. The corresponding quantities of heat and work for the engine ER are Q_h and W_r respectively.

Let us assume that the efficiency of EX is larger than the efficiency of ER. Therefore $W_x > W_r$ because the heat input for both engines is the same. Now when ER is reversed and made to function as a heat pump, the magnitudes of the heat and work interactions remain the same but they occur in the opposite directions.

Therefore the reversed-ER will transfer Q_h to the high temperature reservoir using a work input of W_r. This work input is supplied by EX leaving a quantity of work $(W_x - W_r)$ for delivery to the surroundings. Consider the composite cyclic device enclosed within the broken-line boundary. It receives a quantity of heat $(W_x - W_r)$ from the low temperature reservoir and converts all of it to work on a net basis, violating the K-P-S of the second law. Therefore our assumption that the efficiency of EX is larger than the efficiency of ER is invalid. We conclude that no engine can have an efficiency larger than that of a reversible engine operating between the same reservoirs.

The second statement can be proved by following a similar type of reasoning as for statement 1. Consider the arrangement in Fig. 6.12 where EX is now a second reversible engine whose efficiency is higher than that of ER. We reverse the engine ER with the lower efficiency and operate it as a reversible heat pump with the work input supplied by engine EX with the higher efficiency. The composite device within the broken-line boundary will absorb heat from the low temperature reservoir and convert all of it to work thus violating the K-P-S. The same reasoning will apply no matter which of the two engines is assumed to have the higher efficiency. The conclusion therefore is that both reversible engines must have the same efficiency, thus proving part (ii) of the Carnot principle.

By following a similar procedure (see problem 6.4) we could prove: (i) that no heat pump (reversed heat engine) could have a COP higher than that of a reversible heat pump, and (ii) that all *reversible* heat pumps must have the same COP.

6.5.1 *Efficiency of a Carnot cycle using an ideal gas*

We now derive an expression for the efficiency of a particular reversible heat engine operating with a piston-cylinder arrangement where the working fluid is an ideal gas. The generalization of this expression for any reversible engine will be discussed in the next section.

Consider a fixed mass of an ideal gas contained in a piston-cylinder arrangement similar to that shown in Fig. 6.1. The gas is brought into communication with a high temperature reservoir at temperature T_h whose temperature differs infinitesimally from that of the gas. Work is done by the gas during this process and consequently its temperature is maintained constant.

This process is shown as 1-2 in the P-V and T-V diagrams of Figs. 6.13(a) and (b) respectively. During the process 2-3 the cylinder is insulated and the gas is subjected to a reversible adiabatic expansion delivering work to the surroundings. A low temperature heat sink at T_c is then applied to the cylinder while the gas is compressed isothermally as indicated by process 3-4 in the figures. There is only an infinitesimal difference between the temperatures of the heat sink and the gas to ensure that the process is reversible. Finally, the gas is subjected to a reversible adiabatic compression from 4 to 1 to restore its initial state, thus completing the cycle.

Applying the general expression in Eq. (6.3) we obtain the efficiency of the cycle as

$$\eta = 1 - \frac{Q_c}{Q_h} \tag{6.11}$$

where Q_h is the heat received from the high temperature reservoir during process 1-2 and Q_c is the heat transferred to the low temperature reservoir during process 3-4. There is no heat transfer during the two adiabatic processes 2-3 and 4-1. Hence

$$Q_{23} = 0 \tag{6.12}$$

$$Q_{41} = 0 \tag{6.13}$$

The change in internal energy of an ideal gas during an isothermal process is zero. Hence by applying the first law to the isothermal processes 1-2 and 3-4 we have

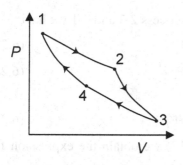

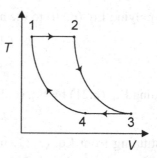

Fig. 6.13(a) Carnot cycle: *P-V* diagram   Fig. 6.13(b) Carnot cycle: *T-V* diagram

$$Q_h = W_{12} \quad \text{and} \quad Q_c = W_{34} \tag{6.14}$$

The expression for the work done by an ideal gas during an isothermal process was obtained earlier in worked example 3.11 in chapter 3. Applying the expression to processes 1-2 and 3-4 we obtain the following

$$W_{12} = mRT_h \ln(V_2/V_1) \tag{6.15}$$

$$W_{34} = mRT_c \ln(V_3/V_4) \tag{6.16}$$

From Eqs. (6.11)–(6.16) it follows that

$$\eta = 1 - \frac{Q_c}{Q_h} = 1 - \frac{T_c \ln(V_3/V_4)}{T_h \ln(V_2/V_1)} \tag{6.17}$$

In worked example 4.10 in chapter 4 we showed that the *P-V* relation for a quasi-static adiabatic process of an ideal gas to be

$$PV^\gamma = C \tag{6.18}$$

where C is a constant, and γ is the ratio of the specific heat capacities.

The ideal gas equation of state is

$$PV = mRT \tag{6.19}$$

Eliminating P between Eqs. (6.18) and (6.19) we obtain the following relation for a quasi-static adiabatic process of an ideal gas

$$TV^{\gamma-1} = C_1 \tag{6.20}$$

where C_1 is a constant.

Applying Eq. (6.20) to the adiabatic process 2-3 and 4-1 we have

$$T_h V_2^{\gamma-1} = T_c V_3^{\gamma-1} \tag{6.21}$$

$$T_h V_1^{\gamma-1} = T_c V_4^{\gamma-1} \tag{6.22}$$

Dividing Eq. (6.21) by Eq. (6.22)

$$V_2/V_1 = V_3/V_4 \tag{6.23}$$

Substituting from Eq. (6.23) in Eq. (6.17) we obtain the expression for the efficiency as:

$$\eta = 1 - \frac{T_c}{T_h} \tag{6.24}$$

Although the derivation of the Eq. (6.24) was somewhat tedious, the final expression, however, involves only the *absolute temperatures* of the two reservoirs. It is noteworthy that the expression for the efficiency of the Carnot cycle is independent of the design parameters of the piston-cylinder set-up and the operating conditions of the cycle.

In the next section we shall obtain an expression similar to Eq. (6.24) following a different route that leads us to the concept of a thermo-dynamic temperature scale.

6.5.2 *Thermodynamic temperature scale*

An important consequence of the reversible heat engine is the develop-ment of a *thermodynamic temperature scale (Absolute temperature scale)* which is independent of the thermometric properties of materials. The Carnot principle showed that the efficiency of all reversible heat engines operating between the same two reservoirs is the same. Therefore the efficiency of the heat engine should only be a function of the two reservoir temperatures. In the preceding section, we demonstrated this for a *particular* reversible heat engine. Hence the efficiency of a reversible engine may be expressed as

$$\eta = 1 - Q_c/Q_h = F(T_h, T_c) \tag{6.25}$$

It follows that: $$Q_c/Q_h = 1 - F(T_h, T_c) = f(T_h, T_c) \tag{6.26}$$

where F and f are functions of the two reservoir temperatures.

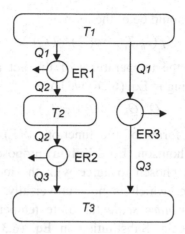

Fig. 6.14 Thermodynamic temperature scale

Consider the arrangement of reversible heat engines shown in Fig. 6.14 where the engine ER1 operates between reservoirs of temperatures T_1 and T_2, while the engine ER2 operates between reservoirs of temperatures T_2 and T_3. A third reversible engine ER3 operates between reservoirs at temperature T_1 and T_3.

Applying Eq. (6.26) to each heat engine in turn we obtain the following relations:

$$Q_2/Q_1 = f(T_1, T_2) \qquad (6.27)$$

$$Q_3/Q_2 = f(T_2, T_3) \qquad (6.28)$$

$$Q_3/Q_1 = f(T_1, T_3) \qquad (6.29)$$

Consider the relation

$$Q_3/Q_1 = (Q_2/Q_1)(Q_3/Q_2) \qquad (6.30)$$

Substituting from Eqs. (6.27) – (6.29) in Eq. (6.30) we have

$$f(T_1, T_3) = f(T_1, T_2)f(T_2, T_3) \qquad (6.31)$$

Now the left hand side of Eq. (6.31) is independent of T_2. Therefore $f(T_1, T_2)$ and $f(T_2, T_3)$ should have the following functional form

$$f(T_1, T_2) = \phi(T_2)/\phi(T_1)$$

and

$$f(T_2, T_3) = \phi(T_3)/\phi(T_2)$$

In general, $f(T_h, T_c)$ should be of the form

$$f(T_h, T_c) = \phi(T_c) / \phi(T_h) \qquad (6.32)$$

where T_h and T_c are the temperatures of the hot and cold reservoirs respectively. Substituting in Eq. (6.26) we have

$$Q_c / Q_h = \phi(T_c) / \phi(T_h) \qquad (6.33)$$

There are several forms of the function, $\phi(T)$ which will satisfy Eq. (6.33). In 1854 Thomson (Lord Kelvin) proposed the simple form, $\phi(T) = T$, which was chosen to agree with the ideal gas temperature scale. This new scale, which is now universally used, is called the *thermodynamic temperature scale* (absolute temperature scale) or the Kelvin temperature scale. Substituting in Eq. (6.33) for the function $\phi(T)$ we have

$$Q_c / Q_h = T_c / T_h \qquad (6.34)$$

Substituting in Eq. (6.25) from Eq. (6.34) we obtain the efficiency of the Carnot engine as:

$$\eta = 1 - T_c / T_h \qquad (6.35)$$

In order to define the absolute temperature completely we need to fix the 'size' of a degree in the thermodynamic temperature scale. The Tenth International General Conference on Weights and Measures in 1954 agreed on the following definition: 'The Kelvin unit of thermodynamic temperature is the fraction 1/273.16 of the thermodynamic temperature T_{tp} of the triple point of water'. Therefore the numerical value assigned to the temperature of the triple point of water is 273.16 K.

Consider the 'thought-experiment' where a Carnot heat engine is operated between a reservoir at T, the temperature to be measured, and a reservoir at T_{tp}, the triple-point temperature of water. Let the respective heat transfers per cycle between the reservoirs and the heat engine be Q and Q_{tp}. If the measured efficiency of the Carnot engine is η_{tp}, then it follows from Eqs. (6.34) and (6.35) that

$$T = T_{tp} Q / Q_{tp} = T_{tp} / (1 - \eta_{tp})$$

We observe that the above equation defines a temperature scale that is independent of thermometric substances, and material properties of temperature sensors. It depends, in concept, only on the efficiency of the reversible heat engine.

The Celsius temperature is then defined by

$$T(^\circ C) = T(K) - 273.15$$

From the analysis of a reversible Carnot cycle, using the ideal gas temperature defined in Sec. 1.5.3, we obtained Eq. (6.24) for its thermal efficiency. This expression is similar to the expression in Eq. (6.35) that is based on the thermodynamic temperature scale. It is the choice of the function, $\phi(T) = T$, and the numerical matching of the temperatures at the triple point of water that has made the ideal gas temperature scale and the thermodynamic temperature scale identical. Note, however, that whereas to use the Celsius degree requires measurements at two fixed points (typically the ice-point and the steam-point) to use the degree defined in the new way requires measurements only at one fixed point (the triple-point of water).

6.5.3 *Cycles interacting with a single thermal reservoir*

We now summarize in analytical form some of the important conse-quences of the second law for cyclic devices operating with a *single thermal reservoir*. Such a device is shown in Fig. 6.15. From the K-P-S it follows that for the cycle

$$\sum_{cycle} W_i \le 0 \qquad (6.36)$$

where W_i the work *output* of process i of the cycle which is taken as *positive*.

Applying the first law to the cycle we have

$$\sum_{cycle} Q_i = \sum_{cycle} W_i \le 0 \qquad (6.37)$$

The integral forms of Eqs. (6.36) and (6.37) are

$$\oint \delta W \le 0 \qquad \oint \delta Q \le 0 \qquad (6.38)$$

Now the K-P-S of the second law does not distinguish explicitly be-tween reversible and irreversible cycles. It is clear that for an *irreversible*

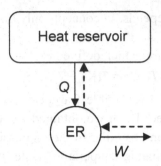

Fig. 6.15 Cyclic device operating with a single heat reservoir

cycle, the net heat input and the net work output are both negative according to the K-P-S.

Now consider a *reversible* heat engine operating as a heat pump, interacting with a single heat reservoir. The heat and work   interactions are shown by the broken-line arrows in Fig. 6.15. Such a device is perfectly feasible because work can be completely converted to heat using a frictional device like a paddle wheel. If we now reverse the heat pump shown in Fig. 6.15, the reversibility of the cycle will ensure that the heat and work interactions are reversed with exactly the same magnitudes, and no other effect on the surroundings. This reversible heat engine cycle, however, violates the K-P-S because it converts the entire heat input to work.

Therefore we conclude that for the K-P-S and reversibility to be satisfied simultaneously, the heat and work interactions must both be zero for a *reversible* cyclic device exchanging heat with a *single* reservoir.

A useful test to determine whether a given process, in which heat is exchanged with a *single reservoir*, is irreversible or otherwise emerges from Eqs. (6.36) and (6.37), and their cyclic integral forms in Eq. (6.38). The procedure can be stated as follows.

Consider the end-states of the process being tested for irreversibility, and introduce a *reversible* process that would take the system from the final state back to the initial state exchanging heat with the *same* reservoir. If the net work transfer of the cycle consisting of the original process and the new reversible process is *negative* then the original process is *irreversible* according to Eq. (6.37).

If, however, the net work transfer is *zero* then the original process is *reversible*. It is clear that if the net work transfer is positive, the K-P-S is violated. The application of the above procedure is demonstrated in worked examples 6.4 to 6.9.

6.5.4 *Cycles interacting with two thermal reservoirs*

Consider the *reversible* heat engine operating between two thermal reservoirs shown schematically in Figs. 6.3 and 6.5. Applying the first law using the sign convention adopted in chapter 4 we have the following conditions.

For a heat engine:

$$\sum_{cycle} W_i > 0 \qquad \text{and} \qquad \sum_{i=1}^{2} Q_i > 0$$

For a reversed heat engine:

$$\sum_{cycle} W_i < 0 \qquad \text{and} \qquad \sum_{i=1}^{2} Q_i < 0$$

where 1 and 2 denote the two reservoirs.

As a consequence of the second law we obtained an additional relation given by Eq. (6.34) for a *reversible* cycle with *two* heat reservoirs. Using the sign convention in chapter 4 we can rewrite Eq. (6.34) in the form:

$$\sum_{i=1}^{2} \frac{Q_i}{T_i} = 0 \qquad\qquad (6.39)$$

for both heat engines and reversed heat engines.

The Carnot principle states that no engine operating between two heat reservoirs can have an efficiency greater then a reversible engine.

Therefore $\eta_{irr} < \eta_{rev}$ $\qquad\qquad (6.40)$

Substituting the expressions for the efficiency of an irreversible cycle from Eq. (6.3), and for a reversible engine from Eq. (6.35) in the above inequality we obtain the condition

$$1 - \frac{Q_{ci}}{Q_{hi}} < 1 - \frac{T_c}{T_h} \qquad (6.41)$$

where the subscript i indicates the heat interactions for the irreversible cycle.

Hence
$$\frac{Q_{hi}}{T_h} - \frac{Q_{ci}}{T_c} < 0 \qquad (6.42)$$

Note that we could show that the condition given by Eq. (6.42) also applies to reversed heat engines or heat pumps by using the fact that

$$COP_{irr} < COP_{rev}$$

Combining Eq. (6.39) and the inequality (6.42) we could conclude that for reversible and irreversible heat engines and heat pumps

$$\sum_{i=1}^{2} \frac{Q_i}{T_i} \leq 0 \qquad (6.43)$$

In applying Eq. (6.43), appropriate signs must be used for the two heat interactions as these are different for heat engines and heat pumps. In summary, to be feasible, a cyclic device operating between *two* heat reservoirs must satisfy Eq. (6.43). Otherwise the device violates the second law.

Having obtained the conditions resulting from the second law for heat engines operating with one and two reservoirs, the next logical step is to apply the second law to reversible and irreversible cyclic devices operating with any number of thermal reservoirs.

6.5.5 *Cycles interacting with any number of thermal reservoirs*

To apply the second law to a system that executes a cycle exchanging heat with any number of reservoirs we consider the arrangement shown in Fig. 6.16.

A closed system S interacts with a series of reservoirs at T_i, which in turn receive heat from a single reservoir at T_o. The cycle executed by the system and the heat transfer interactions between the system and the various heat reservoirs could be reversible or irreversible.

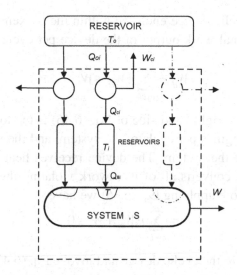

Fig. 6.16 Cyclic device operating with any number of reservoir

However, the heat transfer between the reservoir at T_o and each of the secondary heat reservoirs at T_i are reversible because heat transfer is carried out by operating a reversible heat engine as shown in Fig. 6.16. The heat engines deliver work to the surroundings while the required heat transfer occurs. The temperature T, at the location on the system where heat is exchanged with the secondary reservoir at T_i, may or may not be equal to the latter temperature. If T and T_i are different, then the system would experience an external heat transfer irreversibility in addition to any irreversibilities that are internal to the system S.

Assume that the heat engines execute an integral number of cycles (n_i) per cycle of the system, whose work output is W per cycle. The heat input to and the work output from the various engines are respectively, Q_{oi} and W_{ci} per engine cycle.

Now for each of the reversible heat engines we apply Eq. (6.39) to obtain

$$\frac{Q_{oi}}{T_o} - \frac{Q_{ci}}{T_i} = 0 \qquad (6.44)$$

Since each of the secondary thermal reservoirs operates in a cycle

$$n_i Q_{ci} = Q_{si} \qquad (6.45)$$

Consider the cyclic device enclosed within the broken-line boundary in Fig. 6.16. The total work output of the device per cycle of the system is

$$W_d = \sum_{cycle} n_i W_{ci} + W \tag{6.46}$$

The first term on the right hand side of Eq. (6.46) is the total work done by the reversible engines per cycle of the system, and the second term is the work output of the system. The device receives heat from a single reservoir at T_o and converts all of it to work violating the K-P-S of the second law. Therefore applying Eq. (6.36) we have

$$W_d = \sum_{cycle} n_i W_{ci} + W \le 0 \tag{6.47}$$

Apply the first law to the cyclic device and invoke Eq. (6.47) to obtain

$$\sum_{cycle} n_i Q_{oi} = \sum_{cycle} n_i W_{ci} + W \le 0 \tag{6.48}$$

Substituting for Q_{oi} from Eq. (6.44) in Eq. (6.48) we have

$$T_o \sum \frac{n_i Q_{ci}}{T_i} \le 0 \tag{6.49}$$

Since T_o is positive, the relation between the heat received by the reservoir and its temperature becomes

$$\sum \frac{n_i Q_{ci}}{T_i} \le 0 \tag{6.50}$$

Substituting from Eq. (6.45) in Eq. (6.50) we have

$$\sum \frac{Q_{si}}{T_i} \le 0 \tag{6.51}$$

In the above equation the inequality sign applies when the processes undergone by the system S are irreversible. The irreversibilities could include both internal irreversibilities of the system and external irreversibilities due to the heat transfer across finite temperature differences between the reservoirs at T_i and the location on the system at T. When *all* the processes are reversible the equality sign applies. If the cyclic process does not satisfy the condition expressed by Eq. (6.51) then it violates the second law.

6.6 Worked Examples

Example 6.1 Propose ideal reversible processes to reverse the processes listed below. Hence show that the original processes are irreversible.

(i) A block of metal at 120°C is placed in thermal contact with a heat reservoir at 30°C until their temperatures are equal.

(ii) A well-insulated cylindrical vessel is divided into two compartments with a light insulating piston. Initially, a fixed mass of air at 80°C and 1 bar pressure is trapped in one section, while the other section is evacuated. The piston is held in position by a pin. The pin is removed and the air expands to fill the whole vessel.

(iii) A well-insulated piston-cylinder apparatus contains wet steam at a temperature of 105°C and quality of 0.3. The external pressure on the piston due to the weight of the piston and the surrounding air pressure is constant. Located inside the cylinder is an electrical-resistance heater. The heater is switched on and the steam evaporates until the quality is 0.4.

Solution (i) In the final equilibrium state the metal block will have the same temperature of 30°C as the reservoir. We operate a reversible or Carnot heat pump to transfer heat from the reservoir to the block to raise its temperature to 120°C thus completing the cycle executed by the composite system consisting of the block and heat pump.

The net work interaction of this cycle which exchanges heat with a single reservoir at 30°C is negative because of the work input to the heat pump. Therefore the cycle is irreversible according to Eq. (6.36).

Since the second process introduced by us is reversible, the original process has to be irreversible.

(ii) When the pin is removed, the light piston will be pushed out and the air will expand to fill the vessel. For this process the work done is zero because of the free expansion, and the heat interaction is zero due the insulation. Therefore according to the first law the internal energy is unchanged. Treating the air as an ideal gas we conclude that the temperature of the air at the final state is also 80°C.

In order to reverse the process we compress the air in a quasi-static isothermal process to the original volume maintaining thermal contact with a reservoir at 80°C. The net work interaction of the cycle is negative because of the work of compression, and therefore the cycle irreversible according to Eq. (6.36). Since the compression process is reversible, the original free-expansion process has to be irreversible.

(iii) The process can be reversed by removing some of the insulation and placing a reservoir at 105°C in thermal contact with the steam. An infinitesimal temperature difference between the steam and the reservoir will initiate condensation. The condensation process at constant pressure and temperature is continued until the quality of the steam is 0.3, thereby returning the steam to its original state. The work done by the piston during the expansion is equal in magnitude and opposite in direction to the compression work during condensation. However, the electrical energy input during original evaporation process is a negative work interaction which makes the net work interaction of the cycle negative. Therefore the cycle is irreversible according to Eq. (6.36). The condensation process is reversible and therefore the original heating process is irreversible.

Example 6.2 A mass M attached to a spring and a damper rests on a table between two guides as shown in Fig. E6.2. The damping force is proportional to the relative velocity between the piston and the cylinder, and the spring force is proportional to the displacement.

(i) Describe a method to move the mass in a reversible manner from positions A to B considering the following conditions (a) the damper is detached from the mass and there is no friction between the mass and the guides, (b) the damper is attached and there is no friction between the mass and the guides and (c) the above situations with sliding friction between the mass and the guide.

(ii) When the mass is at B the constraints that ensured equilibrium during the quasi-static process are suddenly removed. Described the ensuing motion of the mass and locate its final equilibrium position for cases (a), (b) and (c) above.

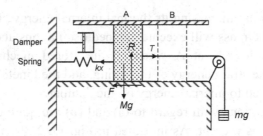

Fig. E6.2 Displacement of a sliding mass

Solution (a) In order to move the mass from A to B in a quasi-static manner we attach to it a cord that runs over a frictionless pulley as shown in Fig. E6.2. Weights are hung at the end of the cord in small steps so that we could control the motion to any degree we desire.

If a small weight is removed, the direction of motion is reversed due to the tension in the spring leaving no permanent change in the system or the surroundings. Now for equilibrium,

$$T = mg = kx$$

where T is the tension, x is the displacement, and k is the spring constant. Therefore for 'small' changes of mass, $g\delta m = k\delta x$.

When the mass is at position B and the cord is cut, the tension in the spring will not be balanced any longer. The mass will execute oscillations about position A where the spring force is zero. Ideally, in the absence of any dissipative mechanisms, such as air resistance, the oscillations will continue steadily.

(b) The situation is similar to (a) above except that the damper is now attached. The force balance on the mass gives,

$$T = mg = kx + cv$$

where v is the velocity of the mass, and c is the force per unit velocity of the damper. Since mass is moved through infinitesimal steps quasi-statically, its velocity is zero. Consequently, the damping force is also zero. The position of the mass can therefore be reversed quasi-statically as in (a). However, when the cord is cut, the mass will gain kinetic energy and the damper will exert a force opposing the motion. In the damper the work done by the damping force is converted to a heat interaction thereby increasing its thermal internal energy. This process is irreversible. The mass will execute damped oscillations about position A until it comes to rest at A when all the stored mechanical energy, consisting of the strain energy of the spring, and the kinetic energy of the mass is converted to internal energy of the damper.

(c) The situations with regard to (a) and (b) are quite different when sliding friction is present. As indicated in Fig. E6.2, the frictional force between the sliding surfaces has a limiting value given by

$$F = \mu R = \mu Mg$$

This maximum value is reached when the mass is *just* about to move. The equilibrium equation for the mass when it is moved towards position B is given by

$$T = kx + \mu Mg$$

Therefore in order for the mass M to move, the tension T in the chord has to exceed the spring force, kx by μMg. On the other hand, to reverse the direction of movement of M, the spring force has to exceed the tension T by μMg because the frictional force F changes direction to oppose the motion. Therefore 'small' changes in the tension of the cord will not produce 'small' changes in the displacement of M in both directions as in the case of (a) and (b). Moreover, any movement of M will involve some frictional work at the sliding surface which is converted to a heat interaction. This process is clearly irreversible.

The heat interaction is equal to the frictional work done, which is given by the decrease in potential energy of the added weight. At position B, when the cord is cut the mass M will execute damped oscillations with mechanical energy dissipation in the damper and at the sliding surface. Eventually the mass will come to rest at a position that is within $(\pm \mu Mg/k)$ about A when the spring force balances the frictional force.

Example 6.3 Figure E6.3 shows an arrangement for the constant-volume heating of an ideal gas A in a reversible manner by supplying work to the ideal gas B trapped in a piston-cylinder set-up. The wall separating the constant volume of gas A from gas B is rigid and made of a highly conducting material. The piston and the cylinder are well-insulated on the outside. (a) Obtain an expression for the P-V relation for the gas B in terms of the quantities indicated in the figure, and (b) suggest two other methods for heating a gas reversibly at constant volume.

Solution The various relevant quantities like the temperatures, pressures and volumes are indicated in Fig. E6.3. Applying the first law to the composite system consisting of the two gases we have

$$\delta Q - dU + \delta W \qquad (E6.3.1)$$

The system is insulated and the compression process is quasi-static. Due to the highly conducting wall between gases A and B their temperatures are equal. Let this temperature be T. With these conditions, Eq. (E6.3.1) becomes

$$0 = (m_B c_{vB} + m_A c_{vA})dT + P_B dV_B \qquad (E6.3.2)$$

where c_v is the specific heat capacity of the gas.

Applying the ideal gas equation to gas B

$$P_B V_B = m_B R_B T \qquad (E6.3.3)$$

From Eqs. (E6.3.2) and (E6.3.3) we obtain the following

$$0 = (m_B c_{vB} + m_A c_{vA})dT/T + m_B R_B dV_B/V_B \qquad (E6.3.4)$$

Integrating Eq. (E6.3.4) we have

$$V_B T^\beta = C \qquad (E6.3.5)$$

where C is a constant and the exponent is given by

$$\beta = (m_B c_{vB} + m_A c_{vA})/m_B R_B$$

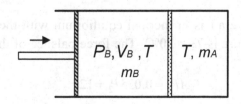

Fig. E6.3 Constant-volume heating of an ideal gas

Eliminating T between Eqs. (E6.3.3) and (E6.3.5) the P-V relation for gas B is obtained as

$$P_B V_B^{(1+1/\beta)} = \text{constant} \qquad\qquad (E6.3.6)$$

If the pressure and volume of gas B are varied according to Eq. (E6.3.6) the heat transfer to gas A will occur in a reversible manner. A force-balancing mechanism consisting of a cam and a set of weights to implement this process is described in detail in Ref. [2].

(b) We can transfer heat reversibly in a constant volume process by supplying heat to the gas using a series of external heat sources as was discussed in worked example 1.6, part (iii).

Alternatively, we could use a series of infinitesimal Carnot heat engines operating between a high temperature reservoir and the gas as the heat sink. The heat flow from the engine to the gas occurs reversibly across an infinitesimal temperature difference.

Example 6.4 A fixed mass of air is trapped in a cylinder behind a frictionless piston of area 0.035 m². The mass of the piston is 12 kg and the external pressure on the piston is zero. Initially, the conditions of the air are: $P_1 = 1.5$ bar, $V_1 = 10^{-4}$ m³, $T_1 = 30°C$. The piston is held in position by a pin. The air, the piston, and the cylinder are in thermal equilibrium with the surroundings, which may be regarded as a thermal reservoir at 30°C. The pin is now removed, and the piston moves up until a final equilibrium state is attained. Calculate (i) the final temperature, pressure, and volume of the air, and (ii) the heat and work interaction between the system and the surroundings.

Describe a reversible process to return the air to its initial state. Hence show that the original expansion process is irreversible.

Solution Since the air is in thermal equilibrium with the surroundings, the final temperature is $T_2 = 30°C$. The force balance of the piston in the final state gives

$$AP_2 = 0.035P_2 = 12g$$

where the acceleration due to gravity, $g = 9.81$ ms⁻².

Hence $P_2 = 3.363 \times 10^3$ Nm⁻²

Applying the ideal gas equation to states 1 and 2 we have

$$mR = P_1V_1/T_1 = P_2V_2/T_2$$

Substituting numerical values in the above equation

$$V_2 = 150 \times 10^{-4}/3.363 = 44.6 \times 10^{-4} \text{ m}^3$$

The change in elevation of the piston is given by

$$\Delta H = (V_2 - V_1)/A = (44.6 - 1) \times 10^{-4}/0.035 = 0.1246 \text{ m}$$

The work done by the air is equal to the potential energy gained by the piston. Hence

$$W_{12} = Mg\Delta H = 12 \times 9.91 \times 0.1246 = 14.67 \text{ J}$$

The change in internal energy of the air is

$$U_2 - U_1 = mC_v(T_2 - T_1) = 0 \text{ J}$$

Applying the first law we have

$$Q_{12} = U_2 - U_1 + W_{12} = 14.67 \text{ J}$$

To reverse the state of the air we use a *reversible isothermal* compression process while the air exchanges heat with a reservoir at 30°C.

The work *input* for this isothermal process is given by

$$W_{23} = mRT_2 \ln(V_2/V_3) = P_2V_2 \ln(V_2/V_1)$$

Substituting numerical values in the above equation we have

$$W_{23} = 3.363 \times 10^3 \times 44.6 \times 10^{-4} \times \ln(44.6/1.0) = 56.96 \text{ J}$$

The net work *output* of the cycle consisting of the original process 1-2 and the isothermal compression process is

$$W_{net} = 14.67 - 56.96 = -42.3 \text{ J}$$

The system interacts only with a single reservoir at 30°C and the net work *output* of the cycle is negative. Therefore the cycle is irreversible according to Eq. (6.36). The isothermal compression process is reversible and therefore the original process 1-2 has to be irreversible.

Example 6.5 An ideal gas undergoes an irreversible adiabatic process from state 1 to state 2. Show that the state of the gas could be returned from 2 to 1 with a combination of a reversible adiabatic process and a reversible isothermal process. Obtain expressions for the work and heat interactions of the two processes.

Solution The original irreversible process 1-2 is shown by the broken-line in Fig. E6.5. A reversible isothermal process 2-3 and a reversible adiabatic process 3-1 are used to reverse the state of the ideal gas from 2 to 1. We then consider the work output of the cycle 1-2-3-1 to determine whether the latter cycle exchanging heat with a single reservoir is reversible by applying the condition in Eq. (6.36).

For the adiabatic process 3-1 of the ideal gas

$$P_3 V_3^{\gamma} = P_1 V_1^{\gamma} \qquad \text{(E6.5.1)}$$

Also for the isothermal process 2-3

$$T_3 = T_2$$

Applying the ideal gas equation

$$P_3 V_3 = P_2 V_2 \qquad \text{(E6.5.2)}$$

From Eqs. (E6.5.1) and (E6.5.2) we obtain

$$V_3 = \left(P_1 V_1^{\gamma} / P_2 V_2 \right)^{1/(\gamma - 1)}$$

Substituting in Eq. (E6.5.2) we have

$$P_3 = P_2 V_2 \left(P_2 V_2 / P_1 V_1^{\gamma} \right)^{1/(\gamma - 1)}$$

The above equation gives the pressure P_3 to which the gas has to be compressed isothermally with the temperature at T_2.

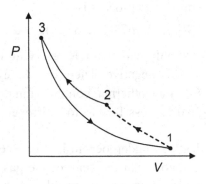

Fig. E6.5 Reversal of the state from 2-1.

The work done in the reversible adiabatic process 3-1 is

$$W_{31} = mR(T_3 - T_1)/(\gamma - 1) = mR(T_2 - T_1)/(\gamma - 1)$$

For the adiabatic process 3-1, $Q_{31} = 0$.

The work done in the reversible isothermal process 2-3 is

$$W_{23} = -P_2 V_2 \ln(P_3/P_2) = -mRT_2 \ln(P_3/P_2)$$

Since the change in internal energy during process 2-3 is zero, $Q_{23} = W_{23}$.

Example 6.6 A closed system consisting of 0.1 kg of air undergoes a process from an initial state of 100 kPa and 303 K to a final state of 380 kPa and 550 K while in thermal communication with a reservoir at 550 K from which it receives 10.5 kJ of heat. Determine the characteristics of a reversible isothermal and a reversible adiabatic process that could be used to return the system from the final state to the initial state. Hence show that the original process is irreversible.

Solution Applying the first law to the given process

$$Q_{12} = U_2 - U_1 + W_{12} \tag{E6.6.1}$$

Substituting numerical values in Eq. (E6.6.1)

$$10.5 = mC_v(T_2 - T_1) + W_{12}$$

$$10.5 = 0.1 \times 0.718 \times (550 - 303) + W_{12}$$

$$W_{12} = -7.23 \text{ kJ}$$

A combination of a reversible isothermal process 2-3, and a reversible adiabatic process 3-1 that should return the system to the initial state is shown in Fig. E6.5. Eliminating V between the ideal gas equation $PV = mRT$ and the adiabatic process law $PV^\gamma = C$ we have

$$P^{(\gamma-1)/\gamma}/T = C1 \tag{E6.6.2}$$

where $C1$ is a constant.

Applying the above relation to the reversible adiabatic process from 3 to 1 we obtain

$$P_3/P_1 = (T_3/T_1)^{\gamma/(\gamma-1)} \tag{E6.6.3}$$

Substituting numerical values in Eq. (E6.6.3)

$$P_3/100 = (550/303)^{\gamma/(\gamma-1)} = (550/303)^{3.5} = 8.058$$

Hence $P_3 = 805.8$ kPa,

which is the pressure to which the air should be compressed iso-thermally. The gas is then expanded in a reversible adiabatic process 3-1 to the original temperature of 303 K.

Work *output* during the adiabatic process 3-1 is

$$W_{31} = mR(T_3 - T_1)/(\gamma - 1)$$

$$W_{31} = 0.1 \times 0.287 \times (550 - 303)/(1.4 - 1) = 17.72 \text{ kJ}$$

The work *output* during the isothermal process 2-3 is

$$W_{23} = mRT_2 \ln(P_3/P_1)$$

$$W_{23} = -0.1 \times 0.287 \times 550 \times \ln(805.7/380) = -11.86 \text{ kJ}$$

The net work *output* of the cycle 1-2-3-1 is

$$W_{net} = W_{12} + W_{23} + W_{31} = -7.23 - 11.86 + 17.72 = -1.37 \text{ kJ}$$

The work *output* of the cycle 1-2-3-1, executed by the air while exchanging heat with a *single* reservoir at 550 K is negative. Therefore the cycle is irreversible according to Eq. (6.36). Since the two processes 2-3 and 3-1 are reversible, the original process 1-2 has to be irreversible.

Example 6.7 A well-insulated tank containing 0.6 kg of air is at an initial temperature and pressure of 323 K and 1500 kPa respectively. The tank is connected to a well-insulated piston-cylinder set-up through a valve that is initially closed. The piston sits at the bottom of the cylinder and its weight and the force of the atmosphere on the outside requires a pressure of 240 kPa to just lift it. The valve is now opened and the air flows into the cylinder to establish a final equilibrium state of the system. (i) Calculate the final temperature of the air, and the work done by the air. (ii) Determine whether the process is reversible or irreversible.

Solution Let the initial and final equilibrium states of the air be denoted by 1 and 2 respectively. Applying the ideal gas equation to state 1

$$P_1 V_1 = mRT_1 \qquad \text{(E6.7.1)}$$

Substituting the numerical values in Eq. (E6.7.1)

$$1500 \times V_1 = 0.6 \times .287 \times 323$$

Hence
$$V_1 = 0.03708 \text{ m}^3$$

The work done by the air is used to raise the piston of mass M and push the atmospheric air at pressure, P_{atm}. Hence

$$W_{12} = Mg(V_2 - V_1)/A + P_{atm}(V_2 - V_1)$$

where A is the area of the piston.

For the final equilibrium state

$$P_2 = Mg/A + P_{atm} = 240 \text{ kPa}$$

which is also the pressure required to just lift the piston.

Applying the first law to the process 1-2 we have

$$Q_{12} = U_2 - U_1 + W_{12} \qquad \text{(E6.7.2)}$$

$$0 = mC_v(T_2 - T_1) + P_2(V_2 - V_1)$$

$$0 = 0.6 \times 0.718 \times (T_2 - 323) + 240 \times (V_2 - 0.03708) \qquad \text{(E6.7.3)}$$

Applying the ideal gas equation to the final state 2

$$240 \times V_2 = 0.6 \times 0.287 \times T_2 \qquad \text{(E6.7.4)}$$

Substituting for V_2 in Eq. (E6.7.3) from Eq. (E6.7.4)

$$0 = 0.6 \times 0.287 \times (T_2 - 323) + 0.6 \times 0.286 \times T_2 - 0.03708 \times 240$$

Therefore the final temperature of the air is, $T_2 = 245.5 \text{ K}$

Substituting in Eq. (E6.7.4), $V_2 = 0.1763 \text{ m}^3$

The work done by the air during process 1-2 is

$$W_{12} = P_2(V_2 - V_1) = 240 \times (0.1763 - 0.03708) = 33.4 \text{ kJ}$$

We use a reversible adiabatic process 2-3 and a reversible isothermal process 3-1, (Fig. E6.7) to reverse the state of the air from 2 to 1. The adiabatic compression will increase the temperature to 323 K and the isothermal compression that follows will increase the pressure to 1500 kPa. Let 3 denote the properties of the intermediate state. Using Eq. (E6.6.2) for the adiabatic process 2-3 we have

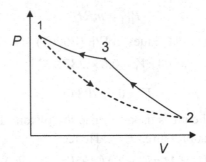

Fig. E6.7 Reversal of the state from 2 to 1

$$P_3 / 240 = (323/245.5)^{\gamma/(\gamma-1)} = 2.61$$

Hence $P_3 = 626.4 \ \text{kPa}.$

The work done during reversible adiabatic compression process 2-3, is the same as that in the adiabatic process 1-2 because for both processes the heat transfer is zero and the temperature change is the same. Therefore $W_{23} = 33.4 \ \text{kJ}$.

During the process 3-1 the air is compressed isothermally to a pressure of 1500 kPa from 626.4 kPa thus returning the air to its original state. During this process the air exchanges heat with a single reservoir at 323 K. The work input for this isothermal process 3-1 is given by

$$W_{31} = mRT \ln(P_1/P_3)$$

$$W_{31} = 0.6 \times 0.287 \times 323 \times \ln(1500/626.4) = 48.6 \ \text{kJ}$$

The net work *output* of the cycle 1-2-3-1 is

$$W_{net} = W_{12} + W_{23} + W_{31} = 33.4 - 33.4 - 48.6 = -48.6 \ \text{kJ}$$

Since the net work *output* of the cycle while exchanging heat with a *single* reservoir at 323 K is negative, the cycle is irreversible according to Eq. (6.36). However, the two processes 2-3 and 3-1 used to reverse the state of the system from 2 to 1 are reversible. Therefore the original process 1-2 is irreversible.

Example 6.8 A paddle wheel of negligible thermal capacity is located inside a well-insulated, rigid cylindrical vessel containing 0.2 kg of air at an initial temperature of 293 K. The paddle wheel is rotated using an external work source until the final temperature of the air becomes 353 K. Assume that the thermal capacity of the wall of the vessel is negligible. (i) Calculate the total work input to the air by the paddle wheel. (ii) Propose a reversible process to return the air to its initial state. (iii) Show that the original process is irreversible.

Solution (i) Applying the first law to the air, treating it as an ideal gas we have

$$Q_{12} = U_2 - U_1 + W_{12}$$

$$0 = mc_v(T_2 - T_1) + W_{12} \tag{E6.8.1}$$

Hence the work input by the paddle-wheel is

$$W_{12} = -0.2 \times 0.718 \times (353 - 293) = -8.616 \text{ kJ}$$

(ii) We use a series of infinitesimal Carnot heat engines operating between the air in the vessel, as the heat source, and a cold reservoir at 293 K, as the heat sink to cool the air reversibly to 293 K. Figure E6.8 shows one such engine.

Applying Eq. (6.39) to the infinitesimal Carnot cycle,

$$\delta Q_1 / T = \delta Q_2 / T_c \tag{E6.8.2}$$

Applying the first law to the constant volume of air in the vessel

$$\delta Q_1 = -mc_v dT \tag{E6.8.3}$$

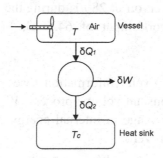

Fig. E6.8 Infinitesimal Carnot cycle

Applying the first law to the infinitesimal Carnot cycle

$$\delta Q_1 - \delta Q_2 = \delta W \qquad (E6.8.4)$$

From Eqs. (E6.8.2), (E6.8.3) and (E6.8.4) we obtain

$$\delta W = -mc_v dT + mc_v T_c dT / T \qquad (E6.8.5)$$

The integration of Eq. (E6.8.5) gives

$$W = -mc_v (T_f - T_i) + mc_v T_c \ln(T_f / T_i) \qquad (E6.8.6)$$

where subscripts i and f denote the initial and final states of the air.

Substituting the numerical values in Eq. (E6.8.6)

$$W = -0.2 \times 0.718 \times (293 - 353) + 0.2 \times 0.718 \times 293 \ln(293 / 353)$$

The work output is $W = 0.78 \, \text{kJ}$

(iii) The air has been returned to its initial state and the system consisting of the air and the heat engines has executed a cycle exchanging heat with a single reservoir at 293 K. The net work output of the cycle is

$$W_{net} = -8.616 + 0.78 = -7.836 \ \text{kJ}.$$

Since the net work output of the cycle is negative, the cycle is irreversible according to Eq. (6.36). The process used to reverse the state of the air is reversible and therefore the original process 1-2 is irreversible.

Example 6.9 A fixed mass of an ideal gas undergoes a cycle 1-2-3 consisting of three processes. In the first constant volume process 1-2 it receives heat from a reservoir at 600 K. During the next adiabatic process 2-3 it delivers a work output of 88 kJ to the surroundings. The gas rejects heat to a reservoir at 280 K during the isothermal process 3-1 while it receives a work input of 64.5 kJ. Show that the cycle is irreversible.

Solution From the given information we draw the following conclusions. For the constant volume process, $W_{12} = 0$. For the adiabatic process $Q_{23} = 0$. The change in internal energy of the ideal gas during the isothermal process is zero.

Hence $Q_{31} = W_{31} = -64.5 \ \text{kJ}.$

Applying the first law to the cycle

$$Q_{12} + Q_{23} + Q_{31} = W_{12} + W_{23} + W_{31}$$

Substituting the numerical values in the above equation we have

$$Q_{12} + 0 - 64.5 = 0 + 88 - 64.5$$

Therefore $\qquad\qquad Q_{12} = 88$ kJ

Applying the condition given by Eq. (6.43) to the cycle operating with two heat reservoirs we have

$$Q_{12}/600 - Q_{31}/280 = 88/600 - 64.5/280 = -0.084 < 0$$

Therefore the cycle is irreversible.

Example 6.10 Ocean thermal energy conversion (OTEC) systems make use of the surface water, and deep water of the ocean as thermal reservoirs to operate a vapor power cycle similar to that shown in Fig. 6.2. (i) If the surface and deep water temperatures are 30°C and 6°C respectively, calculate the maximum thermal efficiency that could be achieved. (ii) If heat transfer between the reservoirs and the working fluid of the cycle requires a temperature difference of 4°C at each reservoir, calculate the thermal efficiency of the internally reversible cycle.

Solution (i) The maximum efficiency that could be achieved is the Carnot efficiency given by

$$\eta = 1 - T_c/T_h$$

The hot and cold reservoir temperatures are:

$$T_h = 273 + 30 = 303 \text{ K} \quad \text{and} \quad T_c = 273 + 6 = 279 \text{ K}$$

Substituting in the above equation we have

$$\eta = 1 - 279/303 = 7.9\%$$

(ii) The temperature difference between the working fluid of the cyclic device and each reservoir is 4°C as indicated in Fig. E6.10. Therefore the effective reservoir temperatures of the *internally reversible* cycle are:

$$T_h' = 273 + 30 - 4 = 299 \text{ K} \quad \text{and} \quad T_c' = 273 + 6 + 4 = 283 \text{ K}.$$

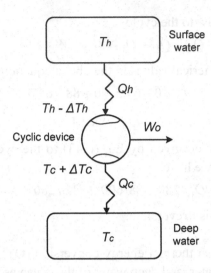

Fig. E6.10 Internally-reversible heat engine cycle

The efficiency of the engine is given by

$$\eta = 1 - T_c'/T_h' = 1 - 283/299 = 5.4\%$$

Note that it is the external thermal irreversibility due to heat transfer across a finite temperature difference at the two reservoirs that causes the decrease in the efficiency of the engine.

Example 6.11 An inventor claims that he has designed a heat engine that operates in a cycle receiving heat at the rate of 120 kW from a waste-heat source at 320°C. The engine delivers work at the rate of 65 kW and rejects heat to the atmosphere at 30°C. Is the claim of the inventor valid?

Solution We shall treat the engine as a cyclic device operating between two heat reservoirs. The hot and cold reservoir temperatures are respectively,

$$T_h = 320 + 273 = 593 \text{ K} \quad \text{and} \quad T_c = 30 + 273 = 303 \text{ K}.$$

The heat transfer rate between the cycle device and the hot reservoirs is

$$Q_h = 120 \text{ kW}$$

Applying the first law to the cyclic device, the heat rejection rate is

$$Q_c = Q_h - W = 120 - 65 = 55 \text{ kW}$$

Apply the condition given by Eq. (6.43) to the cycle operating with two reservoirs

$$Q_h / 593 - Q_c / 303 = 120 / 593 - 55 / 303 = 0.0208 > 0$$

The cycle violates the second law according to Eq. (6.43). Therefore the claim of the inventor is not valid.

Example 6.12 A Carnot-cycle refrigerator, operates between two heat reservoirs at -15°C and 45°C. (i) Calculate the COP of this refrigeration cycle. (ii) Calculate the COP of the cycle if it a operated as a heat pump. (iii) If the heat absorption at the low-temperature reservoir is 25 kW, calculate the power input to the refrigerator and the heat rejection rate. (iv) If the temperature difference between the reservoir and the working fluid of the refrigerator is 4°C at each reservoir, calculate COP of the refrigerator and the heat pump.

Solution The hot and cold reservoir temperatures are respectively,

$$T_h = 45 + 273 = 318 \text{K} \quad \text{and} \quad T_c = -15 + 273 = 258 \text{ K}$$

The COP of the refrigerator is given by Eq. (6.6) as

$$(COP)_r = Q_c / (Q_h - Q_c) \qquad \text{(E6.12.1)}$$

For a fully-reversible cycle Eq. (6.43) gives

$$Q_c / T_c = Q_h / T_h \qquad \text{(E6.12.2)}$$

From Eqs. (E6.12.1) and (E6.12.2) we obtain

$$(COP)_r = T_c / (T_h - T_c) \qquad \text{(E6.12.3)}$$

Substituting numerical values in Eq. (E6.12.3)

(i) $$(COP)_r = 258 / (318 - 258) = 4.3$$

From Eq. (6.10) the COP of the heat pump is

(ii) $$(COP)_h = 1 + (COP)_r = 1 + 4.3 = 5.3 \qquad \text{(E6.12.4)}$$

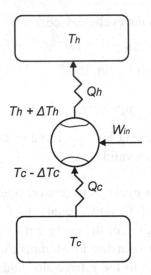

Fig. E6.12 Internally-reversible refrigeration cycle

Now the COP of the refrigerator is

$$(COP)_r = Q_c/W = 25/W = 4.3$$

(iii) Hence the work input is, $W = 5.81$ kW.

Applying the first law the heat rejection rate is

$$Q_h = Q_c + W = 25 + 5.81 = 30.81 \text{ kW}$$

Consider the internally reversible cycle, with a finite temperature difference at each reservoir as shown in Fig. E6.12.

(iv) The effective reservoir temperatures are given by

$$T_h' = 45 + 273 + 4 = 322 \text{ K} \quad \text{and} \quad T_c' = -15 + 273 - 4 = 254 \text{ K}$$

The COP of the refrigerator is

$$(COP)_r = 254/(322 - 254) = 3.74$$

The COP of the heat pump is

$$(COP)_h = 1 + (COP)_r = 1 + 3.74 = 4.74$$

Example 6.13 A Carnot heat engine operates between hot and cold heat reservoirs of temperature T_h and T_c respectively. Determine which of the following changes to the reservoir temperatures will be more effective in increasing the thermal efficiency of the engine: (i) increase T_h to $T_h + \Delta T$ keeping T_c constant, or (ii) decrease T_c to $T_c - \Delta T$ keeping T_h constant.

Solution The efficiency of the Carnot heat engine is given by

$$\eta = 1 - T_c/T_h \qquad (E6.13.1)$$

Substituting the proposed changes (i) and (ii) to the reservoir temperatures in Eq. (E6.13.1) we obtain the following expressions for the efficiency under the new conditions:

$$\eta_h = 1 - T_c/(T_h + \Delta T) = (T_h - T_c + \Delta T)/(T_h + \Delta T) \qquad (E6.13.2)$$

$$\eta_c = 1 - (T_c - \Delta T)/T_h = (T_h - T_c + \Delta T)/T_h \qquad (E6.13.3)$$

Dividing Eq. (E6.13.3) by Eq. (E6.13.2) we have

$$\eta_c/\eta_h = (T_h + \Delta T)/T_h > 1$$

From the above equation it follows that decreasing the cold reservoir temperature leads to a larger increase in the cycle efficiency.

Example 6.14 A reversible heat engine operates between two heat reservoirs at 800 K and 300 K respectively. The work output of the engine is used to run a reversible heat pump that removes heat at the rate of 8 kW from a waste heat source at 310 K, and supplies it to a reservoir at 330 K. Calculate (i) the rate of heat supply to the engine, and (ii) the rate of heat supply to the reservoir at 330 K.

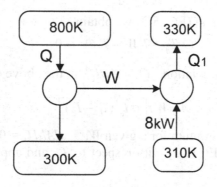

Fig. E6.14 Reversible engine operating a heat pump

Solution Applying the first law and the condition for reversibility given by Eq. (6.43) to the heat pump we have

$$Q_1/330 = (W+8)/330 = 8/310$$

Hence the rate of work transfer from the engine to the heat pump is

$$W = 0.516 \text{ kW}$$

The rate of heat supply to the reservoir is

$$Q_1 = 8 + W = 8.516 \text{ kW}$$

The efficiency of the reversible heat engine is given by

$$\eta = W/Q = 0.516/Q = 1 - 300/800 = 0.625$$

Hence $Q = 0.826 \text{ kW}$

Example 6.15 A reversible heat engine operating in space rejects heat at the rate of, $Q_c = A\sigma T_c^4$ where T_c is the radiator temperature, A is the area of the radiator, and σ is the Stefan-Boltzmann constant. The engine is designed to have the minimum area for the required power output W and a heat source temperature T_h. Obtain a relation between T_c and T_h. What is the efficiency of the engine?

Solution The efficiency of the reversible heat engine is given by

$$\eta = 1 - T_c/T_h = W/Q_h \qquad (E6.15.1)$$

Applying the first law to the cycle

$$Q_h = Q_c + W \qquad (E6.15.2)$$

From Eqs. (E6.15.1) and (E6.15.2) we obtain

$$Q_c T_h - T_c W - T_c Q_c = 0$$

Substituting the given relation, $Q_c = A\sigma T_c^4$ in the above equation

$$A = W/\left[\sigma T_c^3 (T_h - T_c)\right] \qquad (E6.15.3)$$

When the area is a minimum for a given W, $dA/dT_c = 0$
Differentiating Eq. (E6.15.3) with respect to T_c, and applying the above condition we have

$$T_c/T_h = 0.75$$

The efficiency of the engine is

$$\eta = 1 - T_c / T_h = 1 - 0.75 = 25\%$$

Note that Eq. (E6.15.3) also reveals that for a given area of the radiator A, the work output W is a maximum under the same conditions.

Example 6.16 An internally reversible heat engine receives heat from a reservoir at T_1 and rejects heat to a reservoir at T_2. The heat flow rate from the high temperature reservoir is given by, $Q_{in} = AU(T_1 - T_{f1})$, where A and U are the area and heat transfer coefficient of the heat exchanger. T_{f1} is the constant temperature of the working fluid during heat reception. The temperature difference between the working fluid and the low temperature reservoir during heat rejection is negligible. Such a cycle is called an endoreversible cycle. Show that when the work output of the cycle is a maximum, $T_{f1} = (T_1 T_2)^{1/2}$. Obtain an expression for the efficiency of the cycle.

Solution The efficiency of the *internally reversible* heat engine is given by

$$\eta = 1 - T_2 / T_{f1} = W / Q_{in} = W / \left[AU(T_1 - T_{f1}) \right] \quad \text{(E6.16.1)}$$

Rearranging Eq. (E6.16.1) we have

$$W = AU(T_1 - T_{f1})(1 - T_2 / T_{f1}) \quad \text{(E6.16.2)}$$

When the work output of the cycle is a maximum,

$$dW / dT_{f1} = 0$$

Differentiating Eq. (E6.16.2) and applying the above condition we obtain

$$T_{f1} = \sqrt{T_1 T_2}$$

Substituting for T_{f1} in Eq. (E6.16.1), the cycle efficiency at maximum work output becomes

$$\eta = 1 - T_2 / T_{f1} = 1 - \sqrt{T_2 / T_1}$$

Example 6.17 Heat is transferred reversibly between two identical bodies of specific heat capacity C and mass M which are initially at temperatures θ_1 and θ_2 respectively. A series of infinitesimal Carnot heat engines are used for this purpose. The bodies are at constant pressure and undergo no phase change. The final equilibrium temperature of the bodies is θ_3. (i) Show that $\theta_3 = (\theta_1 \theta_2)^{1/2}$. (ii) Obtain an expression for the total work output W of the engines.

Solution Consider the operation of an infinitesimal Carnot heat engine when the temperatures of the hot and cold bodies have reached T_1 and T_2 respectively (see Fig. E6.8). Let the heat interactions between the bodies and the engine at this stage be δQ_1 and δQ_2 respectively. Let the work output of the infinitesimal cycle be δW.

Applying the first law to the two bodies and the cyclic engine we obtain the following three equations.

$$\delta Q_1 = -MCdT_1 \tag{E6.17.1}$$

$$\delta Q_2 = MCdT_2 \tag{E6.17.2}$$

$$\delta W = \delta Q_1 - \delta Q_2 \tag{E6.17.3}$$

Applying the condition for reversibility of the cycle with two heat reservoirs given by Eq. (6.43) we have

$$\delta Q_1 / T_1 = \delta Q_2 / T_2 \tag{E6.17.4}$$

Substituting in Eq. (E6.17.4) from Eqs. (E6.17.1) and (E6.17.2)

$$MCdT_1 / T_1 + MCdT_2 / T_2 = 0 \tag{E6.17.5}$$

Integrating Eq. (E6.17.5) from the initial to the final equilibrium states of the bodies we obtain

$$\ln(\theta_3 / \theta_1) + \ln(\theta_3 / \theta_2) = 0 \tag{E6.17.6}$$

$$\ln(\theta_3^2 / \theta_1 \theta_2) = 0 \tag{E6.17.7}$$

From Eq. (E6.17.7) it follows that:

$$\theta_3 = \sqrt{\theta_1 \theta_2}$$

Substituting in Eq. (E6.17.3) from Eqs. (E6.17.1) and (E6.17.2) we have

$$\delta W = -MCdT_1 - MCdT_2$$

Integrating the above equation from the initial to the final states of the bodies

$$W = MC(\theta_1 - \theta_3) + MC(\theta_2 - \theta_3)$$

$$W = MC(\theta_1 + \theta_2 - 2\theta_3) = 2MC\left|(\theta_1 + \theta_2)/2 - \sqrt{\theta_1\theta_2}\right| > 0$$

Note that $W > 0$ because the arithmetic mean of any two positive numbers is greater then the geometric mean.

Example 6.18 A reversible heat engine receives heat at the rate of Q_h from a high temperature reservoir at T_h and rejects heat to the atmosphere at T_a. The work output of the engine drives a reversible heat pump that extracts heat from the atmosphere and delivers heat at the rate Q_s to a space at T_s. The performance factor for this system is defined by $r = (Q_s/Q_h)$. Obtain an expression for r in terms of the given temperatures.

Solution Consider the reversible cyclic system within broken-line boundary in Fig. E6.18. It exchanges heat with three reservoirs and produces no net work output. The various heat interactions and the temperatures are indicated in the figure.

Applying first law to the cyclic system we obtain

$$Q_h + Q_a - Q_s = 0 \tag{E6.18.1}$$

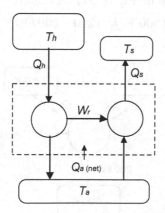

Fig. E6.18 Engine driving a heat -pump

Applying the condition for reversibility of a cycle operating with multiple reservoirs given by Eq. (6.51) we have

$$Q_h/T_h + Q_a/T_a - Q_s/T_s = 0 \qquad \text{(E6.18.2)}$$

Eliminating Q_a between Eqs. (E6.18.1) and (E6.18.2)

$$Q_h(1/T_h - 1/T_a) = Q_s(1/T_s - 1/T_a) \qquad \text{(E6.18.3)}$$

Rearranging Eq. (E6.18.3) we obtain the following expression for r:

$$r = Q_s/Q_h = \frac{T_s(T_h - T_a)}{T_h(T_s - T_a)}$$

Example 6.19 A reversible heat engine receives heat from two reservoirs at 1500 K and 1200 K and it rejects heat to a reservoir at 300 K. The power output of the engine is 700 kW, while the heat rejection rate is 200 kW. Calculate (i) the rate of heat input from the two reservoirs, and (ii) the thermal efficiency of the engine.

Solution Consider the reversible cyclic heat engine shown in Fig. E6.19. The heat interactions between the system and the reservoirs, and the different temperatures are indicated in the figure.

Applying the first law to the cyclic device we have

$$Q_1 + Q_2 - 200 = 700 \qquad \text{(E6.19.1)}$$

Applying the condition for reversibility of a cycle operating with multiple reservoirs given by Eq. (6.51) we obtain

$$Q_1/1500 + Q_2/1200 - 200/300 = 0 \qquad \text{(E6.19.2)}$$

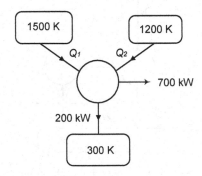

Fig. E6.19 The three-heat-reservoir cycle

Solving Eqs. (E6.19.1) and (E6.19.2) simultaneously for Q_1 and Q_2, the heat interactions are obtained as

$$Q_1 = 500 \text{ kW} \quad \text{and} \quad Q_2 = 400 \text{ kW}$$

The efficiency of the engine is

$$\eta = W/(Q_1 + Q_2) = 700/900 = 78\%$$

Example 6.20 A cyclic device consisting of a reversible engine and a reversible heat pump is used to extract heat from a body of water of mass 2×10^3 kg, initially at 300 K. The device receives heat from a reservoir at 900 K, and rejects heat at a constant rate of 1.6 kW to a reservoir at 330 K. The reservoirs, the device, and the water have no other heat or work interactions with the surroundings. Calculate the time required for the temperature of the water to decrease from 300 K to 275 K. The specific heat capacity of the water is 4.2 kJ kg^{-1} K^{-1}.

Solution The arrangement of the heat engine and the heat pump may be represented schematically as a cyclic device operating with three heat reservoirs as shown in Fig. E6.19. The net work output of the cycle is zero. Applying the first law to the cycle we have

$$\dot{Q}_h + \dot{Q}_c - 1.6 = 0 \tag{E6.20.1}$$

Applying the condition for reversibility, given by Eq. (6.51)

$$\dot{Q}_h/900 + \dot{Q}_c/T_w - 1.6/330 = 0 \tag{E6.20.2}$$

where T_w is the water temperature at time t during the cooling process.

Applying the first law to the water body

$$\dot{Q}_c = -M_w c_w (dT_w/dt) = -4.2 \times 2 \times 10^3 (dT_w/dt) \tag{E6.20.3}$$

Manipulating Eqs. (E6.20.1), (E6.20.2) and (E6.20.3) and substituting the given numerical data we obtain the following relation:

$$-\frac{dT_w}{dt} = \frac{2.764 T_w}{4.2 \times 2 \times 10^3 \times (900 - T_w)}$$

$$\frac{(900 - T_w)}{T_w} dT_w = -0.329 \times 10^{-3} dt \tag{E6.20.4}$$

Integrating Eq. (E6.20.4) from the initial temperature to the final temperature we have

$$900\ln(275/300)-(275-300)=-0.329\times10^{-3}\tau$$

The time to cool the water is given by the above equation as $\tau = 45$ hours.

Example 6.21 A reversible cyclic device consisting of a heat engine and a heat pump is operated between three identical bodies each of mass M and constant thermal capacity C as shown in Fig. E6.21. The initial temperatures of the bodies are 500 K, 400 K and 125 K. The bodies and the cyclic device have no heat or work interactions with the surroundings. Determine (i) the highest temperature to which any one of the bodies can be raised, and (ii) the lowest temperature to which any one of the bodies could be cooled.

Solution Consider the three bodies and the composite cyclic device enclosed within the broken-line boundary shown in Fig. E6.21.

The temperatures of the bodies at some instant, and the infinitesimal heat interactions between the bodies and the cyclic device are indicated in the figure. Applying the first law to the three bodies with the appropriate signs we have

$$\delta Q_1 = -MCdT_1 \qquad\qquad\qquad (E6.21.1)$$

$$\delta Q_2 = MCdT_2 \qquad\qquad\qquad (E6.21.2)$$

$$\delta Q_3 = -MCdT_3 \qquad\qquad\qquad (E6.21.3)$$

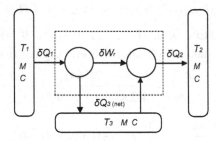

Fig. E6.21 Cyclic device operating with three bodies

Applying the first law to the cyclic device, which has no net work output, we obtain

$$\delta Q_1 - \delta Q_2 + \delta Q_3 = 0 \qquad \text{(E6.21.4)}$$

Manipulating Eqs. (E6.21.1) to (E6.21.4) the following general relation is obtained:

$$dT_1 + dT_2 + dT_3 = 0 \qquad \text{(E6.21.5)}$$

Integrating Eq. (E6.21.5), and applying the given initial conditions for the temperatures of the bodies we have

$$T_1 + T_2 + T_3 = 500 + 400 + 125 = 1025 = S \ (K) \qquad \text{(E6.21.6)}$$

where S is a constant.

Applying the condition for reversibility of the cyclic device given by Eq. (6.51) we obtain

$$\delta Q_1/T_1 - \delta Q_2/T_2 + \delta Q_3/T_3 = 0 \qquad \text{(E6.21.7)}$$

Substituting from Eqs. (E6.21.1) to (E6.21.3) in Eq. (E6.21.7) we have

$$dT_1/T_1 + dT_2/T_2 + dT_3/T_3 = 0 \qquad \text{(E6.21.8)}$$

Integrating Eq. (E6.21.8) and applying the initial conditions for the temperatures

$$T_1 T_2 T_3 = 500 \times 400 \times 125 = 25 \times 10^6 = P \ (K^3) \qquad \text{(E6.21.9)}$$

where P is a constant.

The equations governing the temperatures of the three bodies at any instant are (E6.21.6) and (E6.21.9). We notice that these equations are symmetrical with respect to the three temperatures. The procedure to obtain the maximum temperature can be formulated as an optimization problem as follows. Eliminating T_1 between Eqs. (E6.21.6) and (E6.21.9), and rearranging the resulting equation

$$ST_2 T_3 - (T_2^2 T_3 + T_3^2 T_2) - P = 0 \qquad \text{(E6.21.10)}$$

From Eq. (E6.21.6) we have

$$T_1 = S - T_2 - T_3 \qquad \text{(E6.21.11)}$$

The optimization problem can be stated as follows: Find the maximum value of T_1 in Eq. (E6.21.11), subject to the constraint given by Eq. (E6.21.10). Using the Lagrange multiplier method, the function to be maximized may be written as

$$F = S - T_2 - T_3 + \lambda[ST_2T_3 - (T_2^2T_3 + T_3^2T_2) - P]$$

where λ is the Lagrange multiplier. The conditions at the optimum point are:

$$\partial F/\partial T_2 = 0, \quad \partial F/\partial T_3 = 0 \quad \text{and} \quad \partial F/\partial \lambda = 0$$

Applying the first two conditions and solving the resulting equations simultaneously it is easy to show that $T_2 = T_3$. Applying the condition $\partial F/\partial \lambda = 0$ and substituting $T_2 = T_3$ in the resulting equation we have

$$S - 2T_2 = P/T_2^2$$

Substituting numerical values

$$1025 - 2T_2 = 25 \times 10^6/T_2^2 \tag{E6.21.12}$$

The two roots of the above equation may be obtained by trial-and-error or graphically as $T_2 = 200$ K or 451 K.

Hence the three temperatures at the first optimum point are $T_2 = 200$ K, $T_3 = 200$ K and $T_1 = (1025 - 2 \times 200) = 625$ K. This gives the maximum temperature that could be attained by one of the bodies as 625 K.

At the second optimum point the three temperatures are $T_2 = 451$ K, $T_3 = 451$ K and $T_1 = (1025 - 2 \times 451) = 123$ K. This gives the minimum temperature that one of the bodies could reach as 123 K.

The following practical arrangement may be used to achieve the maximum temperature. Let A, B and C denote the bodies initially at 500 K, 400 K and 125 K respectively. Operate a reversible heat engine between B and C until their temperatures become equal. The entire work output is supplied to a reversible heat pump operating between B and A. The temperature of A will increase while the temperature of B will decrease. Now operate a heat engine between C and B until their temperatures again become equal. The work output is used to operate a heat pump between B and A. The above steps are repeated until the temperatures of B and C become equal in a limiting manner. The temperature of A is then a maximum. A similar procedure could be used to achieve the lowest temperature in C.

Problems

P6.1 Figures 6.2 and 6.4 show schematically the simplified forms of a vapor power cycle, and a compression refrigeration cycle respectively. (i) Identify, using the appropriate terms, the processes that may cause irreversibility in these systems. (ii) Briefly describe these processes. (iii) Suggest possible ideal processes to replace these irreversible processes.

P6.2 A fixed mass of an ideal gas is heated from temperatures T_1 to T_2 by transferring heat from a thermal reservoir at T_2 under the following conditions: (i) constant volume, and (ii) constant pressure. Show that these processes are irreversible. Assume that the temperature of the gas in both cases is uniform. [*Answer*: Reverse the state of the gas using reversible processes]

P6.3 A well-insulated cylinder is divided into two halves by a *thin* piston *of negligible mass*. Initially, the left half of the cylinder contains air at pressure of 2 bar and temperature 25°C. The right half is evacuated and the piston is held in position with a locking pin. The pin is now removed, and the air expands and attains a final equilibrium state. (a) Calculate (i) the work done by the air, and (ii) final pressure and temperature of the air. (b) Describe a reversible process to return the air to its initial state, and hence show that the original process is irreversible. [*Answers*: (a) (i) W = 0, (ii) 1 bar, 25°C]

P6.4 Prove the following statements:
(i) The COP of an irreversible refrigerator is less than that of a reversible refrigerator operating between the same heat reservoirs.
(ii) The COP of an irreversible heat pump is less than that of a reversible heat pump operating between the same heat reservoirs.
(iii) Show that curves representing two different reversible adiabatic processes on a *P-V* diagram cannot intersect. [*Answers*: (i), (ii) and (iii) assume the converse to be true, show that the K-P statement is violated]

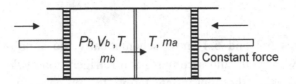

Fig. P6.5 Constant pressure heating of an ideal gas

P6.5 An arrangement for heating an ideal gas A in a reversible constant pressure process by supplying work to the ideal gas B trapped in a piston-cylinder set-up is shown in Fig. P6.5. The wall separating the gas A from gas B is rigid and made of a highly conducting material. Show that the P-V relation for the process in gas B is

$$P_bV_b^{\lambda} = C$$

where $\lambda = (m_a c_{pa} + m_b c_{vb} + m_b R_b)/(m_a c_{pa} + m_b c_{vb})$

P6.6 A series of Carnot heat engines are arranged as in Fig. 6.14, with each intermediate engine absorbing the same amount of heat as that rejected by the preceding engine. The temperature of any one heat reservoir is lower then the one before it by ΔT. (i) Show that the work output all the engines are equal. (ii) As the number of engines is increased, and ΔT is made smaller discuss how the temperature of absolute zero is approached. [*Answer*: (i) $W = Q_1 \Delta T/T_1$]

P6.7 A Carnot heat pump extracts heat from a cold reservoir at T_c and rejects heat to a reservoir at T_h. Determine which of the following changes to the reservoir temperatures will be more effective in increasing the COP of the heat pump: (i) decrease T_h to $(T_h - \Delta T)$ keeping T_c constant, or (ii) increase T_c to $(T_c + \Delta T)$ keeping T_h constant. [see worked example 6.13]

P6.8 A Carnot refrigerator is operated between two identical bodies of mass M and specific heat capacity C, whose initial temperatures are both T_0. Obtain an expression for the total work input to the refrigerator, when one of the bodies has been cooled to a temperature of λT_0 ($\lambda < 1$).
[*Answer*: $W_{in} = MCT_o(1 - \lambda)^2/\lambda$]

P6.9 An inventor claims that he has designed an absorption refrigeration system which extracts heat from a cold space at –10°C at the rate of 100 kW, using the energy from a waste heat source at 110°C. The system rejects heat to the atmosphere at 25°C at the rate of 250 kW. Discuss whether the claim is valid.

[*Answer*: valid claim]

P6.10 In the idealized model of a combined-cycle power plant shown schematically in Fig. P6.10, the reversible engine E1 receives heat from a reservoir at 1250 K, and rejects heat at 500 K to an intermediate heat exchanger, which acts as the heat source for the reversible engine E2. The heat rejection from E2 occurs at 300 K. The heat transfer at the intermediate heat exchanger between the two cyclic engines requires a temperature difference of 20 K.

(i) Calculate the overall thermal efficiency of this combined-cycle.
(ii) If a single reversible engine is operated between the reservoirs at 1250 K and 300 K, what is its thermal efficiency? (iii) Explain the reasons for the difference in the efficiencies obtained in (i) and (ii). (iv) Calculate the efficiency of the combined cycle if the temperature difference in the heat exchanger is negligibly small.

[*Answers*: (i) 75%, (ii) 76%, (iv) 76%]

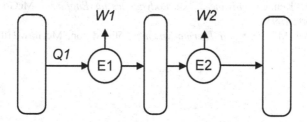

Fig. P6.10 Combined-cycle power plant model

P6.11 An idealized absorption refrigeration system consists of a reversible cyclic device operating with two heat reservoirs, and a cooling load of finite heat capacity. The device receives heat at the rate of 50 kW from a high temperature reservoir at 100°C, and rejects heat to the ambient-reservoir at 30°C. The cooling load of the refrigerator is a tank of water of mass 1800 kg, which is initially at a temperature of 30°C. The water is cooled to a final temperature of 10°C. Calculate (i) the time required to cool the water, (ii) the total heat supplied by the high temperature reservoir, and (iii) the average COP of the refrigerator during the operation.

[*Answers*: (i) 9.1 hours, (ii) 1638 MJ, (iii) 0.092]

References

1. Bejan, Adrian, *Advanced Engineering Thermodynamics*, John Wiley & Sons, New York, 1988.
2. Cravalho, E.G. and J.L. Smith, Jr., *Engineering Thermodynamics*, Pitman, Boston, MA, 1981.
3. Jones, J.B. and G.A. Hawkins, *Engineering Thermodynamics*, John Wiley & Sons, Inc., New York, 1986.
4. Kestin, Joseph, *A Course in Thermodynamics*, Blaisdell Publishing Company, Waltham, Mass., 1966.
5. Reynolds, William C. and Henry C. Perkins, *Engineering Thermodynamics*, 2nd edition, McGraw-Hill, Inc., New York, 1977.
6. Rogers, G.F.C. and Mayhew, Y.R., *Thermodynamic and Transport Properties of Fluids*, 5th edition, Blackwell, Oxford, U.K. 1998.
7. Tester, Jefferson W. and Michael Modell, *Thermodynamics and Its Applications*, 3rd edition, Prentice Hall PTR, New Jersey, 1996.
8. Van Wylen, Gordon J. and Richard E. Sonntag, *Fundamentals of Classical Thermodynamics*, 3rd edition, John Wiley & Sons, Inc., New York, 1985.
9. Wark, Jr., Kenneth, *Advanced Thermodynamics for Engineers*, McGraw-Hill, Inc., New York, 1995.
10. Zemansky, M., *Heat and Thermodynamics*, 5th edition, McGraw-Hill, Inc., New York, 1968.

Chapter 7

Entropy

In the preceding chapter we obtained a number of important results by applying the second law to cyclic processes associated with heat engines and reversed heat engines, operating with one and two thermal reservoirs. The concept of a reversible process, though an idealization of real processes, provided a means to find the upper limit of the efficiency of heat engines operating with two heat reservoirs. The maximum efficiency, called the Carnot-cycle efficiency, is a unique function of the two reservoir temperatures. This lead to the development of the thermodynamic temperature scale. Finally, we generalized these results for cyclic systems experiencing heat interactions with any number of thermal reservoirs.

In this chapter we shall extend the application of the second law to closed systems undergoing any process, reversible or otherwise. The main outcome of this effort is the emergence of a new system property called *entropy* which has a broad significance and very wide applications.

7.1 The Clausius Inequality and Entropy

To apply the second law to a system that executes a cycle exchanging heat over a range of temperatures we consider the arrangement shown in Fig. 6.16. The cycle executed by the system S could be internally reversible or irreversible but the heat interactions between the various thermal reservoirs and the system are *reversible*. This is the main difference between the present development and the analysis in Sec. 6.5.5. For external reversibility the temperature T_i of the secondary reservoir must be equal to the temperature T of the system where heat is received during the process. Since we expect the temperature T to change during the process, we need to have a series of secondary reservoirs with the associated heat engines to achieve reversible heat transfer. Under these new conditions we can rewrite Eq. (6.49) as:

$$\sum \frac{\Delta Q_{si}}{T_i} \le 0 \qquad (7.1)$$

where ΔQ_{si} is the heat flow from the reservoir whose temperature T_i is equal to the temperature T of the system at the location where heat is received.

Considering an infinitesimal heat transfer δQ at temperature T, we can express Eq. (7.1) in the cyclic integral form

$$\oint \frac{\delta Q}{T} \le 0 \qquad (7.2)$$

Equation (7.2), which now involves only the heat input to the system and the temperature of the system at the location where heat is received, is called the *Clausisus inequality*.

As we discussed in Sec. 6.5.3 the equality signs in Eqs. (7.1) and (7.2) apply when the cycle executed by the system is *internally reversible* and the inequality is for *irreversible* cycles. For internally reversible processes the temperature T of the system is uniform because the process occurs quasi-statically. However, for internally irreversible processes there could be temperature gradients within the system because the process is not necessarily quasi-static. For this situation, the term T in Eq. (7.2) is the temperature at the location on the boundary where the heat input δQ enters the system as indicated in Fig. 6.16.

7.1.1 *Entropy – A thermodynamic property*

Consider a closed system executing a *reversible* cycle 1-A-2-B-1 shown in Fig. 7.1. Applying Eq. (7.2) with the equality sign we have

$$\int_{1A}^{2} \frac{\delta Q}{T} + \int_{2B}^{1} \frac{\delta Q}{T} = 0 \qquad (7.3)$$

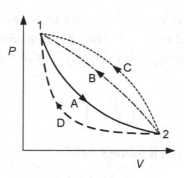

Fig. 7.1 *P-V* diagram of process

We now envisage an alternative reversible process, 2-C-1 that returns the system from states 2 to 1 thus completing the cycle. Applying Eq. (7.2) to the cycle 1-A-2-C-1 we obtain

$$\int_{1A}^{2}\frac{\delta Q}{T}+\int_{2C}^{1}\frac{\delta Q}{T}=0 \qquad (7.4)$$

Subtracting Eq. (7.3) from Eq. (7.4),

$$\int_{2B}^{1}\frac{\delta Q}{T}=\int_{2C}^{1}\frac{\delta Q}{T} \qquad (7.5)$$

Since the process C was selected arbitrarily, it follow from Eq. (7.5) that the quantity, $(\delta Q/T)$ is *path-independent*. It is, therefore, a property of the system, which depends only on state 1 and state 2. Hence we have

$$\int_{2}^{1}\frac{\delta Q}{T}=S_{1}-S_{2} \qquad (7.6)$$

This new property, denoted by S in Eq. (7.6), is called the *entropy* the system. It is an *extensive* property that depends on the mass of the system. We recall that the internal energy of a system arose as a consequence of the first law. In a similar manner, the application of the second law has predicted the existence of this new property, but offers no additional information on its physical meaning. In chapter 15 we shall discuss the microscopic interpretation of entropy.

We now select an *irreversible* process D, shown in Fig. 7.1, to return the system from state 2 to state 1. This choice makes the cycle 1-A-2-D-1 *internally irreversible*. Applying Eq. (7.2) with the inequality sign we have

$$\int_{1A}^{2} \frac{\delta Q}{T} + \int_{2D}^{1} \frac{\delta Q}{T} < 0 \qquad (7.7)$$

From Eqs. (7.4) and (7.7) it follows that

$$\int_{2C}^{1} \frac{\delta Q}{T} > \int_{2D}^{1} \frac{\delta Q}{T} \qquad (7.8)$$

From Eqs. (7.8) and (7.6) we have

$$S_1 - S_2 > \int_{2D}^{1} \frac{\delta Q}{T} \qquad (7.9)$$

For an infinitesimal change of state we can write the differential form of the above relations as

$$dS = \left(\frac{\delta Q}{T} \right)_{rev} \qquad (7.10)$$

$$dS > \left(\frac{\delta Q}{T} \right)_{Irr} \qquad (7.11)$$

It is noteworthy that in Eq. (7.10), T is the system temperature, which is uniform for *internally reversible* processes while in Eq. (7.11), T is the temperature at the location on the boundary where heat enters the system. The above equations constitute the mathematical formulation of the second law.

7.1.2 *The temperature-entropy diagram*

The *P-V* diagram is a useful graphical aid for representing quasi-static processes because the area under the curve is proportional to the work done during the process. In a similar manner, we can use a temperature-entropy diagram (*T-S diagram*) of a *reversible* process to obtain the heat transfer. This follows from Eq. (7.10), which, for a reversible process, may be written as

$$\delta Q_{rev} = TdS \tag{7.12}$$

Integration of Eq. (7.12) gives

$$Q = \int_i^f TdS \tag{7.13}$$

where i and f represent the initial and final states of the system.

A temperature-entropy diagram for a cyclic process consisting of three *reversible* processes is shown in Fig. 7.2. From Eq. (7.13) it follows that the area under the curve 1-2 is the heat *input* to the system during the *reversible* process 1-2. For the process 2-3 the area is traced in the negative S-direction when the system moves from the initial state 2 to the final state 3. We interpret this as a negative area which corresponds to a heat *output* for the process 2-3. For the process 3-1, $dS = 0$. It follows from Eq. (7.12) that $\delta Q_{rev} = 0$, and therefore the process 3-1 is reversible and adiabatic. Such a process is called an *isentropic* process because the entropy remains constant during the process and it can be represented by a straight line parallel to the T-axis as seen in Fig. 7.2. For the cyclic process 1-2-3-1 the area enclosed by the three curves is the net heat *input* to the system during the process.

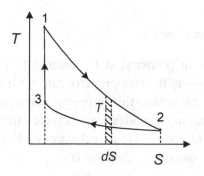

Fig. 7.2 Temperature-entropy (T-S) diagram

7.1.3 *The 'T-ds' equation*

We shall now derive a useful property relation that involves the zeroth law, the first law and the second law for a *simple* thermodynamic system.

Consider an infinitesimal process in a closed system. Applying the first law we have

$$\delta Q = dU + \delta W \tag{7.14}$$

Since the process is reversible

$$\delta W = PdV \tag{7.15}$$

Applying the second law [Eq. (7.10)] to the *reversible* process

$$dS = \left(\frac{\delta Q}{T}\right)_{rev} \tag{7.16}$$

Manipulating Eqs. (7.14), (7.15) and (7.16) we obtain

$$TdS = dU + PdV \tag{7.17}$$

Note that Eq. (7.17) is a relationship between thermodynamic properties of the system and is therefore independent of the process. Moreover, it is independent of the substance constituting the system. This important relation, sometimes called the '*T-dS equation*', involves the zeroth law through the concept of temperature, the first law through the internal energy function, and the second law through the entropy and thermodynamic temperature. A series of important thermodynamics property relations are developed in chapter 14 using the '*T-ds equation*'.

7.1.4 *Entropy of an ideal gas*

Since the entropy is a property, it is possible to express the entropy as $S = S(V, T)$, following the two-property rule for a pure substance. We shall now proceed to obtain this property relation for an ideal gas. Consider a fixed mass m of an ideal gas subjected to a *reversible* process from an initial state (P_o, V_o, T_o) to a final state (P, V, T).

The equation of state of the ideal gas is

$$PV = mRT \tag{7.18}$$

The internal energy of the ideal gas may be expressed as

$$U = mc_v T + U_{ref} \tag{7.19}$$

where U_{ref} is the internal energy of the reference state.

Differentiating Eq. (7.19)

$$dU = mc_v dT \tag{7.20}$$

Substituting in the 'T-dS' equation [Eq. (7.17)] from Eqs. (7.18) and (7.20) we obtain

$$TdS = mc_v dT + mRTdV/V \tag{7.21}$$

$$dS = mc_v dT/T + mRdV/V \tag{7.22}$$

Integrating Eq. (7.22) from the initial state to the final state we have

$$S = S_o + mc_v \ln(T/T_o) + mR \ln(V/V_o) \tag{7.23}$$

Substituting for V in Eq. (7.23) from Eq. (7.18) we obtain a property relation of the type, $S = S(P, T)$.

$$S = S_o + mc_v \ln(T/T_o) + mR \ln(TP_o/T_o P) \tag{7.24}$$

$$S = S_o + m(c_v + R)\ln(T/T_o) - mR \ln(P/P_o)$$

$$S = S_o + mc_p \ln(T/T_o) - mR \ln(P/P_o) \tag{7.25}$$

We note that Eqs. (7.23) and (7.25) are relationships between properties of an ideal gas and therefore applicable to any *equilibrium* state. Moreover, the initial state denoted by o can be regarded as the reference state in a tabulation of data for the entropy.

The T-S diagram for an ideal gas may be drawn by making use of the analytical expressions in Eqs. (7.23) and (7.25). Since these involve three variables we need to draw families of curves keeping one of the variables fixed. For example, for an ideal gas, the constant-pressure lines on the T-S diagram are logarithmic curves according to Eq. (7.25). The relevant expressions for these variations are derived in worked example 7.4.

7.1.5 *Entropy of a pure thermal system*

We shall now obtain a general expression for the entropy change of a pure thermal system which is an idealized model applicable to solids and incompressible fluids. For such systems the specific heat capacity is a constant. Consider an infinitesimal *reversible* heat transfer, δQ to the system. Applying the second law we obtain

$$\delta Q = TdS \qquad (7.26)$$

Applying the first law with zero work transfer

$$\delta Q = dU = MCdT \qquad (7.27)$$

where M and C are the mass and specific heat capacity of the pure thermal system.

From Eqs. (7.26) and (7.27) we obtain

$$TdS = MCdT \qquad (7.28)$$

Integrating Eq. (7.28) from the reference state to the general state

$$S = S_{ref} + MC \ln\!\left(T/T_{ref}\right) \qquad (7.29)$$

where the subscript '*ref*' denotes the properties at the reference state.

7.1.6 *Entropy of a pure substance*

Shown in Fig. 7.3 is the *T-S* diagram for a pure substance undergoing a reversible constant pressure heating process. At A, the fluid is a sub-cooled liquid at a pressure P. As heat is supplied to the fluid at constant pressure, the process on the *T-S* diagram follows the curve A-B up to B where evaporation just begins. The section B-C of the curve corresponds to the evaporation process where S increases while T remains constant

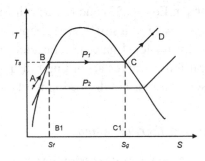

Fig. 7.3 *T-S* diagram for a pure substance

because of the constant pressure. At C evaporation is complete and along the section C-D, the vapor becomes superheated.

For a pure substance like steam, it is not possible to obtain simple analytical expressions for the entropy as we did for an ideal gas. In this case the data has to be obtained from tabulations as for the internal energy and enthalpy. Referring to the data tables in [6], we notice that the entropy per unit mass of a saturated liquid, s_f is chosen to be zero at the triple point of water. The saturated vapor entropy is tabulated under s_g. The change in entropy when a unit mass undergoes phase change from liquid to vapor is given by $(s_f - s_f) = s_{fg}$. The specific entropy of a wet vapor of quality x may be expressed as

$$s = x \, s_g + (1 - x)s_f \qquad (7.30)$$

For superheated vapor, the entropy is tabulated in [6] for different pressures and temperatures. In general, to extract entropy data from the tables we follow the same procedure that was used earlier to obtain the internal energy and the enthalpy of a pure substance.

It follows from Eq. (7.13) that the total heat supplied during the evaporation process B-C in Fig. 7.3 is the area of the rectangle BCC_1B_1. Since the temperature, T_s is constant during evaporation, we obtain the following relation by applying Eq. (7.13) to a unit mass of steam:

$$s_g - s_f = q_{fg}/T_s = (h_g - h_f)/T_s \qquad (7.31)$$

where $q_{fg} = h_g - h_f$, is the heat supplied per unit mass during the evaporation process.

7.2 Principle of Increase of Entropy

The principal of increase of entropy can be deduced by combining Eqs. (7.10) and (7.11) to the form

$$dS \geq \left(\frac{\delta Q}{T}\right) \qquad (7.32)$$

Consider the system A that interacts with a heat reservoir R, and a mechanical energy reservoir MR as shown in Fig. 7.4. MR could be a simple pulley-weight arrangement connected to A with a frictionless

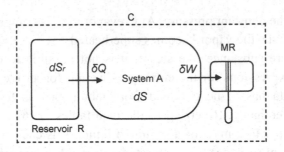

Fig. 7.4 Principle of increase of entropy

shaft. During an infinitesimal change in the state of A its entropy increases by dS while the corresponding change in entropy of the reservoir is dS_r. The heat and work interactions between the system and its surroundings, constituted by R and MR, during the process are δQ and δW respectively. Consider an imaginary boundary C, indicated by the broken-lines in Fig. 7.4, enclosing A, R and MR such that nothing outside of it affects the inside significantly. We call the region within C an *isolated* system for which the heat interaction $\delta Q_c = 0$. Applying Eq. (7.32) to the isolated system C we have

$$(dS)_c \geq 0 \qquad (7.33)$$

Being an extensive property, we can add the entropy changes of the different sub-regions that constitute C to write Eq. (7.33) in the form

$$(dS)_c = dS + dS_r + dS_{mr} \geq 0 \qquad (7.34)$$

Since the entropy change, dS_{mr} of the mechanical energy reservoir MR is zero, Eq. (7.34) becomes

$$dS + dS_r \geq 0 \qquad (7.35)$$

We conclude from the above equation that if all the processes occurring within an *isolated* system are *reversible* the entropy of the system remains constant, otherwise the entropy must increase due the *irreversible* processes. This statement, whose mathematical form is Eq. (7.35), is commonly called the *principle of increase of entropy*. Although the principle of increase of entropy stipulates that the entropy of an *isolated* system can only increase or remain constant, the entropy of some of the individual sub-regions that constitute the isolated system may decrease.

Now the heat transfer to the thermal reservoir R is reversible because its temperature is constant. Therefore by applying Eq. (7.32), with the equality sign we obtain the entropy change of R as

$$dS_r = -\delta Q / T_r \qquad (7.36)$$

Substituting from Eq. (7.36) in (Eq. 7.35) we have

$$dS - \delta Q / T_r \geq 0 \qquad (7.37)$$

Equation (7.37) implies that the entropy of the system A must increase to compensate for the entropy decrease of the reservoir R so that the overall entropy of the isolated system C increases or remains constant. Otherwise the process would violate the second law.

We shall generalize the increase of entropy principle, and make it quantitative by introducing a variable called the *entropy production*. Consider any system and its interacting surroundings that are enclosed in an imaginary boundary and therefore *isolated* (Fig 7.4). The entropy production is given by

$$\delta\sigma = dS_{system} + dS_{surroundings} \geq 0 \qquad (7.38)$$

The equality sign in Eq. (7.30) applies when all the processes within the *isolated* system are reversible and the entropy production, $\delta\sigma$ is therefore zero. On the other hand positive values of $\delta\sigma$ indicate the occurrence of irreversible processes within the isolated system and its magnitude provides an indirect measure of the impact of the irreversibilities on the work output. Note that the entropy production σ is not a property of the system like the entropy S. Therefore σ depends on the type of irreversible process experienced by the system.

7.2.1 *Storage, production and transfer of entropy*

When the principle of increase of entropy, in the form of Eq. (7.38), is applied to real systems we need to distinguish between three different quantities. These are called *stored* entropy, entropy *transfer*, and entropy *production*. We shall illustrate the difference between the above terms using a practical situation.

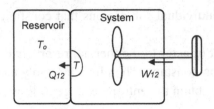

Fig. 7.5 Entropy changes in system and reservoir

Consider as a system the gas contained in the rigid vessel shown in Fig. 7.5 where a rotating paddle wheel supplies work to the system. The vessel is in thermal communication with a single heat reservoir at temperature T_o. The system undergoes a process where the paddle wheel is rotated for a fixed duration of time, and then stopped. Denoting the initial and final equilibrium states by 1 and 2 we apply the first law to the process 1-2 executed by the system to obtain

$$U_2 - U_1 = W_{12} - Q_{12} \qquad (7.39)$$

where $(U_2 - U_1)$ is the *increase* in internal energy, Q_{12} is the heat *output* from the system to the reservoir, and W_{12} is the work *input* to the system through the paddle wheel during the process.

The application of Eq. (7.32) in the integrated-form to the system gives

$$(S_2 - S_1) \geq -\int_1^2 \frac{\delta Q}{T} \qquad (7.40)$$

The negative sign in Eq. (7.40) signifies a heat *output* from the system as indicated in Fig. 7.5.

We convert Eq. (7.40) to an equality by introducing the entropy production term σ. This gives

$$(S_2 - S_1) = \sigma - \int_1^2 \frac{\delta Q}{T} \qquad (7.41)$$

In Eq. (7.41), T is the temperature of the location on the boundary at which heat is transferred from the system to the reservoir.

In order to evaluate the integral in the above equation we need to know how the temperature T varies during the process. Because the irreversible work input by the paddle wheel, shown in Fig 7.5, causes considerable turbulence, it is not possible to determine the variation of T in a straightforward manner. Moreover, T may not be spatially uniform within the system. For the purpose of the present discussion we resolve this difficulty by locating the system boundary in close contact with the reservoir so that T is equal to T_o, the constant reservoir temperature. It is clear that with this arrangement of the system boundary any irreversibility due to heat transfer between the system and the reservoir is now attributed to the system. Hence upon integration, Eq. (7.41) becomes

$$(S_2 - S_1) = \sigma - \frac{Q_{12}}{T_o} \qquad (7.42)$$

In some respects Eq. (7.39), which is an energy balance, is analogous to Eq. (7.42), which could be thought of as an *entropy balance equation*. In both equations the left hand side represents an increase in a 'stored' property of the system. The term Q_{12} is the heat transfer which is a boundary interaction. Therefore we interpret the term (Q_{12}/T_o) as the *entropy transfer* out of the system to the reservoir due to heat transfer at the boundary. Although the terms W_{12} and σ in the two equations are not directly related we are aware that the irreversible frictional dissipation of the work input W_{12} contributes to the entropy production in the system.

It is important to note that σ, is either positive or zero but the *entropy change* or *storage*, $(S_2 - S_1)$ in Eq. (7.42) could be positive or negative depending on the magnitude of the entropy transfer, (Q_{12}/T_o) compared to the entropy production, σ.

The entropy balance equation for the reservoir is given by

$$(S_{r2} - S_{r1}) = \sigma_r + \frac{Q_{12}}{T_o} \qquad (7.43)$$

The heat transfer process in the reservoir is reversible because of its uniform temperature, and therefore the entropy production in the reservoir, $\sigma_r = 0$. Consequently, the increase in the stored entropy of the reservoir is entirely due to the entropy transfer associated with heat transfer from the system across the boundary.

The entropy balance equation for the composite system consisting of the gas and the reservoir is obtained by adding Eqs. (7.42) and (7.43). This gives

$$(S_2 - S_1) + (S_{r2} - S_{r1}) = \sigma \qquad (7.44)$$

We could have written Eq. (7.44) directly by applying the principle of increase of entropy to the composite system which is an isolated system. In summary, the entropy balance equation for a closed system may be expressed in the general form:

$$Entropy_{storage} = Entropy_{production} + Entropy_{transfer} \qquad (7.45)$$

7.2.2 Entropy transfer in a heat engine

Consider the cyclic heat engine operating between two heat reservoirs as shown in Fig. 7.6(a) where the various heat and work interactions and temperatures are indicated. The engine experiences external irreversibilities due heat transfer across finite temperature differences $(T_h - T_{h1})$ and $(T_{c1} - T_c)$ between the cyclic device and the reservoirs. In addition, there are internal mechanical and thermal irreversibilities causing entropy production within the cyclic device.

The T-S diagram for the irreversible heat engine cycle is shown in Fig. 7.6(b). The compression and expansion processes of this cycle are *irreversible adiabatic* processes. However, we assume that during the heat interactions with the reservoirs the temperature of the working fluid of the cyclic device remains constant.

Applying the equation of entropy balance, expressed by Eq. (7.45), to the cyclic device we obtain

$$S_{cycle} = \sigma + \left(\frac{Q_h}{T_h} - \frac{Q_c}{T_c} \right) \qquad (7.46)$$

It should be noted that in the above equation the entropy production due to all irreversibilities, both external and internal, are now included in σ because we use the reservoir temperatures to evaluate the entropy transfers due to heat transfer, and not the temperatures at the boundary of

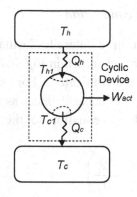

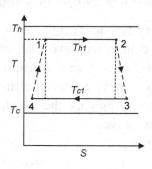

Fig. 7.6(a) Heat engine cycle   Fig. 7.6(b) *T-S* diagram

the cyclic device. Since the heat engine operates in a cycle its entropy change or storage, $S_{cycle} = 0$. Therefore Eq. (7.46) becomes

$$\sigma = -\left(\frac{Q_h}{T_h} - \frac{Q_c}{T_c}\right) \qquad (7.47)$$

Applying the first law to the heat engine, its work output is obtained as

$$W_{act} = Q_h - Q_c \qquad (7.48)$$

Substituting for Q_c from Eq. (7.47) in Eq. (7.48)

$$W_{act} = Q_h - T_c\left(\sigma + \frac{Q_h}{T_h}\right) \qquad (7.49)$$

Now the work output of a *reversible* engine operating between the same heat reservoirs and receiving the same heat input Q_h from the hot reservoir is

$$W_{rev} = Q_h\left(1 - \frac{T_c}{T_h}\right) \qquad (7.50)$$

Subtracting Eq. (7.49) from Eq. (7.50) we have

$$W_{rev} - W_{act} = T_c\sigma \qquad (7.51)$$

We conclude from Eq. (7.51) that the entropy production σ is a measure of the potential loss in work output due to internal and external irreversibilities of the heat engine.

7.2.3 Entropy transfer in steady heat conduction

Consider the *steady* heat conduction in a laterally insulated bar in thermal contact with two heat reservoirs, as shown in Fig. 7.7, where the heat interactions and temperatures are indicated. For steady heat conduction the temperature distribution in the bar is linear as indicated in the figure. Applying the entropy balance equation to the bar as the system we have

$$\dot{S}_{bar} = \dot{\sigma} + \left(\frac{\dot{Q}_h}{T_h} - \frac{\dot{Q}_c}{T_c} \right) \qquad (7.52)$$

where the 'dot' over a quantity stands for differentiation with respect to time.

Since heat conduction is a steady process, the heat flows and the entropy changes are expressed as rates in Eq. (7.52). The entropy storage in the bar is zero because the properties of the bar remain constant under steady conditions. Therefore

$$\dot{S}_{bar} = 0 \qquad (7.53)$$

Applying the first law to the bar we have

$$\dot{Q}_h = \dot{Q}_c = \dot{Q} \qquad (7.54)$$

where $\dot{Q}$ is the constant heat flow rate though the bar. Substituting from Eqs. (7.53) and (7.54) in Eq. (7.52) we obtain

$$\dot{\sigma} = \left(\frac{\dot{Q}}{T_c} - \frac{\dot{Q}}{T_h} \right) > 0 \qquad (7.55)$$

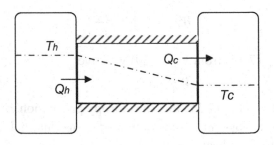

Fig. 7.7 Steady heat conduction in a bar

We could have obtained Eq. (7.55) directly by applying the principle of increase of entropy [Eq. (7.38)] to the composite system consisting of the two reservoirs and the bar which is an isolated system. Then the terms $(\dot{Q}/T_c)$ and $(\dot{Q}/T_h)$ are the rate of increase of entropy in the cold reservoir and the rate of decrease of entropy of the hot reservoir respectively. The entropy production given by Eq. (7.55) is due to the *internal thermal irreversibility* in the bar which results from the heat transfer across a finite temperature difference. If the same heat transfer was carried out reversibly by operating a Carnot heat engine between the two reservoirs, then the rate of work output of the engine is

$$\dot{W}_{rev} = \dot{Q}\left(1 - \frac{T_c}{T_h}\right) \tag{7.56}$$

Manipulating Eqs. (7.55) and (7.56) we have

$$\dot{W}_{rev} = T_c \dot{\sigma} \tag{7.57}$$

The above relation is similar to Eq. (7.51) for the work output of an irreversible engine, except that the potential work available is entirely lost in the case of steady heat conduction.

7.3 Limitations Imposed on Work Output by the Second Law

In this section we shall derive expressions for the upper limits of the heat transfer and work done in a given process, which are important consequences of the second law. Denoting the initial and final states of the closed system by 1 and 2 we apply the first law to obtain

$$Q_{12} = U_2 - U_1 + W_{12} \tag{7.58}$$

Application of the second law to an infinitesimal change of state gives

$$\delta Q \leq TdS \tag{7.59}$$

Integrating Eq. (7.59) from the initial to the final state we have

$$Q_{12} \leq \int_1^2 TdS \tag{7.60}$$

The heat transfer Q_{12} is a maximum for a reversible process for which the equality sign in Eq. (7.60) applies. In order to evaluate the integral on the right hand side of the Eq. (7.60) we need to know the process path. For a reversible process, the path is usually well-defined and the heat transfer is also the area of the T-S diagram of the process. However, for an irreversible process, for which the inequality sign applies, Eq. (7.60) establishes only the upper limit of the heat transfer. Furthermore, the process path required to carry out the integration on the right hand side of Eq. (7.60) is often difficult to determine for an irreversible process.

Eliminating Q_{12} between Eqs. (7.58) and (7.60) we obtain

$$U_2 - U_1 + W_{12} \leq \int_1^2 TdS \qquad (7.61)$$

Hence
$$W_{12} \leq \int_1^2 TdS - (U_2 - U_1) \qquad (7.62)$$

The equality sign in Eq. (7.62) applies for *reversible* processes for which the process path is well defined and the equation establishes the maximum work output of the process. However, for *irreversible* processes the evaluation of the integral on the right hand side of Eq. (7.62) could pose a challenge because the process-path is often not readily available.

7.3.1 *Helmholtz and Gibbs free energy*

We shall now derive expressions for the maximum work output of several special processes that are of considerable practical significance. First consider the application of Eq. (7.62) to an *isothermal* process. The integral in the equation is easily evaluated because the temperature is constant. This gives

$$W_{12} \leq T(S_2 - S_1) - (U_2 - U_1) \qquad (7.63)$$

$$W_{12} \leq -[(U_2 - TS_2) - (U_1 - TS_1)]$$

Hence
$$W_{iso,max} = -[(U_2 - TS_2) - (U_1 - TS_1)] \qquad (7.64)$$

It is seen from Eq. (7.64) that the maximum work output for an *isothermal* process is the change of the function, $(U - TS)$ from state 1 to state 2. This state function is called the *Helmholtz free energy* and usually denoted by F. It is a property of the system because U, T and S are properties. The maximum work output of an isothermal process may be written in terms of the Helmholtz free energy as:

$$W_{iso,max} = -(F_2 - F_1) \qquad (7.65)$$

The physical interpretation of Eq. (7.65) is that the maximum work output of the system is the decrease in the property F. Although the system possesses internal energy of magnitude U, a fraction TS of this internal energy is not free to be used for the production of useful work. Therefore for an *isothermal* process at constant volume the fraction of the internal energy that is *free* for work production is the *Helmholtz free energy* F. It is important to note that Eq. (7.64) for the maximum work output is applicable to situations involving different forms of internal energy and work modes. For example, we could envisage an application like a fuel cell where electrical work is produced isothermally at the expense of chemical internal energy of the system.

We shall now obtain the maximum work output of an *isothermal* process that is carried out under *constant pressure*. In other words, during the isothermal process, the system expands quasi-statically against a *constant* external pressure P. Of the maximum work output of the system a fraction is now used to overcome the external force due to the pressure and therefore not available for *useful* application. Separating the work output into two parts we have

$$W_{iso,max} = W_{useful,max} + P(V_2 - V_1) \qquad (7.66)$$

where V_1 and V_2 are the initial and final volumes of the system.

Substituting for $W_{iso,max}$ from Eq. (7.66) in (7.64)

$$W_{useful,max} + P(V_2 - V_1) = -[(U_2 - TS_2) - (U_1 - TS_1)]$$

$$W_{useful,max} = -[(U_2 + PV_2 - TS_2) - (U_1 + PV_1 - TS_1)]$$

$$W_{useful,max} = -[(H_2 - TS_2) - (H_1 - TS_1)] \qquad (7.67)$$

From Eq. (7.67) it is clear that the maximum work done by a system under simultaneous *isothermal* and *constant pressure* conditions is the change of a property defined as, $G = H - TS$ where H is the enthalpy. The property G is called the *Gibbs free energy*. Therefore Eq. (7.67) can be written in terms of the Gibbs free energy as

$$W_{useful,max} = -[G_2 - G_1] \qquad (7.68)$$

The function G gives the fraction of the enthalpy of the system that could be harnessed to produce *useful non-expansion* work. For example, in an electrolytic cell operating at constant temperature and pressure the amount of energy which is free to produce electrical work is given by the decrease in the Gibbs free energy for the process.

7.3.2 *Availability*

There are numerous engineering applications where a system producing work exchanges heat with a single reservoir like the atmosphere. The arrangement is similar to that shown in Fig. 7.4. In this section we shall derive an expression for the maximum work output of such a system. Let the initial and final states of the closed system and the reservoir be denoted by 1 and 2 respectively. Applying the first law to the system

$$Q_{12} = U_2 - U_1 + W_{12} \qquad (7.69)$$

Consider the *isolated* composite system consisting of the given system and the reservoir. We apply the principle of increase of entropy to obtain

$$(S_2 - S_1) - \frac{Q_{12}}{T_r} \geq 0 \qquad (7.70)$$

where T_r is the reservoir temperature. In Eq. (7.70) the first term is the increase in entropy of the system while the second term is the decrease in entropy of the reservoir due to heat flow from the reservoir to the system. Eliminating Q_{12} between Eqs. (7.70) and (7.69) we have

$$W_{12} \leq -[(U_2 - T_r S_2) - (U_1 - T_r S_1)] \qquad (7.71)$$

The maximum work output is therefore the change in the function $(U - T_r S)$ from the initial to the final state.

Now consider a system interacting with a single reservoir as before, and immersed in an atmosphere at a *uniform* pressure P_{atm}. Of the work output W_{12} of the process, a part is used to move the boundary of the system against the constant pressure of the atmosphere. Therefore the *useful* work produced by the system is given by

$$W_{useful} = W_{12} - P_{atm}(V_2 - V_1) \qquad (7.72)$$

Eliminating W_{12} between Eqs. (7.72) and (7.71) we obtain

$$W_{useful} \leq -[(U_2 + P_{atm}V_2 - T_r S_2) - (U_1 + P_{atm}V_1 - T_r S_1)] \qquad (7.73)$$

It is seen from Eq. (7.73) that the maximum work output is the change of the function, $\Phi = U + P_{atm}V - T_r S$ from the initial to the final state.

If the atmosphere is also the heat reservoir, as it is the case in many engineering systems, then $\Phi = U + P_{atm}V - T_{atm}S$. The latter function is clearly a property of the system-atmosphere combination. We can write Eq. (7.73) in terms of Φ as

$$W_{useful} \leq -[\Phi_2 - \Phi_1] \qquad (7.74)$$

The *minimum* value of Φ occurs when the system establishes equilibrium with the atmosphere and its temperature and pressure are therefore T_{atm} and P_{atm} respectively. If this minimum value of the property Φ is Φ_{min}, then we define the new property called the *availability* as [2]:

$$A = \Phi - \Phi_{min} \qquad (7.75)$$

In terms of the availability, the useful work output given by Eq. (7.74) becomes

$$W_{useful} \leq -[A_2 - A_1] \qquad (7.76)$$

The physical meaning of Eq. (7.76) is that the decrease in the availability of a system-atmosphere combination is the useful work output of a *reversible* process of the system. However, when the system undergoes an *irreversible* process with *same initial and final states* of the system, the work done is less than the change in availability.

7.4 Maximum Work, Irreversibility and Entropy Production

In this section we shall obtain a general relationship between the entropy production and the irreversibility for a *closed* system interacting with a series of thermal reservoirs and producing work as depicted in Fig. 7.8. The atmosphere, shown separately, is chosen as the standard reservoir that undergoes changes of state to accommodate variations of the work output W_o of the system. The system undergoes a process whose initial and final states are denoted by 1 and 2 respectively. The temperatures and the heat interactions between the system and the reservoirs during the process are indicated in the Fig. 7.8.

Applying the first law to the system we have

$$Q_o + \sum Q_i = U_2 - U_1 + W_o \qquad (7.77)$$

where Q_o is the heat flow from the standard reservoir (atmosphere). The second term on the left hand side of Eq. (7.77) is the total heat flow from all the reservoirs *except the standard reservoir*.

Apply the principle of increase of entropy to the isolated composite system consisting of all the thermal reservoirs and the given system. Hence we obtain

$$(S_2 - S_1) - \frac{Q_o}{T_o} - \sum \frac{Q_i}{T_i} = \sigma \qquad (7.78)$$

where σ is the entropy production in the composite system due to all irreversibilities.

The first term in Eq. (7.78) is the change in entropy of the system, while the second and third terms are the changes in the entropy of the atmosphere and the heat reservoirs respectively. Eliminating Q_o between Eqs. (7.77) and (7.78) we have

$$T_o(S_2 - S_1) + \sum Q_i\left(1 - \frac{T_o}{T_i}\right) - (U_2 - U_1) = T_o\sigma + W_o \qquad (7.79)$$

$$\sum Q_i\left(1 - \frac{T_o}{T_i}\right) - [(U_2 - T_oS_2) - (U_1 - T_oS_1)] = T_o\sigma + W_o \qquad (7.80)$$

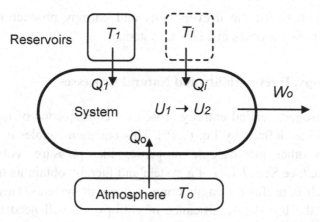

Fig. 7.8 Process in a system interacting with multiple reservoirs

Now consider a *reversible* process that would produce the *same* change of state from 1 to 2 in the system and all the reservoirs *except the atmosphere-reservoir*. The work output for this process, however, would be different from the original process. It is seen from Eq. (7.77) that a change of the work output requires the heat interaction between the system and the atmosphere-reservoir Q_o to adjust to satisfy the latter equation, which is the first law. Now for the *reversible* process the entropy production, $\sigma = 0$. Therefore Eq. (7.80) becomes

$$\sum Q_i \left(1 - \frac{T_o}{T_i}\right) - [(U_2 - T_o S_2) - (U_1 - T_o S_1)] = W_{rev} \qquad (7.81)$$

Equation (7.81) gives the maximum work output of the system for the given change of state from 1 to 2. From Eqs. (7.80) and (7.81) it follows that:

$$W_{rev} - W_o = T_o \sigma \qquad (7.82)$$

The difference between the actual work output of the system W_o and the reversible work output W_{rev} is called the *lost work* or the *irreversibility I.* Substituting in Eq. (7.82) we have

$$I = T_o \sigma \qquad (7.83)$$

We obtained equations similar to Eq. (7.82) for the irreversibility of cyclic heat engines [Eq. (7.51)] and steady heat conduction [Eq. (7.57)] in earlier sections of this chapter. Equation (7.83) is a general relationship between the irreversibility and entropy production that is applicable to any process in a *closed* system.

7.5 Entropy, Irreversibility and Natural Processes

The new property called entropy arose as a consequence of the second law and it was defined by Eq. (7.10). This equation enables us to relate entropy to other macroscopic properties like pressure, volume and temperature (see Sec. 7.1.4) of a system and thereby obtain its numerical value which is useful for engineering analysis of systems. However, to appreciate the broader significance of entropy we still need to seek a more satisfying physical interpretation.

From a microscopic view point entropy can be thought of as a measure of the randomness or disorganization of the constituents of a system. For instance, when a gas is compressed adiabatically the final state confines the molecules of the gas to a smaller volume thereby bringing more organization to the molecules. However, the higher temperature that may result from the compression will make the molecules more agitated and therefore their state more random. This view point is reflected in the expression for the entropy of an ideal gas in Eq. (7.23). A more detailed discussion on the microscopic interpretation of entropy is given in chapter 15.

In our discussion on natural processes in Chapter 6 we noticed that these processes moved in a preferred direction. For instance, heat always flows unaided from hot to cold regions making the latter region more disorganized. Furthermore, gases expand from regions of high pressure to evacuated spaces increasing the level of disorganization of the molecules of the gas. To reverse these processes we need to expend work, which leaves permanent changes in the environment, thus making the processes irreversible.

In view of the above observations we can conclude that *natural* processes take systems from more organized states to less organized states thus increasing their entropy. Often we maintain highly organized states of systems by introducing well designed constraints. For example, we need high quality thermal insulation to prevent heat leaks between a hot body and a cold body. Similarly, we use walls or membranes to confine high pressure gases from expanding to low pressure regions surrounding them. Once the constraints are removed the states of these systems evolve naturally or unaided to less organized states. Since entropy is a measure of the level of disorganization we are led to the conclusion that natural processes cause the entropy of systems to increase and they are, therefore, irreversible.

Now consider a pure work interaction with no frictional effects. The sliding of a metal block down a smooth inclined plane is a possible example. The state of organization of the molecules of the block remains unchanged during the sliding because the kinetic energy gained by the block will affect all the molecules of the block equally in an organized fashion. Therefore there is no change in the entropy of the block due to the motion. However, if sliding friction is present some of the work done is converted to a heat interaction at the sliding surface resulting in a rise in temperature of the block which, in turn, increases the disorganization of its molecules. Therefore the block is at higher entropy in the final state.

An important goal of engineering design to is to develop efficient systems to produce work from stored energy sources. Thermal power plants and combustion engines are examples of such systems. A question that follows naturally is how to relate our knowledge on irreversible processes, level of organization of systems, and entropy to processes in energy conversion systems. The answer with regard to friction is straight forward because friction converts some fraction of the work available or produced directly to heat which cannot be reconverted to work completely even using reversible devices. Now consider the presence of components in an energy conversion system where heat is transferred across finite temperature differences. Ideally we could effect the same heat transfer using a reversible heat engine, thus producing some additional work, as was seen in Sec. 7.2.3. In other words, heat transfer across a finite temperature difference is a 'lost opportunity' for the production of work. Larger the temperature difference, greater would be the potential loss of work. Therefore by allowing the natural process, namely the flow of heat from a higher to a lower temperature in this case, to occur, we lost some of the potential work.

The free expansion of a high-pressure fluid stream through a component like a throttle valve is a similar situation. If the expansion is carried out in a fully-resisted manner, perhaps with the aid of an expander, the 'work potential' in the high pressure stream could have been realized. Moreover, in both instances mentioned above the natural process, if allowed to occur, increases the entropy production of the system.

In summary, differences of intensive properties, like the temperature and pressure between systems, are usually maintained using appropriate constraints. These property gradients offer us opportunities to produce useful work. If the constraints are simply removed, natural processes will level-off these property gradients resulting in entropy production. However, if appropriate devices are introduced to equalize the property differences in a controlled manner some of the work-potential could be realized. Equations (7.82) and (7.83) are useful relationships that enable us to evaluate the potential loss of work by computing the entropy production.

7.6 Worked Examples

Example 7.1 (i) An ideal gas undergoes a polytropic process from state 1 to state 2 according to the process law $PV^n = C$. Obtain an expression for the change in entropy of the gas in terms of the initial and final pressures.

(ii) A unit mass of methane is subjected to a quasi-static compression process which follows the relation $PV^{1.2} = C$. The initial temperature and pressure are 28°C and 88 kPa respectively. The final pressure is 400 kPa. For methane is $R = 0.52$ kJ kg^{-1} K^{-1}, and the ratio of the specific capacities, $\gamma = 1.3$. Calculate the change in entropy of the methane.

Solution We could derive the required expression for the entropy change for a polytropic process from first principles as was done in Sec. 7.1.4. However, it is more convenient to use the expression already obtained in Sec. 7.1.4 for an ideal gas because the change in entropy depends only on the initial and final states of the ideal gas, and not on the process path. The general expression for the change in entropy of an ideal gas is given by Eq. (7.25) as

$$S = S_o + mc_p \ln(T/T_o) - mR \ln(P/P_o) \qquad \text{(E7.1.1)}$$

Manipulating the ideal gas equation of state, $PV = mRT$ and the polytropic process relation, $PV^n = C$ we obtain

$$P^{(n-1)/n}/T = C_1 \qquad \text{(E7.1.2)}$$

where C_1 is a constant.

Hence
$$P_o^{(n-1)/n}/T_o = P^{(n-1)/n}/T \qquad \text{(E7.1.3)}$$

where the subscript o denotes quantities at the initial state.

Substituting for T/T_o from Eq. (E7.1.3) in Eq. (E7.1.1)

$$S = S_o + mc_p(1 - 1/n)\ln(P/P_o) - mR \ln(P/P_o) \qquad \text{(E7.1.4)}$$

Using the ideal gas relations,

$$(c_p - c_v) = R \qquad \text{and} \qquad (c_p/c_v) = \gamma,$$

Equation (E7.1.4) can be expressed in the compact form

$$S = S_o - mR \frac{(\gamma - n)}{n(\gamma - 1)} \ln(P/P_o) \qquad \text{(E7.1.5)}$$

Note that if we substitute, $n = \gamma$ in Eq. (E7.1.5) the entropy change becomes zero because the process then is *isentropic*.

(ii) The numerical data pertinent to the problem are $m = 1$ kg, $R = 0.52$ kJ kg^{-1} K^{-1}, $\gamma = 1.3$ and $n = 1.2$. Substituting in Eq. (E7.1.5)

$$S - S_o = -1 \times 0.52 \times \frac{(1.3 - 1.2)}{1.2(1.3 - 1)} \ln(400/88) = -0.219 \text{ kJ K}^{-1}$$

The entropy of the gas has decreased because heat has been transferred out of the gas during the process.

Example 7.2 A fixed quantity of water of mass 1.3 kg at an initial temperature and pressure of 200°C and 30 bar respectively is contained in a piston-cylinder apparatus. The water undergoes a reversible isothermal expansion to a lower pressure while receiving 3500 kJ of heat. Calculate (i) the change in entropy of the water, and (ii) the final pressure.

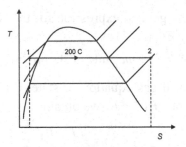

Fig. E7.2 *T-S* diagram for water

Solution At the initial state 1, the water is a compressed liquid because the pressure is higher than the saturation pressure at 200°C. The path of the isothermal heating process is indicated by the horizontal line 1-2 in Fig. E7.2. The area under the line 1-2 gives the heat supplied during the process because for a reversible process

$$Q_{12} = \int_1^2 TdS \qquad\qquad (E.7.2.1)$$

Since T is constant, Eq. (E7.2.1) can be integrated directly to obtain

$$Q_{12} = Tm(s_2 - s_1) \qquad\qquad (E.7.2.2)$$

where s is the entropy per unit mass.

For compressed water we ignore the effect of pressure and find the saturated liquid entropy at 200°C. Form the tabulated data in [6] we obtain by interpolation the liquid entropy as, $s_1 = 2.331$ kJ K^{-1} kg^{-1}.

Substituting the given numerical data in Eq. (E7.2.2) we have

$$3500 = (200 + 273) \times 1.3 \times (s_2 - s_1) \qquad\qquad (E7.3.2)$$

From Eq. (7.3.2) the entropy change of water is given by

$$(s_2 - s_1) = 5.69 \text{ kJ K}^{-1} \text{ kg}^{-1}$$

Hence $s_2 = 8.02$ kJ K^{-1} kg^{-1}

In order to determine the final state of the steam we need to obtain the pressure of superheated steam at 200°C for which $s_2 = 8.02$ kJ K^{-1} kg^{-1}. From the superheated steam data in [6] the pressure is obtained by linear interpolation as 0.68 bar.

Example 7.3 (a) Draw the temperature-entropy diagram for a Carnot refrigeration cycle. Hence obtain an expression for the COP of the cycle. (b) Obtain an expression for thermal efficiency of the reversible heat engine cycle whose *T-S* diagram is shown in Fig. E7.3(b).

Solution (a) The temperature–entropy diagram for a Carnot refrigeration cycle is shown in Fig. E7.3(a). Heat is extracted from the cold reservoir during the isothermal process 1-2. The process 2-3 is an isentropic compression during which the temperature of the cycle becomes equal to the hot reservoir temperature. Heat is rejected isothermally to the hot reservoir during the process 3-4. Finally, the isentropic expansion process 4-1 completes the cycle. Since all the processes are reversible, the area under the process path on the *T-S* diagram is the heat transfer. Using the quantities indicated in the figure we write the following expressions for the heat interactions:

$$Q_c = T_c(S_2 - S_1) \tag{E7.3.1}$$

$$Q_h = T_h(S_3 - S_4) \tag{E7.3.2}$$

Applying the first law to the cycle

$$W_{in} = Q_h - Q_c \tag{E7.3.3}$$

The COP of the refrigerator is defined as

$$COP_{ref} = \frac{Q_c}{W_{in}} = \frac{Q_c}{Q_h - Q_c} \tag{E7.3.4}$$

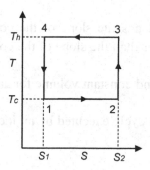

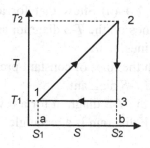

Fig. E7.3(a) *T-S* diagram   Fig. E7.3(b) *T-S* diagram

Because of the rectangular shape of the *T-S* diagram

$$(S_2 - S_1) = (S_3 - S_4) \qquad (E7.3.5)$$

Substituting from Eqs. (E7.3.1) and (E7.3.2) in Eq. (E7.3.4) with the condition in Eq. (E7.3.5) we obtain

$$COP_{ref} = \frac{T_c}{T_h - T_c}$$

(b) The *T-S* diagram for the heat engine cycle, shown in Fig. E7.3(b), is a triangle. The engine receives heat during the process 1-2, undergoes an isentropic expansion from 2-3, and rejects heat during the process 3-1 to complete the cycle. The net heat transfer is the area enclosed by the triangle which by the first law is also equal to the net work output. Therefore

$$W_{net} = Q_{net} = (T_2 - T_1)(S_2 - S_1)/2 \qquad (E7.3.6)$$

The total heat input to the cycle is the area of the trapezium 1-2-b-a, which can be written as

$$Q_{in} = (T_1 + T_2)(S_2 - S_1)/2 \qquad (E7.3.7)$$

The thermal efficiency of the cycle is

$$\eta = \frac{W_{net}}{Q_{in}} = \frac{(T_2 - T_1)}{(T_2 + T_1)} = 1 - \frac{T_1}{(T_1 + T_2)/2} \qquad (E7.3.8)$$

Equation (E7.3.8) shows that the efficiency of the engine is equal to that of a Carnot engine operating with a heat source temperature equal to the mean of the given cycle temperatures.

Example 7.4 (a) Show that for an ideal gas the slope of the constant volume lines on the *T-S* diagram are larger than the slope of the constant pressure lines.
(b) Sketch the lines of constant pressure and constant volume for an ideal gas on the *T-S* diagram.
(c) Draw the *T-S* diagram for a reversible cycle executed by an ideal gas whose *P-V* diagram is a rectangle.

Solution (a) In Sec. 7.1.4 we obtained the following expressions for the entropy of an ideal gas

$$S = S_o + mc_v \ln(T/T_o) + mR\ln(V/V_o) \qquad (E7.4.1)$$

$$S - S_o = mc_p \ln(T/T_o) - mR\ln(P/P_o) \qquad (E7.4.2)$$

The above expressions are used to find the required gradients $(\partial T/\partial S)_V$ and $(\partial T/\partial S)_P$ of the constant volume and constant pressure lines respectively on the T-S diagram.

Differentiate Eq. (E7.4.1) with respect to T keeping V constant. Hence

$$(\partial S/\partial T)_V = mc_v/T \qquad (E7.4.3)$$

Differentiate Eq. (E7.4.2) with respect to T keeping P constant. Hence

$$(\partial S/\partial T)_P = mc_p/T \qquad (E7.4.4)$$

From Eqs. (E7.4.3) and (E7.4.4) (also see chapter 14) we have

$$(\partial T/\partial S)_V = T/mc_v$$

$$(\partial T/\partial S)_P = T/mc_p$$

Now for an ideal gas, $c_p = c_v + R > c_v$. Therefore it follows from the two equations above that:

$$(\partial T/\partial S)_V > (\partial T/\partial S)_P$$

(b) In order to sketch the constant volume and constant pressure lines we first obtain functional forms for the variation of S versus T under these conditions. Rearranging Eq. (E7.4.1)

$$(T/T_o) = \exp((S - S_o)/mc_v - R\ln(V/V_o)/c_v)$$

Hence

$$(T/T_o) = (V_o/V)^{R/c_v} \exp[(S - S_o)/mc_v] \qquad (E7.4.5)$$

Similarly, by rearranging Eq. (E7.4.2)

$$(T/T_o) = (P/P_o)^{R/c_p} \exp[(S - S_o)/mc_p] \qquad (E7.4.6)$$

The exponential curves given by Eqs. (E7.4.5) and (E7.4.6) are sketched in Fig. E7.4(a) where the point O represents the reference state. Notice that the T-V and T-P relations for an isentropic process $(S = S_o)$ could be deduced from these equations.

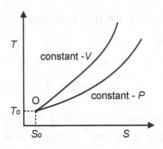

Fig. E7.4(a) *T-S* diagrams of constant-*P* and constant-*V* processes

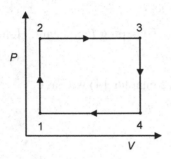

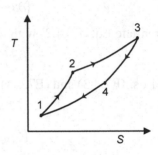

Fig. E7.4(b)   *P-V* diagram   Fig. E7.4(c)   *T-S* diagram

(c) The *P-V* diagram shown in Fig. 7.4(b) has a rectangular shape with two constant-volume processes 1-2 and 3-4 and two constant-pressure processes 2-3 and 4-1. We use the shapes obtained in part (b) for such processes to sketch the cycle on the *T-S* diagram shown in Fig. 7.4(c).

The constant pressure and constant volume processes on the *T-S* diagram are represented by exponential functions according to Eqs. (E7.4.5) and (E7.4.6).

Example 7.5 A fixed quantity of helium gas of mass 0.6 kg undergoes a reversible process that has a linear path on the *T-S* diagram. The temperature and volume at the initial state are 38°C and 0.3 m^3 respectively. The corresponding values are 260°C and 0.73 m^3 at the final state. Calculate the heat transfer during the process. For helium assume that $c_v = 3.12$ kJ kg^{-1} K^{-1} and $R = 2.08$ kJ kg^{-1} K^{-1}.

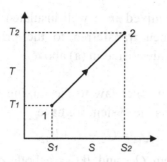

Fig. E7.5 *T-S* diagram

Solution Let the initial and final states of the process be denoted by 1 and 2 respectively (see Fig. E7.5). Since helium is treated as an ideal gas the change in entropy is obtained from the general ideal gas property relation derived in Sec. 7.1.4. Hence

$$S_2 - S_1 = mc_v \ln(T_2/T_1) + mR\ln(V_2/V_1)$$

Substituting numerical values in the above equation

$$S_2 - S_1 = 0.6 \times 3.12 \times \ln\left(\frac{533}{311}\right) + 0.6 \times 2.08 \times \ln\left(\frac{0.73}{0.3}\right) = 2.1178$$

$$S_2 - S_1 = 2.1178 \text{ kJ K}^{-1}$$

Since the process is reversible, the heat transfer is equal to the area of the *T-S* diagram under the straight line 1-2, which has the trapezoidal shape as seen in Fig. E7.5. Therefore the heat transfer is

$$Q_{12} = 0.5 \times (T_1 + T_2)(S_2 - S_1)$$

$$Q_{12} = 0.5 \times (533 + 311) \times 2.1178 = 893.7 \text{ kJ}.$$

Example 7.6 Two separate quantities of water of masses m_a and m_b of the same constant specific heat capacity c are at temperatures T_a and T_b respectively. The two masses of water are mixed in a well-insulated vessel of heat capacity c_v and mass m_v and left to attain equilibrium.
(a) Obtain expressions for (i) the entropy changes of the water and the vessel after the final equilibrium state is reached, and (ii) the entropy production in the composite system.

(b) If the water is first mixed in a well-insulated vessel of negligible thermal capacity and then transferred to the original vessel, obtain expressions for the quantities listed in (a) above.

Solution (a) Applying the first law to the mixing process, considering the water and the vessel as the system we have

$$Q_{12} = U_2 - U_1 + W_{12} \qquad (E7.6.1)$$

For the composite system Q_{12} and W_{12} are both zero. Therefore from Eq. (E7.6.1)

$$m_a c_a T_a + m_b c_b T_b + m_v c_v T_v = (m_a c_a + m_b c_c + m_v c_v)T_f$$

where T_f is the final equilibrium temperature of the system.

$$T_f = \frac{m_a c_a T_a + m_b c_b T_b + m_v c_v T_v}{m_a c_a + m_b c_b + m_v c_v} \qquad (E7.6.2)$$

The water and the vessel may be treated as pure thermal systems because of their low compressibility. Since entropy is a property we use the general expression, Eq. (7.29), derived earlier in Sec. 7.1.5 for a pure thermal system, to obtain the entropy changes as:

$$\Delta S_{water} = m_a c_a \ln(T_f / T_a) + m_b c_b \ln(T_f / T_b) \qquad (E7.6.3)$$

$$\Delta S_{vessel} = m_v c_v \ln(T_f / T_v) \qquad (E7.6.4)$$

The water and the vessel constitute an isolated system that does not exchange heat with the environment. The entropy production of the composite system is therefore equal to the net change in entropy of the water and the vessel.

$$\sigma = \Delta S_{water} + \Delta S_{vessel} \qquad (E7.6.5)$$

Substituting from Eqs. (E7.6.3) and (E7.6.4) in Eq. (E7.6.5)

$$\sigma = m_a c_a \ln(T_f / T_a) + m_b c_b \ln(T_f / T_b) + m_v c_v \ln(T_f / T_v)$$

(b) Applying the first law to the two-step mixing process it is clear that the final equilibrium temperature of the water and the vessel is the same as that given by Eq. (E7.6.2). Since entropy is a property and therefore independent of the process path, the entropy changes and the entropy production for the two-step mixing process are the same as those for part (a) above.

Example 7.7 A rigid tank contains 0.1 kg of air at an initial temperature and pressure of 27°C and 110 kPa respectively. The air receives 8.5 kJ of energy through the following alternative boundary interactions.
(a) A work input of 8.5 kJ by means of a paddle wheel while the tank is well-insulated.
(b) A work input of 8.5 kJ by means of a paddle wheel while the tank looses 1.5 kJ of heat to the surroundings at 30°C.
(c) A heat input of 8.5 kJ from a heat reservoir at 300°C with the tank well-insulated.
(i) Calculate for each of the above processes the change of entropy of the air. (ii) Calculate for the processes (b) and (c) the entropy changes of the reservoirs. (iii) Calculate the entropy production in the universe for each of the three process.

Solution (a) Applying the first law assuming no heat losses from the tank to the surroundings we obtain

$$0 = mc_v(T_2 - T_1) + W_{12} \qquad (E7.7.1)$$

where 1 and 2 denote the initial and final equilibrium states and W_{12} is the work input of the paddle wheel. Substituting numerical values in Eq. (E7.7.1)

$$0 = 0.1 \times 0.718 \times (T_2 - 27) - 8.5$$

Therefore $T_2 = 145.38°C$

Assuming air to be an ideal gas we apply the following property relation to calculate the entropy change

$$S_2 - S_1 = mc_v \ln(T_2/T_1) + mR\ln(V_2/V_1) \qquad (E7.7.2)$$

Since the tank is rigid, the volume of air is unchanged. Substituting numerical values in Eq. (E7.7.2) we have

$$S_{a2} - S_{a1} = 0.1 \times 0.718 \times \ln(418.38/300) = 0.0239 \text{ kJ K}^{-1}$$

Since the tank is well insulated there is no entropy transfer due to heat transfer. The entropy production σ of the system and the surroundings (the universe) is equal to the entropy change of the air. Therefore $\sigma = 0.0239$ kJ K^{-1}. Note that the work source driving the paddle wheel, which is part of the surroundings, undergoes no entropy change because unlike heat transfer, work transfer is not accompanied by any entropy transfer.

(b) Applying the first law to the air we have

$$Q_{12} = mc_v(T_2 - T_1) + W_{12} \qquad (E7.7.3)$$

where Q_{12} is the heat loss from the air to the surroundings.

Substituting numerical values in Eq. (E7.7.3)

$$-1.5 = 0.1 \times 0.718 \times (T_2 - 27) - 8.5$$

Hence $T_2 = 124.49\,°C$

We substitute numerical values in Eq. (E7.7.2) to find the entropy change of the air. Thus

$$S_{a2} - S_{a1} = 0.1 \times 0.718 \times \ln(397.49/300) = 0.0202 \text{ kJ K}^{-1}$$

The entropy change of the surroundings, which may be treated as a reservoir at 30°C is given by

$$S_{r2} - S_{r1} = Q_{12}/T_r = 1.5/303 = 0.00495 \text{ kJ K}^{-1}$$

The entropy production of the system and the surroundings (the universe) is

$$\sigma = 0.0202 + 0.00495 = 0.02515 \text{ kJ K}^{-1}$$

The total entropy production is larger now because in addition to the irreversibility associated with work dissipation by the paddle wheel there is an external irreversibility due to heat transfer between the tank and the surroundings.

(c) The heat input of 8.5 kJ from the reservoir is the same as the work input from the paddle wheel in (a). Therefore the final equilibrium temperature of the air is the same as in (a), that is $T_2 = 145.38°C$. The entropy change of the air is also the same as in (a), which is given by

$$S_{2a} - S_{1a} = 0.0239 \text{ kJ K}^{-1}$$

The entropy change of the high temperature reservoir that supplies the heat input is

$$S_{r2} - S_{r1} = -Q_{12}/T_r = -8.5/573 = -0.0148 \text{ kJ K}^{-1}$$

Note that the reservoir experiences a decrease in entropy because of the heat flow out of it. The entropy production of the composite system consisting of the air and the reservoir (the universe) is

$$\sigma = 0.0239 - 0.0148 = 0.009065 \text{ kJ K}^{-1}$$

The lowest entropy production occurs with the arrangement in (c) because it involves a heat transfer irreversibility that is not as 'dissipative' as the direct conversion of work to heat as happens with the paddle wheel in processes (a) and (b).

Example 7.8 An electrical heater with a resistance of 30 ohms is connected to a DC voltage source of 120 V. The mass and specific heat capacity of the heater are 0.015 kg and 0.8 kJ kg^{-1} K^{-1} respectively. The heater is switched on for 2 s during which time it is maintained at a constant temperature of 30°C by immersing in a heat-sink reservoir. (a) Calculate (i) the change in entropy of the heater, (ii) the change in entropy of the reservoir, (iii) the entropy transfer to the reservoir, and (iv) the entropy production in the universe. (b) If the heater has been well-insulated, and initially at a temperature of 30°C, calculate the (i) the change in entropy of the heater, and (ii) the entropy production in the universe.

Solution The electrical energy (work) input to the heater during the 2 s is

$$W_{12} = V^2 t/R = 120^2 \times 2/30 = 960 \text{ J}$$

Treat the heater as a pure thermal system. Since the heater is maintained at a constant temperature, its change in internal energy is zero. Applying the first law we have

$$Q_{12} = W_{12} = 960 \text{ J}$$

where Q_{12} is the heat transfer from the heater to the heat sink reservoir at 30°C. The change in entropy of the heater is zero because it is a pure thermal system maintained at a constant temperature (see Eq. 7.29). The entropy change in the reservoir is

$$S_{r2} - S_{r1} = Q_{12}/T_r = 0.96/303 = 0.003168 \text{ kJ K}^{-1}$$

The entropy increase of the reservoir is the same as the entropy transfer to it from the heater because the reservoir does not produce any entropy of its own. The entropy production of the composite system consisting of the heater and the reservoir (the universe) is

$$\sigma = (S_{r2} - S_{r1}) + (S_{h2} - S_{h1}) = 0.003168 \text{ kJ K}^{-1}$$

(b) When the heater is insulated the heat transfer $Q_{12} = 0$. Applying the first law we have

$$0 = (U_2 - U_1) + W_{12} = MC(T_2 - T_1) - 960$$

Substituting numerical values

$$0.015 \times 0.8 \times (T_2 - 30) = 0.96$$

The final equilibrium temperature of the heater is, $T_2 = 110°C$.

Since the heater is a pure thermal system its entropy change is given by

$$S_{h2} - S_{h1} = MC \ln(T_2/T_1)$$

$$S_{h2} - S_{h1} = 0.015 \times 0.8 \times \ln(383/303) = 0.00281 \text{ kJ K}^{-1}$$

There is no entropy transfer from the heater because it is thermally insulated. Therefore the entropy production of the universe is equal to the entropy increase of the heater, which is 0.00281 kJ K^{-1}.

Example 7.9 A fixed quantity water of mass 1.5 kg and specific heat capacity 4.2 kJ kg^{-1} K^{-1} is at an initial temperature of 300 K. The following three alternative processes are considered for heating the water to a temperature of 360 K. (a) The water receives heat from a reservoir at 360 K until it attains thermal equilibrium with the reservoir. (b) The water receives heat successively from two reservoirs at 330 K and 360 K attaining equilibrium with each reservoir in turn, and (c) The water receives heat successively from three reservoirs at 320 K, 340 K and 360 K attaining equilibrium with each reservoir in turn. Calculate (i) the entropy change in the water and the reservoirs, and (ii) the entropy production in the universe for each of the three scenarios.

Solution We first obtain the general expressions for the change in entropy of the water, treating it as a pure thermal system, and the reservoirs. For the water we have

$$S_{w2} - S_{w1} = m_w c_w \ln(T_{w2}/T_{w1}) \tag{E7.9.1}$$

Applying the first law to the water with no work transfer

$$Q_{12} = m_w c_w (T_{w2} - T_{w1}) \tag{E7.9.2}$$

Applying the first law to the reservoir

$$Q_{12} = -Q_{res} \tag{E7.9.3}$$

The change in entropy of the reservoir is given by

$$S_{r2} - S_{r1} = Q_{res}/T_r \tag{E7.9.4}$$

Substituting in Eq. (E7.9.4) from Eqs. (E7.9.3) and (E7.9.2) we obtain

$$S_{r2} - S_{r1} = -m_w c_w (T_{w2} - T_{w1})/T_r \tag{E7.9.5}$$

We now apply Eqs. (E7.9.1) and (E7.9.5) to each of the alternative processes to determine the entropy changes.

(a) The change in entropy of the water, from Eq. (E7.9.1) is

$$S_{w2} - S_{w1} = 1.5 \times 4.2 \times \ln(360/300) = 1.148 \text{ kJ K}^{-1}$$

The change in entropy of the reservoir from Eq. (E7.9.5) is

$$S_{r2} - S_{r1} = -1.5 \times 4.2 \times (360 - 300)/360 = -1.05 \text{ kJ K}^{-1}$$

The entropy production in the composite system consisting of the water and the reservoir (the universe) is

$$\sigma = (S_{w2} - S_{w1}) + (S_{r2} - S_{r1})$$

$$\sigma = 1.148 - 1.05 = 0.098 \text{ kJ K}^{-1}$$

(b) For the two-step heating process, the initial and final temperatures of the water are the same as for (a). Therefore the change in entropy of the water is also the same as for (a) because the entropy is a function of the temperature. However, the entropy changes in the two reservoirs are different and these are obtained by applying Eq. (E7.9.5) to each reservoir in turn.

$$(S_{r2} - S_{r1})_1 = -1.5 \times 4.2 \times (330 - 300)/330 = -0.5727 \text{ kJ K}^{-1}$$

$$(S_{r2} - S_{r1})_2 = -1.5 \times 4.2 \times (360 - 330)/360 = -0.525 \text{ kJ K}^{-1}$$

The entropy production in the composite system consisting of the water
and the two reservoirs 1 and 2 (the universe) is

$$\sigma = 1.148 - 0.5727 - 0.525 = 0.0503 \ \text{kJ K}^{-1}$$

(c) We use the same set of equations for the three-step heating process.
Again, the entropy change of the water is the same as in (a). The entropy
changes of three reservoirs, however, are given by

$$(S_{r2} - S_{r1})_1 = -1.5 \times 4.2 \times (320 - 300)/320 = -0.3937 \ \text{kJ K}^{-1}$$

$$(S_{r2} - S_{r1})_2 = -1.5 \times 4.2 \times (340 - 320)/340 = -0.3706 \ \text{kJ K}^{-1}$$

$$(S_{r2} - S_{r1})_3 = -1.5 \times 4.2 \times (360 - 340)/360 = -0.35 \ \text{kJ K}^{-1}$$

The entropy production of the composite system consisting of the water
and the three reservoirs 1, 2 and 3 (the universe) is

$$\sigma = 1.148 - 0.3937 - 0.3706 - 0.35 = 0.034 \ \text{kJ K}^{-1}$$

From the numerical values for the entropy production for parts (a), (b)
and (c) we observe that the entropy production is largest for (a) where,
on average, heat is transferred across the largest temperature difference
between the reservoir and the water. On the other hand, the three-
reservoir heating process, (c) has the smallest entropy production. If the
number of reservoirs is further increased, the entropy production would
be still lower, and eventually with a series of reservoirs with infinitesimal
temperature increments between them, the heating process would
approach the reversible limit. Such a process, though ideal, would have
zero entropy production. Recall that in worked example 1.6 we discussed
a quasi-static heating process based on the same concept.

Example 7.10 A reversible heat engine operating with infinitesimal
cycles receives heat from a block of material of mass m_1 and specific
heat capacity c_1, and rejects heat to a block of mass m_2 and specific
heat capacity c_2. The initial temperatures of the blocks are T_{1i} and T_{2i}
respectively. Finally the two blocks attain equilibrium with their
temperatures equal to T_f.
(a) (i) Obtain expressions for the total change of entropy of the two
blocks and the reversible engine. (ii) Obtain an expression for final

equilibrium temperature T_f. (iii) Obtain an expression for the total work output of the engine.

(b) If an *irreversible* engine is operated between the same blocks with the same initial conditions, (i) obtain an expression for the final equilibrium temperature T_{fi}, and (ii) compare the magnitudes of T_f and T_{fi}.

Solution The two blocks may be treated as pure thermal systems whose entropy change is given by Eq. (7.29). We apply this equation to obtain the following expressions for the entropy change of the two blocks

$$S_{1f} - S_{1i} = m_1 c_1 \ln\left(T_f / T_{1i}\right) \qquad (E7.10.1)$$

$$S_{2f} - S_{2i} = m_2 c_2 \ln\left(T_f / T_{2i}\right) \qquad (E7.10.2)$$

The reversible engine executes an integral number of cycles during the process. Therefore the entropy change of the cyclic device itself is zero.

$$(S_{ef} - S_{ei}) = 0 \qquad (E7.10.3)$$

The entropy production of the composite system consisting of the two blocks and the cyclic engine (the universe) is

$$\sigma = (S_{1f} - S_{1i}) + (S_{2f} - S_{1i}) + (S_{ef} - S_{ei}) \qquad (E7.10.4)$$

Substituting in Eq. (E7.10.4) from Eqs. (E7.10.1) to (E7.10.3) we have

$$\sigma = m_1 c_1 \ln\left(T_f / T_{1i}\right) + m_2 c_2 \ln\left(T_f / T_{2i}\right) + 0 \qquad (E7.10.5)$$

For a reversible engine operating between the two blocks the entropy production in the universe is zero. Hence from Eq. (E7.10.5) it follows that

$$m_1 c_1 \ln\left(T_f / T_{1i}\right) + m_2 c_2 \ln\left(T_f / T_{2i}\right) = 0$$

$$\left(T_f / T_{1i}\right)^{m_1 c_1} \left(T_f / T_{2i}\right)^{m_2 c_2} = 1$$

$$T_f = T_{1i}{}^{\alpha} T_{2i}{}^{\beta} \qquad (E7.10.6)$$

where the exponents α and β are given by

$$\alpha = m_1 c_1 / (m_1 c_1 + m_2 c_2)$$

and

$$\beta = m_2 c_2 / (m_1 c_1 + m_2 c_2)$$

(iii) Applying the first law to the composite system consisting of the two blocks and the cyclic engine we have

$$Q_{12} = U_f - U_i + W_{out} \qquad (E7.10.7)$$

Now the heat transfer between the composite system and the surroundings, Q_{12} is zero. Therefore substituting in Eq. (E7.10.7) we obtain

$$0 = (m_1 c_1 T_f + m_2 c_2 T_f) - (m_1 c_1 T_{1i} + m_2 c_2 T_{2i}) + W_{out}$$

The total work output becomes

$$W_{out} = (m_1 c_1 T_{1i} + m_2 c_2 T_{2i}) - (m_1 c_1 + m_2 c_2) T_f \qquad (E7.10.8)$$

where T_f is given by Eq. (E7.10.6).
(b) If an irreversible heat engine is used the entropy production is positive according to the principle of increase of entropy. Therefore using Eq. (7.10.5) we have

$$\sigma = m_1 c_1 \ln(T_{fi}/T_{1i}) + m_2 c_2 \ln(T_{fi}/T_{2i}) \qquad (E7.10.9)$$

where T_{fi} is the new final equilibrium temperature of the two blocks. Eq. (E7.10.9) may be expressed in the form

$$(T_{fi}/T_{1i})^{m_1 c_1} (T_f/T_{2i})^{m_2 c_2} = e^{\sigma}$$

$$T_{fi} = T_{1i}^{\alpha} T_{2i}^{\beta} e^{\lambda} \qquad (E\ 7.10.10)$$

where the exponent λ is given by

$$\lambda = \sigma/(m_1 c_1 + m_2 c_2) > 0 \qquad (E7.10.11)$$

Dividing Eq. (E7.10.10) by Eq. (E7.10.6) we have

$$T_{fi}/T_f = e^{\lambda} > 1$$

because of the condition in Eq. (E7.10.11).

Therefore the final equilibrium temperature of the blocks is larger when the irreversible engine is used. Now the expression in Eq. (E7.10.8) for the work output, which is based on the first law, is applicable both to the reversible engine and the irreversible engine. Due to the higher final equilibrium temperature, $(T_{fi} > T_f)$ the work output of the irreversible engine is lower as expected. Compare the present solution with the solution of a similar problem in worked example 6.17 where an alternative approach was used.

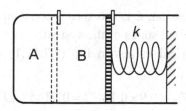

Fig. E7.11 Spring-loaded piston-cylinder set-up

Example 7.11 A non-conducting piston of area 10^{-3} m^2 and negligible mass is free to move in a well-insulated cylinder as shown in Fig. E7.11. Initially, the compartment A of the cylinder with a volume of 10^{-4} m^3 contains helium at 315 K and 15 bar while the piston is held in position by a pin. The compartment B is evacuated. The movement of the piston is resisted by a spring with a spring constant of 0.9 kNm^{-1} that is initially uncompressed. The pin is now removed allowing the gas to expand until the piston hits a second pin at a distance of 0.18 m from the first pin.

Calculate (a) (i) the work done by helium, (ii) the change in entropy of helium, and (iii) the entropy production of the universe.
(b) If the helium undergoes a free expansion without the spring calculate the quantities listed in (a) for the new arrangement.
(c) If the expansion is carried out in a quasi-static manner, calculate the quantities listed in (a).
(d) Comment on the answers to parts (a), (b) and (c).
For helium $c_v = 3.12$ kJ kg^{-1} K^{-1} and $R = 2.08$ kJ kg^{-1} K^{-1}

Solution (a) Applying the equation of state to the initial state

$$P_1 V_1 = mRT_1$$

$$m = 15 \times 100 \times 10^{-4} / (2.08 \times 315) = 2.289 \times 10^{-4} \text{ kg}.$$

The work done by helium is equal to the strain energy stored in the spring because the piston experiences no other external resistance during the expansion.

Hence
$$W_{12} = kx_2^2/2 - kx_1^2/2 \tag{E7.11.1}$$

where x is the compression, k is the spring constant. Subscripts 1 and 2 denote the initial and final states of the system.

Substituting numerical values in Eq. (E7.11.1) we obtain the work done as

$$W_{12} = 0.9 \times 0.18^2 / 2 = 0.01458 \text{ kJ} \qquad (E7.11.2)$$

Applying the first law to the helium gas with, $Q_{12} = 0$ we have

$$0 = U_2 - U_1 + W_{12} \qquad (E7.11.3)$$

Substituting in Eq. (E7.11.3) from Eq. (E7.11.2)

$$0 = mc_v (T_2 - T_1) + 0.01458 \qquad (E7.11.4)$$

Substituting numerical values in Eq. (E7.11.4)

$$0 = 2.289 \times 10^{-4} \times 3.12 \times (T_2 - 315) + 0.01458$$

Therefore $\qquad\qquad\qquad\qquad T_2 = 294.58 \text{ K}$

Assuming that helium is an ideal gas we use the following property relation [Eq. (7.23)] to find the entropy change.

$$S_2 - S_1 = mc_v \ln(T_2/T_1) + mR \ln(V_2/V_1) \qquad (E7.11.5)$$

Substituting numerical values in Eq. (E7.11.5)

$$S_2 - S_1 = 2.289 \times 10^{-4} \times [3.12 \ln(294.58/315) + 2.08 \ln(0.28/0.1)]$$

$$S_2 - S_1 = 4.42 \times 10^{-4} \text{ kJ K}^{-1}$$

The entropy production is also equal to the entropy change of the helium because there is no entropy transfer to the surroundings.

Hence $\qquad\qquad\qquad\qquad \sigma = 4.42 \times 10^{-4} \text{ kJK}^{-1}$

(b) When helium undergoes a free expansion the work done is zero because of the vacuum on the outside of the piston offers no resistance. The heat interaction is also zero because of the thermal insulation. Therefore from the first law it follows that the internal energy is constant. Since helium is an ideal gas, $T_2 = T_1$.

Applying Eq. (7.11.5), the entropy change is

$$S_2 - S_1 = 2.289 \times 10^{-4} [3.12 \ln(315/315) + 2.08 \ln(0.28/0.1)]$$

$$S_2 - S_1 = 4.902 \times 10^{-4} \text{ kJ K}^{-1}$$

The entropy production is the same as the entropy change of the helium.

Therefore $\qquad\qquad \sigma = 4.902 \times 10^{-4}$ kJ K^{-1}

(c) If the helium is expanded to the same final volume in a quasi-static process, with the cylinder insulated, the entropy change is zero according to Eq. (7.10). Applying Eq. (E7.11.5) we have

$$mc_v \ln(T_2 / T_1) + mR \ln(V_2 / V_1) = 0$$

Substituting numerical values in the above equation

$$3.12 \ln(T_2 / 315) + 2.08 \ln(0.28 / 0.1) = 0$$

Hence $\qquad\qquad T_2 = 158.5$ K

Applying the first law

$$0 = mc_v(T_2 - T_1) + W_{12}$$

Substituting numerical values, we have

$$0 = 2.289 \times 10^{-4} \times 3.12 \times (158.5 - 315) + W_{12}$$

Therefore $\qquad\qquad W_{12} = 0.112$ kJ

(d) From the numerical results for the three different expansion processes we observe that the maximum work output is obtained with the reversible adiabatic process, which produces zero entropy. The free expansion, which has no work output produces the largest entropy. The *partially-resisted* expansion using a spring delivers some work output and produces less entropy than the free expansion. We note that entropy production is a measure of the irreversibility of the process.

Example 7.12 A slab of thermal insulation of thickness 0.08 m, area 1.2 m^2 and thermal conductivity 0.05 W m^{-1} K^{-1} is placed between a heat source at 80°C and a heat sink at 25°C, both of which can be treated as reservoirs. There is steady one-dimensional heat flow through the insulation slab.

(a) Calculate (i) the rate of change of entropy of the insulation, (ii) the net rate of entropy transfer from the insulation slab, and (iii) the entropy production rate in the insulation.

(b) If the insulation slab is replaced with one having a thermal conductivity of 0.035 W m^{-1} K^{-1}, what would be the entropy production in the insulation?

Solution (a) Applying Fouriers' law of heat conduction we obtain the steady heat flow rate through the slab as

$$\dot{Q} = kA(T_h - T_c)/\Delta L = 0.05 \times 1.2 \times (80 - 25)/0.08 = 41.25 \ \text{W}$$

Since the slab is in a steady state all its properties are constant with time. Therefore the *change* in entropy of the slab is zero. However, the rate of entropy *production* in steady heat conduction is given by Eq. (7.55). Applying this equation we have

$$\dot{\sigma} = \left(\dot{Q}/T_c - \dot{Q}/T_h \right)$$

$$\dot{\sigma} = (41.25/298 - 41.25/353) = 0.0216 \ \text{WK}^{-1}.$$

We recall that there is entropy transfer at the interfaces between the insulation slab and the two reservoirs. The rate of entropy transfer out of the slab exceeds the entropy transfer into the slab because of the entropy production within the slab. The net entropy transfer rate is equal to the entropy production rate.

(b) With the new insulation slab the heat flow rate is

$$\dot{Q} = kA(T_h - T_c)/\Delta L = 0.035 \times 1.2 \times (80 - 25)/0.08 = 28.88 \ \text{W}$$

The entropy production rate becomes

$$\sigma = (28.88/298 - 28.88/353) = 0.0151 \ \text{WK}^{-1}$$

We note that the use of an insulation slab with a lower thermal conductivity, that is a superior insulation, reduces the entropy production, and therefore the irreversibility.

Example 7.13 Experimental data have been recorded on the work done by a fixed mass of helium gas executing an adiabatic process. Due to an oversight the initial and final states of the gas have not been reliably identified. The available data for two states 1 and 2 are as follows:

State	P bar	V m³
1	6.9	0.085
2	1.01	0.28

(a) Identify correctly the initial and final states of the process.
(b) Calculate the work done by the process.
Assume that helium is an ideal gas for which $c_v = 3.12$ kJ kg⁻¹ K⁻¹ and $R = 2.08$ kJ kg⁻¹ K⁻¹

Solution Let us assume that the initial and final states are 1 and 2 respectively. Since the process is adiabatic, the entropy can only increase or remain constant according to the principle of increase of entropy. Assuming helium to be an ideal gas the entropy change is given by the property relation:

$$S_2 - S_1 = mc_v \ln(T_2/T_1) + mR\ln(V_2/V_1)$$

Substituting for T from the ideal gas equation we have

$$S_2 - S_1 = mc_v \ln(P_2V_2/P_1V_1) + mR\ln(V_2/V_1)$$
$$S_2 - S_1 = mc_v \ln(P_2/P_1) + m(c_v + R)\ln(V_2/V_1)$$
$$S_2 - S_1 = mc_v \ln(P_2/P_1) + mc_p \ln(V_2/V_1)$$

Substituting the given experimental data,

$$(S_2 - S_1)/m = 3.12 \times \ln(101/690) + (3.12 + 2.08)\ln(0.28/0.085)$$
$$(S_2 - S_1)/m = 0.204 > 0$$

Therefore we conclude that 1 and 2, respectively are the initial and final states of helium.

Applying the first law to the adiabatic process,

$$0 = U_2 - U_1 + W_{12}$$

Substituting the ideal gas relations in the above equation we have

$$W_{12} = mc_v(T_1 - T_2) = c_v(P_1V_1 - P_2V_2)/R$$

Substituting numerical values in the above equation,

$$W_{12} = 3.12 \times (690 \times 0.085 - 101 \times 0.28)/2.08 = 45.55 \text{ kJ.}$$

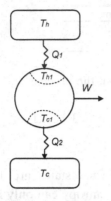
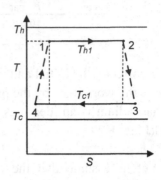

Fig. E7.14(a) Heat engine cycle   Fig. E7.14(b) *T-S* diagram

Example 7.14 An experimental heat engine operates between a heat source reservoir at 1000 K and a heat sink reservoir at 300 K as shown schematically in Fig. E7.14(a). The measured work output and thermal efficiency of the engine are 200 kW and 35 percent respectively. The temperature differences between the working fluid and the heat reservoir are 60°C and 20°C for the hot and cold reservoirs respectively. Assume that the temperature of the working fluid is constant during the heat interactions.

Calculate (i) the rates of entropy transfer, the entropy change, and the entropy production for the cycle and the heat reservoirs, (ii) the entropy production due to heat transfer at the two reservoirs, and (iii) the entropy production in the universe.

Solution The heat flow rates from the reservoirs, the work output and the relevant temperatures are indicated in the energy-flow diagram in Fig. E7.14(a). The *T-S* diagram is shown in Fig. E7.14(b). The pertinent numerical data are:

$$T_c = 300 \, \text{K}, \quad T_{c1} = 300 + 20 = 320 \, \text{K}, \quad T_h = 1000 \, \text{K},$$

$$T_{h1} = 1000 - 60 = 940 \, \text{K}, \quad W = 200 \, \text{kW}, \quad \eta = 35\%.$$

From the given data the efficiency is

$$\eta = W/Q_1 = 200/Q_1 = 0.35$$

Hence the heat input rate is, $Q_1 = 571.4$ kW

Applying the first law to the cycle, the heat rejection rate is

$$Q_2 = Q_1 - W = 571.4 - 200 = 371.4 \text{ kW}$$

The rates of entropy change in the hot and cold reservoirs are:

$$S_{rh} = -Q_1/T_h = -571.4/1000 = -0.5714 \text{ kW K}^{-1}$$

$$S_{rc} = Q_2/T_c = 371.4/300 = 1.238 \text{ kW K}^{-1}$$

For the cyclic engine the entropy *transfer* rates are:

$$S_{e1} = Q_1/T_{h1} = 571.4/(1000 - 60) = 0.608 \text{ kW K}^{-1}$$

$$S_{e2} = Q_2/T_{c1} = 371.4/(300 + 20) = 1.16 \text{ kW K}^{-1}$$

Since the engine operates in a cycle, its entropy *change* is zero. Applying the entropy balance equation to the engine we have

$$S_{eng} = S_{e1} - S_{e2} + \sigma_e$$

Hence $\qquad\qquad \sigma_e = 1.16 - 0.608 = 0.552 \text{ kW K}^{-1}$

which is the entropy production rate in the engine due to *internal* irreversibilities.

The entropy production rates due to heat transfer across the finite temperature differences between the hot and cold reservoirs and the working fluid of the engine are given by Eq. (7.55) as:

$$\sigma_{hot} = Q_1(1/T_{h1} - 1/T_h) = 571.4 \times (1/940 - 1/1000) = 0.0365$$

$$\sigma_{hot} = 0.0365 \text{ kW K}^{-1}$$

$$\sigma_{cold} = Q_2(1/T_c - 1/T_{c1}) = 371.4 \times (1/300 - 1/320)$$

$$\sigma_{cold} = 0.0774 \text{ kW K}^{-1}$$

The above entropy productions are caused by the *external* heat transfer irreversibilities of the engine. The total entropy production rate in the composite system consisting of the two heat reservoirs and the cyclic engine (the universe) is

$$\sigma_{tot} = \sigma_e + \sigma_{hot} + \sigma_{cold}$$

$$\sigma_{tot} = 0.552 + 0.0365 + 0.0774 = 0.666 \text{ kW K}^{-1}$$

Note that by applying the entropy balance for composite system we can verify that

$$S_{rh} + S_{rc} = -0.5714 + 1.238 = 0.666 = \sigma_{tot}$$

Example 7.15 (a) A fixed quantity of dry saturated steam of mass 0.1 kg undergoes a reversible adiabatic expansion from an initial pressure of 600 kPa to a final pressure of 50 kPa. Calculate the work done by the steam.

(b) If an irreversible adiabatic expansion to the same final pressure produces 80% of the work produced by the reversible process, calculate the final equilibrium state of the steam for the irreversible process.

(c) Indicate these processes on a *T-S* diagram for water.

Solution During the reversible adiabatic expansion the entropy of the steam remains constant. From the data for saturated steam at 600 kPa, we obtain [6] for the initial state

$$s_{g1} = 6.761 \text{ kJ K}^{-1} \text{ kg}^{-1}$$

Assume that at the final pressure of 50 kPa the steam is wet. From the saturated steam table [6] at 0.5 bar

$$s_{g2} = 7.593 \text{ kJ K}^{-1} \text{ kg}^{-1} \quad \text{and} \quad s_{f2} = 1.0991 \text{ kJ K}^{-1} \text{ kg}^{-1}$$

Equating the entropies for final and initial states we have

$$ms_1 = ms_2$$

where *m* is the mass of steam.

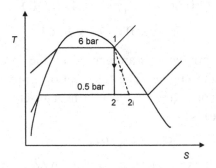

Fig. E7.15 *T-S* diagram for water

Now
$$s_2 = x_2 s_{g2} + (1 - x_2) s_{f2}$$

where x_2 is the steam quality at state 2.

Substituting numerical values in the above equation
$$6.761 = 7.593 x_2 + 1.091(1 - x_2)$$

Therefore
$$x_2 = 0.872$$

Since the final quality of steam is less than one the original assumption that the steam is wet is valid.

Applying the first law to the adiabatic expansion we obtain
$$0 = U_2 - U_1 + W_{12}$$

The internal energies are obtained from the saturated steam table [6]. On substitution in the above equation
$$0 = m[x_2 u_{g2} + (1 - x_2) u_{f2}] - m u_{g1} + W_{12}$$

$$W_{12} = -0.1 \times [0.872 \times 2483 + (1 - 0.872) \times 340] + 0.1 \times 2568$$

Hence
$$W_{12} = 35.93 \text{ kJ}$$

(b) Applying the first law to the irreversible process we have
$$0 = U_2 - U_1 + 0.8 W_{12}$$

Substituting numerical values
$$0 = U_2 - 2568 \times 0.1 + 0.8 \times 35.93$$

Therefore
$$U_2 = 228.056 \text{ kJ}$$

Assume that the steam is wet in the final state. Hence
$$U_2 = 228.056 = 0.1 \times [2483 x_{2i} + 340(1 - x_{2i})]$$

where x_{2i} is the quality of the steam in the final state of the irreversible expansion. From the above equation, $x_{2i} = 0.905$. Note that the steam is in a drier state at the end of the irreversible expansion compared to the reversible expansion.

Example 7.16 An absorption refrigeration system can be represented by a simplified model in which a cyclic device exchanges heat with three reservoirs with no work interaction with the surroundings. The device receives 15kW of heat from a high temperature reservoir at $100\,^{\circ}C$ while absorbing 10 kW from a cold space at $5\,^{\circ}C$. It rejects heat to a heat sink at a temperature of $30\,^{\circ}C$. (a) Calculate (i) the rate of change of entropy in the three reservoirs, and (ii) the rate of entropy production in the universe.

(b) If the device is operated reversibly, what would be the required heat input from the high temperature reservoir to absorb 10 kW from the cold space?

Solution Denote the quantities associated with the high temperature source by 1, the cold space by 2, and the heat sink by 3. We obtain the entropy changes of the hot and cold reservoirs as follows

$$\Delta S_1 = Q_1/T_1 = -15/373 = -0.04021 \text{ kW K}^{-1}$$

$$\Delta S_2 = Q_2/T_2 = -10/278 = -0.03597 \text{ kW K}^{-1}$$

The entropy *change* of the cyclic device is zero because it operates in a cycle.

Applying the first law to the cyclic device with zero work transfer

$$Q_3 = Q_1 + Q_2 = 15 + 10 = 25 \text{ kW}$$

For the heat-sink reservoir

$$\Delta S_3 = Q_3/T_3 = 25/303 = 0.08251 \text{ kW K}^{-1}$$

The entropy production in the composite system consisting of the three reservoirs and the cyclic device is

$$\sigma = \Delta S_1 + \Delta S_2 + \Delta S_3$$

$$\sigma = -0.04021 - 0.03597 + 0.08251 = 0.6328 \times 10^{-2} \text{ kW K}^{-1}$$

Since the entropy production is positive the composite system operates under irreversible conditions. However, the present computation does not reveal the exact source of the irreversibility that causes the entropy production.

(b) If the system is operated reversibly the total entropy production is zero. Applying the same equations as in part (a) for the change of entropy of the reservoirs we obtain the total entropy production as:

$$-Q_1/T_1 - Q_2/T_2 + (Q_1 + Q_2)/T_3 = 0$$

Substituting numerical values in the above equation we have

$$-Q_1/373 - 10/278 + (Q_1 + 10)/303 = 0$$

Hence $\qquad\qquad Q_1 = 4.79\,\text{kW}$

Note that the heat input required to operate the absorption refrigerator increases from a minimum of 4.79 kW to 15 kW due to the irreversibilities of the first system.

Example 7.17 A real refrigeration system extracts heat from a cold space at 10°C and rejects heat to the surroundings at 27°C. Both the cold space and the surroundings may be treated as heat reservoirs as shown schematically in Fig. E7.17. The measured COP and the work input to the refrigerator are 2.1 and 80 kW respectively. The temperature differences between the working fluid and the heat reservoirs are 6°C and 10°C for the cold and hot reservoirs respectively. Assume that the temperature of the working fluid is constant during the heat interactions.

Calculate (i) the rates of entropy transfer, the entropy change, and the entropy production for the cycle and the heat reservoirs, and (ii) the entropy production due to heat transfer at the two reservoirs.

Solution The heat flow rates, the work input and the temperatures of the refrigeration cycle are indicated in the energy flow diagram in Fig. E7.17(a). The T-S diagram of the irreversible refrigeration cycle is shown in Fig. E7.17(b). We assume that during the heat interactions the working fluid of the cyclic device remains at constant temperature. The numerical data pertinent to the problem are as follows:

$$T_c = 273 + 10 = 283 \text{ K}, \qquad T_{c1} = 283 - 6 = 277 \text{ K},$$

$$T_h = 273 + 27 = 300 \text{ K}, \qquad T_{h1} = 300 + 10 = 310 \text{ K}.$$

$$W = 80 \text{ kW}, \qquad\qquad COP = 2.1$$

For the refrigeration cycle,

$$COP_{ref} = Q_c/W = Q_c/80 = 2.1$$

Hence $\qquad\qquad Q_c = 168 \text{ kW}$

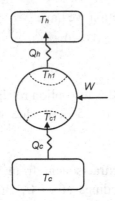

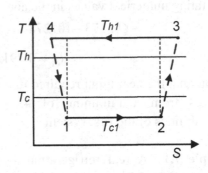

Fig. E7.17(a) Refrigeration cycle   Fig. E7.17(b) *T-S* diagram

Applying the first law to the cycle

$$Q_h = Q_c + W = 80 + 168 = 248 \text{ kW.}$$

The entropy *change* of the cyclic device is zero because it operates in a cycle manner.

The net entropy transfer rate *out* of the cyclic device is

$$\Delta S_{tr,c} = -Q_c/T_{c1} + Q_h/T_{h1}$$

$$\Delta S_{tr,c} = -168/277 + 248/310 = 0.1935 \text{ kW K}^{-1}$$

Applying the entropy balance equation to the cyclic device

$$\Delta S_{change} = \sigma + \Delta S_{trans}$$

$$0 = \sigma - 0.1935$$

Therefore the entropy production rate in the cyclic device is

$$\Delta S_{pro} = 0.1935 \text{ kW K}^{-1}$$

The entropy change of the cold reservoir is

$$\Delta S_{r,cold} = -Q_c/T_c = -168/283 = -0.5936 \text{ kW K}^{-1}$$

The entropy change of the hot reservoir is

$$\Delta S_{r,hot} = Q_h/T_h = 248/300 = 0.8267 \text{ kW K}^{-1}$$

The entropy production of the heat reservoirs are zero because the heat transfer processes in the reservoirs are reversible. Therefore the entropy *transfers* to the two reservoirs are equal to their respective entropy *changes* which were calculated above.

The heat transfer between the cyclic device and the two reservoirs occur across finite temperature differences as indicated in Fig. E7.15(a). These are *external* irreversibilies of the cyclic device. The entropy production due to heat transfer at the two reservoirs is obtained by applying Eq. (7.55),

$$\sigma_{cold} = Q_c\left(1/T_{c1} - 1/T_c\right) = 168 \times \left(1/277 - 1/283\right)$$

$$\sigma_{cold} = 0.01286 \text{ kW K}^{-1}$$

$$\sigma_{hot} = Q_h\left(1/T_h - 1/T_{h1}\right) = 248 \times \left(1/300 - 1/310\right)$$

$$\sigma_{hot} = 0.02667 \text{ kW K}^{-1}$$

Note that it could be easily verified that the sum of the entropy production of the cyclic device and the entropy production due to the two heat transfer irreversibilies is equal to the net entropy change of the two reservoirs.

Example 7.18 A cyclic heat engine is operated between a system and a single thermal reservoir at a temperature T_r as shown in Fig. E7.18. The system does expansion work against an atmosphere at P_o while it undergoes a process from an initial state 1 to a final state 2. The engine executes an integral number of cycles during the process. Obtain an expression for the maximum work output of the engine.

Solution Consider an infinitesimal cycle of the engine whose heat and work interactions are indicated in Fig. E7.18. Apply the first law to the system to obtain

$$-\delta Q = dU + P_o dV \qquad (E7.18.1)$$

Apply the first law to the cyclic engine. Hence

$$\delta Q = \delta W + \delta Q_r \qquad (E7.18.2)$$

Applying the second law to the composite system

$$dS + \delta Q_r / T_r \geq 0 \qquad (E7.18.3)$$

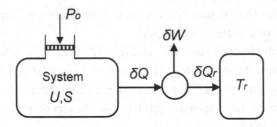

Fig. E7.18 Maximum work output of a closed system

Adding Eqs. (E7.18.1) and (E7.18.2) we have

$$dU + P_o dV + \delta W + \delta Q_r = 0 \qquad \text{(E7.18.4)}$$

Substituting form Eq. (E7.18.4) in Eq. (E7.18.3)

$$-(dU + P_o dV + \delta W) + T_r dS \geq 0$$

$$\delta W \leq T_r dS - dU - P_o dV \qquad \text{(E7.18.5)}$$

Integrating Eq. (E7.18.5) from the initial to the final state of the system

$$W_{12} \leq T_r(S_2 - S_1) - (U_2 - U_1) - P_o(V_2 - V_1)$$

$$W_{12} \leq (U_1 + P_o V_1 - T_r S_1) - (U_2 + P_o V_2 - T_r S_2)$$

Therefore the maximum work output is

$$W_{12,\text{max}} = (U_1 + P_o V_1 - T_r S_1) - (U_2 + P_o V_2 - T_r S_2)$$

Example 7.19 A waste-heat recovery device is to be designed to produce a work output from a fixed mass of hot exhaust air at a temperature and pressure of 1100°C and 4 bar respectively. The air finally attains equilibrium with the atmosphere at 1 bar pressure and 30°C temperature. Calculate the maximum work that could be extracted from the hot air under the given conditions. Assume that air is an ideal gas.

Solution We derived an expression (see Sec. 7.3.2 and worked example 7.18) for the maximum useful work that could be produced by a *closed* system exchanging heat with a single reservoir at T_o, P_o as

$$W_{12,\max} = (U_1 + P_oV_1 - T_oS_1) - (U_2 + P_oV_o - T_oS_2) \qquad \text{(E7.19.1)}$$

where 1 and 2 denote the initial and final states of the system.

The final temperature and pressure of the air are equal to the corresponding values of the atmosphere because the air attains equilibrium with the latter. Hence

$$T_2 = T_o \qquad \text{and} \qquad P_2 = P_o \qquad \text{(E7.19.2)}$$

Assume that air is an ideal gas with the equation state

$$PV = mRT \qquad \text{(E7.19.3)}$$

The entropy change of an ideal gas is obtained from the property relation

$$S - S_o = mc_p \ln(T/T_o) - mR \ln(P/P_o) \qquad \text{(E7.19.4)}$$

Substituting from Eqs. (E7.19.2), (E7.19.3) and (E7.19.4) in Eq. (E7.19.1)

$$\frac{W_{12,\max}}{m} = c_v(T_1 - T_o) + P_o\left(\frac{RT_1}{P_1} - \frac{RT_o}{P_o}\right) - T_o\left[c_p \ln\left(\frac{T_1}{T_o}\right) - R\ln\left(\frac{P_1}{P_o}\right)\right]$$

Substituting the given numerical values and the properties of air in the above equation we have

$$W_{12,\max}/m = 0.718 \times (1373 - 303) + 0.287 \times (1373/4 - 303/1)$$

$$- 303 \times [1.005 \times \ln(1373/303) - 0.287 \times \ln(4/1)]$$

$$W_{12,\max} = 440.2 \text{ kJ per kg of air.}$$

Example 7.20 A process plant has excess steam at 1.5 bar pressure and 200°C temperature. The steam is brought to equilibrium with the atmosphere at 1 bar and 20°C. Calculate the maximum work that could be produced using this change of state of the steam.

Solution We apply the general relation derived earlier for the maximum work available from a process interacting with a single reservoir. Thus

$$W_{12,\max} = (U_1 + P_oV_1 - T_oS_1) - (U_2 + P_oV_o - T_oS_2) \qquad \text{(E7.20.1)}$$

where 1 and 2 represent the initial and final states of the steam.

For superheated steam at 1.5 bar and 200°C we obtain from the steam tables [6]:

$$u_1 = 2656 \text{ kJ kg}^{-1}, \quad v_1 = 1.445 \text{ m}^3 \text{ kg}^{-1}, \quad s_1 = 7.643 \text{ kJ K}^{-1} \text{ kg}^{-1}$$

In the final equilibrium state, the liquid water is sub-cooled. We ignore the effect of pressure and use the saturated liquid data at 20°C. Hence

$$u_2 = 83.9 \text{ kJ kg}^{-1}, v_2 = 1.0018 \times 10^{-2} \text{ m}^3 \text{ kg}^{-1}, \ s_2 = 0.296 \text{ kJ K}^{-1} \text{ kg}^{-1}$$

The ambient conditions are:

$$T_o = 20 + 273 = 293 \text{ K} \quad \text{and} \quad P_o = 100 \text{ kPa}.$$

Substituting the above numerical values in Eq. (E7.20.1) we obtain the maximum work output per unit mass of steam as:

$$W_{12,\text{max}} = (2656 + 100 \times 1.445 - 293 \times 7.643)$$

$$- (83.9 + 100 \times 1.0018 \times 10^{-2} - 293 \times 0.296) = 562.9 \text{ kJ kg}^{-1}$$

Problems

P7.1 Show that any reversible process executed by a system could be replaced with a combination of two reversible adiabatic processes and a reversible isothermal process for which the magnitudes of the work and heat interactions are the same as those of the original process. Hence show that for a reversible cycle, $\oint \delta Q / T = 0$.

P7.2 A fixed quantity of steam of mass 0.12 kg is compressed in a reversible and adiabatic process from an initial pressure of 0.8 bar and quality 0.9 until the steam is just dry saturated. Calculate (i) the pressure in the final state, and (ii) the work done. [*Answers*: (i) 5.126 bar, (ii) 33.04 kJ]

P7.3 A block of copper of mass 1.5 kg at a temperature of 80°C is placed in thermal contact with a block of iron of mass 2 kg at a temperature of 20°C. The two blocks are well insulated from the surroundings. Calculate (i) the final equilibrium temperature of the blocks, (ii) the change in entropy of the blocks, and (iii) the entropy production in the universe.

The specific heat capacities of copper and iron are 0.38 kJ kg^{-1} K^{-1} and 0.5 kJ kg^{-1} K^{-1} respectively.
[*Answers*: (i) 41.8°C, (ii) −0.0653 kJ K^{-1}, 0.0717 kJ K^{-1}, (iii) 0.0064 kJ K^{-1}]

P7.4 A fixed quantity of air of mass 0.12 kg is trapped in a well-insulated cylinder behind a piston. The initial equilibrium pressure and temperature are 1.1 bar and 283 K respectively. A large mass is now placed on the piston and released. After a few oscillations the system attains an equilibrium state with an air pressure of 1.6 bar. (a) Calculate (i) the final temperature of the air, and (ii) the entropy change of the air. (b) Suggest a series of reversible processes to take the air back to its original state. [*Answers*: (a) (i) 46.7°C, (ii) 0.00182 kJ K^{-1}, (b) Reversible adiabatic expansion to 1.0436 bar and 283 K, reversible isothermal compression to 1.1 bar]

P7.5 In an experiment a fixed quantity of steam at 4 bar and 150°C undergoes an adiabatic expansion to a pressure of 1 bar. The final equilibrium quality of the steam has been measured as 0.94. (i) Check whether the process is feasible. (ii) If the expansion was reversible what should be the final quality of the steam? [*Answers*: feasible, $s_f - s_i = 0.066 > 0$, $x_2 = 0.929$]

P7.6 A closed system consisting of 0.15 kg of air, initially at 100 kPa and 303 K, undergoes a process to a final state of 360 kPa and 500 K, while in thermal contact with a reservoir at 500 K, from which it receives 8 kJ of heat.

Calculate (i) the change in entropy of the air, (ii) the change in entropy of the reservoir, (iii) the entropy production in the universe, and (iv) the minimum reversible work input for the same change of state of the system and the reservoir with a standard ambient reservoir at 293 K. [*Answers*: (i) 0.02036 kJ K^{-1}, (ii) −0.016 kJ K^{-1}, (iii) 0.00436 kJ K^{-1}, (iv) 11.8 kJ]

P7.7 A fixed mass of air is trapped in a cylinder behind a frictionless piston of diameter 0.035 m. The mass of the piston and the weights placed on it is 12 kg and the external pressure on the piston is zero.

Initially, the conditions of the gas are: $P_1 = 1.5$ bar, $V_1 = 10^{-4} \, m^3$ and $T_1 = 30°C$. The piston is help in position by a pin. The air, the piston and cylinder are in thermal equilibrium with the surroundings which may be regarded as a thermal reservoir at 30°C. The pin is now removed and the piston moves up until a final equilibrium state is reached. (a) Calculate (i) change in entropy of the air, (ii) the change in entropy of the surroundings, and (iii) the entropy production in the universe. (b) Suggest a reversible process to return the air to the original state. [*Answers*: (a) (i) 1.0×10^{-5} kJ K^{-1}, (ii) -0.912×10^{-5} kJ K^{-1}, (iii) 0.88×10^{-6} kJ K^{-1}, (b) reversible isothermal compression with a work input of 3.055×10^{-3} kJ]

P7.8 A piston-cylinder apparatus contains $0.12 \, m^3$ of nitrogen at pressure of 1.0 bar and temperature 28°C, initially. The nitrogen undergoes a compression process with a work input of 48 kJ until its pressure and temperature become 12 bar and 220°C respectively. During the process the nitrogen exchanges heat with the surroundings at 28°C. Calculate (i) the heat transfer from the nitrogen, (ii) the entropy change of the nitrogen, (iii) the entropy change of the surroundings treating it as a reservoir, and (iv) the entropy production in the universe. For nitrogen, $c_p = 1.04$ kJ kg^{-1} K^{-1}, $c_v = 0.743$ kJ kg^{-1} K^{-1}. [*Answers*: (i) -28.88 kJ, (ii) -0.0301 kJ K^{-1}, (iii) 0.0959 kJ K^{-1}, (iv) 0.0658 kJ K^{-1}]

P7.9 The water in a tank is maintained at a steady uniform temperature of 80°C by an electrical immersion heater. The tank is thermally insulated with a material of thickness 0.05 m and thermal conductivity 0.045 $Wm^{-1}K^{-1}$. The temperature of outer surface of the insulation is equal to the ambient temperature of 20°C. The tank is in a steady state with a constant mass of water.

Calculate, per unit area of insulation: (i) the rate of change of entropy of the water, (ii) the rate of change of entropy of the ambient, which can be regarded as a reservoir at 20°C, and (iii) the rate of entropy production in the tank, the insulation, and the universe. [*Answers*: (i) 0 W K^{-1} m^{-2}, (ii) 0.184 W K^{-1} m^{-2}, (iii) 0.153, 0.031, 0.184 W K^{-1} m^{-2}]

P7.10 A poorly-insulated vessel contains a block of ice of mass 600 kg at a temperature of −20°C and pressure of 1 bar. The ice exchanges heat with the surroundings at 30°C and eventually attains thermal equilibrium with the latter. The pressure remains constant during the process. Calculate (i) the total heat flow between the surroundings and the ice, (ii) the change in entropy of the ice, (iii) the change in entropy of the surroundings, and (iv) the entropy production in the universe. Neglect the heat capacity of the vessel. The specific heat capacity and latent heat of fusion of ice are 2.1 kJ kg^{-1} K^{-1} and 334 kJ kg^{-1} respectively. [*Answers*: (i) 301.2×10^3 kJ, (ii) 1092.6 kJ K^{-1}, (iii) 994.0 kJ K^{-1}, (iv) 98.6 kJ K^{-1}]

P7.11 A real heat engine operates between two reservoirs at 800 K and 290K. The cyclic device of the engine experiences a temperature difference of 15 K and 10 K at the hot and cold reservoirs respectively, due to heat transfer. The efficiency of the engine is 28% and the power output is 600 kW. Calculate (i) entropy change in the two reservoirs, (ii) the entropy production due to heat transfer at the two reservoirs, and (iii) the internal entropy production of the cyclic device.
[*Answers*: (i) −2.68 kW K^{-1}, 5.32 kW K^{-1}, (ii) 0.0512 kW K^{-1}, 0.1773 kW K^{-1}, (iii) 2.411 kW K^{-1}]

P7.12 A fixed quantity of air of mass 0.2 kg is at a pressure of 10 bar and temperature 80°C. The air is to undergo a process exchanging heat with a standard-ambient reservoir at 20°C and 1 bar until the final pressure and temperature are 2 bar and 30°C. Calculate the maximum useful work that could be produced for this change of state of the air. [*Answer*: 18.58 kJ]

P7.13 The specific heat capacity of an ideal gas varies with temperature according to the relationship, $c_v = a + bT$. The equation of state of the ideal gas is, $PV = mRT$. Obtain (i) an expression for the change in entropy from state 1 to state 2, (ii) the T-V relationship for an isentropic expansion of the gas from state 1 to state 2, and (iii) an expression for the efficiency of a Carnot cycle using the ideal gas as the working fluid.
[*Answers*: (i) $S_2 - S_1 = ma\ln(T_2/T_1) + mb(T_2 - T_1) + mR(V_2/V_1)$, (ii) $V_2/V_1 = (T_1/T_2)^{(a/R)}\exp[b(T_1 - T_2)/R]$, (iii) $\eta = 1 - T_c/T_h$]

362 Engineering Thermodynamics

P7.14 An internally irreversible heat engine operating with an ideal gas receives a quantity of heat Q_h per cycle isothermally from a reservoir at T_h, and rejects heat isothermally to a reservoir at T_c. The entropy production during the adiabatic compression and expansion process of the cycle are σ_c and σ_e respectively. Sketch the T-S diagram of the cycle. Obtain an expression for the difference in efficiency between this cycle, and a Carnot cycle operating between the same reservoirs. [*Answer*: $\eta_{Carnot} - \eta = T_c(\sigma_c + \sigma_e)/Q_h$]

References

1. Bejan, Adrian, *Advanced Engineering Thermodynamics*, John Wiley & Sons, New York, 1988.
2. Cravalho, E.G. and J.L. Smith, Jr., *Engineering Thermodynamics*, Pitman, Boston, MA, 1981.
3. Jones, J.B. and G.A. Hawkins, *Engineering Thermodynamics*, John Wiley & Sons, Inc., New York, 1986.
4. Kestin, Joseph, *A Course in Thermodynamics*, Blaisdell Publishing Company, Waltham, Mass., 1966.
5. Reynolds, William C. and Henry C. Perkins, *Engineering Thermodynamics*, 2nd edition, McGraw-Hill, Inc., New York, 1977.
6. Rogers, G.F.C. and Mayhew, Y.R., *Thermodynamic and Transport Properties of Fluids*, 5th edition, Blackwell, Oxford, U.K. 1998.
7. Tester, Jefferson W. and Michael Modell, *Thermodynamics and Its Applications*, 3rd edition, Prentice Hall PTR, New Jersey, 1996.
8. Van Wylen, Gordon J. and Richard E. Sonntag, *Fundamentals of Classical Thermodynamics*, 3rd edition, John Wiley & Sons, Inc., New York, 1985.
9. Wark, Jr., Kenneth, *Advanced Thermodynamics for Engineers*, McGraw-Hill, Inc., New York, 1995.
10. Zemansky, M., *Heat and Thermodynamics*, 5th edition, McGraw-Hill, Inc., New York, 1968.

Chapter 8

Second Law Analysis of Open Systems

The second law and its important consequences were introduced in chapters 6 and 7 where the emphasis was on applications to closed systems. In this chapter we shall apply the entropy balance equation to open systems or control volumes with the aim of developing the tools for availability or exergy analysis of such systems. More recently, this type of performance evaluation of engineering systems has been called second law analysis, although it is the combination of the first and second laws that forms the basis of exergy analysis and the second law efficiency.

8.1 The Entropy Balance Equation

In this section the entropy balance equation, developed in chapter 7 as Eq. (7.45), is extended to control volumes or open systems. The entropy balance equation involves three quantities called entropy transfer, entropy production, and entropy change or storage. Entropy transfer to a control volume occurs due to heat and mass exchange between the control volume and the surroundings. The production of entropy is a result of the irreversibilities within the control volume. The storage of entropy in the control volume is a consequence of the imbalance between the production and transfer of entropy.

To derive the entropy balance equation we shall adopt an approach similar to that used earlier in chapter 5, to derive the mass balance and energy balance equations for a control volume. Consider the control volume depicted schematically in Figs. 8.1(a) and (b) where matter enters through an inlet duct at A and leaves through an exit duct at B. We focus our attention on the changes experienced by a fixed mass of matter

363

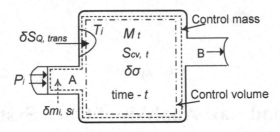

Fig. 8.1(a) Control mass and volume at time t

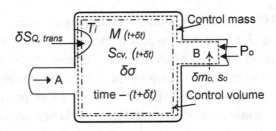

Fig. 8.1(b) Control mass and volume at time $(t + \delta t)$

contained within the control mass boundary shown in Fig. 8.1(a) at time t, and in Fig, 8.1(b) at time, $(t + \delta t)$. It should be noted that although shown separately for clarity, the boundaries of the control volume, the control mass, and the physical enclosure are coincidental, except for the small sections in the inlet port A and the exit port B. At time t, the boundary of the control mass includes the mass M_t in the control volume with a total entropy of $S_{cv,t}$ and the mass δm_i with a specific entropy of s_i in a small section of the inlet duct at A.

During the small time interval δt, the boundary of the control mass deforms to the shape in Fig. 8.1(b) where the mass in duct A has just entered the control volume while a mass δm_o, with a specific entropy of s_o, has been pushed out of the control volume into a small section of the exit duct at B.

At the same time the mass within the control volume has become $M_{(t+\delta t)}$ and the total entropy of the control volume has changed to $S_{cv,(t+\delta t)}$. Let the total entropy transfer to the control volume due to heat transfer and the total entropy production due to irreversibilities during the time δt be $\delta \Delta S_{Q,trans.}$ and $\delta \sigma$ respectively.

The use of single inlet and outlet streams, in the derivation of the entropy balance equation, is not a limitation because it is possible, by straightforward summation, to write the corresponding expressions for multiple streams. The entropy balance equation for the *fixed* mass of matter within the *control mass* boundary can be written as,

$$entropy_{change} = entropy_{production} + entropy_{transfer} \qquad (8.1)$$

$$(S_{cv,(t+\delta t)} + \delta m_o s_o) - (S_{cv,t} + \delta m_i s_i) = \delta \sigma + \delta \Delta S_{Q,trans} \qquad (8.2)$$

We divide Eq. (8.2) by the time interval δt and rearrange it to obtain

$$\frac{(S_{cv,(t+\delta t)} - S_{cv,t})}{\delta t} = \frac{s_i \delta m_i}{\delta t} - \frac{s_o \delta m_o}{\delta t} + \frac{\delta \Delta S_{Q,trans}}{\delta t} + \frac{\delta \sigma}{\delta t} \qquad (8.3)$$

Taking the limit of each term in Eq. (8.3) as $\delta t \to 0$, the rate-form of the entropy balance equation is obtained as

$$\dot{S}_{cv} = \dot{m}_i s_i - \dot{m}_o s_o + \Delta \dot{S}_{Q,trans} + \dot{\sigma} \qquad (8.4)$$

where the 'dot' over a quantity represents, (d/dt) the differentiation with respect to time. The entropy production rate, denoted by $\dot{\sigma}$, is not a property like entropy but is a path-dependent quantity. However, if all processes are reversible, then, $\dot{\sigma} = 0$.

For a control volume with multiple inlet and exit streams we generalize Eq. (8.4) by summing over the different streams. Thus

$$\dot{S}_{cv} = \sum_{inlet} \dot{m}_i s_i - \sum_{outlet} \dot{m}_o s_o + \Delta \dot{S}_{Q,trans} + \dot{\sigma} \qquad (8.5)$$

The entropy transfer rate $\Delta \dot{S}_{Q,trans}$ needs additional consideration before it could be expressed in terms of the various heat interactions between the control volume and the surroundings. We have two possible options for relating the entropy transfer to the corresponding heat transfer.

If the control volume receives a heat *input* of magnitude $\dot{Q}_i$ across a location *on the control volume boundary* where the temperature is T_i (see Fig. 8.1a) then the entropy transfer *to the control volume* is

$$\Delta \dot{S}_{Q,trans} = \dot{Q}_i / T_i \qquad (8.6)$$

When there are multiple locations, say n, over the *control volume* where such heat interactions and entropy transfers occur, the total entropy transfer becomes

$$\Delta \dot{S}_{Q,trans} = \sum_n \dot{Q}_i / T_i \qquad (8.7)$$

In many practical situations the temperature of the control volume where heat is received is not readily available. This is especially true when the processes undergone by the control volume are irreversible and consequently, there are internal temperature gradients within the control volume.

Usually, the heat flow to a control volume occurs from thermal reservoirs with known constant temperatures. In such situations we could express the entropy transfer in terms of the reservoir temperatures. Thus

$$\Delta \dot{S}_{Q,trans} = \sum_n \dot{Q}_i / T_{ri} \qquad (8.8)$$

where T_{ri} is the temperature of reservoir i. Both Eqs. (8.7) and (8.8) are valid expressions for the entropy transfer to the control volume.

However, for the same change of state of the control volume the magnitude of the entropy production term, $\dot{\sigma}$ depends on which expression is used to obtain the entropy transfer. It is clear that if, $T_{ri} = T_i$, then both expressions will give the same entropy production. But if these temperatures are different, then there would be an external irreversibility at each heat transfer location because heat is being transferred across the finite temperature difference $(T_{ri} - T_i)$. This, in turn, would add to the total entropy production of the control volume when the reservoir temperatures are used in the analysis. We shall rewrite Eq. (8.5) using both expressions for entropy transfer for easy reference later in the chapter. These take the forms

$$\dot{S}_{cv} = \sum_{inlet} \dot{m}_i s_i - \sum_{outlet} \dot{m}_o s_o + \sum_n \dot{Q}_i / T_i + \dot{\sigma}_{cv} \qquad (8.9)$$

$$\dot{S}_{cv} = \sum_{inlet} \dot{m}_i s_i - \sum_{outlet} \dot{m}_o s_o + \sum_n \dot{Q}_i / T_{ri} + \dot{\sigma} \qquad (8.10)$$

For the *same change of state* of the control volume, the total entropy production due to heat transfer between the reservoirs and the control volume, $\dot{\sigma}_{ht}$ is obtained by subtracting Eq. (8.9) from Eq. (8.10) as

$$\dot{\sigma}_{ht} = \dot{\sigma} - \dot{\sigma}_{cv} = \sum_{n} \dot{Q}_i (1/T_i - 1/T_{ri}) \qquad (8.11)$$

We obtained an expression similar to Eq. (8.11) in Sec. 7.2.3 for the entropy production due to steady heat conduction in a rod.

The following are some noteworthy conclusions, regarding the nature of the various terms in Eqs. (8.9) and (8.10).

(i) According to the principle of increase of entropy, the entropy production, $\dot{\sigma}$ can never be negative. Moreover, when all the processes are *reversible* the entropy production is zero.

(ii) The entropy transfer due to heat transfer can be positive or negative.

(iii) The net entropy transfer due to matter exchange between the control volume and the surroundings can be positive or negative.

(iv) The entropy change or storage, $\dot{S}_{cv}$ can be positive or negative for *unsteady* operation of the control volume.

(v) If the control volume is in a *steady state* then, $\dot{S}_{cv} = 0$ and the entropy balance expressed by Eq. (8.10) takes the form

$$\sum_{inlet} \dot{m}_i s_i - \sum_{outlet} \dot{m}_o s_o + \sum_{n} \dot{Q}_i / T_{ri} + \dot{\sigma} = 0 \qquad (8.12)$$

8.2 The Combined First and Second Laws

The vast majority of energy conversion systems are designed to produce work in the form of motive power or electricity utilizing the stored energy in fuels. Even under ideal conditions, the maximum work output of these systems is subject to the limits imposed by the second law. Although the first law ensures energy balance it does not distinguish the difference in 'quality' between work and heat.

This difference in quality arises because the second law stipulates that heat cannot be converted *completely* to work while the reverse process is feasible. Although the second law reveals the presence of irreversibilities in a system, it does not relate their impact directly to the work output of

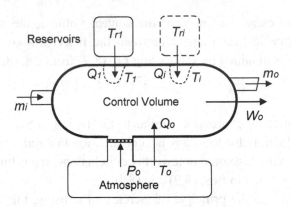

Fig. 8.2 Control volume interacting with reservoirs

the system. For the design and optimization of energy conversion systems it is useful to develop a direct relationship between the magnitude of the irreversibilities and the maximum work that could be harnessed.

We shall now derive such a relationship for a control volume by combining the first and second laws as described below. Consider the control volume shown in Fig. 8.2 where heat interactions occur with a series of reservoirs while delivering a work output W_o.

The first law for the control volume may be expressed as

$$\dot{E}_{cv} = \sum_{in} \bar{h}_i \dot{m}_i - \sum_{out} \bar{h}_o \dot{m}_o + \sum_{n+1} \dot{Q}_i + \dot{Q}_o - \dot{W}_o \qquad (8.13)$$

where, for the sake of brevity, the 'effective-enthalpies' of the inlet and outlet streams are denoted by:

$$\bar{h}_i = (h_i + V_i^2/2 + gz_i) \quad \text{and} \quad \bar{h}_o = (h_o + V_o^2/2 + gz_o)$$

respectively. All other terms in Eq. (8.13) were defined Sec. 5.2 when we applied the first law to a control volume. Application of the second law, in the form of the entropy balance equation, to the control volume gives

$$\dot{S}_{cv} = \sum_{inlet} \dot{m}_i s_i - \sum_{outlet} \dot{m}_o s_o + \sum_{n+1} \dot{Q}_i/T_{ri} + \dot{\sigma} \qquad (8.14)$$

In Eq. (8.14), the entropy transfers to the control volume due to heat transfers are expressed in terms of the reservoir temperatures. These temperatures, however, could be easily replaced with control-volume temperatures if they are available.

In the following analysis we relate the useful work output of the control volume to the entropy production rate for a given *change of state* of the latter. It is seen that when $\dot{W}_o$ is varied for the same change of state of the control volume, the energy balance expressed by Eq. (8.13) can be satisfied only by allowing at least *one* of the heat interactions with the heat reservoirs to vary. The reservoir selected for this purpose is designated by the subscript o in Fig. 8.2. In most practical applications this reservoir is the local atmosphere whose temperature T_o and pressure P_o are constant. If the control volume has a flexible boundary, then a fraction of the work output of the control volume may be needed to overcome the force due to the external pressure P_o of the atmosphere. This is represented schematically by the piston-cylinder arrangement in Fig. 8.2.

We now rewrite Eqs. (8.13) and (8.14) indicating explicitly the interactions between the control volume and the atmosphere-reservoir. Thus

$$\dot{E}_{cv} = \sum_{in} \overline{h}_i \dot{m}_i - \sum_{out} \overline{h}_o \dot{m}_o + \sum_{i=1}^{n} \dot{Q}_i + \dot{Q}_o - \dot{W}_o \tag{8.15}$$

$$\dot{S}_{cv} = \sum_{inlet} \dot{m}_i s_i - \sum_{outlet} \dot{m}_o s_o + \sum_{i=1}^{n} \frac{\dot{Q}_i}{T_{ri}} + \frac{\dot{Q}_o}{T_o} + \dot{\sigma} \tag{8.16}$$

where n is the number of heat reservoirs *excluding* the atmosphere-reservoir. Multiplying Eq. (8.16) by T_o and subtracting the resulting equation from Eq. (8.15) we obtain

$$(\dot{E}_{cv} - T_o \dot{S}_{cv}) = \sum_{in}(\overline{h}_i - T_o s_i)\dot{m}_i - \sum_{out}(\overline{h}_o - T_o s_o)\dot{m}_o$$

$$+ \sum_{i=1}^{n} \dot{Q}_i(1 - T_o/T_{ri}) - \dot{W}_o - T_o\dot{\sigma} \tag{8.17}$$

When a fraction of the work output is used to overcome the *external* pressure of the atmosphere, we may express the *useful* work output as

$$\dot{W}_{use} = \dot{W}_o - P_o \dot{V}_{cv} \tag{8.18}$$

where $\dot{V}_{cv}$ is the rate of change of volume of the control volume.

Substituting for $\dot{W}_o$ from Eq. (8.18) in Eq. (8.17) we have

$$\dot{W}_{use} = -(\dot{E}_{cv} + P_o\dot{V}_{cv} - T_o\dot{S}_{cv}) + \sum_{in}(\bar{h}_i - T_o s_i)\dot{m}_i$$

$$-\sum_{out}(\bar{h}_o - T_o s_o)\dot{m}_o + \sum_{i=1}^{n}\dot{Q}_i(1 - T_o/T_{ri}) - T_o\dot{\sigma} \qquad (8.19)$$

If all processes within the control volume and the heat transfer between the reservoirs and the control volume are *reversible* the entropy production is zero. The work *output* is then a maximum and Eq. (8.19) takes the form

$$\dot{W}_{rev} = -(\dot{E}_{cv} + P_o\dot{V}_{cv} - T_o\dot{S}_{cv}) + \sum_{in}(\bar{h}_i - T_o s_i)\dot{m}_i$$

$$-\sum_{out}(\bar{h}_o - T_o s_o)\dot{m}_o + \sum_{i=1}^{n}\dot{Q}_i(1 - T_o/T_{ri}) \qquad (8.20)$$

For the same *change of state* of the control volume and the same *heat interactions* with the reservoirs, all terms in Eqs. (8.19) and (8.20) are identical *except* the work output and the entropy production. Subtracting Eq. (8.20) from (8.19) we have

$$\dot{W}_{use} = \dot{W}_{rev} - T_o\dot{\sigma} = \dot{W}_{act} \qquad (8.21)$$

Hence $\qquad \dot{I} = (\dot{W}_{rev} - \dot{W}_{act}) = T_o\dot{\sigma} \geq 0 \qquad (8.22)$

The difference between the maximum (reversible) work output and the actual output $(\dot{W}_{act})$, given by Eq. (8.22), is called the *irreversibility rate, $\dot{I}$* or *rate of lost work*.

We note that Eq. (8.22) was obtained for a control volume that produces a work *output*; therefore both $\dot{W}_{act}$ and $\dot{W}_{rev}$ are positive quantities with $\dot{W}_{rev} \geq \dot{W}_{act}$. However, Eq. (8.22) is equally valid for control volumes that absorb work, for which $\dot{W}_{act}$ and $\dot{W}_{rev}$ are both negative quantities.

Since $T_o\dot{\sigma} \geq 0$ it follows from Eq. (8.22) that $\dot{W}_{rev} \leq \dot{W}_{act}$ for work absorbing control volumes.

Recall that in Sec. 7.4, we derived the corresponding expressions for the reversible work and irreversibility for closed systems.

For control volumes where all processes occur under *steady-state* conditions we can rewrite Eqs. (8.19) and (8.20) as:

$$\dot{W}_{use} = \sum_{in} (\overline{h}_i - T_o s_i)\dot{m}_i - \sum_{out} (\overline{h}_o - T_o s_o)\dot{m}_o$$

$$+ \sum_{i=1}^{n} \left(1 - \dot{Q}_i/T_{ri}\right) - T_o \dot{\sigma} \tag{8.23}$$

$$\dot{W}_{rev} = \sum_{in} (\overline{h}_i - T_o s_i)\dot{m}_i - \sum_{out} (\overline{h}_o - T_o s_o)\dot{m}_o + \sum_{i=1}^{n} \left(1 - \dot{Q}_i/T_{ri}\right) \tag{8.24}$$

8.2.1 *Availability or exergy*

A close examination of Eq. (8.20), for the reversible work output, reveals that the different terms of the equation involve properties of the control volume and the inlet and outlet streams. In order to develop a consistent physical interpretation of the various terms of the above equation we first define an atmosphere whose state is called the *standard environmental state* or the *dead state*.

If the control volume and the flow streams are to individually attain equilibrium with the dead state, through a *reversible work device exchanging heat solely with the dead-state-reservoir*, then the work output of the device gives the maximum work-potential of the control volume and the streams. The maximum work-potential is therefore a property, and is called *availability* or *exergy*. When equilibrium is attained with the dead state the kinetic and gravitational potential energies of the control volume and the streams are zero; the temperature and pressure are equal to those of the chosen dead state.

If we denote the dead-state properties by the superscript o, then we can express the maximum work-potential of the control volume, given by the first term on the right hand side of Eq. (8.20), in the form

$$\Phi_{cv} = E_{cv} - E^o + P^o(V_{cv} - V^o) - T^o(S_{cv} - S^o) \tag{8.25}$$

The property Φ_{cv} is called the control-volume *availability* or *exergy*. At equilibrium with the dead-state, $E^o = U(T^o)$ because the kinetic energy and potential energy of the dead state are both zero. The volume and entropy of the control volume at equilibrium with the dead state are V^o and S^o respectively.

The maximum work-potential or flow-exergy of a typical flow-stream, included in the second and third terms of Eq. (8.20), may be expressed in the form

$$\psi_i = (\bar{h}_i - h^o) - T^o(s_i - s^o) \qquad (8.26)$$

The *flow-exergy* is the maximum work that could be produced by the flow-stream when it attains equilibrium with the dead state, exchanging heat solely with the latter. The enthalpy and entropy per unit mass of the flow-stream at equilibrium with the dead state are given by,

$$h^o = u^o + P^o v^o$$

and

$$s^o = s^o(T^o, P^o)$$

respectively. The kinetic and potential energies of the flow-stream at equilibrium with the dead state are zero.

The availability or exergy of a typical heat interaction, included in the last term of Eq. (8.20), is

$$\Phi_{Qi} = Q_i \left(1 - T^o / T_{ri}\right) \qquad (8.27)$$

The *exergy* of the heat input Q_i can be interpreted as the maximum work output of a Carnot heat engine, operating between a reservoir at the heat input temperature, T_{ri} and a reservoir at the dead state temperature T^o, with a heat input of Q_i.

Using the availability or exergy terms defined above we can rewrite Eq. (8.19) in the form

$$\dot{W}_{use} = -\frac{d}{dt} \Phi_{cv} + \sum_{in} \psi_i \dot{m}_i - \sum_{out} \psi_o \dot{m}_o + \sum_n \dot{\Phi}_{Qi} - T^o \dot{\sigma}$$

$$+ \left[-\frac{dM_{cv}}{dt}(u^o + P^o v^o - T^o s^o) + \sum_{in} \dot{m}_i (h^o - T^o s^o) \right]$$

$$+ \left[-\sum_{out} \dot{m}_i (h^o - T^o s^o) \right] \qquad (8.28)$$

It follows from the mass balance equation for the control volume, that the sum of the terms within the two square-brackets in the above equation is zero. Therefore

$$\dot{W}_{use} = -\dot{\Phi}_{cv} + \sum_{in}\psi_i\dot{m}_i - \sum_{out}\psi_o\dot{m}_o + \sum_{n}\dot{\Phi}_{Qi} - T^o\dot{\sigma} \qquad (8.29)$$

Equation (8.29) gives the actual amount of work harnessed from the 'maximum work-potential' or exergy available in the various input streams and heat interactions. A non-zero entropy production, $\dot{\sigma}$ is an indication that some of the original work-potential of the control volume, the flow streams and the heat inputs have been destroyed due to irreversibilities.

For steady-flow situations Eq. (8.29) becomes

$$\dot{W}_{use} = \sum_{in}\psi_i\dot{m}_i - \sum_{out}\psi_o\dot{m}_o + \sum_{n}\dot{\Phi}_{Qi} - T^o\dot{\sigma} \qquad (8.30)$$

8.2.2 *First law and second law efficiency*

The efficiency of a cyclic device, based on the first law, is the ratio of the desired output of the device to the total energy input to the device. We have already adopted this definition of the first law efficiency to evaluate the performance of heat engines and refrigerators. In the case of work devices like turbines the first law efficiency is defined as the ratio of the actual work output to the maximum theoretical work output. For compressors and pumps that absorb work the ratio of the theoretical minimum work required to the actual work input gives the first law efficiency. Though the first law efficiency is used widely in engineering practice, it does not take into account the limits imposed by the second law on the maximum work output of a device or a cycle. The second law emphasizes the fact that the same quantity of energy may have very different 'work-potentials' or availabilities.

Following are two general definitions of the second law efficiency. In the first version the second law efficiency is defined as the ratio of the useful availability (exergy) output to the availability (exergy) input

$$\eta_2 = \frac{useful - availibilty - out}{availability - in} \qquad (8.31)$$

Alternatively, the second law efficiency may be defined as the ratio of the minimum quantity of available energy, A_{min} required to perform a task to the actual available energy A_{act} consumed by the system being used to perform the task. Hence

$$\eta_2 = A_{min} / A_{actual} \tag{8.32}$$

The second law efficiency defined by Eq. (8.31) is specific to the process or device, while the second version given by Eq. (8.32) compares the given system or process with an ideal process or system that requires the least amount of available energy to perform the same task. We need to be cautious when comparing numerical values of the second law efficiency because of possible differences in the definition of the efficiency.

8.3 Worked Examples

Example 8.1 (i) Obtain an expression for the work output of a reversible, adiabatic steady-flow process.
(ii) Obtain an expression for the work input to a *non-adiabatic* reversible compressor.
(iii) Obtain an expression for the reversible work required to compress a liquid from a pressure P_1 to P_2 in a steady flow process.

Solution (i) Consider a packet of fluid of mass δm undergoing an infinitesimal *reversible* change of state through an *adiabatic* work device. Applying the first law to the fluid packet, neglecting changes in kinetic and potential energy, we have

$$\delta Q_{in} = \delta m(h + dh) - \delta m h + \delta W_{out} = 0$$

where h is the specific enthalpy of the fluid.
    Hence

$$- \delta W_{out} / \delta m = dh = d(u + Pv) = du + Pdv + vdP \tag{E8.1.1}$$

We invoke the general thermodynamic property relation [Eq. (7.17)] for a unit mass given by

$$Tds = du + Pdv \tag{E8.1.2}$$

For a reversible adiabatic process, $ds = 0$. Therefore

$$du + Pdv = 0 \qquad \text{(E8.1.3)}$$

From Eqs. (E8.1.1) and (E8.1.3) it follows that

$$\delta W_{out}/\delta m = -vdP \qquad \text{(E8.1.4)}$$

The total work *output* per unit mass of fluid is obtained by integrating Eq. (E8.1.4) as

$$W_{out} = -\int_{in}^{out} vdP$$

(ii) The analysis in (i) above could be applied if the fluid packet is subject to a *non-adiabatic reversible* process. The relevant equations are

First law: $\qquad\qquad \delta Q_{in} = \delta m(h + dh) - \delta mh + \delta W_{out}$

Second law: $\qquad\qquad \delta Q_{in} = \delta mTds$

General property relation [Eq. (7.17)]: $\quad Tds = du + Pdv$

Manipulating the three equations above we have

$$\delta W_{out}/\delta m = -vdP$$

The total work *input* per unit mass of fluid is

$$W_{in} = \int_{in}^{out} vdP \qquad \text{(E8.1.5)}$$

(iii) It was shown in parts (i) and (ii) that the reversible work is given by

$$W_{in} = \int_{in}^{out} vdP \qquad \text{(E8.1.6)}$$

In general, the compressibility of a liquid is very low and therefore the change in specific volume during a compression process is relatively small. Hence it is reasonable to assume that the specific volume is constant and equal to the mean value, $\bar{v} = (v_1 + v_2)/2$. Substituting in Eq. (E8.1.6) we obtain

$$W_{in} = \int_{in}^{out} vdP = \bar{v}\int_{P_1}^{P_2} dP = \bar{v}(P_2 - P_1) \qquad \text{(E8.1.7)}$$

Example 8.2 (i) Air is compressed in a steady-flow, reversible process from 400 kPa and 20°C to 800 kPa. The compressor is cooled externally. The air follows the process-relation, $Pv^{1.2} = C$. Calculate the work input per unit mass of air.

(ii) If the air is compressed from the same inlet conditions to the same exit pressure in an irreversible adiabatic compressor, calculate the work input per unit mass. The process-relation is $Pv^{1.5} = C$.

Solution (i) The process-relation is, $Pv^{1.2} = C$. For a reversible non-adiabatic flow process it was shown in worked example 8.1 above that the work done per unit mass is

$$W_{out} = \int_{in}^{out} vdP$$

Substituting for v from the process-relation in the above equation we obtain the total work input per unit mass as

$$W_{in} = \int_1^2 \frac{C}{P^{1/n}} dP = \frac{n(P_2 v_2 - P_1 v_1)}{(n-1)}$$

$$W_{in} = \frac{nP_1 v_1 (P_2 v_2 / P_1 v_1 - 1)}{(n-1)} \qquad (E8.2.1)$$

Using the ideal gas equation of state, $Pv = RT$ and the process-relation, Eq. (E8.2.1) may be manipulated to obtain the work input per unit mass as

$$W_{in} = \frac{nRT_1[(P_2/P_1)^{(n-1)/n} - 1]}{(n-1)} \qquad (E8.2.2)$$

Substituting numerical values in Eq. (E8.2.2)

$$W_{in} = \frac{1.2 \times 0.287 \times 293 \times [(800/400)^{(1.2-1)/1.2} - 1]}{(1.2-1)} = 61.8$$

Therefore the work input per unit mass is 61.8 kJ kg^{-1}.

(ii) Applying the SFEE to the *adiabatic irreversible* process we have

$$0 = \dot{m}h_2 - \dot{m}h_1 + \dot{W}_{out}$$

$$\dot{W}_{in}/\dot{m} = c_p T_2 - c_p T_1 \qquad (E8.2.3)$$

From the ideal gas equation of state and the process-relation it follows that

$$T_2 = T_1 \left(P_2 / P_1 \right)^{(n-1)/n} = 293 \times 2^{0.333} = 369.15\,\text{K}$$

Substituting numerical values in Eq. (E8.2.3) we obtain the work per unit mass as

$$\dot{W}_{in}/\dot{m} = 1.005 \times (369.15 - 293) = 76.45\ \text{kJ kg}^{-1}$$

Note that the irreversible compressor requires more work than the reversible compressor for the same change of pressure.

Example 8.3 A rigid pressure vessel of volume 1.5 m³ contains air at a pressure of 1 bar and temperature 303 K. A reversible isothermal compressor is used to evacuate the vessel to zero pressure while discharging the air to the ambient at 1 bar. The temperature of the air in the vessel during the process may be assumed constant and equal to the ambient temperature of 303 K. Calculate (i) the work input to the compressor, (ii) the heat transfer, and (iii) the entropy production in the universe.

Solution (i) We showed in worked example 8.1 that the work input per unit mass for a *non-adiabatic, reversible* flow process is

$$W_{in} = \int\limits_{in}^{out} v\,dP \qquad\qquad (E8.3.1)$$

For an ideal gas the equation of state per unit mass is

$$Pv = RT \qquad\qquad (E8.3.2)$$

Substituting for v in Eq. (E8.3.1) from Eq. (E8.3.2) we have

$$W_{in} = \int\limits_{P_1}^{P_2} \frac{RT}{P}\,dP = RT \ln\left(P_2 / P_1 \right) \qquad\qquad (E8.3.3)$$

for an *isothermal* process at temperature T.

It follows from Eq. (E8.3.3) that the work input required to pump a small quantity of air of mass δm out of the vessel (Fig. E8.3) is

$$\delta W_{in} = \delta m RT \ln\left(P_2 / P_1 \right) \qquad\qquad (E8.3.4)$$

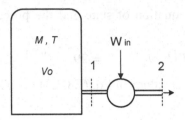

Fig. E8.3 Evacuation of a vessel

where P_1 is the pressure inside the tank and P_2 is the ambient pressure. The constant temperature of the air in the vessel is T.

Applying the mass balance equation for the air in the vessel we have

$$\delta m = -\delta M$$

Substituting from the ideal gas equation for the mass of air, M in the vessel

$$\delta m = -\delta M = -\delta(P_1 V_o / RT) = -(V_o / RT)\delta P_1 \qquad \text{(E8.3.5)}$$

where V_o is the volume of the vessel.

From Eqs. (E8.3.4) and (E8.3.5) we have

$$\delta W_{in} = -V_0 \ln(P_2 / P_1)\delta P_1 \qquad \text{(E8.3.6)}$$

The total work required to evacuate the vessel is obtained by integrating Eq. (E8.3.6) from the initial pressure, $P_o = P_2 = 100$ kPa to zero. Hence

$$W_{in} = -V_0 \int_{P_o}^{0} \ln(P_2 / P_1) dP_1 = V_o P_o = 1.5 \times 100 = 150 \,\text{kJ}$$

(ii) Applying the *unsteady* energy equation to the vessel and the compressor we have

$$\dot{m}h_i + \dot{Q}_{in} = \dot{E} + \dot{m}h_o + \dot{W}_{out} \qquad \text{(E8.3.7)}$$

Integrating Eq. (E8.3.7) from the beginning to the end of the evacuation process

$$Q_{in} = 0 - M_f c_v T_o + M_f c_p T_o - P_o V_o$$

where M_f is the total mass of the air removed from the vessel. The work input to the compressor is $P_o V_o$. Substituting from the ideal gas equation in the above equation we obtain

$$Q_{in} = 0 - M_f c_v T_o + M_f c_p T_o - M_f R T_0$$

$$Q_{in} = -M_f c_v T_o + M_f c_p T_o - M_f T_0 (c_p - c_v) = 0$$

The net heat transfer is zero because the heat flow from the atmosphere to the vessel is equal to the heat flow out of the compressor to the atmosphere during the isothermal evacuation process.

(iii) The entropy balance equation for the control volume consisting of the vessel and the compressor is

$$\dot{S}_{cv} = \dot{m} s_i - \dot{m} s_o + \dot{Q}_{in} / T_o + \dot{\sigma}_{cv} \qquad (E8.3.8)$$

Integrate Eq. (E8.3.8) from the beginning to the end of the evacuation process to obtain

$$S_{cv,final} - S_{cv,initial} = 0 - M_f s_o + 0 + \sigma_{cv}$$

$$0 - M_f s_{cv} = 0 - M_f s_o + \sigma_{cv}$$

$$M_f (s_o - s_{cv}) = \sigma_{cv} \qquad (E8.3.9)$$

The pressure and temperature of the air initially in the vessel and finally in the atmosphere are the same. Hence from the property relation for the change in entropy of an ideal gas [Eq. (7.25)] it follows that

$$(s_o - s_{cv}) = 0$$

From Eq. (E8.3.9), the entropy production in the control volume, $\sigma_{cv} = 0$. The entropy production of the atmosphere, $\sigma_{res} = 0$. The entropy production in the universe is

$$\sigma_{universe} = \sigma_{cv} + \sigma_{res} = 0$$

Since the pumping and heat transfer processes are reversible, we could have concluded that the entropy production in the universe is zero.

Example 8.4 Air from the atmosphere at temperature T_o and pressure P_o flows through a reversible isothermal turbine into a rigid tank of volume V_i, which is initially evacuated. The turbine produces a continuous work output until the tank pressure becomes equal to the ambient pressure of P_o. The wall of the tank is well-insulated. Obtain expressions for: (i) the work output of the turbine, (ii) the final mass of the air in the tank,

(iii) the final temperature of air in the tank, (iv) the total heat transfer between the composite system and the atmosphere, and (v) the total entropy production in the universe. Assume that air is an ideal gas.

Solution (i) The work output per unit mass of the *isothermal* turbine is given by Eq. (E8.3.3) as

$$W'_{out} = -\int_{P_o}^{P_2} \frac{RT_o}{P} dP = -RT_o \ln(P_2/P_o) \qquad \text{(E8.4.1)}$$

where P_2 is the pressure in the tank during the filling process.

Apply the transient energy equation to the *adiabatic* tank for a small change of state, and substitute for mass from the ideal gas equation. Thus

$$\delta m c_p T_o = \delta E = \delta(Mc_v T) = \delta\left(\frac{P_2 V_o c_v T}{RT}\right) \qquad \text{(E8.4.2)}$$

where M and T are the mass and temperature of the air in the tank at any instant, and δm is a small mass flowing into the tank through the turbine. From Eq. (E8.4.2) we have

$$\delta m = (V_o/\gamma RT_o)\delta P_2 \qquad \text{(E8.4.3)}$$

where $\gamma = c_p/c_v$

The work output of the turbine is

$$\delta W_{out} = W'_{out}\delta m = -mRT_o \ln(P_2/P_o)\delta m$$

Substituting for δm from Eq. (E8.4.3) in the above equation and integrating the resulting equation we have

$$W_{out} = -(V_o/\gamma)\int_0^{P_o} \ln(P_2/P_o)dP_2 = P_o V_o/\gamma$$

(ii) Integrating Eq. (E8.4.3), the final mass of air in the tank becomes

$$M_f = \int_0^{P_o} (V_o/\gamma RT_o)dP_2 = P_o V_o/\gamma RT_o \qquad \text{(E8.4.4)}$$

Applying the ideal gas equation to the final state of air in the tank

$$M_f = P_o V_o/RT_f \qquad \text{(E8.4.5)}$$

(iii) From Eqs. (E8.4.4) and (E8.4.5) the final temperature of the air in the tank is obtained as

$$T_f = \gamma T_o$$

(iv) Apply the transient energy equation to the turbine and the tank

$$\dot{m}h_i + \dot{Q}_{in} = \dot{E} + \dot{m}h_o + \dot{W}_{out} \qquad (E8.4.6)$$

Integrating Eq. (E8.4.6) from the initial state to the final state of the filling process we have

$$Q_{in} + M_f c_p T_o = M_f c_v T_f + W_{out} \qquad (E8.4.7)$$

where M_f is the final mass of air in the tank. Substituting for T_f and W_{out} in the above equation

$$Q_{in} + M_f c_p T_o = M_f c_v \gamma T_o + M_f R T_f / \gamma$$

Hence
$$Q_{in} = M_f R T_o = P_o V_o$$

(v) The entropy balance equation for the composite system consisting of the tank, the turbine, and the atmosphere is integrated from the beginning to the end of the filling process to obtain

$$\sigma = (S_{cv,f} - S_{cv,i}) - M_f s_i - Q_{in}/T_o \qquad (E8.4.8)$$

where M_f is the final mass of air in the tank.

Substituting the expression for the entropy of an ideal gas [Eq. (7.25)] in Eq. (E8.4.8) we have

$$\sigma = M_f s_f - M_f s_i - Q_{in}/T_o$$

$$\sigma = M_f c_p \ln(\gamma T_o / T_o) - M_f R T_o / T_o$$

$$\sigma = M_f (c_p \ln \gamma - R) = (P_o V_o / \gamma R T_o)(c_p \ln \gamma - R) \qquad (E8.4.9)$$

Substituting the properties of air in Eq. (E8.4.9),
$$\sigma = M_f (1.005 \times \ln 1.4 - 0.287) = M_f \times 0.051 > 0$$

The total entropy production in the universe is positive and therefore filling the tank through the turbine is an irreversible process. The irreversibility is due to the finite temperature difference between the air entering the tank from the turbine and the air already in the tank during the process.

Example 8.5 Air is compressed in a steady flow adiabatic process from the inlet conditions of 90 kPa, 15°C to the exit conditions of 400 kPa and 200°C. The air velocities at the inlet and exit are 120 ms^{-1} and 50 ms^{-1} respectively. Calculate per kg of air: (i) the entropy production, (ii) the reversible work input for the same change of state of air, assuming the standard ambient conditions of 100 kPa and 27°C, and (iii) the irreversibility.

Solution Applying the entropy balance equation to the steady-flow adiabatic compressor,

$$\dot{m}s_i - \dot{m}s_o + \dot{\sigma} = 0 \qquad\qquad (E8.5.1)$$

Assume that air is an ideal gas and substitute the expression for the change in entropy from Eq. (7.25). Thus

$$\dot{\sigma}/\dot{m} = c_p \ln(T_o/T_i) - R\ln(P_o/P_i) \qquad\qquad (E8.5.2)$$

Substituting numerical values in Eq. (E8.5.2) we have

$$\dot{\sigma}/\dot{m} = 1.005\ln(473/288) - 0.287\ln(400/90) = 0.07 \text{ kJ K}^{-1}\text{ kg}^{-1}$$

Applying the SFEE to the compressor,

$$\dot{Q}_{in} + \dot{m}(h_i + V_i^2/2 + gz_i) = \dot{W}_{out} + \dot{m}(h_o + V_o^2/2 + gz_o)$$

Assume that the change in elevation is negligible and that air is an ideal gas. Therefore

$$(c_pT_i + V_i^2/2) = (\dot{W}_{out}/\dot{m}) + (c_pT_o + V_o^2/2) \qquad\qquad (E8.5.3)$$

Substituting numerical values in Eq. (E8.5.3) we have

$$(\dot{W}_{out}/\dot{m}) = -(1.005 \times 473 + 50^2 \times 10^{-3}/2)$$

$$+ (1.005 \times 288 + 120^2 \times 10^{-3}/2) = 180 \text{ kJ kg}^{-1}$$

The work *input* per unit mass is 180 kJ kg^{-1}.

The availability (exergy) of the inlet air stream per unit mass is

$$\psi_{in} = (h_{in} + V_i^2/2 - h^o) - T^o(s_{in} - s^o)$$

$$\psi_{in} = c_p(T_{in} - T^o) + V_i^2/2 - T^o\left[c_p \ln(T_{in}/T^o) - R\ln(P_{in}/P^o)\right]$$

Note that the kinetic energy and the potential energy of the standard state (dead state) are zero.

Substituting numerical values in the above equation we have

$$\psi_{in} = 1.005 \times (288 - 300) + 120^2 \times 10^{-3}/2$$
$$- 300 \times [1.005 \ln(288/300) - 0.287 \ln(90/100)]$$

Hence
$$\psi_{in} = -1.623 \ kJ \ kg^{-1}$$

Similarly, for the outlet stream

$$\psi_{out} = 1.005 \times (473 - 300) + 50^2 \times 10^{-3}/2$$
$$- 300 \times [1.005 \ln(473/300) - 0.287 \ln(400/100)]$$

$$\psi_{out} = 157.159 \ kJ \ kg^{-1}$$

The reversible work input per unit mass is
$$W_{rev} = \psi_{out} - \psi_{in} = 158.8 \ kJ \ kg^{-1}$$

For the work *absorbing* compressor, the irreversibility is given by
$$I = W_{actal} - W_{rev} = 180 - 158.8 = 21.2 \ kJ \ kg^{-1}$$

We also note that

$$I = T^o \sigma / \dot{m} = 300 \times 0.07 = 21 \ kJ \ kg^{-1}$$

As expected the two results are in agreement. The small difference in the numerical values is due to 'rounding-off'.

Example 8.6 Superheated steam at 10 bar and 300°C is supplied steadily to a turbine at the rate of 1.5 kg s⁻¹. The steam leaves at 0.18 bar. The work output of the turbine is 900 kW. The rate of heat loss from the turbine to the surroundings at 25°C is 18 kW. Assume standard ambient conditions of 10°C and 1 bar. Calculate (i) the entropy production rate in the turbine, the surroundings, and the universe, (ii) the reversible work for the same change of state of the steam, and (iii) the irreversibility rate.

Solution Apply the SFEE to the turbine, neglecting kinetic energy and potential energy changes, to obtain

$$\dot{Q}_{in} + \dot{m}h_i = \dot{m}h_o + \dot{W}_o \qquad \text{(E8.6.1)}$$

At the inlet the steam is superheated. From tabulated data in [6] we have

$$h_i = 3052 \text{ kJ kg}^{-1} \quad \text{and} \quad s_i = 7.124 \text{ kJ K}^{-1} \text{ kg}^{-1}.$$

Substituting numerical values in Eq. (E8.6.1)

$$-18 + 1.5 \times 3052 = 1.5 \times h_o + 900$$

Hence $\qquad\qquad\qquad h_o = 2440 \text{ kJ kg}^{-1}$

The saturated enthalpy of steam at the outlet pressure of 0.18 bar is 2605 kJ kg⁻¹, from data in [6]. Therefore the steam is wet at the outlet.

$$2440 = h_{go}x_o + h_{fo}(1 - x_o) = 2605x_o + 242(1 - x_o)$$

Hence the quality $\qquad\qquad x_o = 0.93$

The entropy at the outlet is

$$s_o = s_{go}x_o + s_{fo}(1 - x_o)$$

$$s_o = 7.944x_o + 0.804(1 - x_o) = 7.444 \text{ kJ K}^{-1} \text{ kg}^{-1}$$

Apply the entropy balance equation to the turbine in the steady-state

$$s_i\dot{m} - s_o\dot{m} + \dot{Q}_{in}/T_{amb} + \dot{\sigma}_{tur} = 0 \qquad \text{(E8.6.2)}$$

Substituting numerical values in Eq. (E8.6.2) we have

$$\dot{\sigma}_{tur} = 7.444 \times 1.5 - 7.124 \times 1.5 + 18/298 = 0.54 \text{ kJ K}^{-1}$$

The entropy production in the surroundings is zero because the latter is a thermal reservoir with a uniform temperature.

The entropy production in the universe is

$$\dot{\sigma}_{univ} = \dot{\sigma}_{tur} + \dot{\sigma}_{sur} = 0.54 \text{ kJ K}^{-1}$$

The standard ambient or dead-state conditions are 10°C and 1 bar. We assume that in the dead state the water is a sub-cooled liquid at 10°C. Ignoring the effect of pressure, we obtain the following properties at 10°C form [6],

$$h^o = 42 \text{ kJ kg}^{-1} \quad \text{and} \quad s^o = 0.151 \text{ kJ K}^{-1} \text{ kg}^{-1}$$

The availabilities or exergies per unit mass of the inlet and outlet streams are given by:

$$\psi_{in} = (h_i - h^o) - T_o(s_i - s^o)$$

$$\psi_{out} = (h_o - h^o) - T_o(s_o - s^o)$$

Substituting numerical values in the above expressions we have

$$\psi_{in} = (3052 - 42) - 283 \times (7.124 - 0.151) = 1036.64 \text{ kJ kg}^{-1}$$

$$\psi_{out} = (2440 - 42) - 283 \times (7.444 - 0.151) = 334.08 \text{ kJ kg}^{-1}$$

The exergy outflow due to the heat loss is

$$\Phi_Q = \dot{Q}(1 - T^o / T_{sur}) = 18 \times (1 - 283/298) = 0.906 \text{ kW}$$

The maximum reversible work output is

$$\dot{W}_{rev} = \dot{m}\psi_{in} - \dot{m}\psi_{out} - \Phi_Q$$

$$\dot{W}_{rev} = 1.5 \times (1036.64 - 334.081) - 0.906 = 1052.9 \text{ kW}$$

The irreversibility rate is

$$\dot{I} = \dot{W}_{rev} - \dot{W}_{act} = 1052.9 - 900 = 152.9 \text{ kW}$$

Also, $$\dot{I} = T^o \dot{\sigma} = 283 \times 0.54 = 152.8 \text{ kW}$$

Example 8.7 A heat exchanger for cooling oil using cold water is shown in Fig. E8.7. Oil flows through the inner tube at the rate of 0.6 kg s^{-1} while water flows through the annulus at the rate of 0.5 kg s^{-1}. The oil is to be cooled from 75°C to 50°C. The temperature of the entering water is 28°C. Oil and water may be treated as pure substances with constant specific heat capacities of 1.65 kJ kg^{-1} K^{-1} and 4.2 kJ kg^{-1} K^{-1} respectively. The surroundings are at a temperature of 25°C. Calculate (i) the entropy

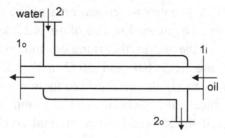

Fig. E8.7 Heat exchanger for cooling oil

production rate in the heat exchanger, and (ii) the irreversibility rate assuming the standard ambient to be the surroundings.

Solution (i) Apply the SFEE to the heat exchanger assuming no heat losses to the surroundings.

$$\dot{m}_o c_o (T_{oi} - T_{oo}) = \dot{m}_w c_w (T_{wo} - T_{wi}) \qquad \text{(E8.7.1)}$$

where subscripts o and w denote oil and water respectively.
   Substituting numerical values in Eq. (E8.7.1) we have

$$0.6 \times 1.65 \times (75 - 50) = 0.5 \times 4.2 \times (T_{wo} - 28)$$

Therefore the water outlet temperature is, $T_{wo} = 39.78\,^{\circ}\text{C}$
   Apply the entropy balance equation to the heat exchanger.

$$\dot{\sigma} = \sum_{out} \dot{m}_o s_o - \sum_{in} \dot{m}_i s_i \qquad \text{(E8.7.2)}$$

Neglecting any pressure losses of the two streams we have

$$\dot{\sigma} = \dot{m}_o c_o \ln(T_{oo}/T_{oi}) + \dot{m}_w c_w \ln(T_{wo}/T_{wi}) \qquad \text{(E8.7.3)}$$

Substituting numerical values in Eq. (E8.7.3) we obtain

$$\dot{\sigma} = 0.6 \times 1.65 \times \ln(323/348) + 0.5 \times 4.2 \times \ln(312.8/301)$$

$$\dot{\sigma} = -0.0738 + 0.0806 = 0.00695 \ \text{kW K}^{-1}$$

which is the entropy production rate.
(ii) The standard ambient is at a temperature of 25°C. Therefore the irreversibility rate is

$$\dot{I} = T^o \dot{\sigma} = (273 + 25) \times 0.00695 = 2.07 \ \text{kW}$$

Example 8.8 Air flowing through a converging duct receives heat from a reservoir at 300°C. The cross-sectional area of the duct changes from 0.25 m² to 0.075 m². The mass flow rate of air is 2.2 kg s⁻¹. The pressure and temperature of the air at the entrance and exit sections of the converging duct are 1.15 bar and 30°C and 1.35 bar and 120°C respectively. Calculate (i) the entropy production rates in the control volume of the duct, the reservoir, and the universe, and (ii) the irreversibility rate if the standard environmental conditions are 1.0 bar and 25°C.

Solution Let the subscripts 1 and 2 denote quantities at the entrance and exit of the duct. Applying the mass balance equation we have

$$\dot{m} = A_1 V_1 \rho_1 \qquad (E8.8.1)$$

Substituting for density from the ideal gas equation in Eq. (E8.8.1)

$$\dot{m} = A_1 V_1 P_1 / RT_1 \qquad (E8.8.2)$$

Substituting numerical values in Eq. (E8.8.2), we have for the entrance section

$$V_1 = \dot{m} RT_1 / A_1 P_1 = 2.2 \times 0.287 \times 303 / 0.25 \times 115 = 6.65 \text{ ms}^{-1}$$

Similarly, for the exit section

$$V_2 = \dot{m} RT_2 / A_2 P_2 = 2.2 \times 0.287 \times 393 / 0.075 \times 135 = 24.5 \text{ ms}^{-1}$$

Applying the SFEE to a control volume surrounding the duct we have

$$\dot{Q}_{in} = \dot{m}(h_2 + V_2^2/2) - \dot{m}(h_1 + V_1^2/2) \qquad (E8.8.3)$$

Note that in Eq. (E8.8.3) the difference in potential energy is neglected and the work output is zero.

Assuming air to be an ideal gas, Eq. (E8.8.3) can be written as

$$\dot{Q}_{in} = \dot{m}(c_p T_2 + V_2^2/2) - \dot{m}(c_p T_1 + V_1^2/2) \qquad (E8.8.4)$$

Substituting numerical values in Eq. (E8.8.4)

$$\dot{Q}_{in} = 2.2 \times 1.005 \times (393 - 303) + 2.2 \times 10^{-3} \times (24.5^2/2 - 6.65^2/2)$$

Hence
$$\dot{Q}_{in} = 199.6 \text{ kW}$$

Apply the entropy balance equation to the control volume

$$\dot{S}_{stor,cv} = \dot{m}s_1 - \dot{m}s_2 + \dot{Q}_{in}/T_r + \dot{\sigma}_{cv} \qquad (E8.8.5)$$

Since the flow is steady, the rate of entropy storage in the control volume is zero. Substituting the expression for the entropy change of an ideal gas in Eq. (E8.8.5)

$$\dot{\sigma}_{cv} = \dot{m}\left(c_p \ln(T_2/T_1) - R\ln(P_2/P_1)\right) - \dot{Q}_{in}/T_r$$

Substituting numerical values in the above equation

$$\dot{\sigma}_{cv} = 2.2 \times \left(1.005 \times \ln(393/303) - 0.287 \times \ln(135/100)\right) - 199.6/573$$

$$\dot{\sigma}_{cv} = 0.3855 - 0.3483 = 0.0372 \text{ kW K}^{-1}$$

The entropy production rate of the reservoir is zero because the heat transfer process in the reservoir is reversible. Note, however, that the rate of entropy change of the reservoir is, $199.6/573 = 0.348$ kW K^{-1}.

Therefore entropy production rate of the universe is

$$\dot{\sigma}_{uni} = \dot{\sigma}_{cv} + \dot{\sigma}_{res} = 0.0372 \text{ kW K}^{-1}$$

The irreversibility rate is

$$\dot{I} = \dot{W}_{rev} - \dot{W}_{act} = T_o \dot{\sigma}_{uni} = 298 \times 0.0372 = 11.09 \text{ kW}$$

Example 8.9 Two liquids with specific heat capacities c_1 and c_1 flow in parallel through a control volume with flow rates $\dot{m}_1$ and $\dot{m}_2$ as shown in Fig. E8.9. At the inlet the temperatures of the fluids are T_{i1} and T_{i2} respectively. A series of infinitesimal reversible work devices, operating on the Carnot cycle, extract heat from the hot stream 1 and reject heat to stream 2. At the exit the fluids leave with the same temperature T_f. There is no heat interaction between the control volume and the surroundings. Obtain an expression for the maximum work output of the device.

Solution Consider the operation of an infinitesimal Carnot engine which receives a quantity of heat δQ_1 from the hot stream and rejects δQ_2 to the cold stream while producing a work output of δW_o as shown in Fig. E8.9.

Applying the SFEE to the liquid streams we have

$$\delta Q_1 = -\dot{m}_1 c_1 dT_1 \qquad (E8.9.1)$$

$$\delta Q_2 = \dot{m}_2 c_2 dT_2 \qquad (E8.9.2)$$

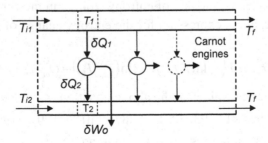

Fig. E8.9 Maximum work from two streams

where T_1 and T_2 are respectively the temperatures of the hot and cold streams at the locations where an engine receives and rejects heat.

Apply the first law to the Carnot engine to obtain

$$\delta W_o = \delta Q_1 - \delta Q_2 \qquad \text{(E8.9.3)}$$

Applying the entropy balance equation to the Carnot engine we have

$$\delta Q_1 / T_1 = \delta Q_2 / T_2 \qquad \text{(E8.9.4)}$$

Substituting form Eqs. (E8.9.1) and (E8.9.2) in Eq. (E8.9.4)

$$-\dot{m}_1 c_1 dT_1 / T_1 = \dot{m}_2 c_2 dT_2 / T_2 \qquad \text{(E8.9.5)}$$

Integrating Eq. (E8.9.5) and substituting the given conditions at the inlet of the streams we obtain

$$\dot{m}_1 c_1 \ln T_1 + \dot{m}_2 c_2 \ln T_2 = \dot{m}_1 c_1 \ln T_{1i} + \dot{m}_2 c_2 \ln T_{2i} \qquad \text{(E8.9.6)}$$

The work output is a maximum when the both streams reach the same final exit temperature. Thereafter no more work could be extracted from the streams. Let the final equilibrium exit temperature of the two streams be T_f. Then from Eq. (E8.9.6) it follows that

$$(\dot{m}_1 c_1 + \dot{m}_2 c_2) \ln T_f = \dot{m}_1 c_1 \ln T_{1i} + \dot{m}_2 c_2 \ln T_{2i}$$

Hence

$$T_f = T_{1i}{}^\lambda T_{2i}{}^\beta \qquad \text{(E8.9.7)}$$

where the exponents in Eq. (E8.9.7) are given by:

$$\lambda = \dot{m}_1 c_1 / (\dot{m}_1 c_1 + \dot{m}_2 c_2) \quad \text{and} \quad \beta = \dot{m}_2 c_2 / (\dot{m}_1 c_1 + \dot{m}_2 c_2)$$

Integrating Eq. (E8.8.3) from the entrance to the exit we obtain the maximum total work output as

$$\dot{W}_o = \dot{m}_1 c_1 (T_{1i} - T_f) - \dot{m}_2 c_2 (T_f - T_{2i})$$

where T_f is given by Eq. (E8.9.7).

Example 8.10 A vortex tube (Fig. E8.10) is a device that receives a stream of air tangentially at location 1 at 500 kPa, 27°C and delivers two streams of air at 100 kPa, –23°C and 100 kPa, 37°C at the exit locations 2 and 3 respectively. The mass flow rate of air at 1 is 0.01 kg s^{-1}. The device has no heat or work interactions with the surroundings. Calculate (i) the mass flow rates at 2 and 3, (ii) the entropy production in the

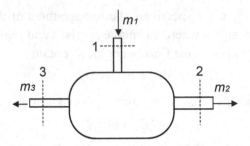

Fig. E8.10 Vortex tube cooler

device, and (iii) for the same inlet and outlet conditions of the air, the maximum work output if the device is allowed to exchange heat with a reservoir at 20°C. Assume that air is an ideal gas.

Solution The vortex-tube is shown schematically in Fig. E8.10. It receives high pressure air in a tangential manner and produces a cold stream and a hot stream. The vortex tube has no moving parts and it is being used for special cooling applications.

Applying the mass balance equation to the control volume 1-2-3

$$\dot{m}_1 = \dot{m}_2 + \dot{m}_3 \qquad (E8.10.1)$$

Applying the SFEE to the control volume, neglecting kinetic and potential energy changes we have

$$\dot{m}_1 h_1 = \dot{m}_2 h_2 + \dot{m}_3 h_3 \qquad (E8.10.2)$$

Substituting the expressions for the enthalpy of an ideal gas in Eq. (E8.10.2)

$$\dot{m}_1 c_p T_1 = \dot{m}_2 c_p T_2 + \dot{m}_3 c_p T_3 \qquad (E8.10.3)$$

Substituting numerical values in Eqs. (E8.10.1) and (E8.10.3) we obtain

$$\dot{m}_2 + \dot{m}_3 = 0.01$$

and $$250\dot{m}_2 + 310\dot{m}_3 = 300 \times 0.01$$

Solving the two equations above simultaneously

$$\dot{m}_2 = 1.667 \times 10^{-3} \text{ kg s}^{-1} \quad \text{and} \quad \dot{m}_3 = 8.333 \times 10^{-3} \text{ kg s}^{-1}$$

Applying the entropy balance equation to the control volume 1-2-3 we have

$$\dot{S}_{stor} = \dot{m}_1 s_1 - \dot{m}_2 s_2 - \dot{m}_3 s_3 + \dot{\sigma} \qquad (E8.10.4)$$

Since the system is in a steady state the rate of entropy storage in the control volume is zero. Hence

$$\dot{\sigma} = -(\dot{m}_2 + \dot{m}_3)s_1 + \dot{m}_2 s_2 + \dot{m}_3 s_3$$

$$\dot{\sigma} = \dot{m}_2 (s_2 - s_1) + \dot{m}_3 (s_3 - s_1) \qquad (E8.10.5)$$

Substituting the expression for the entropy change of an ideal gas [Eq. (7.25)] in Eq. (E8.10.5) we have

$$\dot{\sigma} = \dot{m}_2 \left(c_p \ln(T_2/T_1) - R \ln(P_2/P_1) \right)$$

$$+ \dot{m}_3 \left(c_p \ln(T_3/T_1) - R \ln(P_3/P_1) \right)$$

Substituting numerical values in the above equation

$$\dot{\sigma} = 1.667 \times 10^{-3} (1.005 \ln(250/300) - 0.287 \ln(100/500))$$

$$+ 8.33 \times 10^{-3} (1.005 \ln(310/300) - 0.287 \ln(100/500))$$

$$\dot{\sigma} = 4.587 \times 10^{-3} \ \text{kW K}^{-1}$$

The irreversibility rate is given by

$$\dot{I} = \dot{W}_{rev} - \dot{W}_{act} = T_o \dot{\sigma} = 293 \times 4.587 \times 10^{-3} = 1.34 \ \text{kW}.$$

The actual work output of the device is zero. Therefore the maximum reversible work is

$$\dot{W}_{rev} = 293 \times 4.587 \times 10^{-3} = 1.34 \ \text{kW}.$$

Example 8.11 Air enters an ideal reversible adiabatic turbine with a mass flow rate of $\dot{m}$ at pressure P_i and temperature T_i. The air leaving the turbine at a pressure P_o, is divided into two streams each with a mass flow rate of $\dot{m}/2$. A series of infinitesimal Carnot heat pumps are used to transfer energy from one stream to the other (see Fig. E8.11) so that one stream is cooled while the other is heated. The heat pumps consume the entire work output of the turbine. Obtain expressions for the maximum and minimum temperatures of the streams that could be achieved by

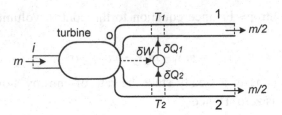

Fig. E8.11 Turbine operating reversible heat pumps

this system. Calculate these temperatures for the following conditions: $P_i = 400$ kPa, $T_i = 300$ K, $P_o = 100$ kPa. There are on other heat or work interactions between the system and the surroundings. Assume that air is an ideal gas.

Solution As the air expands through the reversible adiabatic (isentropic) turbine the entropy change is zero. Using the expression for the entropy change of an ideal gas [Eq. (7.25)] we have

$$c_p \ln(T_o/T_i) - R\ln(P_o/P_i) = 0 \qquad (E8.11.1)$$

It follows from Eq. (E8.11.1) that

$$T_o = T_i(P_o/P_i)^{(\gamma-1)/\gamma} \qquad (E8.11.2)$$

where $\gamma = c_p/c_v$.

Apply the SFEE to the control volume i-1-2, neglecting kinetic and potential energy changes.

$$\dot{m}c_pT_i = \dot{m}c_pT_{1e}/2 + \dot{m}c_pT_{2e}/2$$

$$T_i = (T_{1e} + T_{2e})/2 \qquad (E8.11.3)$$

where e denotes the exit conditions of streams 1 and 2.

Apply the SFEE to the turbine control volume i-o (Fig. E8.11)

$$\dot{W}_o = \dot{m}c_p(T_i - T_o) \qquad (E8.11.4)$$

where $\dot{W}_o$ is the work output of the turbine.

From Eqs. (E8.11.3) and (E8.11.4) we have

$$\dot{W}_o = \dot{m}c_p[(T_{1e} + T_{2e})/2 - T_o] \qquad (E8.11.5)$$

Consider the operation of an infinitesimal Carnot heat pump shown in Fig. E8.11. Applying the SEFF to the hot and cold streams at the locations where the heat pump receives and rejects heat we have

$$\delta \dot{Q}_1 = \dot{m} c_p dT_1 / 2 \qquad (E8.11.6)$$

$$\delta \dot{Q}_2 = -\dot{m} c_p dT_2 / 2 \qquad (E8.11.7)$$

Applying the first law to the Carnot heat pump

$$\delta \dot{W} = \delta \dot{Q}_1 - \delta \dot{Q}_2 \qquad (E8.11.8)$$

Entropy balance for the Carnot heat pump gives

$$\delta \dot{Q}_1 / T_1 = \delta \dot{Q}_2 / T_2 \qquad (E8.11.9)$$

From Eqs. (E8.11.6), (E8.11.7) and (E8.11.9) we have

$$dT_1 / T_1 + dT_2 / T_2 = 0 \qquad (E8.11.10)$$

Integrating Eq. (E8.11.10) and applying the entry condition, $T_1 = T_2 = T_0$ and the exit conditions, $T_1 = T_{1e}$ and $T_2 = T_{2e}$ for the hot and cold streams we obtain

$$T_{1e} T_{2e} = T_o^2 \qquad (E8.11.11)$$

The total work input to the heat pumps is obtained by integrating Eq. (E8.11.8). Hence

$$\dot{W}_{in} = \dot{m} c_p (T_{1e} - T_o)/2 - \dot{m} c_p (T_o - T_{2e})/2$$

$$\dot{W}_{in} = \dot{m} c_p [(T_{1e} + T_{2e})/2 - T_o] \qquad (E8.11.12)$$

It is given that the work output of the turbine is equal to the total work consumed by the heat pumps. Hence $\dot{W}_o = \dot{W}_{in}$. The exit temperatures of the hot and cold streams are obtained by solving Eqs. (E8.11.2), (E8.11.3) and (E8.11.11) simultaneously. The given numerical values are:

$P_i = 400$ kPa, $T_i = 300$ K, and $P_o = 100$ kPa. For air $\gamma = 1.4$.

Substituting these numerical values in Eqs. (E8.11.2), (E8.11.3) and (E8.11.11) we have

$$T_o = T_i(P_o/P_i)^{(\gamma-1)/\gamma} = 300(100/400)^{0.286} = 201.88$$

$$(T_{1e} + T_{2e}) = 600 \tag{E8.11.13}$$

$$T_{1e}T_{2e} = T_o^2 = 201.88^2 \tag{E8.11.14}$$

From Eqs. (E8.11.13) and (E8.11.14) we obtain the quadratic equation

$$T_{1e}^2 - 600T_{1e} + 201.88^2 = 0$$

The two roots of the above equation give the highest and lowest exit temperatures as:

$$T_{1e} = 521.9 \text{ K} \quad \text{and} \quad T_{2e} = 78.1 \text{ K}$$

Notice the similarity between the operation of the above *ideal* device (Fig. E8.11), and the operation of the *actual* vortex tube (Fig. E8.10), considered in worked example 8.10.

Example 8.12 An adiabatic steam ejector pumps saturated water from a pressure of 10 kPa to 400 kPa using an input of superheated steam at 2000 kPa and 300°C. The mass flow rate of water is 3 times that of the steam. The water leaves the ejector at 400 kPa. Calculate (i) the entropy production rate of the ejector, and (ii) the maximum reversible work for the same inlet and outlet conditions, assuming the standard state to be saturated water at 30°C.

Solution A schematic of the steam ejector pump is shown in Fig. E8.12. Consider the control volume 1-2-3 for which the heat and work interactions are zero. Neglect the kinetic and potential energy of the streams.

Mass conservation gives

$$\dot{m}_1 + \dot{m}_2 = \dot{m}_3 = 4\dot{m}_1 \tag{E8.12.1}$$

Applying the SFEE to the control volume

$$\dot{m}_1 h_1 + \dot{m}_2 h_2 = \dot{m}_3 h_3 \tag{E8.12.2}$$

Substituting the property data obtained from the steam tables [6] in Eq. (E8.12.2)

$$3025\dot{m}_1 + 192 \times 3\dot{m}_1 = 4\dot{m}_1 h_3$$

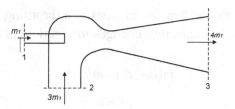

Fig. E8.12 Steam ejector pump

Hence $\qquad h_3 = 900.25 \ \text{kJ kg}^{-1}$

At the exit 3

$$P_3 = 400 \ \text{kPa}, \quad h_{g3} = 2739 \ \text{kJ kg}^{-1} \quad \text{and} \quad h_{f3} = 605 \ \text{kJ kg}^{-1}$$

Since $h_{g3} > h_3 > h_{f3}$, the steam leaving at 3 is wet. Hence

$$h_3 = 900.25 = x_3 \times 2739 + (1 - x_3) \times 605$$

Therefore the quality, $\qquad x_3 = 0.138$

Apply the entropy balance equation to the control volume.

$$\dot\sigma = \dot m_3 s_3 - (\dot m_1 s_1 + \dot m_2 s_2)$$

$$\dot\sigma = \dot m_1 (4 s_3 - s_1 - 3 s_2) \qquad\qquad \text{(E8.12.3)}$$

Substituting numerical data from [6] in Eq. (E8.12.3) we have

$$\dot\sigma = \dot m_1 (4 \times 2.4827 - 6.768 - 3 \times 0.649) = \dot m_1 \times 1.215 > 0$$

(i) The entropy production per unit mass is 1.215 kJ kg^{-1} K^{-1}.

(ii) The actual work output of the ejector is zero. Therefore the maximum reversible work output per unit mass is

$$\dot W_{rev} / \dot m_1 = I = T_o \dot\sigma / \dot m_1 = (273 + 30) \times 1.215 = 368.14 \ \text{kJ kg}^{-1}$$

Example 8.13 Find the irreversibility per unit mass for the following adiabatic throttling processes. Assume the standard ambient conditions as 20°C and 1 bar.

(i) Air at 20 bar and 40°C is throttled to 5 bar.

(ii) Steam at 10 bar and quality 0.95 is throttled to 1.1 bar.

(iii) Water at a temperature of 30°C and 5 bar is throttled to 1 bar.

Solution The work and heat interactions of a throttling process are zero. Applying the SFEE, neglecting changes in kinetic and potential energy, we have

$$\dot{m}(h+dh)-\dot{m}h=0$$

Hence
$$dh=0 \tag{E8.13.1}$$

We invoke the general property relationship [Eq. (7.17)].
$$Tds=du+Pdv=dh-vdP \tag{E8.13.2}$$

From Eqs. (E8.13.1) and (E8.13.2) we obtain
$$ds=-vdP/T \tag{E8.13.3}$$

The entropy balance equation for the throttling process is
$$\dot{\sigma}=\dot{m}\int_{in}^{out}ds=-\dot{m}\int_{in}^{out}vdP/T \tag{E8.13.4}$$

(i) Assuming air to be an ideal gas we have $Pv=RT$ per unit mass. Substituting in Eq. (E8.13.4),
$$\dot{\sigma}=\dot{m}\int_{in}^{out}ds=-\dot{m}\int_{in}^{out}RdP/P=\dot{m}R\ln\left(P_{in}/P_{out}\right) \tag{E8.13.5}$$

Therefore the irreversibility per unit mass is
$$\dot{I}/\dot{m}=T_o\dot{\sigma}/\dot{m}=T_oR\ln\left(P_{in}/P_{out}\right)$$

$$\dot{I}/\dot{m}=293\times0.287\ln(20/5)=116.57\ \text{kJ kg}^{-1}$$

(ii) For the throttling process, $dh=0$.

Hence
$$h_{out}=h_{in}=x_1h_{g1}+(1-x_1)h_{f1} \tag{E8.13.6}$$

Substituting tabulated data from [6] in Eq. (E8.13.6)
$$h_{out}=0.95\times2778+(1-0.95)\times763=2677.25\ \text{kJ kg}^{-1}$$

Assuming that the steam is wet at the exit pressure of 1.1 bar
$$h_{out}=2680x_2+429\times(1-x_2)=2677.25\ \text{kJ kg}^{-1}$$

Hence
$$x_2=0.998$$

Using tabulated data from [6], the entropy at the inlet and outlet are:

$$s_1 = 6.586 \times 0.95 + (1 - 0.95) \times 2.138 = 6.3636 \text{ kJ K}^{-1} \text{ kg}^{-1}$$

$$s_2 = 7.327 \times 0.998 + (1 - 0.998) \times 1.333 = 7.3149 \text{ kJ K}^{-1} \text{ kg}^{-1}$$

The entropy balance equation gives

$$\dot{\sigma} = \dot{m}(s_2 - s_1) = \dot{m} \times 0.9514$$

Therefore the irreversibility per unit mass is

$$\dot{I} / \dot{m} = T_o \dot{\sigma} / \dot{m} = 293 \times 0.9514 = 278.76 \text{ kJ kg}^{-1}$$

(iii) For a liquid, the changes in specific volume and temperature during a throttling process are very small. Therefore from Eq. (E8.13.4)

$$\dot{\sigma} = \dot{m} \int_{in}^{out} ds = -\dot{m}(\bar{v} / T) \int_{in}^{out} dP = \dot{m}\bar{v}(P_{in} - P_{out}) / T$$

$$\dot{\sigma} = \dot{m} \times 0.10044 \times 10^{-2} \times (500 - 100) / 303 = 0.1326 \times 10^{-2} \dot{m}$$

Therefore the irreversibility per unit mass is

$$\dot{I} / \dot{m} = T_o \dot{\sigma} / \dot{m} = 293 \times 0.1326 \times 10^{-2} = 0.388 \text{ kJ kg}^{-1}$$

Example 8.14 A closed feed water heater of a steam power plant receives saturated steam at 150 kPa from the turbine. Feed water enters the heater at 2800 kPa and 40°C. The saturated liquid condensate leaving the heater at 150 kPa is pumped to a pressure of 2650 kPa and mixed with the heated feed water to attain a final mixed temperature of 90°C. Calculate (i) the entropy production in the heater per unit mass of water leaving the heater, and (ii) the irreversibility if the standard state is saturated water at 30°C.

Solution Consider the control volume 1-2-3 shown in Fig. E8.14. The steam enters the feed water heater at 1 and condenses to water by transferring heat to the feed water entering at 2. The condensate is pumped and mixed with the feed water before the mixture leaves at 3. Mass balance gives

$$\dot{m}_1 + \dot{m}_2 = \dot{m}_3 \qquad (E8.14.1)$$

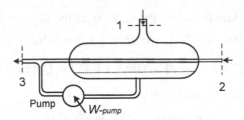

Fig. E8.14 Closed feed-water heater

Apply the SFEE, neglecting heat losses from the heater. Thus

$$\dot{m}_1 h_1 + \dot{m}_2 h_2 + \dot{W}_{pump} = \dot{m}_3 h_3 \qquad \text{(E8.14.2)}$$

Let $r = \dot{m}_1 / \dot{m}_3$. From Eqs. (E8.14.1) and (E8.14.2) we have

$$h_3 = (1-r)h_2 + rh_1 + \dot{W}_{pump} / \dot{m}_3 \qquad \text{(E8.14.3)}$$

The ideal work input to the liquid pump is given by Eq. (E8.1.7) as

$$\dot{W}_{pump} = \dot{m}_1 \bar{v}(P_3 - P_1) \qquad \text{(E8.14.4)}$$

where $\bar{v}$ is the mean specific volume of the liquid.

Substituting from Eq. (E8.14.4) in Eq. (E8.14.3) we obtain

$$h_3 = (1-r)h_2 + rh_1 + r\bar{v}(P_3 - P_1) \qquad \text{(E8.14.5)}$$

Substituting numerical data from [6] in Eq. (E8.14.5),

$$376.9 = (1-r) \times 167.5 + r \times 2693 + r \times 0.1052 \times 10^{-2}(2650 - 150)$$

Hence $r = 0.0828$

Applying the entropy balance equation,

$$\dot{\sigma} = \dot{m}_3 s_3 - \dot{m}_1 s_1 - \dot{m}_2 s_2$$

Substituting numerical data from [6] in the above equation

$$\dot{\sigma} / \dot{m}_3 = 1.192 - 7.223r - 0.572 \times (1-r) = 0.0693 \text{ kJ K}^{-1} \text{ kg}^{-1}$$

The irreversibility per unit mass is given by

$$\dot{I} / \dot{m}_3 = T_o \dot{\sigma} / \dot{m}_3 = (273 + 30) \times 0.0693 = 21 \text{ kJ kg}^{-1}$$

Example 8.15 Two streams of waste-water at 80°C and 20°C are discharged from a process plant. The flow rates of the hot and cold streams are 20 kg s^{-1} and 35 kg s^{-1} respectively. A series of infinitesimal reversible heat engines are to be used to extract the maximum work-potential available in the two streams. (i) Calculate the maximum work output of the reversible engines. (ii) The work produced is to be used to operate a reversible heat pump to heat a stream of cold water entering the plant at 15°C with a flow rate of 20 kg s⁻¹. The heat pump extracts heat from the local ambient at 20°C. Calculate the temperature of the water after the heating process. Specific heat capacity of water is 4.2 kJ kg⁻¹ K⁻¹.

Solution In worked example 8.9 we derived general expressions for the exit temperature of two liquid streams when a series of reversible heat engines operate between them. For the sake of brevity we shall use these expressions in the present example.

$$T_f = T_{1i}^{\lambda} T_{2i}^{\beta}$$ (E8.15.1)

where the exponents in Eq. (E8.15.1) are given by

$$\lambda = \dot{m}_1 c_1 / (\dot{m}_1 c_1 + \dot{m}_2 c_2) \quad \text{and} \quad \beta = \dot{m}_2 c_2 / (\dot{m}_1 c_1 + \dot{m}_2 c_2)$$

Substituting numerical values in the above expressions we have

$$\lambda = 20 \times 4.2 / (20 \times 4.2 + 35 \times 4.2) = 0.3636$$

$$\beta = 35 \times 4.2 / (20 \times 4.2 + 35 \times 4.2) = 0.6364$$

$$T_f = (353)^{0.3636} (293)^{0.6364} = 313.5 \text{ K}$$

Applying the first law to the Carnot engines, the total work output is

$$W_o = 20 \times 4.2 \times (353 - 313.5) - 35 \times 4.2 \times (313.5 - 293)$$

$$W_o = 304.5 \text{ kW}$$

Now consider a Carnot heat pump that extracts heat from the *constant* temperature ambient and heats the cold stream (see Fig. E8.11). Applying the SFEE to the cold stream we have

$$\delta \dot{Q}_3 = \dot{m} c d T_3$$ (E8.15.2)

where T_3 is the *variable* temperature of the cold stream.

Applying the first law to the heat pump

$$\delta \dot{W} = \delta \dot{Q}_3 - \delta \dot{Q}_o$$ (E8.15.3)

Applying the entropy balance equation to the heat pump

$$\delta \dot{Q}_o/T_o = \delta \dot{Q}_3/T_3 = \dot{m}cdT_3/T_3 \qquad (E8.15.4)$$

where T_o is the local ambient temperature and $\delta \dot{Q}_o$ is the heat extracted from the ambient.

Manipulating Eqs. (E8.15.2), (E8.15.3) and (E8.15.4)

$$\delta \dot{W} = \dot{m}cdT_3 - T_o\dot{m}cdT_3/T_3 \qquad (E8.15.5)$$

Integrating Eq. (E8.15.5) we obtain

$$\dot{W} = \dot{m}c(T_f - T_{3i}) - \dot{m}cT_o \ln\left(T_f/T_{3i}\right) \qquad (E8.15.6)$$

where T_f is the final temperature of the cold stream.

Substituting numerical values in Eq. (E8.15.6)

$$\dot{W} = 20 \times 4.2 \times (T_f - 288) - 20 \times 4.2T_o \ln\left(T_f/288\right)$$

The work output from the Carnot engines is equated to the total work input to the heat pumps. Thus

$$20 \times 4.2(T_f - 288) - 20 \times 4.2 \times 293 \ln\left(T_f/288\right) = 304.5$$

The above equation is solved by trial-and-error to obtain, $T_f = 342$ K.

Example 8.16 Ambient air at a temperature T_o flows through a reversible adiabatic turbine into an evacuated rigid tank of volume V_t, producing a work output, until the tank pressure becomes equal to the ambient pressure of P_o. The tank is well insulated. Obtain expressions for (i) the work output of the turbine, (ii) the final temperature of the air in the tank, and (iii) the total entropy production in the universe. Assume that air is an ideal gas.

Solution As air fills the tank, the pressure difference across the turbine changes with time. Consider the situation when the tank pressure is P_1 and the temperature of the air leaving the turbine is T_1. The instantaneous work output of the turbine per unit mass is obtained by applying the SFEE as

$$W' = c_p(T_o - T_1) \qquad (E8.16.1)$$

Since the air undergoes a reversible adiabatic process in the turbine, the entropy change is zero. From the expression for the entropy change of an ideal gas [Eq. (7.25)] it follows that:

$$T_o/T_1 = (P_o/P_1)^{(\gamma-1)/\gamma} \tag{E8.16.2}$$

where $\gamma = c_p/c_v$

From Eqs. (E8.16.1) and (E8.16.2) we obtain

$$W' = -c_p T_o \left[(P_1/P_o)^{(\gamma-1)/\gamma} - 1 \right] \tag{E8.16.3}$$

Consider the entry of a small mass of air, δm into the tank from the turbine. Energy balance for the tank gives:

$$\delta m c_p T_1 = \delta(M_t c_v T_t) \tag{E8.16.4}$$

Substituting for the mass of air in the tank from the ideal gas equation

$$\delta m c_p T_1 = \delta(P_1 V_t c_v T_t / RT_t)$$

where T_t is the temperature of the air *in* the tank.

$$\delta m = (V_t/\gamma RT_1)\delta P_1 \tag{E8.16.5}$$

The total work output of the turbine is

$$W_{tot} = \int_o^{P_o} \delta m W' = -\int_0^{P_o} c_p T_o \left[(P_1/P_o)^{(\gamma-1)/\gamma} - 1 \right] (V_t/\gamma RT_1) dP_1$$

$$W_{tot} = -\frac{c_v V_t}{R} \int_0^{P_o} \left[1 - (P_o/P_1)^{(\gamma-1)/\gamma} \right] dP_1 = V_t P_o$$

Applying the energy equation to the filling process [Eq. (E5.15.7)]

$$M_f c_p T_o = W_{tot} + M_f c_v T_f$$

where M_f is the final mass of air in the tank. Substituting for mass from the ideal gas equation

$$P_o V_t c_p T_o / RT_f = P_o V_t + P_o V_t c_v / R$$

Simplifying the above equation we obtain the final temperature as

$$T_f = T_o$$

The temperature and pressure of the air at the initial and final states are the same. Therefore from the expression for the entropy change of an ideal gas [Eq. (7.25)] it follows that the entropy change of the air from the beginning to the end of the filling process is zero. The heat interaction with the surroundings is also zero. Consequently, the entropy production in the universe is zero.

Example 8.17 A fixed mass of air is at a pressure of 50 bar and 300 K. The air undergoes a series of reversible process which reduces its pressure to 1 bar while interacting *only* with atmospheres-reservoir at 300 K and 1 bar. It has no other interactions with the surroundings. Calculate the maximum temperature that the air could attain. What combination of ideal processes could be used to achieve this maximum temperature?

Solution (i) Consider the overall change of state of the fixed mass of air, from an initial state 1 to a final state 2. The air interacts *only* with the atmosphere-reservoir at pressure P_o and temperature T_o. Applying the first law to the air as a closed system we have

$$Q_{in} = U_2 - U_1 + P_o(V_2 - V_1) \qquad (E8.17.1)$$

The second term on the right hand side of Eq. (E8.17.1) represents the work done against the constant pressure of the atmosphere. The heat interaction between the air and the atmosphere-reservoir is Q_{in}. Applying the entropy balance equation to the composite system

$$\sigma = S_2 - S_1 - Q_{in}/T_o \geq 0 \qquad (E8.17.2)$$

From Eqs. (E8.17.1) and (E8.17.2) we obtain

$$U_2 - U_1 + P_o(V_2 - V_1) \leq T_o(S_2 - S_1) \qquad (E8.17.3)$$

Assume that air is an ideal gas and substitute the expression [Eq. (7.25)] for the change in entropy in Eq. (E8.17.3) to obtain

$$mc_v(T_2 - T_1) + P_o(V_2 - V_1) \leq mT_o\left[c_p \ln(T_2/T_1) - R\ln(P_2/P_1)\right]$$

Substituting the given conditions, $P_2 = P_o$ and $T_1 = T_o$, we have

$$mc_v(T_2 - T_o) + mP_o(RT_2/P_o - RT_o/P_1)$$

$$\leq mT_o\left[c_p \ln(T_2/T_1) - R\ln(P_o/P_1)\right]$$

$$T_2 - T_o \ln(T_2/T_o) \le (c_v/c_p + RP_o/P_1 c_p)T_o + \frac{RT_o}{c_p}\ln(P_1/P_o)$$

Substituting the given numerical values and the properties of air in the above equation we obtain

$$T_2 - 300\ln(T_2/300) \le (216 + 335.15)$$

From the above equation it follows that $T_2 \le 871\,\mathrm{K}$. Hence the maximum temperature is 871 K.

(ii) The following idealized processes may be used to achieve the maximum temperature in the air. The air at 50 bar and 300 K is first expanded in a reversible adiabatic process to the reservoir pressure of 1 bar. The total work output of this process is used to operate a Carnot heat pump that extracts heat from the atmosphere at temperature at 300 K, and supplies it to the air to heat it at a constant pressure of 1 bar to the maximum temperature of 871 K.

Example 8.18 A steam turbine receives superheated steam at 40 bar and 350°C. The steam expands to the condenser pressure of 0.2 bar. The mass-flow rate of steam is 2.0 kg s^{-1}. The work output of the turbine is 1500 kW. Calculate (i) the second law efficiency if the standard ambient is at 20°C, and (ii) the isentropic efficiency of the turbine. Neglect heat losses from the turbine to the surroundings.

Solution The following pertinent data for steam are obtained from [6]. At 350°C and 40 bar,

$$h_1 = 3094\ \mathrm{kJ\ kg^{-1}} \quad \text{and} \quad s_1 = 6.584\ \mathrm{kJ\ K^{-1}\ kg^{-1}}$$

At 0.2 bar

$$h_{g2} = 2609\ \mathrm{kJ\ kg^{-1}}, \qquad h_{f2} = 251\ \mathrm{kJ\ kg^{-1}},$$

$$s_{g2} = 7.907\ \mathrm{kJ\ K^{-1}\ kg^{-1}}, \qquad s_{f2} = 0.832\ \mathrm{kJ\ K^{-1}\ kg^{-1}}$$

Applying the SFEE to the turbine, neglecting kinetic and potential energy changes, we have

$$\dot{W}_{act} = \dot{m}(h_1 - h_2)$$

Substituting numerical values in the above equation

$$1500 = 2.0 \times (3094 - h_2)$$

Hence $h_2 = 2344$ kJ kg^{-1}. Assume that the steam is wet at the exit 2.

$$h_2 = 2344 = 2609 x_2 + (1 - x_2)251$$

Hence $x_2 = 0.888$

$$s_2 = 7.907 x_2 + (1 - x_2) \times 0.832 = 7.1146$$

Applying the entropy balance equation to the turbine

$$\dot{\sigma} = \dot{m}(s_2 - s_1) = 2 \times (7.1146 - 6.584) = 1.061 \text{ kW K}^{-1}$$

The irreversibility rate for the *work-producing* turbine is given by

$$\dot{I} = \dot{W}_{rev} - \dot{W}_{act} = T_o \dot{\sigma} = 293 \times 1.061 = 310.87 \text{ kW}$$

$$\dot{W}_{rev} = \dot{W}_{act} + \dot{I} = 1500 + 310.87 = 1810.87 \text{ kW}$$

The second-law efficiency is defined as

$$\eta_2 = \dot{A}_{min} / \dot{A}_{act} = \dot{W}_{act} / \dot{W}_{rev} = 1500 / 1810.87 = 82.8\%$$

Note that the task is to produce the required work output.

The isentropic efficiency is a first-law performance parameter. The steam is now expanded in an isentropic process to the *same final pressure* of 0.2 bar. For the expansion from state 1 to the new state 2s we have

$$s_1 = s_{2s} = 6.584 = 7.907 x_{2s} + 0.832 \times (1 - x_{2s})$$

Hence $x_{2s} = 0.813$

$$h_{2s} = 2609 x_{2s} + 251 \times (1 - x_{2s}) = 2168.05$$

Applying the SFEE, the work output of the isentropic expansion is

$$\dot{W}_{iso} = \dot{m}(h_1 - h_{2s}) = 2.0 \times (3094 - 2168.05) = 1851.9 \text{ kW}$$

The isentropic efficiency of the turbine is given by

$$\eta_{iso} = \dot{W}_{act} / \dot{W}_{iso} = 1500 / 1851.9 = 81\%$$

It should be noted that the numerical value of the second-law efficiency depends on the chosen standard ambient temperature, T_o while the isentropic efficiency depends on the final pressure.

Example 8.19 Atmospheric air is compressed in an air compressor from 100 kPa and 27°C to 500 kPa. The air flow rate through the compressor is 1.2 kg s^{-1}. The work input to the compressor is 280 kW. Calculate (i) the isentropic efficiency of the compressor, and (ii) the second law efficiency if the standard ambient conditions are 100 kPa and 20°C. Assume that air is an ideal gas. Neglect any heat exchange with the surroundings.

Solution Applying the SFEE to the compressor, neglecting kinetic and potential energy changes, we have

$$\dot{W}_{in} = \dot{m}c_p(T_o - T_i) = 1.2 \times 1.005(T_o - 300) = 280 \text{ kW}$$

Hence

$$T_o = 532.17 \text{ K}$$

The entropy balance equation for the compressor is

$$\dot{\sigma} = \dot{m}(s_o - s_i)$$

Assuming air to be an ideal gas we use the expression in Eq. (7.25) for the change in entropy. Hence

$$\dot{\sigma} = \dot{m}\left[c_p \ln(T_o/T_1) - R\ln(P_o/P_1)\right]$$

Substituting numerical values in the above equation

$$\dot{\sigma} = 1.2 \times \left[1.005\ln(532.17/300) - 0.287\ln(5/1)\right] = 0.137 \text{ kW K}^{-1}$$

The irreversibility rate for the *work-absorbing* compressor is

$$\dot{I} = \dot{W}_{act} - \dot{W}_{rev} = T_o\dot{\sigma} = 293 \times 0.137 = 40.1 \text{ kW}$$

$$\dot{W}_{rev} = \dot{W}_{act} - T_o\dot{\sigma} = 280 - 40.1 = 239.9 \text{ kW}$$

The second-law efficiency is defined as

$$\eta_2 = \dot{A}_{min}/\dot{A}_{act} = \dot{W}_{rev}/\dot{W}_{act} = 239.9/280 = 85.7\%$$

The task in this case is to compress the air to the desired pressure.

Consider an *isentropic process* in which the air is compressed to the same final pressure. The entropy change is zero. From the expression for the entropy change of an ideal gas [Eq. (7.25)] it follows that

$$T_{o,iso}/T_i = (P_o/P_i)^{(\gamma-1)/\gamma}$$

where $\gamma = c_p/c_v = 1.4$, for air.

Substituting numerical values in the above equation

$$T_{o,iso}/300 = (5/1)^{0.2857}$$

Hence $T_{o,iso} = 475.13 \text{ K}$

Applying the SFEE to the compressor, neglecting changes in kinetic and potential energy, we obtain the work input to the isentropic compressor as

$$\dot{W}_{iso} = \dot{m}c_p(T_{o,iso} - T_i) = 1.2 \times 1.005(475.13 - 300) = 211.2 \text{ kW}$$

The isentropic efficiency of the *work-absorbing* compressor is given by

$$\eta_{iso} = \dot{W}_{iso}/\dot{W}_{act} = 211.2/280 = 75.4\%$$

Note that the second law efficiency depends on the temperature of the chosen standard ambient.

Example 8.20 A stream of air at a pressure P_i and temperature T_i flows into a control volume with a mass flow rate of $\dot{m}$. The stream is to attain equilibrium with an ambient at a pressure P_o and temperature T_o. Suggest a series of ideal processes to extract the maximum work-potential in the stream before equilibrium is established.

Solution We could use two flow processes in series to extract work from the stream and bring it to equilibrium with the surroundings. First, a series of Carnot engines are operated between the stream and the atmosphere to reduce its temperature from T_i to T_o at a constant pressure P_i. In the second process, in series with above process, the air expands through an isothermal turbine to decrease the pressure from P_i to P_o at constant temperature T_o.

The work output of the first process is obtained as follows. Consider an infinitesimal Carnot engine receiving heat from the stream at T_1 and rejecting heat to the standard atmosphere at T_o (see Fig. E8.11).

Applying the SFEE to the stream where the engine receives heat

$$\delta\dot{Q}_1 = -\dot{m}c_p dT_1 \qquad\qquad \text{(E8.20.1)}$$

Applying the first law to the Carnot engine, the work output is

$$\delta \dot{W}_o = \delta \dot{Q}_1 - \delta \dot{Q}_o \qquad \text{(E8.20.2)}$$

where $\delta \dot{Q}_o$ is the heat rejected to the atmosphere.

The entropy balance equation for the Carnot engine gives

$$\delta \dot{Q}_1 / T_1 = \delta \dot{Q}_o / T_o \qquad \text{(E8.20.3)}$$

Manipulating Eqs. (E8.20.1)–(E8.20.3) we obtain

$$\delta \dot{W}_o = -\dot{m} c_p dT_1 + \dot{m} c_p T_o dT_1 / T_1 \qquad \text{(E8.20.4)}$$

Integrating Eq. (E8.20.4) from inlet, i to outlet, o

$$\dot{W}_o = \dot{m} c_p (T_i - T_o) - \dot{m} c_p T_o \ln(T_i / T_o) \qquad \text{(E8.20.5)}$$

The work output of the *reversible* turbine is given by Eq. (E8.3.1),

$$\dot{W}_{tur} = -\dot{m} \int_{P_i}^{P_o} v dP \qquad \text{(E8.20.6)}$$

For an isothermal process, substitute for v from the ideal gas equation in Eq. (E8.20.6), and integrate to obtain

$$\dot{W}_{tur,iso} = \dot{m} RT_o \ln(P_i / P_o) \qquad \text{(E8.20.7)}$$

The total reversible work output is

$$\dot{W}_{rev} = \dot{W}_o + \dot{W}_{tur,iso}$$

Substituting from Eqs. (E8.20.5) and (E20.8.7) in the above equation

$$\dot{W}_{rev} = \dot{m} c_p (T_i - T_o) - \dot{m} c_p T_o \ln(T_i / T_o) + \dot{m} RT_o \ln(P_i / P_o)$$

$$\dot{W}_{rev} = \dot{m} c_p (T_i - T_o) - \dot{m} T_o (s_i - s_o)$$

$$\dot{W}_{rev} = \dot{m}(h_i - T_o s_i) - \dot{m}(h_o - T_o s_o) \qquad \text{(E8.20.8)}$$

$$\dot{W}_{rev} = \dot{m}(\psi_i - \psi_o)$$

We note that the expression for the maximum reversible work given by Eq. (E8.20.8) agrees with the general expression in Eq. (8.26) for the exergy change of a flow stream, which is also the change in the 'work- potential' of the stream.

Engineering Thermodynamics

Problems

P8.1 Air enters an adiabatic nozzle at 1.5 bar with a mass flow rate of 0.2 kg s^{-1}. At the exit of the nozzle the air is at 1.2 bar and 27°C. The exit area of the nozzle is 1.2×10^{-3} m^2. The kinetic energy at the entrance section is negligible. Calculate (i) the entropy production rate, and (ii) the maximum reversible work for the same change of state of the air if the standard ambient is at 1 bar and 20°C.
[*Answers*: (i) $\dot{\sigma} = 8.09\times10^{-3}$ kW K^{-1}, (ii) $\dot{W}_{rev} = 2.37$ kW]

P8.2 A steam turbine is designed to have a power output of 10 MW for a steam flow rate of 20 kg s^{-1}. At the entrance to the turbine the conditions of the steam are 40 bar and 450°C. At the exit the steam is dry saturated at 5 bar. The velocities of the steam at the inlet and the exit are 250 ms^{-1} and 70 ms^{-1} respectively. The surroundings are at a temperature of 20°C. Calculate (i) the entropy production rate in the universe, and (ii) the irreversibility rate.
[*Answers*: $\dot{\sigma} = 5.23$ kW K^{-1}, $\dot{I} = 1532.4$ kW]

P8.3 A stream of high pressure air at 250 kPa and 307.5 K, flowing with a mass flow rate of 1 kg s^{-1}, enters an adiabatic device. The device produces a cold stream at 100 kPa and 115 K with a flow of 0.5 kg s^{-1} and a hot stream at 100 kPa and 500 K with a flow rate of 0.5 kg s^{-1}. The device and the streams have no other interactions with the surroundings. (i) Does the device violate any of the laws of thermodynamics? (ii) What combination of ideal processes could be used to operate such a device?
[*Answers*: (i) $\dot{\sigma} = 0.013$ kW K^{-1}, (ii) see worked example 8.11]

P8.4 A fixed mass of air is at a pressure of 20 bar and temperature 300 K. The air undergoes a series of reversible processes which reduces its pressure to 1 bar while interacting with the atmosphere-reservoir at 300 K and 1 bar. The air has no other interactions with the surroundings. (i) Calculate the minimum temperature that the air could attain. (ii) What combination of ideal processes could be used to achieve the minimum temperature?
[*Answer*: 80.5 K, see worked example 8.17]

P8.5 An adiabatic turbine receives air at 500 kPa and 250°C with a steady mass flow rate of 0.5 kg s⁻¹. The isentropic efficiency of the turbine is 85 percent. The turbine discharges air to the surroundings at 95 kPa and 27°C. Assume that the standard ambient is at 27°C. Calculate (i) the actual work output of the turbine, (ii) the actual outlet temperature, (iii) the irreversibility rate, (iv) the reversible work for the same end-states of the air, and (v) the second-law efficiency. [*Answers*: (i) 84.4 kW, (ii) 355.3 K, (iii) 13.2 kW, (iv) 97.6 kW, (v) 86.5%]

P8.6 Steam at 50 bar and 400°C is supplied to an adiabatic turbine through a throttle valve that reduces the inlet pressure to the turbine inlet pressure of 40 bar. In the turbine the steam is expanded to 0.5 bar. The work output of the turbine per unit mass of steam is 760 kJ kg⁻¹. Assume the standard ambient conditions to be 1 bar and 27°C. Calculate the (i) irreversibility of the valve, (ii) the isentropic efficiency of the turbine, (iii) the irreversibility of the turbine, (iv) the second law efficiency of the turbine, and (v) the second law efficiency of the system consisting of the valve and the turbine. [*Answers*: (i) 28.6 kJ kg⁻¹, (ii) 88.1%, (iii) 78.6 kJ kg⁻¹, (iv) 90.7%, (v) 87.6%]

P8.7 A series of Carnot engines are operated between a stream of steam and a stream of water flowing in parallel through a control volume. The steam at entry is dry saturated at 1.5 bar. The water enters at a temperature of 27°C. At the exit the quality of steam is 0.2. The mass flow rates of steam and water are 0.25 kg s⁻¹ and 2.1 kg s⁻¹ respectively. Calculate (i) the outlet temperature of water, and (ii) the total work output of the engines. [*Answers*: (i) 69°C, (ii) 74.8 kW]

P8.8 Steam enters an adiabatic turbine at 90 bar and 400°C and expands to 10 bar and 190°C. The mass flow rate of steam is 1.3 kg s⁻¹. The standard ambient conditions are 1 bar and 30°C. Calculate (i) the work output of the turbine, (ii) the isentropic efficiency, and (iii) the second law efficiency. [*Answers*: (i) 408.7 kW, (ii) 66 %, (iii) 74.5%]

P8.9 A process plant discharges a stream of air at 500°C and 100 kPa with a mass flow rate of 1.8 kg s⁻¹. A *reversible* device is to be used to reduce the temperature of the stream to 40°C at constant pressure. The device exchanges heat with the surroundings at 27°C and 100 kPa. The standard ambient conditions are 100 kPa and 20°C. Calculate (i) the maximum work output of the device, and (ii) the exergy (availability) of the air stream at the inlet and the exit of the device. Verify that the maximum reversible work is equal to the net change in exergy. [*Answers*: (i) 341.5 kW, (ii) 196.75 kJ kg^{-1}, 0.656 kJ kg^{-1}]

P8.10 Two rigid tanks with equal volumes of 0.15 m^3 contain air at an initial pressure of 200 kPa and a temperature of 10°C. A reversible compressor connected between the tanks sucks air from one tank and discharges it to the other tank. The tanks and the compressor may be assumed isothermal with the temperature at 10°C. Calculate (i) the total work input when the pressure in the high pressure tank reaches 300 kPa, (ii) the total heat transfer during the process, and (iii) the change in entropy of the air. [*Answers*: (i) $W_{in} = 7.85$ kJ, (ii) $Q_{out} = 7.85$ kJ, (iii) $S_f - S_i = -0.0277$ kJ kg^{-1}]

References

1. Bejan, Adrian, *Advanced Engineering Thermodynamics*, John Wiley & Sons, New York, 1988.
2. Cravalho, E.G. and J.L. Smith, Jr., *Engineering Thermodynamics*, Pitman, Boston, MA, 1981.
3. Jones, J.B. and G.A. Hawkins, *Engineering Thermodynamics*, John Wiley & Sons, Inc., New York, 1986.
4. Kestin, Joseph, *A Course in Thermodynamics*, Blaisdell Publishing Company, Waltham, Mass., 1966.
5. Reynolds, William C. and Henry C. Perkins, *Engineering Thermodynamics*, 2nd edition, McGraw-Hill, Inc., New York, 1977.
6. Rogers, G.F.C. and Mayhew, Y.R., *Thermodynamic and Transport Properties of Fluids*, 5th edition, Blackwell, Oxford, U.K. 1998.
7. Tester, Jefferson W. and Michael Modell, *Thermodynamics and Its Applications*, 3rd edition, Prentice Hall PTR, New Jersey, 1996.
8. Van Wylen, Gordon J. and Richard E. Sonntag, *Fundamentals of Classical Thermodynamics*, 3rd edition, John Wiley & Sons, Inc., New York, 1985.
9. Wark, Jr., Kenneth, *Advanced Thermodynamics for Engineers*, McGraw-Hill, Inc., New York, 1995.

Chapter 9

Vapor Power Cycles

In this chapter we shall consider the analysis of vapor power cycles that convert the stored energy of fuels to mechanical work. The main distinguishing feature of these power cycles is that the working fluid undergoes phase change during the operation of the cycle. Steam power plants, which use water as the working fluid, generate much of the electrical power in the world. Although the design and construction of such power plants involve many engineering considerations, our focus in this chapter will be the thermodynamic aspects that impact on the efficiency of the plant.

9.1 The Carnot Cycle Using a Vapor

In chapter 6 we discussed the operation of the Carnot heat engine cycle and derived an expression for its efficiency when the working fluid is an ideal gas. The feasibility of operating a Carnot cycle using a vapor is considered in this section. Figure 9.1 shows the T-s diagram of a Carnot cycle in which all processes occur within the liquid-vapor region. The working fluid enters the evaporator at 4 as a saturated liquid where it is heated to a saturated vapor state and exits at 1. The vapor undergoes an isentropic expansion 1-2 to produce a work output. During the condensation process 2-3 the wet vapor rejects heat to a heat sink at constant temperature. Finally, the vapor is compressed from 3 to 4 in an isentropic compression process to complete the cycle.

We now apply the SFEE to each of the steady-flow processes of the cycle, neglecting the kinetic and potential energy of the fluid, to obtain the following expressions.

411

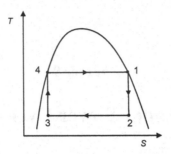

Fig. 9.1 Carnot cycle using a vapor

For the process 4-1 in the evaporator:

$$\dot{Q}_{41} = \dot{m}(h_1 - h_4) \qquad (9.1)$$

where $\dot{m}$ is the steady mass flow rate of the working fluid.

For the heat removal process 2-3:

$$\dot{Q}_{23} = \dot{m}(h_2 - h_3) \qquad (9.2)$$

Applying the first law to the cycle, the net work output is

$$\dot{W}_{net} = \dot{Q}_{41} - \dot{Q}_{23} \qquad (9.3)$$

The efficiency of the cycle 1-2-3-4 is given by

$$\eta = \frac{\dot{W}_{net}}{\dot{Q}_{41}} \qquad (9.4)$$

Manipulating Eqs. (9.1) to (9.4) we have

$$\eta = 1 - \frac{(h_2 - h_3)}{(h_1 - h_4)} \qquad (9.5)$$

We relate the enthalpy changes in Eq. (9.5) to entropy changes by integrating the general thermodynamic property relation

$$dh = Tds - vdP \qquad (9.6)$$

Note that the above relation follows directly from the 'T-ds' equation [Eq. (7.17)] when we substitute for u from the relation, $h = u + Pv$.

The pressure and temperature during the phase change processes 4-1 and 2-3 are constant. Therefore $dP = 0$ and $T = T_{sat}$, the saturation temperature. Integrating Eq. (9.6) and applying the resulting relationship to processes 4-1 and 2-3 we have

$$h_1 - h_4 = T_1(s_1 - s_4) \tag{9.7}$$

$$h_2 - h_3 = T_2(s_2 - s_3) \tag{9.8}$$

From the rectangular shape of the *T-s* diagram we see that

$$(s_1 - s_4) = (s_2 - s_3)$$

Substituting in Eq. (9.5) from Eqs. (9.7) and (9.8) we obtain the familiar expression for the efficiency of the Carnot cycle,

$$\eta = 1 - \frac{T_2}{T_1} \tag{9.9}$$

There are several difficulties in constructing a real heat engine that operates on the Carnot cycle using a vapor as the working fluid. The isothermal process 4-1 in the boiler could be approximated in practice because vaporization occurs at constant temperature. However, during the expansion 1-2 in the turbine, the vapor is wet and therefore could cause erosion of the blades of the turbine due to the impact of the liquid drops. In the case of the condensation process 2-3 it is difficult to control the final quality of the vapor at 3 where it enters the compressor. Moreover, practical problems are encountered in compressing the wet vapor during the process 3-4.

When operating the Carnot cycle in the vapor-liquid region, the maximum temperature of heat supply has to be below the critical temperature of the working fluid. This in turn limits the achievable cycle efficiency, which is a function of the heat supply temperature as seen from Eq. (9.9). Several modifications have been done to the Carnot cycle to overcome the aforementioned factors and develop a practically workable heat engine cycle that operates using vapors. This cycle is known as the Rankine cycle.

9.2   The Rankine Cycle

The ideal cycle on which the vapor power plant operates is called the Rankine cycle. In the flow diagram of the Rankine cycle shown in Fig. 9.2, vapor in a superheated state enters the turbine at 1. During the expansion process 1-2 through the turbine a part of the initial vapor enthalpy is converted to work. At 2 the wet vapor enters the condenser, in which heat is rejected to a cooling medium. The fully condensed liquid exiting from the condenser at 3 is pumped into the boiler by the feed pump, to be heated in a constant pressure process from 4 to 1.

There are several aspects in which the Rankine cycle differs from the Carnot cycle shown in Fig. 9.1. These help to overcome some of practical difficulties associated with the Carnot cycle.

In the more common version of the Rankine cycle, the vapor is superheated to a temperature much higher than the saturation temperature corresponding to the boiler pressure. This process (1a-1) is carried out in a separate heater called the *superheater* as shown in Fig. 9.2. Superheating results in a higher average temperature of heat supply to the cycle, which in turn increases its thermal efficiency.

The liquid entering the boiler is usually in a sub-cooled state and the section of the boiler in which it is heated to the saturation temperature (4-4a) is sometimes termed the *economizer*.

In a typical turbine, the vapor expands over a series of blades attached to a wheel mounted on a shaft. During this process (1-2) the vapor transfers a part of its enthalpy to the shaft as mechanical work. In the early versions of the Rankine cycle power plant using water as the working fluid, the wet vapor exiting from the work-device (the steam

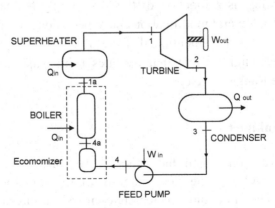

Fig. 9.2 The Rankine cycle

engine) was discharged to the atmosphere. This limited the exit temperature of the steam to about 100°C. In the modern version of the Rankine cycle, the exit temperature of the steam from the turbine is lowered by introducing a *condenser* in which the steam rejects heat to a cooling medium whose temperature is much closer to the ambient temperature.

From our studies of the Carnot cycle, we note that by lowering the heat rejection temperature of the cycle we could increase the cycle efficiency. In the Rankine cycle, this potential increase in efficiency is achieved practically by using the condenser. However, the pressure of the vapor in the condenser is typically below ambient pressure, and therefore the condenser is prone to air leakage into it from the ambient.

In the Rankine cycle, the working fluid entering the feed pump is in a liquid state and it could therefore be pumped to the boiler pressure (3-4) conveniently with a relatively small work input. By condensing the vapor completely to a liquid in the condenser we eliminate the practical difficulties of pumping a wet vapor to the boiler, which was the case with the Carnot cycle discussed earlier.

9.2.1 *Temperature-entropy and enthalpy-entropy diagrams*

The *T-s* diagram of the ideal Rankine cycle is shown in Fig. 9.3. The process 4-1 where heat is supplied and the process 2-3 during which heat is rejected are both constant pressure processes. The expansion 1-2 in the turbine and the pumping process 3-4 are reversible adiabatic processes.

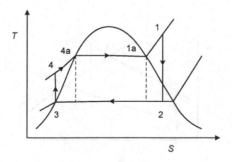

Fig. 9.3 Rankine cycle: *T-s* Diagram

With these ideal processes the ideal Rankine cycle is *internally* reversible. However, the temperature of the working fluid varies from T_4 to T_1 during heat reception in the boiler and the superheater. Therefore if heat is supplied from an external source at constant temperature, the cycle will be *externally* irreversible due to heat transfer across a finite temperature difference. In contrast, the heat rejection during process 2-3 occurs at a constant temperature and could therefore be *idealized* as an externally reversible process if heat is rejected to a reservoir. The area enclosed by the cycle 1-2-3-4 is equal to the net heat interaction, which by virtue of the first law is also the net work output of the cycle.

We observe that the state of the vapor during the expansion in the turbine (1-2) is much drier when the initial state is superheated (state 1) compared to the expansion from the initially saturated state 1a. This is desirable in practice because wet vapor tends to cause erosion of the turbine blades due to the impact of liquid droplets present in the expanding vapor.

A property diagram that is especially useful for the analysis of vapor power cycles is the enthalpy-entropy (*h-s*) diagram shown in Fig. 9.4. In Sec. 9.1 we found that when the kinetic and potential energy terms in the SFEE are neglected, the heat and work interactions are given by the change in enthalpy of the working fluid. Therefore the magnitude of the heat and work interactions of the various sub-components of the Rankine cycle could be obtained directly from the vertical distances between the state points in the *h-s* diagram in Fig. 9.4. The comprehensive form of the *h-s* diagram including additional constant property lines is known

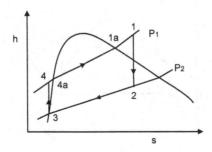

Fig. 9.4 Rankine cycle: *h-s* Diagram

as the Mollier chart and it is a useful practical tool for analyzing vapor power cycles.

The general shape of the constant pressure lines in the h-s diagram may be obtained by applying the property relation given by Eq. (9.6). For a constant pressure process, $dP = 0$.Therefore from Eq. (9.6) it follows that

$$(\partial h/\partial s)_P = T \qquad (9.10)$$

In Fig. 9.4 we have shown two constant pressure lines for the heat supply and heat rejection pressures at P_1 and P_2 respectively. From Eq. (9.10) we note that in the vapor-liquid region, the slope of these constant pressure lines is equal to the saturation temperature which is constant.

In the superheat region the slope of the constant pressure lines increase with temperature while in the sub-cooled liquid region the slopes decrease with the degree of sub-cooling. The change in quality, dx of the vapor within the vapor-liquid region is proportional to the change in enthalpy, dh because, $dh = h_{fg}dx$, where h_{fg} is the enthalpy of vaporization, which may be assumed constant.

9.2.2 *Analysis of the Rankine cycle*

We now apply the SFEE to each component of the Rankine cycle heat engine depicted in Fig. 9.2. The important assumptions are: (i) the kinetic energy and potential energy of the working fluid are negligible compared to the enthalpy, (ii) the heat interaction between the working fluid and the surroundings is negligible, and (iii) the pressure drop due to fluid friction is negligible.

The work output for the reversible adiabatic expansion $(s_1 = s_2)$ in the turbine 1-2 is

$$\dot{W}_{1-2} = \dot{m}(h_1 - h_2) \qquad (9.11)$$

where $\dot{m}$ is the mass flow rate of the working fluid.

The constant-pressure heat input in the boiler is

$$\dot{Q}_{4-1a} = \dot{m}(h_{1a} - h_4) \qquad (9.12)$$

The constant-pressure heat input in the super-heater is

$$\dot{Q}_{1a-1} = \dot{m}(h_1 - h_{1a}) \qquad (9.13)$$

The constant-pressure heat rejection in the condenser is

$$\dot{Q}_{2-3} = \dot{m}(h_2 - h_3) \qquad (9.14)$$

The work input to the adiabatic feed pump is

$$\dot{W}_{3-4} = \dot{m}(h_4 - h_3) \qquad (9.15)$$

Assuming the compression process in the pump to be isentropic ($ds = 0$) we use the property relation in Eq. (9.6) to simplify Eq. (9.15) to the form

$$\dot{W}_{3-4} = \dot{m}(h_4 - h_3) = \dot{m}\int_{P_3}^{P_4} v\,dP \qquad (9.16)$$

Since the liquid has a very low compressibility, its specific volume may be assumed constant. Integrating Eq. (9.16) we have

$$\dot{W}_{3-4} = \dot{m}\int_{P_3}^{P_4} v\,dP = \dot{m}\bar{v}(P_4 - P_3) \qquad (9.17)$$

where $\bar{v}$ is the mean specific volume.

The thermal efficiency of the cycle is

$$\eta = \frac{\dot{W}_{net}}{\dot{Q}_{4-1}} = \frac{(\dot{W}_{1-2} - \dot{W}_{3-4})}{(\dot{Q}_{4-1a} + \dot{Q}_{1a-1})} \qquad (9.18)$$

The above analysis of the ideal Rankine cycle can be easily extended to cover deviations from ideal conditions such as pressure drops in the piping, heat losses to the surroundings, and the inefficiencies in the turbine and the pump. We shall consider these practical situations in the worked examples to follow in this chapter.

9.3   The Reheat Cycle

The reheat cycle is a modified version of the basic Rankine cycle that is usually employed in large power generation systems. The main objective

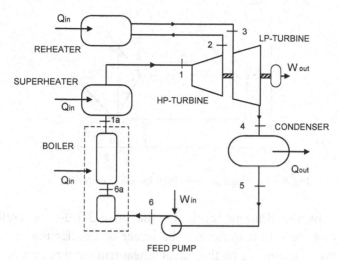

Fig. 9.5 Rankine cycle with reheating

of the reheat cycle is to ensure that the quality of the steam during its expansion through the turbine does not become too low for the safe operation of the turbine blades. Excessive wetness of the steam can result in the erosion of turbine blades due to the impact of liquid drops. A schematic diagram of the reheat cycle is shown in Fig. 9.5.

Superheated steam entering the high-pressure section of the turbine is first expanded to an intermediate pressure 2. The steam then passes through a reheater where it receives heat in a constant pressure process to attain a superheated state 3.

After expanding in the low-pressure section of the turbine, the steam is discharged into the condenser at 4. The condensate is pumped to the boiler in the process 5-6.

A *T-s* diagram of the reheat cycle is shown in Fig. 9.6. If the steam was expanded from the superheated state 1 to the condenser pressure in a single process the final state would be 2a. With the inclusion of a reheater the quality of the steam at the end of the expansion in the low-pressure turbine (4) is much higher.

It is clear from the enclosed area of the *T-s* diagram that the work output of the reheat cycle 1-2-3-4-5-1 is higher than the basic Rankine cycle 1-2a-5-1. However, the heat input to the reheat cycle is also larger

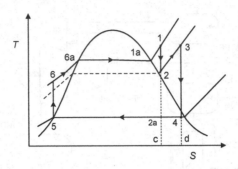

Fig. 9.6 *T-s* diagram of the Rankine cycle with reheating

than that for the Rankine cycle by the area 2-3-d-c. Therefore the efficiency of the reheat cycle may be larger or smaller than that of the Rankine cycle depending on the actual reheat temperature range.

9.3.1 *Analysis of the reheat cycle*

We now apply the SFEE to the various components of the reheat cycle making the same assumptions as in Sec. 9.2. The energy equations for the boiler, the condenser and the feed pump are similar to the corresponding equations obtained in Sec. 9.2 for the basic Rankine cycle. The work output of the high-pressure turbine is

$$\dot{W}_{1-2} = \dot{m}(h_1 - h_2) \tag{9.19}$$

where $\dot{m}$ is the mass flow rate of the working fluid.

The constant-pressure heat input in the reheater is

$$\dot{Q}_{2-3} = \dot{m}(h_3 - h_2) \tag{9.20}$$

The work output of the low-pressure turbine is

$$\dot{W}_{3-4} = \dot{m}(h_3 - h_4) \tag{9.21}$$

The thermal efficiency of the reheat cycle is given by

$$\eta = \frac{\dot{W}_{net}}{\dot{Q}_{in}} = \frac{(\dot{W}_{1-2} + \dot{W}_{3-4} - \dot{W}_{5-6})}{(\dot{Q}_{6-1a} + \dot{Q}_{1a-1} + \dot{Q}_{2-3})} \tag{9.22}$$

9.4 The Regenerative Power Cycle

One of the main deficiencies of the basic Rankine cycle is the external heat transfer irreversibility in the boiler due to the relatively large temperature difference between the heat source and the working fluid. This is particularly acute in the economizer section of the boiler where the sub-cooled liquid is heated to the saturation temperature corresponding to the boiler pressure.

The regenerative cycle is a modified version of the Rankine cycle that is designed to reduce the external heat input in the economizer by raising the inlet temperature of the working fluid at entry to the boiler. The heat required for this purpose is obtained by extracting a fraction of the expanding steam from the turbine, and mixing it with the feed water from the condenser in a series of feed water heaters (feed heaters).

A schematic diagram of the regenerative cycle with two feed heaters is shown in Fig. 9.7. Steam is extracted at points 2 and 3 at the end of the high-pressure and intermediate-pressure stages of the turbine. Mixing between the steam and the feed water from the condenser occurs in direct contact heat exchangers called *open feed heaters*. Since the pressures in the two feed heaters are equal to the turbine pressures at the corresponding steam extraction locations, additional pumps need to be installed in the feed water line to pump the water from one heater to the next.

Although we have shown two feed heaters in Fig. 9.7, large commercial power plants may have many more feed heaters. It should be noted that by extracting steam from the turbine for feed water heating we reduce the steam flow rate through the turbine stages. Therefore the fraction of the steam extracted must be carefully optimized for maximum thermal efficiency since increasing this fraction causes a corresponding decrease in the turbine work output.

The T-s diagram of the regenerative cycle is shown in Fig. 9.8. Under the ideal conditions depicted in the Fig. 9.8, the mass flow rates of steam into the feed heaters $\dot{m}_3$ and $\dot{m}_2$ are such that at the exit locations 7 and 9 from the feed heaters the water is in a saturated liquid state. We notice that external heat input is only needed to heat the water from temperature T_{10} to T_{10a} in the economizer. Moreover, the average temperature of heat supply that is now between T_{10} and T_1 is higher than the average temperature without feed heating, which is between T_6 and T_1. This higher average temperature of heat supply of the regenerative cycle results in an increase in the thermal efficiency of the cycle.

As the number of feed heaters is increased the saw-tooth-like heating lines from 5 to 10 will approach the liquid saturation curve, and the feed water inlet temperature to the boiler will approach the saturation temperature of the boiler. In the absence of superheating, the temperature of heat supply will then be constant. Therefore the efficiency of the cycle with the above ideal feed heater arrangement will approach the Carnot efficiency.

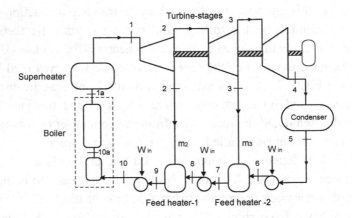

Fig. 9.7 The regenerative Rankine cycle

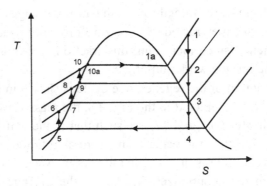

Fig. 9.8 *T-s* diagram of the regenerative Rankine cycle

9.4.1 *Analysis of the regenerative cycle with open-feed-heaters*

We shall apply the mass balance equation and the SFEE to the various components of the regenerative cycle using open feed heaters shown in Fig. 9.7. We assume that the kinetic and potential energy of the streams, the frictional pressure loss in the piping and components and the heat losses to the surrounding to be negligible.

Let the mass flow rates of steam through the boiler and super-heater be $\dot{m}_1$ and the two feed heaters be $\dot{m}_2$ and $\dot{m}_3$ respectively.

The work output of the high-pressure turbine is

$$\dot{W}_{1-2} = \dot{m}_1 (h_1 - h_2) \tag{9.23}$$

The work output of the intermediate-pressure turbine is

$$\dot{W}_{2-3} = (\dot{m}_1 - \dot{m}_2)(h_2 - h_3) \tag{9.24}$$

The work output of the low-pressure turbine is

$$\dot{W}_{3-4} = (\dot{m}_1 - \dot{m}_2 - \dot{m}_3)(h_3 - h_4) \tag{9.25}$$

For the isentropic expansions through the turbines we have

$$s_1 = s_2 = s_3 = s_4$$

The constant pressure heat removal rate in the condenser is

$$\dot{Q}_{4-5} = (\dot{m}_1 - \dot{m}_2 - \dot{m}_3)(h_4 - h_5)$$

The constant pressure heat input in the super-heater is

$$\dot{Q}_{a1-1} = \dot{m}_1 (h_1 - h_{1a}) \tag{9.26}$$

The constant pressure heat input in the boiler is

$$\dot{Q}_{10-1a} = \dot{m}_1 (h_{1a} - h_{10}) \tag{9.27}$$

Applying the SFEE to the feed heater 1 we have

$$\dot{m}_1 h_9 - \dot{m}_2 h_2 - (\dot{m}_1 - \dot{m}_2)h_8 = 0 \tag{9.28}$$

Applying the SFEE to the feed heater 2 we have

$$(\dot{m}_1 - \dot{m}_2)h_7 - \dot{m}_3 h_3 - (\dot{m}_1 - \dot{m}_2 - \dot{m}_3)h_6 = 0 \tag{9.29}$$

The ideal work input to the feed pumps is obtained by applying Eq. (9.17).

$$\dot{W}_{5-6} = (\dot{m}_1 - \dot{m}_2 - \dot{m}_3)\bar{v}_{5-6}(P_6 - P_5) \qquad (9.30)$$

$$\dot{W}_{7-8} = (\dot{m}_1 - \dot{m}_2)\bar{v}_{7-8}(P_8 - P_7) \qquad (9.31)$$

$$\dot{W}_{9-10} = \dot{m}_1\bar{v}_{9-10}(P_{10} - P_9) \qquad (9.32)$$

where $\bar{v}$ is the mean specific volume of the liquid.

The thermal efficiency of the regenerative cycle is given by

$$\eta = \frac{\dot{W}_{net}}{\dot{Q}_{10-1}} = \frac{(\dot{W}_{1-2} + \dot{W}_{2-3} + \dot{W}_{3-4} - \dot{W}_{5-6} - \dot{W}_{7-8} - \dot{W}_{9-10})}{(\dot{Q}_{10-1a} + \dot{Q}_{1a-1})} \qquad (9.33)$$

9.4.2 Closed-feed-heaters

A disadvantage of using open feed heaters in a Rankine cycle is the need to install separate feed water pumps before and after each heater to boost the feed water pressure to the pressure of the extracted steam.

Closed feed heaters which are essentially shell and tube heat exchangers may be arranged to operate with a single feed water pump as shown in Fig. 9.9. Notice that in Fig. 9.9 we have shown only the feed water heating section of the complete Rankine cycle depicted in Fig. 9.7.

The feed water flowing in the tubes of the heaters receive heat from the extracted steam condensing on the outside of the tubes. Ideally, the condensate leaving the heaters at 9 and 7 is in a saturated state. Steam traps that allow only liquid to pass through are installed in the condensate line between the heaters to throttle the liquid to the lower pressure in a constant-enthalpy process. Hence

$$h_9 = h_{9a} \qquad \text{and} \qquad h_7 = h_{7a}$$

Applying the mass balance equation and the SFEE to the two feed heaters 1 and 2 and the condenser we obtain the following equations.

$$\dot{m}_1(h_{10} - h_8) - \dot{m}_2(h_2 - h_9) = 0 \qquad (9.34)$$

$$\dot{m}_1(h_8 - h_6) - \dot{m}_3(h_3 - h_7) - \dot{m}_2(h_9 - h_7) = 0 \qquad (9.35)$$

$$\dot{Q}_{con} + \dot{m}_1 h_5 - (\dot{m}_1 - \dot{m}_2 - \dot{m}_3)h_4 - (\dot{m}_2 + \dot{m}_3)h_7 = 0 \qquad (9.36)$$

where $\dot{Q}_{con}$ is the rate of heat removal in the condenser.

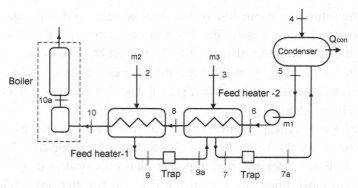

Fig. 9.9 Closed feed water heating arrangement

A significant advantage of closed feed heaters is that they do not require a separate pump for each heater but a drawback of the closed heaters is that they do not heat the feed water to the saturation temperature of the extracted steam as open heaters do. Despite their differences in performance, many power plants incorporate a combination of both open and closed feed water heaters.

9.5 The Choice of Working Fluid

Based on our discussion of the Rankine cycle thus far, we can list the desirable characteristics of a working fluid to be used in a vapor-power cycle.

(i) The critical temperature of the fluid is well above the metallurgical limit of the boiler so that most of the heat could be supplied at the highest practical temperature without the need for a super-heater.

(ii) Saturation pressure at the metallurgical limit is moderate so that the construction cost of the boiler is low.

(iii) The specific heat capacity of the liquid is low, so that the energy required to heat the liquid to the saturation state in the economizer is low.

(iv) The latent enthalpy of vaporization of the fluid is high so that the fluid flow rate per unit work output is low. The overall plant size will therefore be low.

(v) The saturation vapor line of the fluid is steep so that the quality of the expanding vapor in the turbine remains close to unity. Reheating could be dispensed with under such conditions.

(vi) The saturation pressure at the lowest available heat sink temperature is above ambient pressure so that vacuum is not needed in the condenser.

(vii) The fluid is cheap, non-toxic, non-corrosive and chemically stable.

There is no fluid that has all these desirable properties. Water is widely used as a working fluid in power plants because its properties are better than most other fluids available. However, water has a relatively low critical temperature of 374°C compared to the limiting temperatures of boiler materials. Moreover, even at moderate temperatures the saturation pressure of water is quite high. These deficiencies of water as a working fluid at the high temperature part of the Rankine cycle may be partially overcome by combining two fluids in a binary vapor cycle.

9.5.1 *Binary vapor cycle*

Mercury and water are the most commonly used working fluids in the binary vapor cycle. The binary vapor cycle shown in Fig. 9.10, consists essentially of two Rankine cycles with mercury and water as the working fluids. The heat rejected by the condensing mercury is used to evaporate the water. Mercury is heated in a constant pressure process 4-1 and expands through the mercury turbine in the process 1-2.

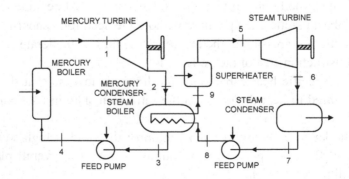

Fig. 9.10 Binary vapor cycle

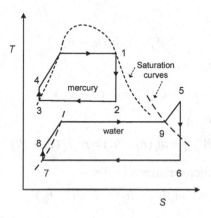

Fig. 9.11 *T-s* diagram of the binary cycle

Condensation of the mercury occurs in the mercury condenser/steam boiler in which sub-cooled water is evaporated in the process 8-9. After being heated in the superheater in the constant pressure process 9-5, the steam expands in the turbine from 5 to 6. The liquid mercury and water are circulated by the respective feed pumps. In the *T-s* diagram of the binary vapor cycle, shown in Fig. 9.11, we have assumed that the expansion processes in the two turbines, and the compression processes in the two feed pumps are isentropic.

9.5.2 *Analysis of the binary vapor cycle*

Let the mass flow rates of the mercury and steam be $\dot{m}_m$ and $\dot{m}_s$ respectively. Applying the SFEE to the various components of the two Rakine cycles (Fig. 9.10), making the same assumptions as in Sec. 9.2.2, we obtain following equations:

For the mercury turbine:

$$\dot{W}_{t-m} = \dot{m}_m (h_1 - h_2) \tag{9.37}$$

For the mercury boiler:

$$\dot{Q}_{b-m} = \dot{m}_m (h_1 - h_4) \tag{9.38}$$

For the mercury feed pump:

$$\dot{W}_{p-m} = \dot{m}_m (h_4 - h_3) = \dot{m}_m \bar{v}_m (P_4 - P_3) \tag{9.39}$$

For the steam turbine:

$$\dot{W}_{t-s} = \dot{m}_s (h_5 - h_6) \tag{9.40}$$

For the steam superheater:

$$\dot{Q}_{sh-s} = \dot{m}_s (h_5 - h_9) \tag{9.41}$$

For the water feed pump:

$$\dot{W}_{p-s} = \dot{m}_s (h_8 - h_7) = \dot{m}_s \bar{v}_w (P_8 - P_7) \tag{9.42}$$

For the mercury condenser/steam boiler:

$$\dot{m}_m (h_2 - h_3) = \dot{m}_s (h_9 - h_8) \tag{9.43}$$

The thermal efficiency of the binary vapor cycle is

$$\eta = \frac{\dot{W}_{t-m} + \dot{W}_{t-s} - \dot{W}_{p-m} - \dot{W}_{p-s}}{\dot{Q}_{b-m} + \dot{Q}_{sh-s}} \tag{9.44}$$

9.5.3 *Supercritical vapor power cycle*

Another method of increasing the thermal efficiency of the Rankine cycle is to operate under supercritical conditions. Here the working fluid receives heat at a pressure in excess of the critical pressure. Unlike the standard Rankine cycle, the fluid in a supercritical system does not undergo phase change during the heating process. The temperature of the fluid increases in a continuous manner.

We have shown in Fig. 9.13 the temperature distribution of the furnace gases in a boiler, which is the typical heating medium in a Rankine cycle power plant. During evaporation, the temperature difference between the working fluid and the heating medium become progressively larger thus increasing the external heat transfer irreversibility. However, with a supercritical fluid the latter temperature difference is less divergent (see Fig. E9.13) and therefore the heat transfer irreversibility could be reduced. In the case of water the critical pressure is about 220 bar, and therefore the supercritical power plant using water as the working fluid operates under very high pressures.

9.6 Combined-Heat and Power (CHP) Cycles

Thus far in this chapter we have discussed vapor cycles whose useful output is work. However, there are many industrial plants or factories that require both heat and power as energy inputs to the various processes. The heat input may be used to heat a building, operate an absorption cooling system, or carry out a thermal process. Such power and heating needs can be combined efficiently using combined heat and power (CHP) cycles.

Although the actual configuration of the CHP system may depend on the specific details of the plant we illustrate this important application by considering the arrangement shown in Fig. 9.12. The steam supplied to the steam heater is controlled by installing a pressure regulating valve. The steam flow rates through the turbines and the heater are adjusted to meet the required heating and power needs.

Let the mass flow rates of steam through the boiler, the regulating valve, and the low-pressure turbine be $\dot{m}_1$, $\dot{m}_2$ and $\dot{m}_4$ respectively. We now apply the mass balance equation and the SFEE to the different components to obtain the following equations.

For the heater:

$$\dot{Q}_h = \dot{m}_2 h_3 + \dot{m}_3 h_2 - (\dot{m}_3 + \dot{m}_2)h_7 \qquad (9.45)$$

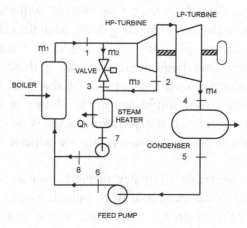

Fig. 9.12 Combined heat and power (CHP) cycle

For the low-pressure turbine:

$$\dot{W}_{lp} = \dot{m}_4(h_2 - h_4) \tag{9.46}$$

For the high-pressure turbine:

$$\dot{W}_{hp} = (\dot{m}_3 + \dot{m}_4)(h_1 - h_2) \tag{9.47}$$

The mass balance equation gives

$$\dot{m}_1 = \dot{m}_2 + \dot{m}_3 + \dot{m}_4 \tag{9.48}$$

The flow through the valve may be assumed a constant enthalpy process. Under ideal design conditions the expansion processes in the two turbines are isentropic.

9.7 Deviations Between Actual and Ideal Cycles

In our analysis of the Rankine cycle and its modified forms we assumed that ideal processes occurred in the various components. The impact of the deviation of actual cycles from ideal conditions is considered briefly in this section. The inevitable frictional pressure losses in the piping and the valves will cause entropy production with a resulting decrease in the thermal efficiency of the plant.

These frictional losses are present in the piping between the boiler and the turbine, in the boiler and to a lesser extent in the condenser. Heat losses to the surroundings from the piping surfaces will also result in net entropy production with a corresponding decrease in the efficiency.

The actual expansion process in the turbine, and the compression process in the pump are both non-isentropic. If these processes are assumed to be adiabatic then the entropy of the fluid at the end of the process will exceed the initial entropy as shown in Fig. 9.13. The final states of the fluid for the isentropic expansion in the turbine and the isentropic compression in the pump are indicated by 2s and 4s respectively.

We define the isentropic efficiency of the turbine as the ratio of the actual work output to the isentropic work output. The efficiency of the pump is the ratio the isentropic work input to the actual work input.

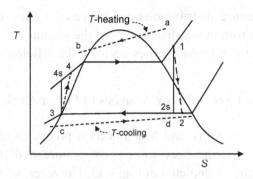

Fig. 9.13 Irreversibilities of the Rankine cycle

These efficiencies may be expressed in terms of the enthalpies of the fluid by applying the SFEE to the turbine and the pump respectively. Hence we have

$$\eta_t = \frac{w_{act}}{w_{ideal}} = \frac{h_1 - h_2}{h_1 - h_{2s}} \qquad (9.49)$$

$$\eta_p = \frac{w_{ideal}}{w_{act}} = \frac{h_{4s} - h_3}{h_4 - h_3} \qquad (9.50)$$

The irreversible processes in the turbine and the pump that occur within the Rankine cycle are usually called internal irreversibilities.

In addition, there are external irreversibilities that are caused by the heat interactions between the working fluid, and the heating and cooling media. In the idealized analysis of cycles we assume that these heat interactions occur with reservoirs at constant temperatures. However, this is not the case with actual power plants. In a fossil fuel power plant, heat is supplied to the working fluid by the high temperature furnace gases, while in a nuclear power plant, the primary reactor coolant constitutes the heat source. In both instances, the heating medium is a single phase fluid that flows counter to the working fluid in the steam generator. In Fig. 9.13, we have indicated the variation of the heating medium temperature by the line a-b. The finite temperature difference between the lines a-b and 4-1 constitutes an external heat transfer irreversibility.

A similar irreversibility occurs due to the temperature difference between the cooling medium and the working fluid in the condenser.

These temperature distributions are indicated by the lines c-d and 2-3 respectively. From our earlier studies of the second law in chapters 5 to 7 it is clear that these irreversibilies decrease the efficiency of the plant.

9.8 Simplified Second Law Analysis of Power Cycles

During the heat supply and heat rejection processes of an actual vapor power cycle both the heat source temperature and the working fluid temperature vary as depicted in Fig. 9.13. However, in order to perform a simplified second-law analysis, we could assume that heat is supplied from a high temperature reservoir and rejected to a heat sink reservoir while the temperature of working the fluid varies as shown in Fig. 9.13. This idealized model is depicted in Fig. 9.14.

Let the heat input, the actual work output and the heat rejection rate of the cycle per unit mass of working fluid be q_h, w_a and q_c respectively.

Since the working fluid undergoes a cyclic process in a steady manner, the entropy change and entropy storage in the working fluid are both zero. Therefore the only components of the cycle where entropy changes occur are the two heat reservoirs whose temperatures are T_{rh} and T_{rc}. In order to generalize our second law analysis of the cycle we introduce a standard ambient or 'dead-state' reservoir at a temperature T_o as shown in Fig. 9.14. Applying first law to the cycle we obtain

$$w_a = q_h - q_c - q_o \tag{9.51}$$

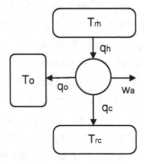

Fig. 9.14 Second law analysis of a power cycle

Application of the second law in the form of the principle of increase of entropy gives

$$-\frac{q_h}{T_{rh}}+\frac{q_c}{T_{rc}}+\frac{q_o}{T_o}\geq 0 \qquad (9.52)$$

We convert Eq. (9.52) to an equality by introducing the entropy production per unit mass of working fluid, σ. This gives

$$-\frac{q_h}{T_{rh}}+\frac{q_c}{T_{rc}}+\frac{q_o}{T_o}=\sigma \qquad (9.53)$$

Multiplying Eq. (9.53) by T_o and adding it to Eq. (9.51) we obtain

$$q_h\left(1-\frac{T_o}{T_{rh}}\right)-q_c\left(1-\frac{T_o}{T_{rc}}\right)-w_a=T_o\sigma \qquad (9.54)$$

When the work output is a maximum, $\sigma=0$ and Eq. (9.54) becomes

$$q_h\left(1-\frac{T_o}{T_{rh}}\right)-q_c\left(1-\frac{T_o}{T_{rc}}\right)-w_{\max}=0 \qquad (9.55)$$

It is interesting to note that the first and second terms on the left hand side of Eq. (9.55) are the exergies of the heat input and the heat output respectively. Therefore the maximum work is the difference in the exergy input and output to the cycle.

From Eqs. (9.54) and (9.55) it follows that

$$w_{\max}=w_a+T_o\sigma \qquad (9.56)$$

We define the second-law effectiveness of the actual cycle as

$$\eta_2=\frac{w_a}{w_{\max}}=\frac{w_a}{w_a+T_o\sigma} \qquad (9.57)$$

We shall use Eq. (9.57) to compare the second-law effectiveness of various modifications to the Rankine cycle in the worked examples that follow.

9.9 Worked Examples

Example 9.1 The vaporization and condensation pressures of an ideal engine operating on the Carnot cycle with steam as the working fluid are 40 bar and 0.04 bar respectively. Calculate (i) the heat and work transfers, (ii) the work ratio, (iii) the specific steam consumption in kg per kWh, and (iv) the cycle efficiency. Sketch the cycle on the *T-s* and *h-s* diagrams for water.

Solution The *T-s* and *h-s* diagrams of the Carnot cycle are shown in Figs. E9.1(a) and (b) respectively. The following properties at points 1 and 4 are obtained from the saturated steam tables at 40 bar in [5].

$$h_1 = 2801 \text{ kJ kg}^{-1}, \qquad s_1 = 6.07 \text{ kJ K}^{-1} \text{ kg}^{-1},$$

$$h_4 = 1087 \text{ kJ kg}^{-1}, \qquad s_4 = 2.797 \text{ kJ K}^{-1} \text{ kg}^{-1}.$$

For the isentropic processes 1-2, $s_1 = s_2$. Hence

$$6.07 = 8.473x_2 + 0.422(1 - x_2).$$

Therefore the quality at 2 is $x_2 = 0.7015$.

For the isentropic process 3-4, $s_3 = s_4$. Hence

$$2.797 = 8.473x_3 + 0.422(1 - x_3).$$

The quality at 3 is $x_3 = 0.295$

$$h_2 = 2554x_2 + 121(1 - x_2) = 1827.8 \text{ kJ kg}^{-1}$$

$$h_3 = 2554x_3 + 121(1 - x_3) = 838.72 \text{ kJ kg}^{-1}$$

The heat and work interactions are obtained by applying the SFEE to the different units of the cycle neglecting the kinetic and potential energy terms.

Turbine work output is

$$w_t = h_1 - h_2 = 2801 - 1827.8 = 973.2 \text{ kJ kg}^{-1}$$

Compressor work input is

$$w_c = h_4 - h_3 = 1087 - 838.72 = 248.29 \text{ kJ kg}^{-1}$$

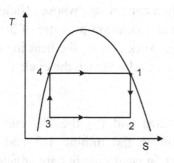

Fig. E9.1(a) *T-s* diagram

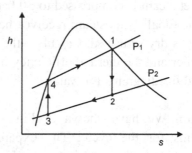

Fig. E9.1(b) *h-s* diagram

Net work output is

$$w_{net} = w_t - w_c = 973.2 - 248.29 = 724.9 \text{ kJ kg}^{-1}$$

Heat input in the boiler is

$$q_b = h_1 - h_4 = 2801 - 1087 = 1714 \text{ kJ kg}^{-1}$$

Heat rejected in the condenser is

$$q_c = h_2 - h_3 = 1827.8 - 838.72 = 989.08 \text{ kJ kg}^{-1}$$

Thermal efficiency of the cycle is

$$\eta = \frac{w_{net}}{q_b} = \frac{724.9}{1714} = 42.3\%$$

The specific steam consumption (SSC) is defined as the kg of steam flow required to generate a kW-hour of work. Since 1 kg of steam produces 724.9 kJ of work, the SSC is given by

$$SSC = 3600 / 724.9 = 4.97 \text{ kg (kWh)}^{-1}$$

Note that the efficiency of the Carnot cycle which is independent of the properties of the working fluid is also given by

$$\eta = 1 - \frac{T_2}{T_1} = 1 - \frac{(273 + 29)}{(273 + 250.3)} = 42.3\%$$

Example 9.2 Dry saturated steam at 30 bar expands to a pressure of 0.04 bar through a turbine whose efficiency is 80%. The steam is then condensed until its entropy is equal to that of saturated water at 30 bar.

The wet steam is compressed to 30 bar in a compressor whose efficiency is 85%. Finally, the steam receives heat at a constant pressure of 30 bar until it is dry saturated. Calculate the net work output, the heat input in the boiler and the thermal efficiency of the cycle. Sketch the cycle on the *T-s* and *h-s* diagrams for water.

Solution We have shown in Fig. E9.2(a) and (b) the *T-s* and *h-s* diagrams for the cycle. The expansion in the turbine 1-2 and the compression in the compressor are both non-isentropic but are adiabatic. The corresponding isentropic processes are indicated by 1-2s and 3-4s respectively. The following data have been obtained from the saturated steam data tables in [5].

$$h_1 = 2803 \text{ kJ kg}^{-1}, \quad s_1 = 6.186 \text{ kJ K}^{-1} \text{ kg}^{-1}, \quad s_{4s} = 2.645 \text{ kJ K}^{-1} \text{ kg}^{-1}.$$

For the isentropic expansion we have $s_1 = s_{2s}$.

$$6.186 = 8.473 x_{2s} + 0.422(1 - x_{2s}). \quad \text{Hence} \quad x_{2s} = 0.716$$

$$h_{2s} = 121(1 - x_2) + 2554 x_2 = 1863 \text{ kJ kg}^{-1}$$

The isentropic efficiency is defined as

$$\eta_i = \frac{w_{12}}{h_1 - h_{2s}} \tag{E9.2.1}$$

Substituting numerical values in Eq. (E9.2.1) we have

$$w_{12} = h_1 - h_2 = 0.8 \times (2803 - 1863) = 752 \text{ kJ kg}^{-1}$$

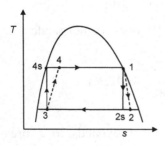

Fig. E9.2(a) *T-S* diagram

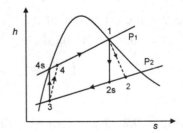

Fig. E9.2(b) *h-s* diagram

Now for the isentropic compression, $s_3 = s_{4s} = 2.645$

$$2.645 = 8.473x_3 + 0.422(1 - x_3), \quad \text{Hence} \quad x_3 = 0.276$$

$$h_3 = 121(1 - x_3) + 2554x_3 = 792.5 \text{ kJ kg}^{-1}$$

The isentropic efficiency of the compressor is defined as

$$\eta_{ic} = \frac{h_{4s} - h_3}{w_{34}} \qquad (E9.2.2)$$

Substituting numerical data in Eq. (E9.2.2) we obtain

$$w_{34} = (1008 - 792.5)/0.85 = 253.5 \text{ kJ kg}^{-1}$$

The net work output of the cycle is

$$w_{net} = w_{12} - w_{34} = 498.5 \text{ kJ kg}^{-1}$$

$w_{34} = h_4 - h_3 = 253.5 \text{ kJ kg}^{-1}$. Hence $h_4 = 1046 \text{ kJ kg}^{-1}$

The heat supplied in the boiler is given by

$$q_{41} = h_1 - h_4 = 1757 \text{ kJ kg}^{-1}$$

Thermal efficiency of the cycle is

$$\eta = \frac{w_{net}}{q_{41}} = \frac{498.5}{1757} = 28.4\%$$

Note the that the Carnot efficiency for the same heat supply and rejection temperatures is

$$\eta = 1 - \frac{T_2}{T_1} = 1 - \frac{(273 + 29)}{(273 + 233.8)} = 40.4\%$$

The decrease in efficiency is due to the internal irreversibilities in the turbine and the compressor which are caused by fluid friction.

Example 9.3 The boiler and condenser pressures of a steam plant operating on the ideal Rankine cycle are 30 bar and 0.04 bar respectively. The steam is superheated to 350°C in a super-heater before it is expanded isentropically in the turbine. The steam is condensed to a saturated liquid and then pumped isentropically to the boiler pressure to complete the cycle. (a) Calculate (i) the work output of the turbine (kJ per kg), (ii) the feed pump work, (iii) the heat rejected in the condenser, and

(iv) the thermal efficiency of the plant. Sketch the cycle on a *T-s* diagram and a *h-s* diagram for water.

(b) If heat is supplied to the boiler and superheater from a reservoir at 400°C and heat is rejected in the condenser to a heat sink reservoir at 20°C, calculate (i) the entropy change in the two reservoirs, and (ii) the second law effectiveness using a standard ambient temperature of 20°C.

Solution The *T-s* and *h-s* diagrams of the ideal Rankine cycle are shown in Figs. E9.3(a) and (b) respectively. The expansion process 1-2 and compression process 3-4 are isentropic. The following data have been obtained from the steam tables in [5].

$$h_1 = 3117 \text{ kJ kg}^{-1}, \quad s_1 = 6.744 \text{ kJ K}^{-1} \text{ kg}^{-1}, \quad h_3 = 121 \text{ kJ kg}^{-1}.$$

For the isentropic process $s_1 = s_2$

$$6.744 = 8.473x_2 + 0.422(1 - x_{2s}). \quad \text{Hence} \quad x_2 = 0.785$$

$$h_2 = 121(1 - x_2) + 2554x_2 = 2030.9 \text{ kJ kg}^{-1}$$

Work output of the turbine is

$$w_{12} = h_1 - h_2 = 3117 - 2030.9 = 1086.1 \text{ kJ kg}^{-1}$$

Work input to the pump is

$$w_{34} = h_4 - h_3 = \bar{v}(P_4 - P_3)$$

Substituting numerical values in the above equation

$$w_{34} = h_4 - h_3 = (30 - 0.04) \times 100 \times 10^{-3} = 3.0 \text{ kJ kg}^{-1}$$

We have assumed that the specific volume of water is $10^{-3} \text{ m}^3 \text{ kg}^{-1}$

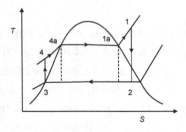

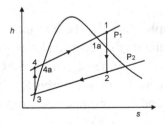

Fig. E9.3(a) *T-s* diagram   Fig. E9.3(b) *h-s* diagram

Hence
$$h_4 = 121 + 3 = 124 \text{ kJ kg}^{-1}$$

Net work output is
$$w_{net} = w_{12} - w_{34} = 1086.1 - 3 = 1083.1 \text{ kJ kg}^{-1}$$

Heat input in the boiler and superheater is
$$q_{41} = h_1 - h_4 = 3117 - 124 = 2993 \text{ kJ kg}^{-1}$$

Heat rejected in the condenser is
$$q_{23} = h_2 - h_3 = 2030.9 - 121 = 1909.9 \text{ kJ kg}^{-1}$$

Net heat input to the cycle is
$$q_{net} = q_{41} - q_{23} = 2993 - 1909.9 = 1083.1 \text{ kJ kg}^{-1}$$

As expected, for the cycle
$$q_{net} = w_{net}$$

The thermal efficiency of the cycle is
$$\eta = \frac{w_{net}}{q_{41}} = \frac{1083.1}{2993} = 36\%$$

(b) Since the engine operates in a cyclic manner there is no change in entropy of the working fluid during the complete cycle. Also because conditions are steady no entropy is stored in the working fluid. The only components of the plant that experience entropy changes are the two reservoirs. The entropy change of the high temperature reservoir per unit mass of steam is given by
$$\Delta S_h = -\frac{Q_{in}}{T_h} = -\frac{2993}{(273 + 400)} = -4.447 \text{ kJ K}^{-1} \text{ kg}^{-1}$$

For the low temperature reservoir the entropy change is
$$\Delta S_c = \frac{Q_{out}}{T_c} = \frac{1909.9}{(273 + 20)} = 6.518 \text{ kJ K}^{-1} \text{ kg}^{-1}$$

Entropy production in the universe is
$$\sigma = \Delta S_h + \Delta S_c = -4.447 + 6.518 = 2.07 \text{ kJ K}^{-1} \text{ kg}^{-1}$$

The second-law effectiveness is given by Eq. (9.57) as
$$\eta_2 = \frac{1083.1}{1083.1 + 293 \times 2.07} = 0.64$$

Example 9.4 A steam plant operating on the Rankine cycle has a boiler pressure of 30 bar and a condenser pressure of 0.04 bar. The steam is superheated to 350°C before it enters the turbine. The isentropic efficiencies of the turbine and the feed-pump are 80% and 85% respectively. (a) Calculate (i) the work output of the turbine, (ii) the heat input to the cycle, (iii) the work input to the pump, (iv) the heat rejected in the condenser, and (v) the thermal efficiency.

(b) If heat is supplied to the boiler and superheater from a reservoir at 400°C and heat is rejected in the condenser to a heat sink reservoir at 20°C, (i) calculate the entropy change in the two reservoirs, and (ii) the second-law effectiveness of the cycle using a standard ambient temperature of 20°C.

Solution The T-s and h-s diagrams of the Rankine cycle are shown in Figs. E9.4(a) and (b) respectively. The isentropic expansion and compression processes are indicated by 1-2s and 3-4s respectively. The ideal cycle 1-2s-3-4s is identical to that in worked example 9.3. Therefore we use the data obtained earlier in that example.

The isentropic efficiency of the turbine is defined by

$$\eta_t = \frac{w_{12}}{h_1 - h_{2s}} = \frac{w_{12}}{1086.1} = \frac{h_1 - h_2}{1086.1} = 0.8$$

Therefore the work output is

$$w_{12} = 868.9 \text{ kJ kg}^{-1}. \quad \text{Hence} \quad h_2 = 2248.1 \text{ kJ kg}^{-1}$$

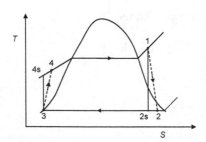

Fig. E9.4(a) T-s diagram

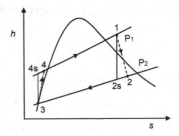

Fig. E9.4(b) h-s diagram

The isentropic efficiency of the pump is defined by

$$\eta_p = \frac{h_{4s} - h_3}{w_{34}} = \frac{3.0}{w_{34}} = \frac{3.0}{h_4 - h_3} = 0.85$$

Therefore the work input is

$$w_{34} = 3.53 \text{ kJ kg}^{-1}. \quad \text{Hence} \quad h_4 = 124.53 \text{ kJ kg}^{-1}$$

The net work output of the cycle is

$$w_{net} = w_{12} - w_{34} - 868.9 - 3.53 = 865.4 \text{ kJ kg}^{-1}$$

Heat input in the boiler and superheater is

$$q_{41} = h_1 - h_4 = 3117 - 124.53 = 2992.5 \text{ kJ kg}^{-1}$$

Heat rejected in the condenser is

$$q_{23} = h_2 - h_3 = 2248.1 - 121 = 2127.1 \text{ kJ kg}^{-1}$$

Net heat input to the cycle is

$$q_{net} = q_{41} - q_{23} = 2992.5 - 2127.1 = 865.4 \text{ kJ kg}^{-1}$$

As expected for the cycle, $\qquad q_{net} = w_{net}$

The thermal efficiency of the cycle is

$$\eta = \frac{w_{net}}{q_{41}} = \frac{865.4}{2992.5} = 28.9\%$$

We note that the decrease in thermal efficiency of the cycle is due to the internal frictional irreversibilities of the adiabatic turbine and pump.

(b) For the high temperature reservoir, the entropy change per unit mass of steam is given by

$$\Delta S_h = -\frac{Q_{in}}{T_h} = -\frac{2992.5}{(273 + 400)} = -4.446 \text{ kJ K}^{-1} \text{ kg}^{-1}$$

For the low temperature reservoir the entropy change is

$$\Delta S_c = \frac{Q_{out}}{T_c} = \frac{2127.1}{(273 + 20)} = 7.259 \text{ kJ K}^{-1} \text{ kg}^{-1}$$

Entropy production in the universe is

$$\sigma = \Delta S_h + \Delta S_c = -4.446 + 7.259 = 2.813 \text{ kJ K}^{-1} \text{ kg}^{-1}$$

The second-law effectiveness is given by Eq. (9.57)

$$\eta_2 = \frac{865.4}{865.4 + 293 \times 2.813} = 0.51$$

We note that the internal entropy produced in the adiabatic turbine and pump due to friction is transferred out to the low temperature reservoir. This causes a significant increase in the entropy of the universe which, in turn, decreases the second law effectiveness of the cycle.

Example 9.5 The condenser and boiler pressures of an ideal Rankine cycle with reheat are 0.04 bar and 30 bar respectively. The temperature of the steam at entry to the turbine is 350°C. In the high-pressure turbine, the steam expands until it is dry saturated before being reheated to 350°C. The steam is then expanded to the condenser pressure in the low-pressure turbine. Calculate (i) the total work output, (ii) the total heat supplied, and (iii) the cycle efficiency. Compare the performance of the cycles with and without reheating.
(b) If heat is supplied to the boiler, superheater and reheater from a reservoir at 400°C, and heat is rejected in the condenser to a heat sink reservoir at 20°C, calculate the entropy change in the two reservoirs, and the second-law effectiveness of the cycle using a standard ambient temperature of 20°C.

Solution The *T-s* diagram for the reheat cycle is depicted in Fig. E9.5. The cycle is ideal and the boiler and condenser conditions are the same as those in example 9.3. Now the expansion in the high pressure turbine 1-2 is isentropic and therefore

$$s_2 = s_1 = 6.744 \text{ kJ K}^{-1} \text{ kg}^{-1}$$

The steam is saturated in state 2. From the saturated steam tables [5] we observe that the saturation vapor pressure, P_2 when the vapor entropy is 6.744 kJ K^{-1} kg^{-1} is 6.33 bar. The corresponding saturated vapor enthalpy at 2 is $h_2 = 2759.3$ kJ kg^{-1}. The steam is reheated at a constant pressure of 6.33 bar to state 3 where the temperature is 350°C.

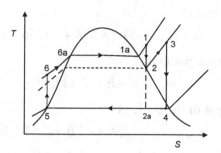

Fig. E9.5 Reheat cycle: *T-s* diagram

From the superheated steam tables we obtain by interpolation the following data,

$$h_3 = 3165.3 \, \text{kJ kg}^{-1} \quad \text{and} \quad s_3 = 7.5219 \, \text{kJ K}^{-1} \text{kg}^{-1}$$

The expansion in the low-pressure turbine 3-4 is isentropic, therefore $s_4 = s_3 = 7.5219$ kJ K^{-1} kg^{-1}. Assume that the state of steam at 4 is wet.

$$7.5219 = 8.473 x_4 + 0.422(1 - x_4). \quad \text{Hence} \quad x_4 = 0.882$$

$$h_4 = 121(1 - x_4) + 2554 x_4 = 2266.9 \, \text{kJ kg}^{-1}$$

Heat input in the boiler and superheater is the same as in worked example 9.3.

Hence $\quad\quad\quad\quad q_{61} = h_1 - h_6 = 2993 \, \text{kJ kg}^{-1}$

Heat input in the reheater is

$$q_{23} = h_3 - h_2 = 3165.3 - 2759.3 = 406 \, \text{kJ kg}^{-1}$$

Total heat input to the cycle is

$$q_{tot} = q_{61} + q_{23} = 3399 \, \text{kJ kg}^{-1}$$

Heat rejected in the condenser is

$$q_{45} = h_4 - h_5 = 2266.9 - 121 = 2145.9 \, \text{kJ kg}^{-1}$$

Work output of the high-pressure turbine is

$$w_{12} = h_1 - h_2 = 3117 - 2759.3 = 357.7 \, \text{kJ kg}^{-1}$$

Work output of the low-pressure turbine is

$$w_{34} = h_3 - h_4 = 3165.3 - 2266.9 = 898.4 \, \text{kJ kg}^{-1}$$

The total work output is

$$w_{tot} = w_{12} + w_{34} = 357.7 + 898.4 = 1256.1 \text{ kJ kg}^{-1}$$

The work input to the pump is

$$w_{35} = h_6 - h_5 = 3 \text{ kJ kg}^{-1}$$

The net work output of the cycle is

$$w_{net} = w_{tot} - w_{56} = 1256.1 - 3.0 = 1253.1 \text{ kJ kg}^{-1}$$

The net heat input to the cycle is

$$q_{net} = q_{tot} - q_{45} = 3399.0 - 2145.9 = 1253.1 \text{ kJ kg}^{-1}$$

As expected for the cycle

$$q_{net} = w_{net}$$

The thermal efficiency of the cycle is

$$\eta = \frac{w_{net}}{q_{tot}} = \frac{1253.1}{3399.0} = 36.9\%$$

The efficiency of the cycle without reheat for the same boiler and condenser conditions in worked example 9.3 was 36%. The improvement in efficiency due to reheat is therefore marginal for the present case. However, the moisture content of the steam expanding in the turbine is reduced significantly due to reheating. With reheat, the quality of steam at the end of the expansion is 0.882 compared with 0.785 without reheat.
(b) For the high temperature reservoir, the entropy change per unit mass of steam is given by

$$\Delta S_h = -\frac{Q_{in}}{T_h} = -\frac{3399}{(273 + 400)} = -5.05 \text{ kJ K}^{-1} \text{ kg}^{-1}$$

For the low temperature reservoir the entropy change is

$$\Delta S_c = \frac{Q_{out}}{T_c} = \frac{2145.9}{(273 + 20)} = 7.323 \text{ kJ K}^{-1} \text{ kg}^{-1}$$

Entropy production in the universe is

$$\sigma = \Delta S_h + \Delta S_c = -5.05 + 7.323 = 2.274 \text{ kJ K}^{-1} \text{kg}^{-1}.$$

The second-law effectiveness is given by Eq. (9.57)

$$\eta_2 = \frac{1253.1}{1253.1 + 293 \times 2.274} = 0.65$$

The increase in the second law effectiveness due to reheat is marginal for the present case.

Example 9.6 In an ideal Rankine cycle, superheated steam enters the turbine at 30 bar and 350°C. In the high-pressure turbine, the steam expands until it is dry saturated before a fraction of the steam is extracted and supplied to an open feed water heater. The rest of the steam is expanded in the low-pressure turbine to the condenser pressure of 0.04 bar. (a) Calculate (i) the total work output, (ii) the total heat supplied, and (iii) the cycle efficiency. Compare the performance of the cycles with and without feed water heating.
(b) If heat is supplied to the boiler and superheater from a reservoir at 400°C and heat is rejected in the condenser to a heat sink reservoir at 20°C, calculate the entropy change in the two reservoirs, and the second-law effectiveness of the cycle using a standard ambient temperature of 20°C.

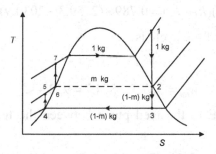

Fig. E9.6 *T-s* diagram

Solution Consider 1 kg of steam flowing through the boiler and superheater. Since the conditions of the boiler, superheater and condenser are the same as example 9.3 we shall use the relevant property data obtained in that example.

Applying the SFEE to the feed water heater (see Fig. 9.7), neglecting the kinetic and potential energy changes

$$mh_2 + (1-m)h_5 = h_6 \qquad (E9.6.1)$$

where m is the mass of steam extracted per kg of steam flowing through the boiler. The steam is dry saturated at state 2 and the feed water is a saturated liquid at state 6. Hence we use the data obtained in worked example 9.5.

Now the work input per kg of fluid to the feed pump between the condenser and the feed heater is

$$h_5 - h_4 = \bar{v}(P_2 - P_3) = 10^{-3} \times (6.33 - 0.04) \times 100 = 0.63$$

Hence $\qquad h_5 = 121 + 0.63 = 121.63 \text{ kJ kg}^{-1}$

Substituting numerical values in Eq. (E9.6.1) we have

$$2759.28m + 121.63(1 - m) = 678.9$$

Hence $\qquad m = 0.211$ kg per kg of steam through the boiler

Work output of the high-pressure turbine is

$$w_{12} = h_1 - h_2 = 3117 - 2759.3 = 357.7 \text{ kJ}$$

Work output of the low-pressure turbine is

$$w_{23} = (1-m)(h_2 - h_3) = 0.789 \times (2759.3 - 2030.9) = 574.7 \text{ kJ}$$

Total work output of the cycle is

$$w_{tot} = w_{12} + w_{23}$$

$$w_{tot} = 357.7 + 574.7 = 932.4 \text{ kJ}$$

Applying the SFEE to the feed pump between the feed heater and the boiler

$$w_{67} = h_7 - h_6 = \bar{v}(P_7 - P_6) = 10^{-3} \times (30 - 6.33) \times 100 = 2.37 \text{ kJ}$$

Hence $\qquad h_7 = 678.9 + 2.37 = 681.27 \text{ kJ kg}^{-1}$

Heat input in the boiler and the superheater is

$$q_{71} = h_1 - h_7 = 3117 - 681.27 = 2435.73 \text{ kJ}$$

Heat rejected in the condenser is

$$q_{34} = (1-m)(h_3 - h_4) = 0.789 \times (2030.9 - 121) = 1506.9 \text{ kJ}$$

Total feed pump work input is

$$w_{fp} = (1-m)(h_5 - h_4) + (h_7 - h_6)$$

$$w_{fp} = (1-0.221) \times 0.63 + 2.37 = 2.86 \text{ kJ}$$

Net work output of the cycle is

$$w_{net} = w_{tot} - w_{fp}$$

$$w_{net} = 932.4 - 2.86 = 929.5 \text{ kJ}$$

Net heat input

$$q_{net} = q_{71} - q_{34} = 2435.73 - 1506.9 = 928.8 \text{ kJ}$$

The thermal efficiency is

$$\eta = \frac{w_{net}}{q_{tot}} = \frac{929.5}{2435.73} = 38.2\%$$

(b) For the high temperature reservoir, the entropy change per unit mass of steam is given by

$$\Delta S_h = -\frac{Q_{in}}{T_h} = -\frac{2435.73}{(273 + 400)} = -3.619 \text{ kJ K}^{-1} \text{ kg}^{-1}$$

For the low temperature reservoir the entropy change is

$$\Delta S_c = \frac{Q_{out}}{T_c} = \frac{1506.9}{(273 + 20)} = 5.143 \text{ kJ K}^{-1} \text{ kg}^{-1}$$

Entropy production in the universe is

$$\sigma = \Delta S_h + \Delta S_c = -3.619 + 5.143 = 1.524 \text{ kJ K}^{-1} \text{ kg}^{-1}.$$

The second-law effectiveness is given by Eq. (9.57).

$$\eta_2 = \frac{929.5}{929.5 + 293 \times 1.524} = 0.675$$

The second law effectiveness without the feed water heater was obtained as 0.64 in worked example 9.3. Therefore the inclusion of feed heating improves the second law effectiveness of the cycle.

Example 9.7 In an ideal regenerative Rankine cycle, superheated steam enters the turbine at 30 bar and 350°C. A fraction of the steam is extracted from the turbine at a location where it is dry saturated, and at a second location where the pressure is 3 bar, and supplied to two open feed water heaters. In each heater the feed water is heated to the saturation temperature of the extracted steam. Pumps are installed upstream of the condenser and the two feed heaters. Calculate (i) the work output per kg of steam supplied by the boiler, and (ii) the thermal efficiency of the cycle.

Solution Since the boiler and condenser conditions of this example are the same as those in worked examples 9.3 and 9.5, we shall use the relevant steam property data obtained earlier.

Let 1 kg of steam flow through the boiler and the superheater. The mass of steam extracted for the two feed heaters are m_1 and m_2 as shown in Fig. E9.7. From worked examples 9.3 and 9.5 we have

$$h_1 = 3117 \text{ kJ kg}^{-1}, \qquad\qquad h_2 = 2759.3 \text{ kJ kg}^{-1},$$

$$h_4 = 2030.9 \text{ kJ kg}^{-1}, \qquad \text{and} \qquad x_4 = 0.785$$

The expansion in each turbine stage is ideal, therefore

$$s_1 = s_2 = s_3 = s_4$$

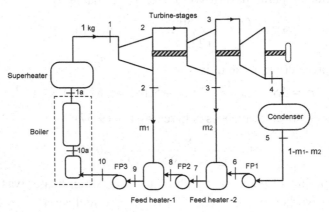

Fig. E9.7 Regenerative Rankine cycle

For the expansion 2-3 we have

$$6.744 = 1.672(1 - x_3) + 6.993x_3. \qquad \text{Hence} \qquad x_3 = 0.953$$

$$h_3 = 561(1 - x_3) + 2725x_3 = 2623.3 \text{ kJ kg}^{-1}$$

For the feed pump FP1

$$h_6 - h_5 = \overline{v}(P_6 - P_5) = 10^{-3}(3 - 0.04) \times 100 = 0.3$$

Hence $\qquad h_6 = 121.3 \text{ kJ kg}^{-1}$ and $\qquad h_7 = 561 \text{ kJ kg}^{-1}$

For feed pump FP2

$$h_8 - h_7 = \overline{v}(P_8 - P_7) = 10^{-3}(6.33 - 3.0) \times 100 = 0.33$$

Hence $\qquad h_8 = 561.33 \text{ kJ kg}^{-1}$ and $\qquad h_9 = 678.9 \text{ kJ kg}^{-1}$

For feed pump FP3

$$h_{10} - h_9 = \overline{v}(P_{10} - P_9) = 10^{-3}(30 - 6.33) \times 100 = 2.37$$

Hence $\qquad\qquad h_{10} = 681.3 \text{ kJ kg}^{-1}$

Applying the SFEE to feed heater 1

$$m_1 h_2 + (1 - m_1)h_8 = h_9$$

$$2759.28 m_1 + 561.33(1 - m_1) = 678.9$$

Hence $\qquad\qquad m_1 = 0.0535$

Applying the SFEE to feed heater 2

$$m_2 h_3 + (1 - m_1 - m_2)h_6 = (1 - m_1)h_7$$

$$2623.3 m_2 + 121.3(0.9465 - m_2) = 0.9465 \times 561$$

Hence $\qquad\qquad m_2 = 0.166$

Work output of the high-pressure turbine is

$$w_{12} = h_1 - h_2 = 3117 - 2759.3 = 357.7 \text{ kJ}$$

Work output of the intermediate-pressure turbine is

$$w_{23} = (1 - m_1)(h_2 - h_3) = 0.9465 \times (2759.3 - 2623.3) = 128.7 \text{ kJ}$$

Work output of the low-pressure turbine is

$$w_{34} = (1 - m_1 - m_2)(h_3 - h_4)$$

$$w_{34} = 0.7805 \times (2623.3 - 2030.9) = 462.37 \text{ kJ}$$

The total feed pump work input is

$$w_{fp} = (h_{10} - h_9) + (1 - m_1)(h_8 - h_7) + (1 - m_1 - m_2)(h_6 - h_5)$$

$$w_{fp} = 2.37 + 0.31 + 0.23 = 2.91 \text{ kJ}$$

Net work output of the cycle is

$$w_{net} = w_{12} + w_{23} + w_{34} - w_{fp}$$

$$w_{net} = 357.7 + 128.71 + 462.37 - 2.91 = 945.87 \text{ kJ}$$

Total heat input in the boiler and the superheater is

$$q_{10-1} = h_1 - h_{10} = 3117 - 681.3 = 2435.7 \text{ kJ}$$

Heat rejected in the condenser is

$$q_{45} = (1 - m_1 - m_2)(h_4 - h_5) = 0.7805 \times (2030.9 - 121) = 1490.7$$

Net heat input to the cycle is

$$q_{net} = q_{10-1} - q_{45} = 2435.7 - 1490.7 = 945.0 \text{ kJ}$$

The thermal efficiency is

$$\eta = \frac{w_{net}}{q_{tot}} = \frac{945.87}{2435.7} = 38.8\%$$

Example 9.8 Superheated steam at 30 bar and 350°C is supplied to the turbine of an ideal Rankine cycle that includes both reheating and feed water heating. At the exit of the high-pressure turbine where the pressure is 6 bar, a fraction of the steam is extracted and supplied to an open feed heater, while the rest of the steam is reheated to 350°C before entering the low-pressure turbine. Steam is extracted from the low-pressure turbine at a location where the pressure is 1 bar to supply a second open feed heater. The condenser pressure is 0.04 bar. Calculate (i) the work output per kg of steam supplied by the boiler, and (ii) the thermal efficiency of the cycle.

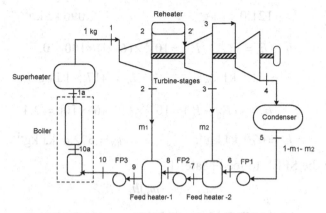

Fig. E9.8 Rankine cycle with reheating and feed heating

Solution A schematic diagram of the cycle is shown in Fig. E9.8. Since the expansion in the turbine is ideal

$$s_1 = s_2 \qquad \text{and} \qquad s_{2'} = s_3 = s_4$$

$$6.744 = 1.931(1 - x_2) + 6.761x_2 \qquad \text{Hence} \quad x_2 = 0.996$$

$$h_2 = 670(1 - x_2) + 2757x_2 = 2748.65 \text{ kJ kg}^{-1}$$

At the exit of the reheater $P_{2'} = 6$ bar and $T_{2'} = 350\,°C$. From the superheated steam tables [5] we obtain

$$s_{2'} = 7.546 \text{ kJ K}^{-1} \text{ kg}^{-1} \qquad \text{and} \qquad h_{2'} = 3166 \text{ kJ kg}^{-1}$$

For the expansion from 2'-3, $s_{2'} = s_3 > s_{g3}$. Therefore at 3 where the pressure is 1 bar, the steam is superheated. From the superheated steam tables [5] we obtain by interpolation $h_3 = 2749.96$ kJ kg^{-1}.

For the expansion from 3-4, $s_{2'} = s_3 = s_4$

$$7.546 = 0.422(1 - x_4) + 8.473x_4, \qquad \text{Hence} \quad x_4 = 0.8848$$

$$h_4 = 121(1 - x_4) + 2554x_4 = 2273.7 \text{ kJ kg}^{-1}$$

For the three feed pumps FP1, FP2 and FP3 we have

$$h_6 - h_5 = \bar{v}(P_6 - P_5) = 10^{-3} \times (1 - 0.04) \times 100 = 0.096$$

Since $\qquad h_5 = 121.0$ kJ kg^{-1}, $\qquad\qquad h_6 = 121.096$ kJ kg^{-1}

$$h_8 - h_7 = \bar{v}(P_8 - P_7) = 10^{-3} \times (6-1) \times 100 = 0.5$$

Since $\qquad h_7 = 417$ kJ kg^{-1}, $\qquad\qquad h_8 = 417.5$ kJ kg^{-1}

$$h_{10} - h_9 = \bar{v}(P_{10} - P_9) = 10^{-3} \times (30-6) \times 100 = 2.4$$

Since $\qquad h_9 = 670$ kJ kg^{-1}, $\qquad\qquad h_{10} = 672.4$ kJ kg^{-1}

Applying the SFEE to feed heater 1

$$m_1 h_2 + (1-m_1)h_8 = h_9 \qquad\qquad\qquad\text{(E9.8.1)}$$

$$2748.65 m_1 + 417.5(1-m_1) = 670$$

Therefore $\qquad\qquad\qquad\qquad m_1 = 0.108$

Applying the SFEE to feed heater2

$$m_2 h_3 + (1-m_1-m_2)h_6 = (1-m_1)h_7 \qquad\qquad\text{(E9.8.2)}$$

$$2749.96 m_2 + 121.096(0.892-m_2) = 0.892 \times 417$$

Therefore $\qquad\qquad\qquad\qquad m_2 = 0.10$

Work output of the high-pressure turbine is

$$w_{12} = h_1 - h_2 = 3117 - 2748.65 = 368.35 \text{ kJ}$$

Work output of the intermediate-pressure turbine is

$$w_{23} = (1-m_1)(h_{2'} - h_3) = 0.892 \times (3166 - 2749.96) = 371.1 \text{ kJ}$$

Work output of the low-pressure turbine is

$$w_{34} = (1-m_1-m_2)(h_3 - h_4)$$

$$w_{34} = 0.792 \times (2749.96 - 2273.7) = 377.2 \text{ kJ}$$

Total work output of the cycle is

$$w_{tot} = w_{12} + w_{23} + w_{34} = 1116.65 \text{ kJ}$$

The total feed pump work input is

$$w_{fp} = (h_{10} - h_9) + (1-m_1)(h_8 - h_7) + (1-m_1-m_2)(h_6 - h_5)$$

$$w_{fp} = 2.4 + 0.892 \times 0.5 + 0.096 \times 0.792 = 2.922 \text{ kJ}$$

Net work output is

$$w_{net} = w_{tot} - w_{fp}$$

$$w_{net} = 1116.65 - 2.92 = 1113.73 \text{ kJ}$$

Heat input in the boiler and the superheater is

$$q_{10-1} = h_1 - h_{10} = 3117 - 672.4 = 2444.6 \text{ kJ}$$

Heat input in the reheater is

$$q_{22'} = (1 - m_1)(h_{2'} - h_2) = 0.892 \times (3166 - 2748.65) = 372.27 \text{ kJ}$$

Total heat input

$$q_{tot} = q_{10-1} + q_{22'} = 2444.6 + 372.27 = 2816.9 \text{ kJ}$$

Heat rejected in the condenser is

$$q_{45} = (1 - m_1 - m_2)(h_4 - h_5) = 0.792 \times (2273.7 - 121) = 1704.94 \text{ kJ}$$

Net heat input

$$q_{net} = q_{10-1} + q_{22'} - q_{45} = 2444.6 + 372.27 - 1704.94 = 1112.0 \text{ kJ}$$

The thermal efficiency is

$$\eta = \frac{w_{net}}{q_{tot}} = \frac{1113.7}{2816.9} = 39.5\%$$

Example 9.9 Superheated steam at 60 bar and 400°C is supplied to the turbine of an ideal Rankine cycle incorporating a closed feed heater at 11 bar and an open feed heater at 1.7 bar. The feed water leaving the closed feed heater is 5°C below the saturation temperature of the steam entering the heater. The condensate from the closed feed heater is pumped and mixed with the feed water leaving the heater. The condenser pressure is 0.04 bar. Calculate (i) the work output per kg of steam supplied by the boiler, and (ii) the thermal efficiency of the cycle.

Solution A schematic diagram of the cycle is shown in Fig. E9.9. From the superheated steam tables [5] we obtain

$$s_1 = 6.541 \text{ kJ K}^{-1} \text{kg}^{-1} \quad \text{and} \quad h_1 = 3177 \text{ kJ kg}^{-1}.$$

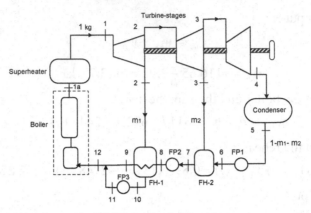

Fig. E9.9 Regenerative Rankine cycle

For the expansion 1-2, $s_1 = s_2$

$$6.541 = 2.179(1 - x_2) + 6.554x_2.$$ Hence $x_2 = 0.997$

$$h_2 = 718(1 - x_2) + 2781x_2 = 2775 \text{ kJ kg}^{-1}$$

For the expansion 2-3, $s_2 = s_3$

$$6.541 = 1.475(1 - x_3) + 7.182x_3.$$ Hence $x_3 = 0.8877$

$$h_3 = 483(1 - x_3) + 2699x_3 = 2450.1$$

For the expansion 3-4, $s_3 = s_4$

$$6.541 = 0.422(1 - x_4) + 8.473x_4.$$ Hence $x_4 = 0.76$

$$h_4 = 121(1 - x_4) + 2554x_4 = 1970.08 \text{ kJ kg}^{-1}$$

For the compression in the pump FP1

$$h_6 - h_5 = \bar{v}(P_6 - P_5) = 10^{-3} \times (1.7 - 0.04) \times 100 = 0.17$$

Hence $$h_6 = 121.17 \text{ kJ kg}^{-1}$$

For the compression in the pump FP2

$$h_8 - h_7 = \bar{v}(P_8 - P_7) = 10^{-3} \times (60 - 1.7) \times 100 = 5.83$$

Hence $$h_8 = 488.83 \text{ kJ kg}^{-1}$$

At location 10, the condensate is a saturated liquid with

$$T_{10} = 184.1\,°C \qquad \text{and} \qquad h_{10} = 781 \text{ kJ kg}^{-1}$$

From the given data,

$$T_9 = (T_{10} - 5) = 179.1\,°C.$$

The liquid enthalpy at 9, ignoring the effect of pressure, is $h_9 = 763$ kJ kg^{-1}

Applying the SFEE to feed heater, FH-1,

$$m_1 h_2 + (1 - m_1)h_8 = (1 - m_1)h_9 + m_1 h_{10}$$

Substituting numerical values

$$2775 m_1 + 488.83(1 - m_1) = 763(1 - m_1) + 781 m_1$$

Therefore $m_1 = 0.1208$ kg per kg of steam flow through the boiler.

Applying the SFEE to feed heater FH-2

$$m_2 h_3 + (1 - m_1 - m_2)h_6 = (1 - m_1)h_7$$

Substituting numerical values

$$2450.1 m_2 + 121.17(1 - m_1 - m_2) = 483(1 - m_1)$$

Therefore $m_2 = 0.1365$ kg per kg of steam flow through the boiler.

Work output of the high-pressure turbine is

$$w_{12} = h_1 - h_2 = 3177 - 2775 = 402 \text{ kJ}$$

Work output of the intermediate-pressure turbine is

$$w_{23} = (1 - m_1)(h_2 - h_3) = 0.879 \times (2775 - 2450.1) = 285.58 \text{ kJ}$$

Work output of the low-pressure turbine is

$$w_{34} = (1 - m_1 - m_2)(h_3 - h_4)$$

$$w_{34} = 0.7425 \times (2450.1 - 1970.08) = 356.4 \text{ kJ}$$

Total work output of the cycle is

$$w_{tot} = w_{12} + w_{23} + w_{34} = 1043.98 \text{ kJ}$$

The total feed pump work input is

$$w_{fp} = m_1(h_{11} - h_{10}) + (1 - m_1)(h_8 - h_7) + (1 - m_1 - m_2)(h_6 - h_5)$$

$$w_{fp} = 0.59 + 5.12 + 0.12 = 5.83 \text{ kJ}$$

Net work output of the cycle is

$$w_{net} = w_{tot} - w_{fp}$$

$$w_{net} = 1043.98 - 5.83 = 1038.15 \text{ kJ}$$

Applying the SFEE to the mixing junction 9-11-12

$$(1 - m_1)h_9 + m_1 h_{11} = h_{12}$$

Substituting numerical values

$$h_{12} = 0.879 \times 763 + 0.1208 \times 785.9 = 765.61 \text{ kJ kg}^{-1}$$

Total heat input in the boiler and the superheater is

$$q_{12-1} = h_1 - h_{12} = 3177 - 765.61 = 2411.38 \text{ kJ}$$

Heat rejected in the condenser is

$$q_{45} = (1 - m_1 - m_2)(h_4 - h_5)$$

$$q_{45} = 0.7425 \times (1970.08 - 121) = 1372.94 \text{ kJ}$$

Net heat input is

$$q_{net} = q_{12-1} - q_{45} = 2411.38 - 1372.94 = 1038.4 \text{ kJ}$$

The thermal efficiency is

$$\eta = \frac{w_{net}}{q_{tot}} = \frac{1038.15}{2411.38} = 43\%$$

Example 9.10 The turbine of an ideal regenerative Rankine cycle, with a closed feed heater at 7 bar and an open feed heater at 1.4 bar, is supplied with superheated steam at 30 bar and 400°C. The feed water leaving the closed heater is 5°C below the saturation temperature of the extracted steam. The condensate from the closed heater is drained to the open feed heater through a steam trap. The condenser pressure is 0.04 bar. Calculate (i) the work output per kg of steam supplied by the boiler, and (ii) the thermal efficiency of the cycle.

Solution A schematic diagram of the cycle is shown in Fig. E9.10. From the superheated steam tables [5]

$$s_1 = 6.921 \text{ kJ K}^{-1} \text{ kg}^{-1} \quad \text{and} \quad h_1 = 3231 \text{ kJ kg}^{-1}.$$

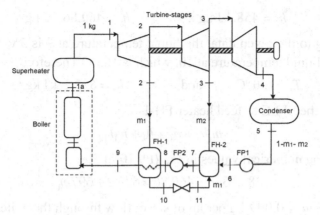

Fig. E9.10 Regenerative Rankine cycle

For the expansion 1-2, $s_1 = s_2$. Assume that the steam at 2 is wet.

$$6.921 = 1.992(1 - x_2) + 6.709x_2$$

Since, $x_2 > 1$, the steam is superheated at 2. From the superheated steam tables [5]

$$h_2 = 2862.5 \text{ kJ kg}^{-1}, \quad s_2 = 6.921 \text{ kJ K}^{-1} \text{ kg}^{-1}, \quad T_{2,sat} = 165 \,^\circ\text{C}$$

For the expansion 2-3, $s_2 = s_3$

$$6.921 = 1.411(1 - x_3) + 7.246x_3, \quad \text{Hence} \quad x_3 = 0.9443$$

$$h_3 = 458(1 - x_3) + 2690x_3 = 2565.67$$

For the expansion 3-4, $s_3 = s_4$

$$6.921 = 0.422(1 - x_4) + 8.473x_4, \quad \text{Hence} \quad x_4 = 0.8072$$

$$h_4 = 121(1 - x_4) + 2554x_4 = 2084.9$$

For the compression in the pump FP1

$$h_6 - h_5 = \overline{v}(P_6 - P_5) = 10^{-3} \times (1.4 - 0.04) \times 100 = 0.14$$

Since $\qquad h_5 = 121.0 \text{ kJ kg}^{-1}, \qquad h_6 = 121.14 \text{ kJ kg}^{-1}$

For the compression in the pump FP2

$$h_8 - h_7 = \overline{v}(P_8 - P_7) = 10^{-3} \times (30 - 1.4) \times 100 = 2.86$$

Since $\qquad h_7 = 458$ kJ kg^{-1}, $\qquad h_8 = 460.86$ kJ kg^{-1}

According to the given data, the liquid temperature at 9 is 5°C below the saturated liquid temperature at 10, which is 165°C. Therefore

$$T_9 = 160\,°C \qquad \text{and} \qquad h_9 = 675.2 \text{ kJ kg}^{-1}$$

Applying the SFEE to feed heater FH-1

$$m_1 h_2 + h_8 = h_9 + m_1 h_{10} \qquad\qquad (E9.10.1)$$

Substituting numerical values in Eq. (E9.10.1)

$$2862.5 m_1 + 460.86 = 675.2 + 697 m_1$$

Therefore $m_1 = 0.099$ kg per kg of steam flow through the boiler.
   Applying the SFEE to feed heater FH-2

$$m_2 h_3 + (1 - m_1 - m_2) h_6 + m_1 h_{11} = h_7 \qquad\qquad (E9.10.2)$$

The flow through the steam trap from 10 to 11 is a throttling process for which

$$h_{11} = h_{10} = 697 \text{ kJ kg}^{-1}$$

Substituting numerical values in Eq. (E9.10.2)

$$2565.67 m_2 + (1 - 0.099 - m_2) \times 121.14 + 0.099 \times 697 = 458$$

Therefore $m_2 = 0.1144$ kg per kg of steam flow through the boiler.
   Work output of the high-pressure turbine is

$$w_{12} = h_1 - h_2 = 3231 - 2862.5 = 368.5 \text{ kJ}$$

Work output of the intermediate-pressure turbine is

$$w_{23} = (1 - m_1)(h_2 - h_3)$$

$$w_{23} = 0.901 \times (2862.5 - 2565.67) = 267.44 \text{ kJ}$$

Work output of the low-pressure turbine is

$$w_{34} = (1 - m_1 - m_2)(h_3 - h_4)$$

$$w_{34} = 0.7866 \times (2565.67 - 2084.9) = 378.17 \text{ kJ}$$

Total work output is

$$w_{tot} = w_{12} + w_{23} + w_{34} = 1014.11 \text{ kJ}$$

The total feed pump work input is

$$w_{fp} = (h_8 - h_7) + (1 - m_1 - m_2)(h_6 - h_5)$$

Net work output is

$$w_{net} = w_{tot} - w_{fp}$$

$$w_{net} = 1014.11 - 2.87 = 1011.1 \text{ kJ}$$

Total heat input in the boiler and the superheater is

$$q_{91} = h_1 - h_9 = 3231 - 675.2 = 2555.8 \text{ kJ}$$

Heat rejected in the condenser is

$$q_{45} = (1 - m_1 - m_2)(h_4 - h_5)$$

$$q_{45} = 0.7866 \times (2084.9 - 121) = 1544.8 \text{ kJ}$$

$$q_{net} = q_{91} - q_{45} = 2555.8 - 1544.8 = 1011.0 \text{ kJ}$$

The thermal efficiency of the cycle is

$$\eta = \frac{w_{net}}{q_{tot}} = \frac{1011.1}{2555.8} = 39.6\%$$

Example 9.11 Figure E9.11 shows a schematic diagram of a binary–vapor Rankine cycle power plant with steam and Freon-12 as the two working fluids. Steam enters the turbine at 40 bar and 400°C. The condensing pressure of steam is 0.8 bar. The heat rejected by the steam as it condenses is used to generate saturated Freon-12 at 80°C. The Freon-12 condenses to a saturated liquid at 30°C. The expansions in the steam and Freon turbines may be assumed isentropic. The work input to the two feed pumps are negligible. Calculate (i) the ratio of the steam flow rate to the Freon flow rate, (ii) the ratio of the power output of the steam turbine to that of the Freon turbine, and (iii) the overall thermal efficiency of the plant.

Solution A schematic diagram of the steam-Freon binary-vapor cycle is shown in Fig. E9.11.

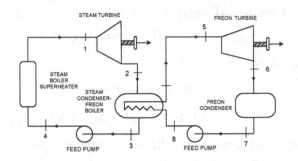

Fig. E9.11 Binary vapor cycle

The following data for steam are obtained from the superheated steam table [5].

$$h_1 = 3214 \text{ kJ kg}^{-1} \quad \text{and} \quad s_1 = 6.769 \text{ kJ K}^{-1} \text{ kg}^{-1}.$$

For the isentropic expansion in the steam turbine $s_1 = s_2$. Therefore

$$6.769 = 1.233(1 - x_2) + 7.434x_2, \quad \text{Hence} \quad x_2 = 0.8928$$

$$h_2 = 392(1 - x_2) + 2665x_2 = 2421.3$$

For Freon we obtain the property data from page 14 in [5].

$$h_5 = 212.83 \text{ kJ kg}^{-1} \quad \text{and} \quad s_5 = 0.6673 \text{ kJ K}^{-1} \text{ kg}^{-1}$$

For the isentropic expansion in the Freon turbine,

$$s_5 = s_6$$

$$0.6673 = 0.2399(1 - x_6) + 0.6853x_6, \quad \text{Hence} \quad x_6 = 0.9596$$

$$h_6 = 64.59(1 - x_6) + 199.62x_6 = 194.16 \text{ kJ kg}^{-1}$$

Applying the SFEE to the steam condenser-Freon boiler we obtain

$$\dot{m}_s(h_2 - h_3) = \dot{m}_f(h_5 - h_8) \qquad (\text{E9.11.1})$$

where $\dot{m}_s$ and $\dot{m}_f$ are the steam and Freon flow rates respectively.
Since the feed pump work is negligible,

$$h_8 = h_7 \quad \text{and} \quad h_4 = h_3.$$

Substituting numerical data in Eq. (E9.11.1)

$$\dot{m}_s(2421.3 - 392) = \dot{m}_f(212.83 - 64.59)$$

Hence $$\dot{m}_s/\dot{m}_f = 0.073$$

The work output of the steam turbine is

$$\dot{W}_s = \dot{m}_s(h_1 - h_2) = \dot{m}_s(3214 - 2421.3) = 792.7\dot{m}_s$$

The work output of the Freon turbine is

$$\dot{W}_f = \dot{m}_f(h_5 - h_6) = \dot{m}_f(212.83 - 194.16) = 18.67\dot{m}_f$$

Therefore

$$\frac{\dot{W}_s}{\dot{W}_f} = \frac{792.7\dot{m}_s}{18.67\dot{m}_f} = \frac{792.7 \times 0.073}{18.67} = 3.1$$

The total work output is

$$\dot{W}_{tot} = \dot{W}_s + \dot{W}_f = 792.7\dot{m}_s + 18.67\dot{m}_f$$

Since the feed pump work is negligible, the total work output is equal to the net work output.

The total heat input is

$$\dot{Q}_{tot} = \dot{Q}_s = \dot{m}_s(h_1 - h_4) = \dot{m}_s(3214 - 392) = 2822\dot{m}_s$$

The overall thermal efficiency of the plant is

$$\eta = \frac{\dot{W}_{net}}{\dot{Q}_{tot}} = \frac{792.7\dot{m}_s + 18.67\dot{m}_f}{2822\dot{m}_s} = \frac{792.7\dot{m}_s + 255.2\dot{m}_s}{2822\dot{m}_s} = 37\%$$

Example 9.12 A binary vapor power plant uses mercury and steam as the working fluids. Saturated mercury vapor enters the turbine at 10 bar and expands to the condenser pressure of 0.08 bar. The heat rejected by the condensing mercury is used to evaporate water from a saturated liquid to a dry saturated vapor at 25 bar. The steam is superheated to 450°C in a superheater before entering the turbine where it expands to the condenser pressure 0.05 bar.

A separate economizer heats the condensate to the saturation temperature of the boiler. Calculate (i) the ratio of the mercury flow rate to the steam flow rate, and (ii) the thermal efficiency of the plant.

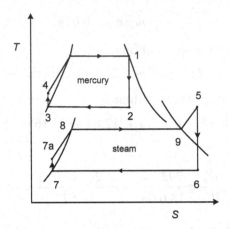

Fig. E9.12 Mercury-steam binary cycle

Solution The *T-s* diagram of the binary vapor cycle is shown in Fig. E9.12. The following property data for mercury have been obtained from page 12 in [5].

$$h_1 = 359.1 \text{ kJ kg}^{-1} \quad \text{and} \quad s_1 = 0.5089 \text{ kJ K}^{-1} \text{ kg}^{-1}.$$

For the isentropic expansion in the mercury turbine $s_1 = s_2$. Therefore

$$0.5089 = 0.087(1 - x_2) + 0.6591x_2, \quad \text{Hence} \quad x_2 = 0.7374$$

$$h_2 = 33.21(1 - x_2) + 327.91x_2 = 250.52$$

Also $\quad h_3 = 33.21 \text{ kJ kg}^{-1}, \quad h_9 = 2802.5 \text{ kJ kg}^{-1}, \quad h_8 = 962 \text{ kJ kg}^{-1}$

Applying the SFEE to the mercury condenser-steam boiler we have

$$\dot{m}_s(h_9 - h_8) = \dot{m}_m(h_2 - h_3)$$

where $\dot{m}_s$ and $\dot{m}_m$ are the steam and mercury flow rates respectively. Substituting numerical values in the above equation

$$\dot{m}_s(2802.5 - 962) = \dot{m}_m(250.52 - 33.2)$$

$$\frac{\dot{m}_s}{\dot{m}_m} = \frac{217.31}{1840.5} = 0.118$$

Therefore the ratio of mercury flow rate to the steam flow rate is 8.47.

For the mercury feed pump

$$h_4 - h_3 = \bar{v}(P_4 - P_3) = \frac{(10 - 0.08) \times 100}{13.6 \times 10^3} = 0.073$$

$$h_4 = 33.21 + 0.073 = 33.28 \text{ kJ kg}^{-1}$$

Heat input in the mercury boiler is

$$\dot{Q}_{41} = \dot{m}_m (h_1 - h_4) = (359.1 - 33.28)\dot{m}_m = 325.82\dot{m}_m$$

Work output of the mercury turbine is

$$\dot{W}_{12} = \dot{m}_m (h_1 - h_2) = (359.1 - 250.52)\dot{m}_m = 108.58\dot{m}_m$$

Consider the steam cycle 5-6-7-8-5. From the superheated steam tables [5],

$$h_5 = 3350 \text{ kJ kg}^{-1}, \quad s_5 = 7.182 \text{ kJ K}^{-1} \text{ kg}^{-1}, \quad h_7 = 138 \text{ kJ kg}^{-1}$$

For the isentropic expansion in the steam turbine $s_5 = s_6$. Therefore

$$7.182 = 0.476(1 - x_6) + 8.394x_6, \quad x_6 = 0.8469$$

$$h_6 = 138(1 - x_6) + 2561x_6 = 2190.0$$

For the feed water pump

$$h_{7a} - h_7 = (25 - 0.05) \times 10^{-3} \times 100 = 2.5$$

$$h_{7a} = h_7 + 2.5 = 140.5 \text{ kJ kg}^{-1}$$

Heat supplied in the economizer is

$$\dot{Q}_{7a-8} = \dot{m}_s (h_8 - h_{7a}) = (962 - 140.5)\dot{m}_s = 821.5\dot{m}_s$$

Heat supplied in the steam superheater is

$$\dot{Q}_{95} = \dot{m}_s (h_5 - h_9) = (3350 - 2802.5)\dot{m}_s = 547.5\dot{m}_s$$

Heat rejected in the steam condenser is

$$\dot{Q}_{67} = \dot{m}_s (h_6 - h_7) = (2190 - 138)\dot{m}_s = 2052\dot{m}_s$$

Work output of the steam turbine is

$$\dot{W}_{56} = \dot{m}_s (h_5 - h_6) = (3350 - 2190)\dot{m}_s = 1160\dot{m}_s$$

Total heat supplied is

$$\dot{Q}_{tot} = \dot{Q}_{7a-8} + \dot{Q}_{95} + \dot{Q}_{41}$$

$$\dot{Q}_{tot} = 821.5\dot{m}_s + 547.5\dot{m}_s + 325.82\dot{m}_m = 1369\dot{m}_s + 325.82\dot{m}_m$$

$$\dot{Q}_{tot} = 1369 \times 0.118\dot{m}_m + 325.82\dot{m}_m = 487.36\dot{m}_m$$

Net heat input is

$$\dot{Q}_{net} = \dot{Q}_{tot} - \dot{Q}_{67} = 487.36\dot{m}_m - 2052 \times 0.118\dot{m}_m = 245.22\dot{m}_m$$

Net work output is

$$\dot{W}_{net} = \dot{W}_{12} + \dot{W}_{56} - \dot{W}_{34} - \dot{W}_{7-7a}$$

$$\dot{W}_{net} = 108.58\dot{m}_m + 1160\dot{m}_s - 0.073\dot{m}_m - 2.5\dot{m}_s = 245.1\dot{m}_m$$

The overall thermal efficiency of the binary cycle is

$$\eta = \frac{\dot{W}_{net}}{\dot{Q}_{tot}} = \frac{245.1\dot{m}_m}{487.36\dot{m}_m} = 50\%$$

Example 9.13 In a supercritical Rankine cycle, saturated water at 40°C is pumped to the boiler where the pressure is 275 bar. The temperature of the fluid leaving the boiler and entering the turbine is 500°C. The power output of the cycle is 50 MW and the isentropic efficiencies of the turbine and feed pump are 0.8 and 0.7 respectively. Calculate (i) the thermal efficiency of the cycle, and (ii) the mass flow rate of steam.

Solution A *T-s* diagram of the supercritical Rankine cycle is shown in Fig. E9.13. The working fluid is heated in the process 4-1 at a constant pressure which is above the critical pressure of 221.2 bar for water. The isentropic expansion between the turbine inlet and exit pressures is represented by 1-2s while the actual expansion process, assumed adiabatic, is indicated by 1-2. The corresponding compression processes in the pump are indicated as 3-4s and 3-4 respectively.

The following data are obtained from the supercritical steam data on page 9 in [5]. For $P_1 = 275$ bar and $T_1 = 500$°C,

$$s_1 = 5.878 \text{ kJ K}^{-1} \text{ kg}^{-1} \qquad \text{and} \qquad h_1 = 3125 \text{ kJ kg}^{-1}$$

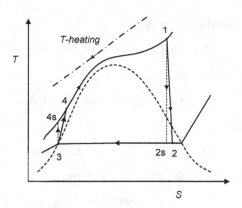

Fig. E9.13 Supercritical cycle

For the isentropic process 1-2s, $s_1 = s_{2s}$

$$5.878 = 0.572(1 - x_{2s}) + 8.256x_{2s}, \quad \text{Hence} \quad x_{2s} = 0.6905$$

$$h_{2s} - 167.5(1 - x_{2s}) + 2573.7x_{2s} = 1828.98 \text{ kJ kg}^{-1}$$

Since the isentropic efficiency is 0.8, we have

$$h_1 - h_2 = 0.8(h_1 - h_{2s}) = 1036.82. \quad \text{Hence} \quad h_2 = 2088.18 \text{ kJ kg}^{-1}$$

For the isentropic compression process 3-4s

$$h_{4s} - h_3 = \overline{v}(P_4 - P_3) = (275 - 0.073) \times 10^{-3} \times 100 = 27.49$$

Since the isentropic efficiency is 0.7, we have

$$h_4 - h_3 = (h_{4s} - h_3)/0.7 = 39.275. \quad \text{Hence} \quad h_4 = 206.77 \text{ kJ kg}^{-1}$$

The work output of the turbine is

$$\dot{W}_{12} = \dot{m}(h_1 - h_2) = 1036.82\dot{m}$$

where $\dot{m}$ is the mass flow rate of steam.

The work input to the pump is

$$\dot{W}_{34} = \dot{m}(h_4 - h_3) = 39.275\dot{m}$$

The net work output is

$$\dot{W}_{net} = \dot{W}_{12} - \dot{W}_{34} = 997.525\dot{m} = 50 \times 10^3 \text{ kW}$$

Therefore the mass flow rate of steam is, $\dot{m} = 50.1 \text{ kg s}^{-1}$.

The heat input in the boiler is

$$\dot{Q}_{41} = \dot{m}(h_1 - h_4) = 2918.23\dot{m} \text{ kW}$$

The heat rejected in the condenser is

$$\dot{Q}_{23} = \dot{m}(h_2 - h_3) = 1920.68\dot{m} \text{ kW}$$

The net heat input is

$$\dot{Q}_{net} = \dot{Q}_{41} - \dot{Q}_{23} = 997.52\dot{m} \text{ kW}$$

The thermal efficiency of the plant is given by

$$\eta = \frac{\dot{W}_{net}}{\dot{Q}_{41}} = \frac{997.5\dot{m}}{2918.23\dot{m}} = 34\%$$

Example 9.14 In a modified Rankine cycle power plant, saturated steam at 40 bar expands in the high pressure turbine to 3.5 bar. The steam then enters a steam separator where liquid is separated and pumped to the boiler while the dry saturated steam is sent to the low pressure turbine in which it expands to the condenser pressure of 0.05 bar. The condensate in the form of a saturated liquid is pumped to the boiler.

The processes in the two turbines and the two feed pumps are isentropic. Calculate (i) the work output of the turbines, (ii) the work input to the pumps, and (iii) the thermal efficiency of the plant. Draw the *T-s* diagram for the cycle.

Solution The *T-s* diagram of the modified Rankine cycle is shown in Fig. E9.14. From the saturated steam tables [5] we obtain

$$s_1 = 6.07 \text{ kJ K}^{-1} \text{ kg}^{-1} \quad \text{and} \quad h_1 = 2801 \text{ kJ kg}^{-1}$$

Consider 1 kg of steam discharged from the boiler.

For the expansion 1-2, $s_1 = s_2$

$$6.07 = 1.727(1 - x_2) + 6.941x_2. \quad \text{Hence} \quad x_2 = 0.833$$

$$h_2 = 584(1 - x_2) + 2732x_2 = 2373.28 \text{ kJ kg}^{-1}$$

Of the 1 kg of steam that enters the steam separator, 0.833 kg is sent to the low-pressure turbine (state 3) while 0.167 kg (state 7) is pumped to the boiler. For the isentropic expansion 3-4 in the low pressure turbine $s_3 = s_4$

$$6.941 = 0.476(1 - x_4) + 8.394x_4 \qquad \text{Hence} \qquad x_4 = 0.816$$

$$h_4 = 138(1 - x_4) + 2561x_4 = 2115.17 \text{ kJ kg}^{-1}$$

For the pumping process 5-6

$$h_6 - h_5 = \overline{v}(P_6 - P_5) = (40 - 0.05) \times 10^{-3} \times 100 = 4.0$$

Since $\qquad h_5 = 138 \text{ kJ kg}^{-1}, \qquad h_6 = 142 \text{ kJ kg}^{-1}$

For the pumping process 7-8

$$h_8 - h_7 = \overline{v}(P_8 - P_7) = (40 - 3.5) \times 10^{-3} \times 100 = 3.65$$

Since $\qquad h_7 = 584 \text{ kJ kg}^{-1}, \qquad h_8 = 587.65 \text{ kJ kg}^{-1}$

Total work output per kg of steam through the boiler is

$$w_{tot} = (h_1 - h_2) + 0.833(h_3 - h_4)$$

$$w_{tot} = (2801 - 2373.28) + 0.833(2732 - 2115.17) = 941.54 \text{ kJ}$$

Total pump work input per kg of steam is

$$w_{tot} = 0.167(h_8 - h_7) + 0.833(h_6 - h_5)$$

$$w_{tot} = 0.167 \times (587.65 - 584) + 0.833 \times (142 - 138) = 3.94 \text{ kJ}$$

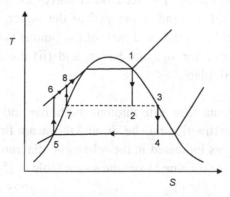

Fig. E9.14 Modified Rankine cycle

Net work output is

$$W_{net} = 941.54 - 3.94 = 937.6 \text{ kJ}$$

Heat input per kg of steam is

$$q_{in} = 0.833(h_1 - h_6) + 0.167(h_1 - h_8)$$

$$q_{in} = 0.833(2801 - 142) + 0.167(2801 - 587.65) = 2584.58 \text{ kJ}$$

Heat rejected in the condenser per kg of steam is

$$q_{45} = 0.833(h_4 - h_5) = 0.833(2115.17 - 138) = 1646.98 \text{ kJ}$$

Net heat input per kg of steam is

$$q_{net} = q_{in} - q_{45} = 937.6 \text{ kJ}$$

The thermal efficiency is given by

$$\eta = \frac{W_{net}}{q_{in}} = \frac{937.6}{2584.58} = 36\%$$

Example 9.15 Superheated steam at 15 bar and 250°C is supplied to the turbine of a combined-heat and power (CHP) system with a net work output of 3000 kW. Some steam is extracted from the turbine at a location where the pressure is 1.4 bar and the rest expands to the condenser pressure of 0.05 bar. Part of the extracted steam is sent to an open feed heater while the rest is sent to a heating system with a heat load of 8000 kW. The condensate leaving the heating system at 55°C enters the open feed heater. The feed water leaves the feed heater as a saturated liquid at 1.4 bar and is pumped to the boiler. The isentropic efficiency of the turbine is 0.8, and each of the pumps is 0.75. Calculate (i) the steam flow rate through the boiler, and (ii) the overall thermal efficiency of the CHP plant.

Solution Let the steam flow rate through the boiler and superheater be $\dot{m}_1$, the total steam extraction rate be $\dot{m}_2$ and the steam flow rate through the heat load be $\dot{m}_3$ as indicated in the schematic diagram in Fig. E9.15. The following data are obtained from the steam table in [5],

$$s_1 = 6.711 \text{ kJ K}^{-1} \text{ kg}^{-1} \qquad \text{and} \qquad h_1 = 2925 \text{ kJ kg}^{-1}$$

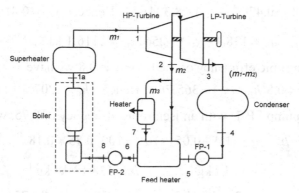

Fig. E9.15 Combined heat and power system

For an ideal expansion in the turbine between pressures at 1 and 2, $s_1 = s_{2s}$ where the subscript s represents the state under isentropic conditions.

$$6.711 = 1.411(1 - x_{2s}) + 7.246x_{2s} \quad \text{Hence} \quad x_{2s} = 0.9083$$

$$h_{2s} = 458(1 - x_{2s}) + 2690x_{2s} = 2485.32 \text{ kJ kg}^{-1}$$

Since the isentropic efficiency of the turbine is 0.8, we have

$$h_1 - h_2 = 0.8(h_1 - h_{2s}) = 351.74 \quad \text{Hence} \quad h_2 = 2573.3 \text{ kJ kg}^{-1}$$

The condensate leaving the heater at 7 is a sub-cooled liquid at 55°C. Ignoring effect of pressure, $h_7 = 230.2 \text{ kJ kg}^{-1}$.

Applying the SFEE to the heater we obtain

$$8000 = \dot{m}_3(h_2 - h_7) = (2573.7 - 230.2)\dot{m}_3$$

Hence

$$\dot{m}_3 = 3.414 \text{ kg s}^{-1}$$

$$h_2 = 2573.3 = 458(1 - x_2) + 2690x_2$$

where x_2 is the actual quality at 2.

Hence

$$x_2 = 0.9477$$

$$s_2 = 1.411(1 - x_2) + 7.246x_2 = 6.9408$$

For the ideal expansion of the steam from the pressure at 2 to 3,

$$s_2 = s_{3s}$$

$$6.9408 = 0.476(1 - x_{3s}) + 8.394x_{3s}, \quad \text{Hence} \quad x_{3s} = 0.8164$$

$$h_{3s} = 138(1 - x_{3s}) + 2561x_{3s} = 2116.1 \text{ kJ kg}^{-1}$$

Since the isentropic efficiency of the turbine is 0.8 we have

$$h_2 - h_3 = 0.8(h_2 - h_{3s}) = 365.76, \quad \text{Hence} \quad h_3 = 2207.5 \text{ kJ kg}^{-1}$$

For the feed pump FP-1, with an isentropic efficiency of 0.75, we have

$$h_5 - h_4 = (1.4 - 0.05) \times 10^{-3} \times 100/0.75 = 0.18$$

Since $h_4 = 138 \text{ kJ kg}^{-1}$, $h_5 = 138.18 \text{ kJ kg}^{-1}$

For the feed pump FP-2, with an isentropic efficiency of 0.75, we have

$$h_8 - h_6 = (15 - 1.4) \times 10^{-3} \times 100/0.75 = 1.81$$

Since $h_6 = 458 \text{ kJ kg}^{-1}$, $h_8 = 459.81 \text{ kJ kg}^{-1}$

The total work output of the turbines is

$$\dot{W}_{tot} = \dot{m}_1(h_1 - h_2) + (\dot{m}_1 - \dot{m}_2)(h_2 - h_3)$$

$$\dot{W}_{tot} = 351.7\dot{m}_1 + 365.8(\dot{m}_1 - \dot{m}_2)$$

The total work input to the two feed pumps is

$$\dot{W}_{fp} = \dot{m}_1(h_8 - h_6) + (\dot{m}_1 - \dot{m}_2)(h_5 - h_4)$$

$$\dot{W}_{fp} = 1.81\dot{m}_1 + 0.18(\dot{m}_1 - \dot{m}_2)$$

The net work output of the system is

$$\dot{W}_{net} = \dot{W}_{tot} - \dot{W}_{fp} = 349.89\dot{m}_1 + 365.62(\dot{m}_1 - \dot{m}_2)$$

$$3000 = 715.51\dot{m}_1 - 365.62\dot{m}_2$$

$$8.205 = 1.957\dot{m}_1 - \dot{m}_2 \qquad (E9.15.1)$$

Applying the SEFF to the open feed heater

$$\dot{m}_3 h_7 + (\dot{m}_2 - \dot{m}_3)h_2 + (\dot{m}_1 - \dot{m}_2)h_5 = \dot{m}_1 h_6$$

Substituting numerical values in the above equation

$$230.2\dot{m}_3 + 2573.3(\dot{m}_2 - \dot{m}_3) + 138.18(\dot{m}_1 - \dot{m}_2) = 458\dot{m}_1$$

$$- 2343.1\dot{m}_3 + 2435.1\dot{m}_2 - 319.82\dot{m}_1 = 0$$

$$- 25.01 + 7.614\dot{m}_2 - \dot{m}_1 = 0 \qquad (E9.15.2)$$

Solving Eqs. (E9.15.1) and (E9.15.2) simultaneously we obtain

$$\dot{m}_1 = 6.293 \text{ kg s}^{-1} \quad \text{and} \quad \dot{m}_2 = 4.11 \text{ kg s}^{-1}$$

The steam flow rate through the boiler is, 6.293 kg s^{-1}.

The heat input to the system is

$$\dot{Q}_{in} = \dot{m}_1(h_1 - h_8) = 6.293(2925 - 459.81) = 15513.4 \text{ kW}$$

We define the overall efficiency of the CHP plant as:

$$\eta = \frac{\dot{Q}_h + \dot{W}_{net}}{\dot{Q}_{in}} = \frac{8000 + 3000}{15513.40} = 70.9\%$$

We note that the CHP plant has a high thermal efficiency because the heat delivered to the load is also included in the numerator of the expression for the efficiency. In general, the first law efficiency is the ratio of the desired output of the plant to the heat input. However, the heat delivered, which now forms a part of the desired output, is not subject to the limitations imposed by the second law on the conversion of heat to work.

Problems

P9.1 The boiler and condenser pressures of a Carnot cycle with steam as the working fluid are 60 bar and 0.06 bar respectively. The steam flow rate is 18 kg s^{-1}. Calculate (i) the power output of the turbine, (ii) the power input to the compressor, and (iii) the thermal efficiency of the plant. Sketch the *T-s* and *h-s* diagrams of the cycle. [*Answers*: (i) 17.48 MW, (ii) 5.17 MW, (iii) 43.6%]

P9.2 Saturated steam at 60 bar is supplied to the turbine of a Rankine cycle. The condenser pressure is 0.06 bar. The turbine and pump are both ideal. The steam flow rate is 18 kg s^{-1}. Calculate (i) the power output of the turbine, (ii) the power input to the pump, (iii) the thermal efficiency of the plant, and (iv) the second law effectiveness of the plant if heat is supplied from a reservoir at 350°C, and rejected to a reservoir at 25°C. Assume that the dead-state temperature is 20°C. Draw the *T-s* and *h-s* diagrams of the cycle. [*Answers*: (i) 17.48 MW, (ii) 108 kW, (iii) 36.7%, (iv) 0.708]

P9.3 A superheat-Rankine-cycle steam plant operates with a boiler pressure of 60 bar and a condenser pressure of 0.06 bar. The temperature of the steam at the exit from the superheater is 450°C. The steam flow rate is 18 kg s^{-1}. The turbine and feed pump are both ideal. Calculate (i) the power output of the turbine, (ii) the power input to the pump, (iii) the thermal efficiency of the plant, and (iv) the second law effectiveness of the plant if heat is supplied from a reservoir at 525°C and rejected to a reservoir at 25°C. Assume that the dead-state temperature is 20°C. Draw the *T-s* and *h-s* diagrams of the cycle. [*Answers*: (i) 22.18 MW, (ii) 108 kW, (iii) 39%, (iv) 0.626]

P9.4 The boiler and condenser pressures of a Rankine cycle power plant are 60 bar and 0.06 bar respectively. The temperature of the steam at the exit from the superheater is 450°C. The steam flow rate is 18 kg s^{-1}. The isentropic efficiencies of the turbine and feed pump are 0.8 and 0.75 respectively. Calculate (i) the power output of the turbine, (ii) the power input to the pump, (iii) the thermal efficiency of the plant, and (iv) the second law effectiveness of the plant if heat is supplied from a reservoir at 525°C, and rejected to a reservoir at 25°C. Assume that the dead-state temperature is 20°C. Draw the *T-s* and *h-s* diagrams of the cycle. [*Answers*: (i) 17.74 MW, (ii) 144 kW, (iii) 31%, (iv) 0.5]

P9.5 Superheated steam is supplied at 60 bar and 450°C to the high pressure turbine of a Rankine cycle power plant. The steam leaves the high pressure turbine at 20 bar and enters a reheater where it is heated to 450°C. In the low pressure turbine the steam expands to the condenser pressure of 0.06 bar. The steam flow rate is 18 kg s^{-1}. The isentropic efficiency of each turbine is 80% and the feed pump is 75%. Calculate (i) the power output of the turbines, (ii) the power input to the pump, (iii) the thermal efficiency of the plant, and (iv) the second law effectiveness of the plant if heat is supplied from a reservoir at 525°C and rejected to a reservoir at 25°C. Assume that the dead-state temperature is 20°C. Draw the *T-s* and *h-s* diagrams of the cycle. [*Answers*: (i) 20.4 MW, (ii) 144 kW, (iii) 32.7%, (iv) 0.526]

P9.6 In a regenerative Rankine cycle power plant superheated steam at 60 bar and 450°C is supplied to the high pressure turbine with a mass flow rate of 18 kg s^{-1}. After the expansion in the high pressure turbine to 20 bar, steam is bled at the rate of 2 kg s^{-1} and sent to an open feed water heater.

The rest of the steam is reheated at constant pressure to 450°C and then enters the low pressure turbine where it expands to the condenser pressure of 0.06 bar.

The isentropic efficiency of each turbine is 80% and each pump is 75%. Calculate (i) the total power output of the turbines, (ii) the power input to the pumps, (iii) the thermal efficiency of the plant, and (iv) the second law effectiveness of the plant if heat is supplied from a reservoir at 525°C and rejected to a reservoir at 25°C. Assume that the dead-state temperature is 20°C. Draw the *T-s* and *h-s* diagrams of the cycle. [*Answers*: (i) 18.6 MW, (ii) 138 kW, (iii) 33.3%, (iv) 0.536]

P9.7 The turbine of an ideal Rankine cycle power plant is supplied with superheated steam at 60 bar and 450°C. The condenser pressure is 0.06 bar. The steam flow rate through the boiler is 18 kg s^{-1}. Steam is bled from the turbine at three locations where the pressures are 20 bar, 8 bar and 1 bar and sent to three open feed heaters. The water leaving each heater is at the saturation temperature of the bled steam. Calculate (i) the total power output of the turbine, (ii) the power input to the pumps, (iii) the thermal efficiency of the plant, and (iv) the second law effectiveness of the plant if heat is supplied from a reservoir at 525°C and rejected to a reservoir at 25°C. Assume that the dead-state temperature is 20°C. Draw the *T-s* and *h-s* diagrams of the cycle. [*Answers*: (i) 18.6 MW, (ii) 108 kW, (iii) 43.1%, (iv) 0.69]

P9.8 The turbine of a supercritical Rankine cycle is supplied with steam at 275 bar and 500°C. The isentropic efficiencies of the turbine and the feed pump are 80% and 65% respectively. The condenser pressure is 0.08 bar and the steam flow rate is 65 kg s^{-1}. Calculate (i) the net power output, and (ii) the thermal efficiency of the cycle. Draw the *T-s* and *h-s* diagrams of the cycle. [*Answers*: (i) 64.3 MW, (ii) 34%]

P9.9 Mercury and steam are used as working fluids in a binary vapor cycle with a net work output of 10 MW. The mercury boiler and condenser operate at pressures of 12 bar and 0.08 bar respectively.

The heat rejected by the condensing mercury is used to evaporate water from a saturated liquid to a dry saturated vapor at 30 bar.

The steam is superheated to 500°C in a superheater before entering the turbine where it expands to the condenser pressure of 0.08 bar. A separate economizer heats the condensate to the saturation temperature of the boiler. Calculate (i) the mercury and steam flow rates, and (ii) the thermal efficiency of the plant. [*Answers*: (i) 39.1 kg s^{-1}, 4.68 kg s^{-1}, (ii) 50.6%]

P9.10 Superheated steam at 15 bar and 300°C is supplied to the turbine of a combined-heat and power (CHP) plant. The steam discharged from the turbine at 1 bar is sent to a heating system with a heat load of 1400 kW. The water leaving the heating system at 55°C is pumped back to the boiler. The isentropic efficiencies of the turbine and the pump are 75% and 70% respectively. Calculate (i) the power output of the turbine, and (ii) the efficiency of the plant. [*Answers*: (i) 229 kW, (ii) 14%]

P9.11 In the CHP plant described in worked example 9.15, heat is supplied to the boiler and superheater from a reservoir at 300°C and rejected to a reservoir at 26°C. The heat load may assumed as a reservoir at 85°C. Calculate (i) the rate of entropy production in the universe, and (ii) the second law effectiveness of the system. Assume that the standard ambient is at 20°C. [*Answers*: (i) 10.38 kW K^{-1}, (ii) 0.59]

P9.12 A Rankine cycle steam plant is to have a single open-feed water heater where the water is heated to the temperature of the bled steam. It is to be assumed that the difference between the enthalpy of steam at any point in the turbine, *H* and the saturation liquid enthalpy *h* at the same pressure, is constant. The condition of the steam at entry to the turbine and the condenser pressure are fixed. Show that the thermal efficiency of the plant is a maximum when, $\Delta H/\Delta H_b = 0.5$ where ΔH is the actual enthalpy rise of the feed water in the heater, and ΔH_b is the enthalpy rise of the feed water if heated to the saturation temperature at the boiler pressure.

References

1. Bejan, Adrian, *Advanced Engineering Thermodynamics*, John Wiley & Sons, New York, 1988.
2. Cravalho, E.G. and J.L. Smith, Jr., *Engineering Thermodynamics*, Pitman, Boston, MA, 1981.
3. Jones, J.B. and G.A. Hawkins, *Engineering Thermodynamics*, John Wiley & Sons, Inc., New York, 1986.
4. Reynolds, William C. and Henry C. Perkins, *Engineering Thermodynamics*, 2nd edition, McGraw-Hill, Inc., New York, 1977.
5. Rogers, G.F.C. and Mayhew, Y.R., *Thermodynamic and Transport Properties of Fluids*, 5th edition, Blackwell, Oxford, U.K. 1998.

Chapter 10

Gas Power Cycles

In the last chapter we analyzed vapor power cycles with working fluids that undergo phase change during operation. However, many common prime movers like petrol engines, diesel engines, and gas turbines use gases as working fluids and their operating cycles are known as gas power cycles. In this chapter we shall analyze gas power cycles considering their performance mainly under ideal conditions.

10.1 Internal-Combustion Engine Cycles

Reciprocating gasoline engines and diesel engines, widely used in motor cars, trucks, buses and other transport applications, are commonly called *internal combustion* engines. There are two broad categories of such engines which use different mechanisms of combustion. These are known as spark-ignition engines and compression-ignition engines.

10.1.1 *Spark-ignition (SI) engines*

In this section we shall consider briefly the basic mode of operation of a typical spark-ignition (SI) engine. Figure 10.1 depicts a schematic of the piston cylinder arrangement that forms the core of the engine. The reciprocating motion of the piston is converted to a rotation of the shaft with the aid of a connecting rod, that links the piston to a crank attached to the shaft. The basic form of the engine consists of an inlet valve through which a mixture of fuel and air is admitted into the cylinder, and an exhaust valve through which the products of combustion are discharged to the atmosphere.

476

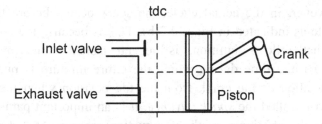

Fig. 10.1 Reciprocating internal combustion engine

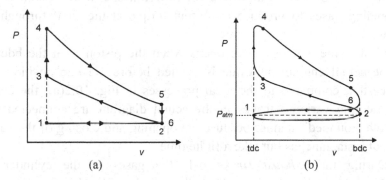

Fig. 10.2 SI engine: (a) Ideal *P-v* diagram, (b) Actual *P-v* diagram

The ideal and actual pressure–volume (*P-v*) diagrams for the gases in the cylinder during operation are shown in Figs. 10.2(a) and 10.2(b) respectively. At point 1 in the *P-v* diagrams the exhaust valve closes and the inlet valve opens to admit a fresh mixture of fuel and air. At this stage the piston is at the top-dead-centre (tdc) where it is closest to the cylinder head. As the piston moves out the pressure in the cylinder is ideally equal to the atmospheric pressure as seen in Fig. 10.2(a), whereas the actual pressure is slightly below as indicated in Fig. 10.2(b). At 2, when the piston is at the bottom-dead-centre (bdc), the furthest position from the cylinder head, the inlet valve closes.

The volume swept by the piston from the tdc to bdc is called the *stroke volume* while the volume at the tdc is the *clearance volume*. During the return stroke from 2 to 3, known as the compression stroke, the air-fuel mixture is compressed. In the spark-ignition engine under consideration, ideally an electric spark is introduced at 3 to ignite the fuel. However, in the actual cycle, the spark occurs before the piston reaches tdc as indicated in Fig. 10.2(b). This is because the combustion process that follows ignition, is not an instantaneous process but occupies a finite time during which the entire mixture is burned. The angular position of the crank from the tdc position when the spark is introduced is called the *spark advance*. It is an important parameter that determines the efficiency of the combustion process. At 4 the cylinder contains the products of combustion at a high pressure and temperature. During the process 4-5, known as the power stroke, work is done by the expanding gases to produce an output torque at the shaft through the crank.

Ideally, the exhaust valve opens when the piston is at the bdc but in the actual engine the value is opened before the bdc position. We notice that compared to the ideal processes in Fig. 10.2(a), the curves representing various strokes in the actual diagram are connected in a smooth 'rounded' manner because the opening and closing of the valves are not instantaneous but take a finite time.

During the *exhaust stroke* 6-1, the gases in the cylinder are discharged to the atmosphere. The exhaust valve then closes and the inlet valve opens to begin a new cycle. The cycle described above is broadly called a *four-stroke cycle* and occupies two complete rotations of the crank. An alternative version of the SI engine where the various processes described above are carried out in a single rotation of the crank is known as the *two-stroke cycle*.

10.1.2 *Compression-ignition (CI) engines*

The compression-ignition (CI) engine differs from the spark-ignition engine due mainly to the difference in the mode of ignition of the fuel. During the intake stroke 1-2 (Fig. 10.3) of the CI engine only air is admitted to the cylinder, unlike the SI engine. The air is compressed in the compression stroke 2-3. Under ideal conditions, at 3, the fuel is sprayed into the high temperature air with the aid of a fuel injector. The fuel undergoes *auto-ignition* thus initiating the combustion process. In general, the temperature and pressure of the air at the end of the

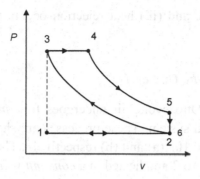

Fig. 10.3 CI Engine: Ideal *P-v* diagram

compression stroke are much higher in the case of a CI engine compared to a SI engine. In the actual CI engine, the fuel injection takes place after the piston has begun to move away from the tdc in stroke 3-4. The ideal combustion process is therefore more close to a constant pressure process as shown in Fig. 10.3. The power and exhaust strokes of the CI engine are similar to those of the SI engine.

It should be noted that compared to the Rankine cycle studied in chapter 9, the cycle in the actual internal combustion engine does not retain the same working fluid throughout its operation. Moreover, due to the combustion process, the chemical constitution of the working fluid changes during the cycle, and therefore does not remain as a pure substance. These and other practical considerations make the analysis of actual gas power cycles very complex. Therefore we resort to a more simplified approach know as *air-standard cycle analysis* to obtain analytical expressions for various parameters like the work output and the thermal efficiency of gas power cycles.

10.2 Air Standard Cycles

The ideal air standard cycles on which the SI engine and the CI engine operate are called the *Otto cycle* and the *Diesel Cycle* respectively. We now develop the energy analysis of these cycles making the following idealizations: (i) the working fluid is air which behaves as an ideal gas, (ii) the combustion process is replaced by simple heat addition from an external heat source, and (iii) heat rejection occurs to an external heat sink.

10.2.1 *Analysis of the Otto cycle*

In the air-standard Otto cycle, air undergoes four *internally reversible* processes in a closed system. These processes are shown in the *P-v* and *T-s* diagrams in Figs. 10.4(a) and (b) respectively. The air is compressed *adiabatically* from 1 to 2 and heated in a *constant volume* process from 2 to 3. The air undergoes an *adiabatic* expansion from 3 to 4 followed by a *constant volume* heat rejection from 4 to 1.

Applying the first law to the cycle

$$W_{net} = Q_{in} - Q_{out} \qquad (10.1)$$

where W_{net}, Q_{in} and Q_{out} are the net work output, the heat input and the heat output per cycle respectively. Assuming that air is an ideal gas with constant specific heat capacities, Eq. (10.1) can be written as

$$W_{net} = mc_v(T_3 - T_2) - mc_v(T_4 - T_1) \qquad (10.2)$$

where m is the mass of air in the system.

The thermal efficiency of the cycle is given by

$$\eta = \frac{W_{net}}{Q_{in}} = 1 - \frac{Q_{out}}{Q_{in}} \qquad (10.3)$$

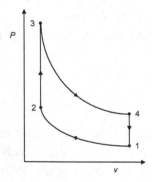

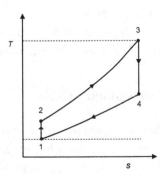

Fig. 10.4(a) Otto cycle: *P-v* diagram   Fig. 10.4(b) Otto cycle: *T-s* diagram

Substituting the ideal gas expressions in Eq. (10.3) we have

$$\eta = 1 - \frac{mc_v(T_4 - T_1)}{mc_v(T_3 - T_2)} \tag{10.4}$$

For the two reversible adiabatic processes, 1-2 and 3-4, the entropy is constant. Using the expression for the entropy change of an ideal gas [Eq. (7.23)] we have

$$S_2 - S_1 = mc_v \ln\left(\frac{T_2}{T_1}\right) + mR\ln\left(\frac{V_2}{V_1}\right) = 0 \tag{10.5}$$

Rearranging Eq. (10.5) we obtain

$$\frac{T_2}{T_1} = \left(\frac{V_1}{V_2}\right)^{(\gamma-1)} \tag{10.6}$$

where $\gamma = c_p/c_v$

Similarly for the isentropic process 3-4

$$\frac{T_3}{T_4} = \left(\frac{V_4}{V_3}\right)^{(\gamma-1)} \tag{10.7}$$

Since $V_1 = V_4$ and $V_2 = V_3$ we have

$$\frac{T_2}{T_1} = \frac{T_3}{T_4} = \left(\frac{V_1}{V_2}\right)^{(\gamma-1)} = r^{(\gamma-1)} \tag{10.8}$$

where $r = V_1/V_2$, is the *compression ratio*.

Substituting from Eq. (10.8) in Eq. (10.4) we obtain the efficiency as

$$\eta = 1 - \frac{1}{r^{(\gamma-1)}} \tag{10.9}$$

The above expression shows that the efficiency of the air-standard Otto cycle increases with the compression ratio. Although we draw this conclusion based on the analysis of the air-standard cycle, it is found that the efficiency of a real SI engine also increases with the compression ratio. However, the maximum compression ratio of actual engines is limited by a phenomenon associated with the combustion process.

In a SI-engine, the combustion is initiated by the spark and the flame front advances rapidly to consume the unburned fuel in the cylinder. However, if the compression ratio is too high, the resulting high temperature and pressure in the cylinder could cause the unburned fuel to undergo *auto-ignition* at some locations before the flame front could reach that location. This phenomenon is called '*knocking*', and it produces pressure waves which will eventually damage the valves and the piston of the engine. The *octane rating* of a fuel, given by its *octane number*, reflects the resistance of the fuel to knocking. Recent advances in the development of high-octane fuels have enabled the compression ratios of SI engines to be increased from earlier values of about 7 to about 10.

10.2.2 *Analysis of the Diesel cycle*

The compression ignition engine operates on the ideal air-standard cycle known as the *Diesel cycle*. The *P-v* and *T-S* diagrams of the Diesel cycle are shown in Figs. 10.5 (a) and (b) respectively. The compression process 1-2 and the expansion process 3-4 are *reversible adiabatic* processes. Heat is supplied in the *constant pressure* process 2-3 while the heat rejection process 4-1 occurs at *constant volume*.

Applying the first law to the cycle

$$W_{net} = Q_{in} - Q_{out} \qquad (10.10)$$

where W_{net}, Q_{in}, and Q_{out} are the net work output, the heat input, and the heat output per cycle respectively. Assuming that air is an ideal gas with constant specific heat capacities, Eq. (10.10) can be written as

$$W_{net} = mc_p(T_3 - T_2) - mc_v(T_4 - T_1) \qquad (10.11)$$

where m is the mass of air in the system.

The thermal efficiency of the cycle is given by

$$\eta = \frac{W_{net}}{Q_{in}} = 1 - \frac{Q_{out}}{Q_{in}} \qquad (10.12)$$

$$\eta = 1 - \frac{mc_v(T_4 - T_1)}{mc_p(T_3 - T_2)} \qquad (10.13)$$

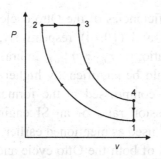 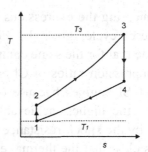

Fig. 10.5(a) Diesel cycle: *P-v* diagram   Fig. 10.5(b) Diesel cycle: *T-s* diagram

We define the compression ratio r and the cut-off ratio r_c as,

$$r = V_1/V_2 \qquad \text{and} \qquad r_c = V_3/V_2$$

Since the process 1-2 is isentropic and air is an ideal gas,

$$\frac{T_1}{T_2} = \left(\frac{V_2}{V_1}\right)^{(\gamma-1)} = \frac{1}{r^{\gamma-1}}.$$

For the constant pressure process 2-3,

$$\frac{T_3}{T_2} = \frac{V_3}{V_2} = r_c.$$

For the isentropic process, 3-4

$$\frac{T_4}{T_3} = \left(\frac{V_4}{V_3}\right)^{\gamma-1} = \left(\frac{r_c}{r}\right)^{\gamma-1}$$

Now

$$\frac{T_4}{T_1} = \frac{T_4 T_3 T_2}{T_3 T_2 T_1} = \left(\frac{r_c}{r}\right)^{\gamma-1} r_c r^{\gamma-1} = r_c^{\gamma}$$

Substituting the above relations in Eq. (10.13) we obtain the efficiency as

$$\eta = 1 - \frac{mc_v(T_4 - T_1)}{mc_p(T_3 - T_2)} = 1 - \frac{1}{r^{\gamma-1}}\left[\frac{r_c^{\gamma} - 1}{\gamma(r_c - 1)}\right] \qquad (10.14)$$

With $\gamma = 1.4$ for air and $r_c > 1$, we can verify numerically that

$$\left[\frac{r_c^{\gamma} - 1}{\gamma(r_c - 1)}\right] > 1$$

Comparing the expressions for the efficiencies of the Otto cycle and the Diesel cycle, given by Eqs. (10.9) and (10.14) respectively, we conclude that for the same compression ratio, r, $\eta_{otto} > \eta_{diesel}$. In practice, the compression ratios of CI engines could be significantly higher than those of SI engines because only air is compressed in the former. In contrast, the highest achievable compression ratio of an SI engine is limited by the knock-resistance of the fuel used as mentioned earlier.

It is clear that the thermal efficiencies of both the Otto cycle and the Diesel cycle are less than that of a Carnot cycle operating between the same maximum and minimum temperatures T_3 and T_1 respectively. This is because heat supply and rejection in the Otto and Diesel cycles occur over a range of temperatures. Therefore if heat is supplied from a reservoir at T_3 and rejected to a heat sink at T_1, these processes will involve external heat transfer ireversibilities due to the finite temperature difference between the reservoirs and the working fluid. However, the Otto and Diesel cycles are both *internally* reversible.

10.2.3 *The dual cycle*

The *P-v* diagrams of actual SI engines and CI engines deviate significantly from the Otto and Diesel cycles. Therefore modifications are made to these air-standard cycles in order to make the shapes of their *P-v* diagrams closer to those of actual engines.

One such modified cycle is the dual cycle, whose *P-v* diagram is shown in Fig. 10.6. In the dual cycle part of the heat is supplied at

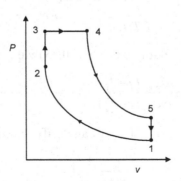

Fig. 10.6 Dual cycle: *P-v* diagram

constant volume while the rest is supplied at constant pressure. The heat rejection occurs at constant volume. The analysis of the duel cycle can be carried out using the methods and equations similar to those used above to analyze the Otto and Diesel cycles. The procedure is illustrated in worked example 10.6.

10.3 Gas Turbine Engine Cycles

Gas turbine engines are used for electricity generation and transport applications, especially in aircraft propulsion. Unlike the SI and CI engines, the gas turbine engine operates in a continuous manner with the working fluid flowing steadily through the system.

The simple gas turbine plant, shown schematically in Fig. 10.7, consists of a compressor, a turbine and a combustion chamber. Atmospheric air flows in steadily through the compressor where it is pressurized and then enters the combustion chamber. Fuel flows steadily to the combustion chamber where continuous combustion occurs. As the products of combustion, with a high pressure and temperature, flow through the turbine a part of its enthalpy is converted to work. Of the total work produced by the turbine a significant fraction is used to drive the compressor. The ratio of the compressor work input to the total turbine work output is called the *back-work-ratio*.

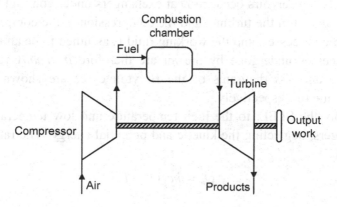

Fig. 10.7 Simple gas turbine plant

In the case of a steam power plant, the work required by the feed-pump is only a very small fraction of the work output of the turbine. In contrast, the compressor work input in a gas turbine plant could be more than half the work produced by the turbine. This is because the compression of a gas requires much more work than pressurizing a liquid as it happens in the steam plant. The efficiency of the compressor is therefore of prime importance in the efficient operation of gas turbine plants.

The actual gas turbine plant does not operate in a cycle. Moreover, the working fluid passing through the compressor and the turbine is not the same pure substance due to the combustion process. The analysis of the actual gas turbine plant is therefore very complex, and requires detailed knowledge of the processes in the different components. As with SI and CI engines, we shall analyze the gas turbine cycles by resorting to the somewhat simplified approach based on *air-standard* cycles.

10.3.1 *Analysis of the Brayton cycle*

The ideal air-standard cycle on which the gas turbine operates is known as the Brayton cycle. Figure 10.8(a) shows a schematic diagram of the air standard cycle which is a closed system with air as the working fluid, where heat is supplied from a high temperature reservoir, and heat is rejected to a heat sink reservoir. These heat interactions between the air and the two reservoirs occur in heat exchangers under constant pressure. The expansion in the turbine, and the compression in the compressor are isentropic processes, and the working fluid is assumed to be an ideal gas. All processes undergone by the air are therefore *internally* reversible. The *P-v* and *T-s* diagrams of the Brayton cycle are shown in Figs. 10.8(b) and (c) respectively.

Apply the SFEE, to the high temperature and low temperature heat exchangers, neglecting the kinetic and potential energy. The rate of heat input is

$$\dot{Q}_h = \dot{m}c_p(T_3 - T_2) \qquad (10.15)$$

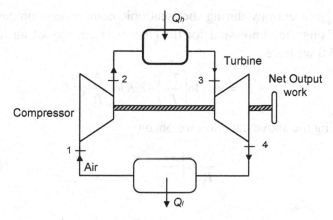

Fig. 10.8(a) Air standard gas turbine cycle: Brayton cycle

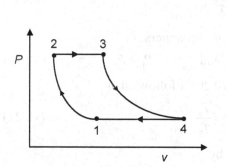

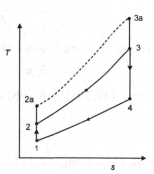

Fig. 10.8(b) Brayton cycle: *P-v* diagram   Fig. 10.8(c) Brayton cycle: *T-s* diagram

The heat rejection rate is

$$\dot{Q}_c = \dot{m}c_p(T_4 - T_1) \qquad (10.16)$$

where $\dot{m}$ is the steady air flow rate through the system.

The work output of the turbine is given by

$$\dot{W}_t = \dot{m}c_p(T_3 - T_4) \qquad (10.17)$$

The work input to the compressor is

$$\dot{W}_c = \dot{m}c_p(T_2 - T_1) \qquad (10.18)$$

The change in entropy during the isentropic compression processes 1-2 is zero. Using the expression for the change in entropy of an ideal gas [Eq. (7.25)] we have

$$S_2 - S_1 = mc_p \ln\left(\frac{T_2}{T_1}\right) - mR \ln\left(\frac{P_2}{P_1}\right) = 0$$

Rearranging the above equation we obtain

$$\frac{P_2}{P_1} = \left(\frac{T_2}{T_1}\right)^{\gamma/(\gamma-1)} \tag{10.19a}$$

where $\gamma = c_p/c_v$.

Similarly for the isentropic expansion 3-4 in the turbine

$$\frac{P_3}{P_4} = \left(\frac{T_3}{T_4}\right)^{\gamma/(\gamma-1)} \tag{10.19b}$$

Neglecting pressure losses in the heat exchangers,

$$P_3 = P_2 \qquad \text{and} \qquad P_4 = P_1 \tag{10.20}$$

From Eqs. (10.19a), (10.19b) and (10.20) it follows that

$$\frac{T_4}{T_1} = \frac{T_3}{T_2} \tag{10.21}$$

The efficiency of the cycle is given by

$$\eta = 1 - \frac{\dot{Q}_c}{\dot{Q}_h} = 1 - \frac{\dot{m}c_p(T_4 - T_1)}{\dot{m}c_p(T_3 - T_2)} \tag{10.22}$$

Substituting in Eq. (10.22) from Eqs. (10.21) and (10.19) we obtain the efficiency as

$$\eta = 1 - \frac{T_1}{T_2} = 1 - \left(\frac{P_1}{P_2}\right)^{(\gamma-1)/\gamma}$$

$$\eta = 1 - \frac{1}{r^{(\gamma-1)/\gamma}} \tag{10.23}$$

where $r = P_2/P_1$, is the pressure ratio of the cycle.

Substituting from Eqs. (10.19) to (10.21) in Eqs. (10.17) and (10.18) we obtain the turbine work output, and the compressor work input as

$$\dot{W}_t = \dot{m}c_pT_3\left(1 - \frac{1}{r^{(\gamma-1)/\gamma}}\right)$$ (10.24)

and $$\dot{W}_c = \dot{m}c_pT_1\left(r^{(\gamma-1)/\gamma} - 1\right)$$ (10.25)

The net work output of the Brayton cycle is given by

$$\dot{W}_{net} = \dot{W}_t - \dot{W}_c$$ (10.26)

We note from Eq. (10.23) that the efficiency of the Brayton cycle increases with the pressure ratio. This is also clear from the *T-s* diagrams in Fig. 10.8(c) where the cycle 1-2a-3a-4 is for a higher pressure than the cycle 1-2-3-4. Since these cycles are internally reversible, the area under the constant pressure lines gives the heat transfer. The heat rejection rates of the two cycles are equal while for the cycle 1-2a-3a-4 the heat input rate is larger. Therefore the efficiency, given by Eq. (10.22), is larger for the latter cycle.

Form Eq. (10.24) we see that the work output of the turbine increases with the pressure ratio, r and the turbine inlet temperature, T_3. The work input to the compressor also increases with the pressure ratio and decreases with decreasing air inlet temperature, T_1 according to Eq. (10.25). However, the net work output of the cycle, given by Eq. (10.26), increases with the pressure ratio, r up to a maximum value and then decreases due to the increase of the compressor work input.

Substituting from Eqs. (10.24) and (10.25) in Eq. (10.26) and differentiating the resulting expression it is easy to show that for a fixed minimum temperature T_1, and a fixed maximum temperature T_3, the net work output of the Brayton cycle is a maximum for a pressure ratio given by

$$r_{opt} = \left(\frac{T_3}{T_1}\right)^{\gamma/2(\gamma-1)}$$

10.3.2 *Gas turbine cycle with regeneration*

In the simple gas turbine plant, the temperature of the fluid leaving the turbine is appreciably higher than the temperature of the fluid leaving the compressor. Therefore it is possible to preheat the air from the compressor using the exhaust gases from the turbine, thus reducing the required energy input in the combustion chamber. In practice, the preheating of air is carried out in a counter-flow heat exchanger know as a regenerator.

Figure 10.9 depicts a schematic of an air standard gas turbine cycle including regenerative heating. The *T-s* diagram in Fig. 10.10 shows that the air is heated in the regenerator to temperature T_a while the exhaust air from the turbine is cooled to T_b. It is clear that, the maximum possible value of T_a is equal to T_4. However, to approach this ideal condition in practice, the required heat exchanger area would be very large, and

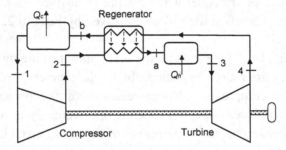

Fig.10.9 Regenerative gas turbine cycle

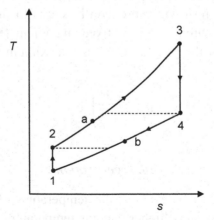

Fig. 10.10 Regenerative cycle: *T-s* diagram

consequently the fluid pressure drop in the heat exchanger would be excessive. Therefore in arriving at the optimum size of the regenerator several factors including the heat transfer area, the fluid pressure drop, and the cost have to be considered. The effectiveness of a regenerator is defined as the ratio of the actual enthalpy rise of the air to the maximum possible enthalpy rise.

Therefore
$$\varepsilon_{reg} = \frac{h_a - h_2}{h_1 - h_2}$$

Assuming that the specific heat capacity is constant, the above relation for the effectiveness becomes

$$\varepsilon_{reg} = \frac{T_a - T_2}{T_4 - T_2} \qquad (10.27)$$

10.3.3 *Analysis of the ideal regenerative cycle*

Consider the air-standard cycle of a gas turbine with a regenerator, shown in Fig. 10.9. The rate of heat input is

$$\dot{Q}_h = \dot{m}c_p(T_3 - T_a) \qquad (10.28)$$

The rate of heat rejection is

$$\dot{Q}_c = \dot{m}c_p(T_b - T_1) \qquad (10.29)$$

Applying the SFEE to the counter-flow regenerator we have

$$\dot{m}c_p(T_a - T_2) = \dot{m}c_p(T_4 - T_b) \qquad (10.30)$$

Now for the ideal regenerator, $T_a = T_4$. Therefore from Eq. (10.30) it follows that $T_b = T_2$. The thermal efficiency of the cycle is given by

$$\eta = 1 - \frac{\dot{Q}_c}{\dot{Q}_h} \qquad (10.31)$$

Substituting from Eqs. (10.28) and (10.29) in Eq. (10.31) and applying the conditions for the ideal regenerator we obtain

$$\eta = 1 - \frac{(T_2 - T_1)}{(T_3 - T_4)} \qquad (10.32)$$

For the isentropic compression process 1-2 and the isentropic expansion process 3-4 we have

$$\frac{T_2}{T_1} = \left(\frac{P_2}{P_1}\right)^{(\gamma-1)/\gamma} = r^{(\gamma-1)/\gamma} \qquad (10.33)$$

and
$$\frac{T_3}{T_4} = \left(\frac{P_2}{P_1}\right)^{(\gamma-1)/\gamma} = r^{(\gamma-1)/\gamma} \qquad (10.34)$$

where r is the pressure ratio, which, in the absence of pressure drops in the heat exchangers and the regenerator, is constant. Substituting in Eq. (10.32) from Eqs. (10.33) and (10.34) we obtain the thermal efficiency as

$$\eta = 1 - \left(\frac{T_1}{T_3}\right) r^{(\gamma-1)/\gamma} \qquad (10.35)$$

The above expression shows that the efficiency of the cycle with an ideal regenerator depends on the pressure ratio as well as the ratio of the minimum to maximum temperature of the cycle. In contrast, the efficiency of the Brayton cycle, given by Eq. (10.23), is only a function of the pressure ratio. Moreover, for a given temperature ratio (T_1/T_3), the efficiency of the ideal regenerative cycle decreases with the pressure ratio while the Brayton cycle efficiency increases with the pressure ratio.

10.4 Gas Turbine Cycles with Intercooling and Reheating

The net work output of a gas turbine cycle given by,

$$\dot{W}_{net} = \dot{W}_t - \dot{W}_c$$

can be increased by increasing the turbine work output $\dot{W}_t$ or by decreasing the compressor work input $\dot{W}_c$. These two options for increasing the net work output can be achieved in practice by inter-cooling, where the average compressor inlet temperature is decreased, and by reheating, where the average turbine inlet temperature is increased.

10.4.1 *Staged-compression with intercooling*

In chapter 5, we developed expressions for the ideal reversible work required to compress a fluid across a given pressure ratio (also see worked example 8.1). We shall now apply these expressions to the compressor of the gas turbine cycle to determine the minimum work input required.

The ideal rate of work input to compress a fluid in a steady flow process from a pressure P_1 to P_2 is given by

$$\dot{W}_c = \dot{m}\int_{P_1}^{P_2} v\,dP \tag{10.36}$$

where $\dot{m}$ is the fluid flow rate through the compressor.

For an ideal gas that is compressed in a *reversible* polytropic process, $Pv^n = C$. Hence Eq. (10.36) may be integrated by substituting for v from the above polytropic relation. After some manipulation we obtain the following expression for the rate of work input to the compressor:

$$\dot{W}_c = \dot{m}RT_1 n\left[r^{(n-1)/n} - 1\right]/(n-1), \quad (n \neq 1) \tag{10.37}$$

where, $r = (P_2/P_1)$, is the pressure ratio and T_1 is the air temperature at the inlet of the compressor. We notice that Eq. (10.37) does not apply for an isothermal process for which, $n = 1$ and $Pv = mRT_1$.

The expression for the work input rate for an isothermal process is obtained by integrating Eq. (10.36), after substituting for v from the ideal gas equation, $Pv = mRT_1$. This gives

$$\dot{W}_c = \dot{m}RT_1 \ln(r) \tag{10.38}$$

Several useful conclusions on the rate of work input to the compressor can be drawn from Eqs. (10.37) and (10.38). For an ideal gas, the polytropic index, n ranges from 1 to γ where $n = 1$ applies to an isothermal process, and $n = \gamma$ applies to an isentropic process. We observe from Eq. (10.37) that the work input increases with the pressure ratio r and the polytropic index n. Therefore for an ideal gas the work required is a minimum when $n = 1$, the lowest polytropic index, and the compression process is then isothermal.

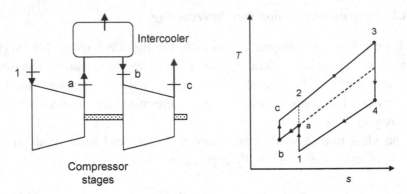

Fig. 10.11(a) Compressor with intercooling   Fig. 10.11(b) *T-s* diagram

In practice, the isothermal process may be approached by compressing the gas using a series of compressors at intermediate pressures with intercooling of the air at the exit of each compressor. An arrangement for two-stage compression with intercooling is shown schematically in Fig. 10.11(a). The *T-s* diagram for the corresponding gas turbine cycle is depicted in Fig. 10.11(b).

In the low pressure stage of the compressor the gas is compressed from P_1 to P_i, the pressure in the intercooler. In the intercooler the gas is cooled from T_a to T_b at constant pressure. In the second stage of the compressor the gas is compressed from P_i to P_2. The work input for the second stage of compression from b to c is less than the work required in the original cycle 1-2-3-4, for compression from a to 2 because the compressor inlet temperature is lower for process b-c [see Eq. (10.25)]. This is also evident from the vertical distance between the constant pressure lines in the *T-s* diagram, which diverge as the temperature increases (see worked example 7.4).

Although the total work of compression is reduced by intercooling, the efficiency of the cycle is less than that of the original cycle 1-2-3-4. The cycle with intercooling could be visualized as a composite cycle consisting of the original cycle 1-2-3-4 and a cycle a-b-c-2 attached to it. The ratio of the average heat supply temperature to the average heat rejection temperature is smaller for the cycle a-b-c-d compared to the original cycle. Consequently, the efficiency of the composite cycle is lower than that of the original cycle. We note that the cycle with intercooling requires an additional heat input for heating the gas from c to 2. However, if intercooling is combined with regenerative heating, then this need for an additional external heat input may be eliminated by heating the air regeneratively from c to 2. Moreover, the potential for regenerative heating is more in the cycle with intercooling because the outlet temperature of the air from the compressor at c is lower than at 2 in the original cycle.

Clearly, the inclusion of several stages of compression with intercooling will reduce the temperature variations during the overall compression process. In the ideal limiting compression process consisting of a series of infinitesimal increments in pressure with intercooling, the compression process will approach an isothermal process. As was mentioned earlier, for such a process the work input will be a minimum for the given overall pressure ratio.

10.4.2 *Multi-staged expansion with reheating*

The second option to increase the net work output of the gas turbine cycle is to include multi-stage expansion in the turbine with reheating between the turbine stages. A schematic of the arrangement for two expansion stages with reheating is shown in Fig. 10.12(a). The *T-s* diagram of the corresponding gas turbine cycle is depicted in Fig. 10.12(b).

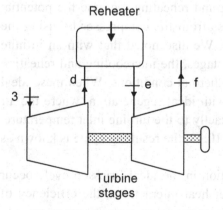

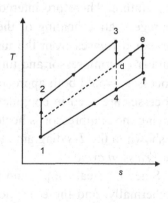

Fig 10.12(a) Turbine with reheating   Fig 10.12(b) *T-s* diagram

The work output of the expansion process in the second turbine-stage from e to f is larger than the expansion work from d to 4 in the original cycle 1-2-3-4 because the turbine inlet temperature is higher for process e-f [see Eq. (10.24)]. This is also evident from the distance between the constant pressure lines in Fig. 10.12(b), which diverge with increasing temperature. Therefore the total work output of the cycle 1-2-3-d-e-f-1 with reheat is larger than that of the original cycle 1-2-3-4. Moreover, the final turbine outlet temperature at f is higher than the temperature at 4 in the original cycle, which enhances the potential for regenerative heating in the reheat cycle. The efficiency of the reheat cycle, however, is lower than that of the original cycle 1-2-3-4. This is because the ratio of the average heat supply temperature to the average heat rejection temperature is smaller for the cycle 1-2-3-4-a-f-1 compared to the original cycle 1-2-3-4. We observe that by increasing the number of turbine stages with reheating, the temperature variation during the expansion in the turbines could be lowered. The ideal limiting expansion process consisting of infinitesimal pressure drops in the turbines combined with reheating between the expansions, will be isothermal.

10.4.3 *The Ericsson cycle*

The introduction of intercooling with multi-stage compression lowers the final compressor outlet temperature, while the use of multi-stage expansion in turbines with reheating increases the final turbine exhaust temperature. Therefore intercooling and reheating increase the potential for regenerative heating of the gases from the compressor by using the hot exhaust gases from the turbine. We also noted that with an infinite number of compressor and turbine stages, the intercooling and reheating processes, would both approach isothermal conditions. When these ideal processes are used in conjunction with ideal regeneration, where the air leaving the compressor is heated exactly to the turbine inlet temperature, as shown in the T-s diagram in Fig. 10.13, the resulting cycle is known as the *Ericsson cycle*.

Since the heat supply and rejection in the ideal Ericsson cycle occur isothermally, and there are no other heat interactions, the efficiency of the cycle approaches the Carnot efficiency.

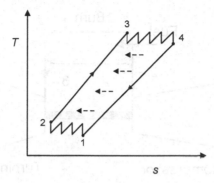

Fig. 10.13 *T-s* diagram of the Ericsson cycle

10.5 Air-Standard Cycle for Jet Propulsion

One of the main applications of gas turbines in transportation is in the propulsion systems of aircrafts. We shall consider briefly the air-standard cycle of such an aircraft gas turbine, shown schematically in Fig. 10.14.

The compressor admits atmospheric air through the intake at 1. The compressed air leaving at 2 flows into the combustion chamber, usually called the burner. In the real engine fuel is sprayed steadily into the air in the burner where continuous combustion occurs. The hot exhaust gases leaving the burner at 3 expands through the turbine. The pressure at the exit 4 of the turbine is such that the work output of the turbine is just sufficient to supply the work required by the compressor.

Consequently, the pressure of the exhaust gases from the turbine will be significantly higher than the surrounding atmospheric pressure. The gases then expand through a nozzle which converts a part of its enthalpy to kinetic energy at the exit section 5 of the nozzle. The change in momentum of the exiting gases produces the thrust to propel the jet aircraft.

Ideally, the expansion in the nozzle is isentropic. Therefore $s_4 = s_5$. Applying the SFEE to the nozzle, neglecting the kinetic energy at he inlet, and any heat losses, we obtain

$$\dot{m}h_4 = \dot{m}\left(h_5 + V_5^2/2\right) \tag{10.39}$$

Assuming the working fluid to be an ideal gas, Eq. (10.39) may be written as

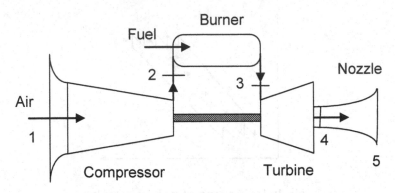

Fig. 10.14 Gas turbine cycle for jet propulsion

$$c_p T_4 = \left(c_p T_5 + V_5^2 / 2 \right) \qquad (10.40)$$

Due to limitations on space and weight regenerators and intercoolers are not used in typical aircraft gas turbine engines. However, evaporative cooling of the compressor may be carried out by injecting water into the compressor inlet. An equivalent of reheating is included in some jet engines by employing a process known as afterburning. Since the air-fuel ratio of jet engines is relatively high there is sufficient oxygen available in the exhaust gases to burn additional fuel before expansion occurs in the nozzle. The resulting increase in the temperature of the gases entering the nozzle increases the kinetic energy at exit, and consequently the thrust.

10.6 Idealizations in Air-Standard Cycles

Our analysis of gas turbine engines thus far was based on air-standard cycles which involved several idealizations. These enabled us to developed simple analytical expressions for most of the important performance parameters of gas turbine cycles. Although, the conclusions drawn from the analysis of air-standard cycles agree well with the trends observed in the performance of actual engines, to obtain better quantitative predictions, we need to eliminated some of the assumptions made.

The pressure losses in the regenerators, the intercoolers, and reheaters could be easily included in the analysis, which will result in different pressure ratios for the compressor and the turbine. For a more realistic analysis, different values can be used for the specific heat capacity of the gas flowing through the compressor and the turbine, based on the respective average temperatures.

In actual engines, the expansion through the turbine and the compression in the compressor are not isentropic. If these irreversible processes are assumed to be adiabatic, then the entropy at the end of the process will be higher than at the beginning as indicated in Fig. 10.15. The isentropic efficiency of the turbine is defined as the actual work output to the ideal work output when the expansion occurs across the same pressure ratio. Hence we have

$$\eta_t = \frac{h_3 - h_4}{h_3 - h_{4s}} \qquad (10.41)$$

where 4s is the final state for the isentropic expansion. Assuming that the gas flowing through the turbine is an ideal gas with constant specific heat capacities, Eq. (10.41) may be written as

$$\eta_t = \frac{T_3 - T_4}{T_3 - T_{4s}} \qquad (10.42)$$

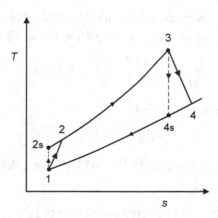

Fig. 10.15 T-s diagram with ideal and actual processes

The corresponding expression for the isentropic efficiency of the compressor, where the ideal work input is less than the actual work input is given by

$$\eta_c = \frac{h_{2s} - h_1}{h_2 - h_1} \qquad (10.43)$$

For an ideal gas with constant specific heat capacities,

$$\eta_c = \frac{T_{2s} - T_1}{T_2 - T_1} \qquad (10.44)$$

10.7 Worked Examples

Example 10.1 An ideal air-standard Otto cycle has a compression ratio of 9. The pressure, temperature and volume of the air at the beginning of the isentropic compression are 100 kPa, 20°C, and 8×10^{-4} m^3 respectively. At the end of the isentropic expansion the temperature is 800°C. Calculate (i) the highest temperature and pressure in the cycle, (ii) the heat input, (iii) the net work output, and (iv) the thermal efficiency of the cycle. For all numerical computations involving air standard cycles assume the following constant properties of air:

$$c_v = 0.718, \ c_p = 1.005, \ \overline{R} = 0.287 \ [\text{kJ kg}^{-1} \text{ K}^{-1}] \text{ and } \gamma = 1.4$$

Solution The *P-v* diagram of the air-standard Otto cycle is shown in Fig. 10.4(a). Applying the ideal gas equation to the air at 1 we obtain

$$m = \frac{P_1 V_1}{RT_1} = \frac{100 \times 8 \times 10^{-4}}{0.287 \times 293} = 9.513 \times 10^{-4} \text{ kg}$$

For the isentropic expansion 3-4

$$T_4 V_4^{(\gamma-1)} = T_3 V_3^{(\gamma-1)}$$

Therefore		$$T_3 = T_4 (V_4 / V_3)^{(\gamma-1)} = 1073 \times 9^{0.4} = 2584 \text{ K}$$

Since air is an ideal gas

$$P_3 = \frac{mRT_3}{V_3} = \frac{9.513 \times 10^{-4} \times 0.287 \times 2584 \times 9}{8 \times 10^{-4}} = 7937 \text{ kPa}$$

For the isentropic compression 1-2

$$T_1 V_1^{(\gamma-1)} = T_2 V_2^{(\gamma-1)}$$

Therefore $\quad T_2 = T_1 (V_1/V_2)^{(\gamma-1)} = 293 \times 9^{0.4} = 705.6\,\text{K}$

The heat supplied is

$$Q_{23} = mc_v(T_3 - T_2)$$

$$Q_{23} = 9.513 \times 10^{-4} \times 0.718 \times (2584 - 705.6) = 1.283\ \text{kJ}$$

The heat rejected is

$$Q_{41} = mc_v(T_4 - T_1)$$

$$Q_{41} = 9.513 \times 10^{-4} \times 0.718 \times (1073 - 293) = 0.533\ \text{kJ}$$

The net work output of the cycle is

$$W_{net} = Q_{23} - Q_{41} = 1.283 - 0.533 = 0.75\ \text{kW}$$

Thermal efficiency of the cycle is

$$\eta = \frac{W_{net}}{Q_{23}} = \frac{0.75}{1.283} = 58\%$$

Using Eq. (10.9) we obtain the efficiency as

$$\eta = 1 - \frac{1}{r^{(\gamma-1)}} = 1 - \frac{1}{9^{0.4}} = 58\%$$

As expected, the use of the formula for the efficiency of the Otto cycle gives the same answer as the detailed calculation.

Example 10.2 The maximum and minimum cycle temperatures of an ideal air-standard Otto cycle are T_3 (K) and T_1 (K) respectively. Obtain an expression for the compression ratio, r for the maximum net work done per cycle.

Solution Referring to the P-v diagram of the Otto cycle in Fig. 10.4(a) we have

$$W_{net} = Q_{23} - Q_{41} = mc_v(T_3 - T_2) - mc_v(T_4 - T_1) \qquad \text{(E10.2.1)}$$

Rearranging Eq. (E10.2.1)

$$W_{net} = mc_v(T_3 - T_4) - mc_v(T_2 - T_1) \qquad (E10.2.2)$$

For the isentropic processes 1-2 and 3-4 we have the following relations:

$$\frac{T_4}{T_3} = \left(\frac{V_3}{V_4}\right)^{(\gamma-1)} = \frac{1}{r^{\gamma-1}}$$

$$\frac{T_2}{T_1} = \left(\frac{V_1}{V_2}\right)^{(\gamma-1)} = r^{\gamma-1}$$

Substituting the above relations in Eq. (E10.2.2) we obtain

$$W_{net} = mc_v T_3(1 - 1/r^{\gamma-1}) - mc_v T_1(r^{\gamma-1} - 1) \qquad (E10.2.3)$$

When the net work output is a maximum, $dW_{net}/dr = 0$. Differentiating Eq. (E10.2.3) with respect to r we obtain the condition for maximum net work output as:

$$r_{opt} = \left(\frac{T_3}{T_1}\right)^{1/2(\gamma-1)}$$

Example 10.3 An actual four-stroke SI engine has a total cylinder volume of 1.5×10^{-3} m^3 and a compression ratio of 9. The engine runs at 3000 rpm and the conditions of the air at the beginning of the compression stroke are 100 kPa and 30°C. The ratio of the actual thermal efficiency of the engine to the air-standard cycle efficiency is 45%, and the air-fuel ratio is 13. The internal energy of combustion of the fuel is 47×10^3 kJ kg^{-1}. Calculate (i) the thermal efficiency of the engine, and (ii) the power output in kW.

Solution The efficiency of the ideal air-standard cycle is

$$\eta = 1 - \frac{1}{r^{(\gamma-1)}} = 1 - \frac{1}{9^{0.4}} = 0.585$$

The thermal efficiency of the actual engine is

$$\eta_{act} = 0.585 \times 0.45 = 0.263$$

The mass of air in the cylinder at the end of the intake stroke is

$$m_a = \frac{P_1 V_1}{R T_1} = \frac{100 \times 1.5 \times 10^{-3}}{0.287 \times 303} = 1.725 \times 10^{-3} \text{ kg}$$

The mass of fuel that is mixed with air per cycle is

$$m_f = m_a / 13 = 1.327 \times 10^{-4} \text{ kg}$$

Now for a four-stroke engine each complete cycle occupies 2 revolutions. Therefore the number of cycles per minute is 1500. The rate of fuel consumption is

$$(1500/60) m_f = 3.317 \times 10^{-3} \text{ kg s}^{-1}$$

The rate of energy input due to combustion is

$$\dot{Q}_{in} = \dot{m}_f U_f = 3.317 \times 10^{-3} \times 47 \times 10^3 = 155.9 \text{ kW}$$

The rate of work output is

$$\dot{W}_{out} = \dot{Q}_{in} \eta_{act} = 155.9 \times 0.263 = 41 \text{ kW}.$$

Example 10.4 A Diesel engine operating on the ideal air standard cycle has a cylinder with a diameter of 0.06 m and a stroke of 0.05 m. The piston has a flat top and the clearance volume is 10^{-5} m³. At the beginning of the compression stroke the air is at 100 kPa and 40°C. The highest temperature of the cycle is 1500°C. Calculate (i) the compression ratio, (ii) the cut-off ratio, (iii) the heat input per cycle, (iv) heat rejected per cycle, (v) the net work output per cycle, and (vi) the thermal efficiency of the cycle.

Solution The *P-v* diagram of the Diesel cycle is shown in Fig. 10.5(a). The stroke volume of the engine is

$$V_s = L_s \pi D_s^2 / 4 = 0.05 \times \pi (0.06)^2 / 4 = 1.4137 \times 10^{-4} \text{ m}^3$$

The clearance volume is $\quad V_c = 0.1 \times 10^{-4} \text{ m}^3$

The compression ratio is given by

$$r = \frac{V_{cyl}}{V_c} = \frac{V_s + V_c}{V_c} = \frac{1.5137 \times 10^{-4}}{0.1 \times 10^{-4}} = 15.137$$

For the isentropic compression 1-2

$$T_1 V_1^{(\gamma-1)} = T_2 V_2^{(\gamma-1)}$$

$$T_2 = T_1 r^{(\gamma-1)} = 313 \times 15.137^{0.4} = 928.02 \text{ K}$$

For the constant pressure process 2-3

$$\frac{V_3}{V_2} = \frac{T_3}{T_2} = \frac{1773}{928.02} = 1.91$$

The cut-off ratio is given by, $r_c = \dfrac{V_3}{V_2} = 1.91$

Now

$$\frac{V_4}{V_3} = \frac{V_1}{V_2} \frac{V_2}{V_3} = \frac{r}{r_c} = 7.92$$

For the isentropic expansion 3-4

$$T_4 V_4^{(\gamma-1)} = T_3 V_3^{(\gamma-1)}$$

Therefore

$$T_4 = T_3 \left(V_3 / V_4\right)^{(\gamma-1)} = 1773/(7.923)^{0.4} = 774.7 \text{ K}$$

The mass of air in the cylinder at the beginning of the compression stroke is

$$m = \frac{P_1 V_1}{R T_1} = \frac{100 \times 1.5137 \times 10^{-4}}{0.287 \times 313} = 1.685 \times 10^{-4} \text{ kg}$$

Heat supplied per cycle is

$$Q_{in} = m c_p (T_3 - T_2)$$

$$Q_{in} = 1.685 \times 10^{-4} \times 1.005 \times (1773 - 928.0) = 0.143 \text{ kJ}$$

Heat rejected per cycle is

$$Q_{out} = m c_v (T_4 - T_1)$$

$$Q_{out} = 1.685 \times 10^{-4} \times 0.718 \times (774.73 - 313) = 0.05586 \text{ kJ}$$

Work output per cycle is

$$W_{net} = Q_{in} - Q_{out} = 0.1431 - 0.05586 = 0.0872 \text{ kJ}$$

The thermal efficiency is given by

$$\eta = \frac{W_{net}}{Q_{in}} = \frac{0.0872}{0.1431} = 60.9\%$$

Applying Eq. (10.14) we obtain the efficiency directly as

$$\eta = 1 - \frac{1}{15.137^{0.4}} \left[\frac{1.9105^{1.4} - 1}{1.4 \times (1.9105 - 1)} \right] = 60.9\%$$

As expected the detailed calculation gives the same efficiency as the general expression.

Example 10.5 The compression ratio and the cut-off ratio of an air-standard Diesel cycle are 16 and 2 respectively. The cylinder volume is 0.015 m³. The conditions of the air at the beginning of the compression stroke are 95 kPa and 10°C. Calculate (i) the heat input per cycle, and (ii) the thermal efficiency.

Solution The *P-v* diagram of the Diesel cycle is shown in Fig. 10.5(a). The mass of air at the beginning of the compression stroke is

$$m = \frac{P_1 V_1}{RT_1} = \frac{95 \times 0.015}{0.287 \times 283} = 0.0175 \text{ kg}$$

For the isentropic process 1-2

$$T_1 V_1^{(\gamma-1)} = T_2 V_2^{(\gamma-1)}$$

Therefore

$$T_2 = T_1 (V_1 / V_2)^{(\gamma-1)} = 283 \times 16^{0.4} = 857.89 \text{ K}$$

For the constant pressure process 2-3

$$\frac{T_3}{T_2} = \frac{V_3}{V_2} = r_c = 2$$

Therefore
$$T_3 = 1715.8 \text{ K}$$

For the isentropic process 3-4

$$T_4 = T_3 (V_3 / V_4)^{(\gamma-1)} = T_3 / (r / r_c)^{\gamma-1}$$

Hence $T_4 = 1715.8/(8)^{0.4} = 746.8$ K

Heat supplied per cycle is

$$Q_{in} = mc_p(T_3 - T_2)$$

$$Q_{in} = 0.0175 \times 1.005 \times (1715.8 - 857.9) = 15.088 \text{ kJ}$$

Heat rejected per cycle is

$$Q_{out} = mc_v(T_4 - T_1)$$

$$Q_{out} = 0.0175 \times 0.718 \times (746.8 - 283) = 5.828 \text{ kJ}$$

Work output per cycle is

$$W_{net} = Q_{in} - Q_{out} = 15.088 - 5.828 = 9.26 \text{ kJ}$$

The thermal efficiency is

$$\eta = \frac{W_{net}}{Q_{in}} = \frac{9.26}{15.088} = 61.4\%$$

Applying Eq. (10.14) we obtain the efficiency as

$$\eta = 1 - \frac{1}{16^{0.4}}\left[\frac{2^{1.4} - 1}{1.4 \times (2 - 1)}\right] = 61.4\%$$

Example 10.6 The total heat input to an air-standard dual cycle of compression ratio 14, is 1130 kJ kg^{-1}. At the beginning of compression the air is at 100 kPa and 27°C. The highest temperature of the air is 1900°C. Calculate (i) the heat rejected, (ii) the net work output, and (iii) the thermal efficiency of the cycle.

Solution The P-v diagram of the duel cycle is shown in Fig. 10.6 where heat is supplied at constant volume from 2 to 3 and at constant pressure from 3 to 4. Therefore the total heat input per cycle is

$$Q_{in} = mc_v(T_3 - T_2) + mc_p(T_4 - T_3) \tag{E10.6.1}$$

For the isentropic process 1-2

$$T_1 V_1^{(\gamma-1)} = T_2 V_2^{(\gamma-1)}$$

Hence $\qquad T_2 = T_1(V_1/V_2)^{(\gamma-1)} = 300 \times 14^{0.4} = 862.13$ K

The highest temperature is given as, $T_4 = 2173$ K

Consider a unit mass of air in the cylinder. Substituting numerical values in Eq. (E10.6.1) we have

$$0.718(T_3 - 862.13) + 1.005(2173 - T_3) = 1130$$

Therefore $\qquad\qquad\qquad T_3 = 1515.18$ K

For the constant pressure process 3-4

$$\frac{V_4}{V_3} = \frac{T_4}{T_3} = \frac{2173}{1515.18} = 1.434$$

Now $\qquad \dfrac{V_4}{V_1} = \dfrac{V_4}{V_3}\dfrac{V_3}{V_1} = 1.434/r = 1.434/14 = 0.1024$

For the isentropic process 4-5

$$T_5 = T_4(V_4/V_5)^{(\gamma-1)} = 2173 \times (0.1024)^{0.4} = 873.3 \text{ K}$$

Heat rejected per unit mass of air is

$$Q_{out} = c_v(T_5 - T_1)$$

$$Q_{out} = 0.718 \times (873.3 - 300) = 411.63 \text{ kJ kg}^{-1}$$

Net work output per unit mass of air is

$$W_{net} = Q_{in} - Q_{out} = 1130 - 411.63 = 718.37 \text{ kJ kg}^{-1}$$

The thermal efficiency is

$$\eta = \frac{W_{net}}{Q_{in}} = \frac{718.37}{1130} = 63.5\%$$

Example 10.7 In a simple air-standard Brayton cycle air enters the compressor at 1 bar and 27°C. The pressure ratios in the compressor and the turbine are each 5. The temperature of the air entering the turbine is 1150°C. Calculate the thermal efficiency of the cycle, (i) if the processes in the turbine and the compressor are ideal, and (ii) if the isentropic efficiencies of the turbine and the compressor are 0.75 and 0.65 respectively.

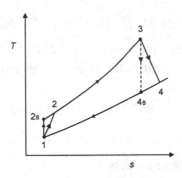

Fig. E10.7 *T-s* diagram

Solution The *T-s* diagram of the simple air-standard Brayton cycle is shown in Fig. E10.7. The following data are given:

$$T_1 = 300 \text{ K}, \qquad T_3 = 1423 \text{ K}, \qquad r = 5.$$

(i) Consider the cycle with an ideal turbine and an ideal compressor. For the isentropic compression process, 1-2s we have

$$\frac{T_{2s}}{T_1} = \left(\frac{P_2}{P_1}\right)^{(\gamma-1)/\gamma} = r^{(\gamma-1)/\gamma} \qquad (E10.7.1)$$

Substituting numerical values in Eq. (E10.7.1), with $\gamma = 1.4$ we obtain

$$T_{2s} = 475.13 \text{ K}$$

For the isentropic expansion process 3-4s in the turbine

$$\frac{T_3}{T_{4s}} = \left(\frac{P_2}{P_1}\right)^{(\gamma-1)/\gamma} = r^{(\gamma-1)/\gamma}$$

Therefore $\qquad\qquad\qquad T_{4s} = 898.48 \text{ K}$

The rate of heat input is

$$\dot{Q}_{in} = \dot{m}c_p(T_3 - T_{2s})$$

$$\dot{Q}_{in} = \dot{m}c_p(1423 - 475.13) = 947.87\dot{m}c_p$$

The rate of heat rejection is

$$\dot{Q}_{out} = \dot{m}c_p(T_{4s} - T_1)$$

$$\dot{Q}_{out} = \dot{m}c_p(898.48 - 300) = 598.48\dot{m}c_p$$

Thermal efficiency is

$$\eta = 1 - \frac{\dot{Q}_{out}}{\dot{Q}_{in}} = 1 - \frac{598.48}{947.87} = 36.8\%$$

(ii) Consider the cycle with a non-ideal turbine and compressor. For the irreversible, adiabatic compression process 1-2, the isentropic efficiency of the compressor is defined as:

$$\eta_c = \frac{T_{2s} - T_1}{T_2 - T_1}$$

Hence

$$T_2 = T_1 + (T_{2s} - T_1)/\eta_c$$

$$T_2 = 300 + (475.13 - 300)/0.65 = 569.43 \text{ K}$$

For the irreversible, adiabatic expansion process 3-4 in the turbine, the isentropic efficiency is defined as:

$$\eta_t = \frac{T_3 - T_4}{T_3 - T_{4s}}$$

Hence

$$T_4 = T_3 - \eta_t (T_3 - T_{4s})$$

$$T_4 = 1423 - 0.75 \times (1423 - 898.48) = 1029.6 \text{ K}$$

The rate of heat input is

$$\dot{Q}_{in} = \dot{m} c_p (T_3 - T_2)$$

$$\dot{Q}_{in} = \dot{m} c_p (1423 - 569.43) = 853.57 \dot{m} c_p$$

The rate of heat rejection is

$$\dot{Q}_{out} = \dot{m} c_p (T_4 - T_1)$$

$$\dot{Q}_{out} = \dot{m} c_p (1029.1 - 300) = 729.1 \dot{m} c_p$$

The thermal efficiency of the cycle is

$$\eta = 1 - \frac{\dot{Q}_{out}}{\dot{Q}_{in}} = 1 - \frac{729.1}{853.6} = 14.6\%$$

We notice that the *internal* ireversibilities of the turbine and the compressor cause a dramatic reduction in the thermal efficiency of the Brayton cycle.

Example 10.8 Air enters a compressor of an air-standard Brayton cycle at 100 kPa and 27°C. The cycle includes an ideal regenerator, and has a pressure ratio of 5. The temperature of the air entering the turbine is 1150°C. Calculate the thermal efficiency of the cycle, (i) if the processes in the turbine and the compressor are ideal, and (ii) if the isentropic efficiencies of the turbine and the compressor are 0.75 and 0.65 respectively.

Solution Notice that in this example we have used the same conditions as the previous worked example so that we could make meaningful comparisons of the effect of the regenerator on the thermal efficiency of the cycle. In the regenerator, the air leaving the compressor is heated form T_2 to T_a using the hot air leaving the turbine as shown in Fig. E10.8. (i) Consider the case when the compressor and turbine are ideal. The following data obtained in worked example 10.7 are applicable to this example:

$$T_3 = 1423K, \quad T_1 = 300\,\text{K}, \quad T_{4s} = 898.48\,\text{K}, \quad T_{2s} = 475.13\,\text{K}$$

For the ideal regenerator,

$$T_a = T_{4s} \qquad \text{and} \qquad T_b = T_{2s}$$

The heat input rate is

$$\dot{Q}_{in} = \dot{m}c_p(T_3 - T_a) = \dot{m}c_p(T_3 - T_{4s})$$

$$\dot{Q}_{in} = \dot{m}c_p(1423 - 898.48) = 524.52\dot{m}c_p$$

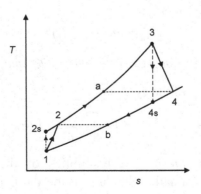

Fig. E10.8 Regenerative cycle: *T-s* diagram

The heat rejection rate is

$$\dot{Q}_{out} = \dot{m}c_p(T_b - T_1) = \dot{m}c_p(T_{2s} - T_1)$$

$$\dot{Q}_{out} = \dot{m}c_p(475.13 - 300) = 175.13\dot{m}c_p$$

Thermal efficiency is given by

$$\eta = 1 - \frac{\dot{Q}_{out}}{\dot{Q}_{in}} = 1 - \frac{175.13}{524.52} = 66\%$$

which agrees with result obtained using Eq. (10.35).

(ii) Consider the cycle in which the turbine and compressor are non-ideal. However, the regenerator is *ideal*, and therefore $T_a = T_4$ and $T_b = T_2$ (see Fig. E10.8). The temperatures T_2 and T_4 were obtained in worked example 10.6 for the same conditions as the present example. The heat input rate is

$$\dot{Q}_{in} = \dot{m}c_p(T_3 - T_4)$$

$$\dot{Q}_{in} = \dot{m}c_p(1423 - 1029.6) = 393.4\dot{m}c_p$$

The heat rejection rate is

$$\dot{Q}_{out} = \dot{m}c_p(T_2 - T_1)$$

$$\dot{Q}_{out} = \dot{m}c_p(569.43 - 300) = 269.43\dot{m}c_p$$

The thermal efficiency is

$$\eta = 1 - \frac{\dot{Q}_{out}}{\dot{Q}_{in}} = 1 - \frac{269.43}{393.4} = 31.5\%$$

Example 10.9 Air enters the compressor of an air-standard Brayton cycle at 100 kPa and 27°C. The cycle includes a regenerator with an effectiveness of 0.85. The pressure ratio is 5 and the temperature of the air entering the turbine is 1150°C. Calculate the thermal efficiency of the cycle, (i) if the processes in the turbine and the compressor are ideal, and (ii) if the isentropic efficiencies of the turbine and the compressor are 0.75 and 0.65 respectively.

Solution Note that in this example we have used the same conditions as the two previous examples so that we could make meaningful comparisons of the effect of the *non-ideal regenerator* on the thermal efficiency of the cycle. In the regenerator, the air leaving the compressor is heated form T_2 to T_a using the hot air leaving the turbine as shown in Fig. E10.8. The regenerator is not ideal and its effectiveness is given by Eq. (10.27) as

$$\varepsilon_{reg} = \frac{T_a - T_2}{T_4 - T_2}$$

(i) Consider the cycle with an ideal turbine and an ideal compressor.

$$T_{as} = \varepsilon_{reg}(T_{4s} - T_{2s}) + T_{2s}$$

where the second subscript s denotes quantities for the ideal turbine and compressor. Substituting the numerical values obtained in the previous example in the above equation we have

$$T_{as} = 0.85 \times (898.48 - 475.13) + 475.13 = 834.98 \text{ K}$$

The application of the SFEE to the regenerator gives

$$\dot{m}c_p(T_{as} - T_{2s}) = \dot{m}c_p(T_{4s} - T_{bs})$$

$$(834.98 - 475.13) = (898.48 - T_{bs})$$

Therefore $T_{bs} = 538.63 \text{ K}$

The rate of heat input is

$$\dot{Q}_{in} = \dot{m}c_p(T_3 - T_{as})$$

$$\dot{Q}_{in} = \dot{m}c_p(1423 - 834.98) = 588.02\dot{m}c_p$$

The rate of heat rejection is

$$\dot{Q}_{out} = \dot{m}c_p(T_{bs} - T_1)$$

$$\dot{Q}_{out} = \dot{m}c_p(538.63 - 300) = 238.63\dot{m}c_p$$

The thermal efficiency is

$$\eta = 1 - \frac{\dot{Q}_{out}}{\dot{Q}_{in}} = 1 - \frac{238.63}{588.02} = 59.4\%$$

(ii) Consider the cycle with a non-ideal turbine and compressor. For the non-ideal regenerator we have

$$T_a = \varepsilon_{reg}(T_4 - T_2) + T_2$$

Substituting the numerical values obtained in the previous example in the above equation

$$T_a = 0.85 \times (1029.6 - 569.43) + 569.43 = 960.57 \text{ K}$$

The application of the SFEE to the regenerator gives

$$\dot{m}c_p(T_a - T_2) = \dot{m}c_p(T_4 - T_b)$$

$$(960.57 - 569.43) = (1029.6 - T_b)$$

Therefore $\qquad T_b = 638.48 \text{ K}$

The rate of heat input is

$$\dot{Q}_{in} = \dot{m}c_p(T_3 - T_a)$$

$$\dot{Q}_{in} = \dot{m}c_p(1423 - 960.57) = 462.43\dot{m}c_p$$

The rate of heat rejection is

$$\dot{Q}_{out} = \dot{m}c_p(T_b - T_1)$$

$$\dot{Q}_{out} = \dot{m}c_p(638.48 - 300) = 338.48\dot{m}c_p$$

The thermal efficiency is

$$\eta = 1 - \frac{\dot{Q}_{out}}{\dot{Q}_{in}} = 1 - \frac{338.48}{462.43} = 26.8\%$$

Example 10.10 Air enters a two-stage compressor of an air-standard Brayton cycle at 100 kPa and 27°C. The pressure ratio is 5. Air leaves the low-pressure compressor, at 250 kPa and enters an intercooler where it is cooled to 60°C before entering the high-pressure compressor. The temperature of the air entering the turbine is 1150°C. The turbine and compressor are ideal. Calculate (i) the thermal efficiency of the cycle, and (ii) the work input to the compressor.

Solution The T-s diagram of the air-standard cycle with intercooling is shown in Fig. E10.10. The pressure ratio for the low-pressure compressor is 2.5. For the isentropic compression 1-as, we have

$$T_{as} = T_1 r_{cl}^{(1-\gamma)/\gamma} = 300 \times 2.5^{0.2857} = 389.77 \text{ K}$$

where the second subscript s denotes quantities under ideal conditions. The pressure ratio for the high-pressure compressor is

$$r_{ch} = 500/250 = 2$$

For the isentropic compression in the high-pressure compressor

$$T_{cs} = T_b r_{ch}^{(1-\gamma)/\gamma} = 333 \times 2.0^{0.2857} = 405.93 \text{ K}$$

Heat rejection rate in the intercooler is

$$\dot{Q}_{ic} = \dot{m} c_p (T_{as} - T_b)$$

$$\dot{Q}_{ic} = \dot{m} c_p (389.77 - 333) = 56.77 \dot{m} c_p$$

Heat rejection rate to the heat sink is

$$\dot{Q}_{out} = \dot{m} c_p (T_{4s} - T_1)$$

From the data obtained in worked example 10.7, the temperature at 4 is

$$T_{4s} = 898.4 \text{ K}$$

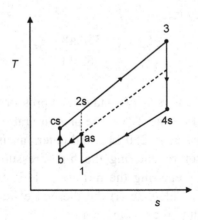

Fig. E10.10 T-s diagram of cycle with intercooling

Hence $\dot{Q}_{out} = \dot{m}c_p(898.48 - 300) = 598.48\dot{m}c_p$

$$\dot{Q}_{out,tot} = \dot{Q}_{ic} + \dot{Q}_{out} = 56.77\dot{m}c_p + 598.48\dot{m}c_p = 655.25\dot{m}c_p$$

Therefore the total heat rejection rate is $655.25\,\dot{m}c_p$
The rate of heat input is

$$\dot{Q}_{in} = \dot{m}c_p(T_3 - T_{cs})$$

$$\dot{Q}_{in} = \dot{m}c_p(1423 - 405.93) = 1017.07\dot{m}c_p$$

The thermal efficiency is

$$\eta = 1 - \frac{\dot{Q}_{out,tot}}{\dot{Q}_{in}} = 1 - \frac{655.25}{1017.07} = 35.6\%$$

The total work input rate to the compressors is

$$\dot{W}_c = \dot{m}\,c_p(T_{as} - T_1) + \dot{m}c_p(T_{cs} - T_b)$$

$$\dot{W}_c = \dot{m}\,c_p(389.77 - 300) + \dot{m}c_p(405.93 - 333) = 162.7\dot{m}c_p$$

The compressor work input rate without intercooling is obtained using the data in worked example 10.7 as

$$\dot{W}_c = \dot{m}\,c_p(475.13 - 300) = 175.13\dot{m}c_p$$

We note that the compressor work input is decreased by about 7 per cent due to the introduction of intercooling. However, the thermal efficiency of the cycle has decreased from 36.8% to 35.6% due to intercooling. Note that the above analysis could be easily modified to deal with a non-ideal compressor and turbine.

Example 10.11 Air enters the compressor of an air-standard Brayton cycle at 100 kPa and 27°C. The pressure ratio is 5. The temperature of the air entering the two-stage turbine is 1150°C. The air expands in the high-pressure turbine to a pressure of 250 kPa before entering a reheater where it is heated to 1050°C. The air expands to 100 kPa in the low-pressure turbine. The compressor and turbine are ideal. Calculate (i) the thermal efficiency of the cycle, and (ii) the work output of the turbine.

Solution The *T-s* diagram of the air-standard cycle with reheating is shown in Fig. E10.11. The pressure ratio of the high-pressure turbine is

$$r_{th} = 500/250 = 2$$

For the isentropic expansion in the high-pressure turbine

$$T_{ds} = T_3/r_{th}^{(\gamma-1)/\gamma} = 1423/2^{0.2857} = 1167.3 \text{ K}$$

where the second subscript s denotes quantities under ideal conditions.

The temperature of the air leaving the reheater is $T_e = 1323$K.

The pressure ratio for the low-pressure turbine is, $r_{tl} = 250/100 = 2.5$.

For the isentropic expansion in the low-pressure turbine,

$$T_{fs} = T_e/r_{tl}^{(\gamma-1)/\gamma} = 1323/2.5^{0.2857} = 1018.28 \text{ K}$$

Rate of heat input from the high temperature reservoir is

$$\dot{Q}_{in} = \dot{m}c_p(T_3 - T_{2s})$$

$$\dot{Q}_{in} = \dot{m}c_p(1423 - 475.13) = 947.87\dot{m}c_p$$

Rate of heat input in the reheater is

$$\dot{Q}_{rh} = \dot{m}c_p(T_e - T_{ds})$$

$$\dot{Q}_{rh} = \dot{m}c_p(1323 - 1167.35) = 155.65\dot{m}c_p$$

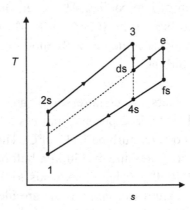

Fig. E10.11 *T-s* diagram of cycle with reheating

Total rate of heat input is

$$\dot{Q}_{in,tot} = \dot{Q}_{in} + \dot{Q}_{rh} = 1103.52 \dot{m} c_p$$

Heat rejection rate is

$$\dot{Q}_{out} = \dot{m} c_p (T_{fs} - T_1)$$

$$\dot{Q}_{out} = \dot{m} c_p (1018.28 - 300) = 718.28 \dot{m} c_p$$

The thermal efficiency is

$$\eta = 1 - \frac{\dot{Q}_{out}}{\dot{Q}_{in,tot}} = 1 - \frac{718.28}{1103.52} = 34.9\%$$

The total rate of work output of the turbine is

$$\dot{W}_t = \dot{m} c_p (T_3 - T_{ds}) + \dot{m} c_p (T_e - T_{fs})$$

$$\dot{W}_t = \dot{m} c_p (1423 - 1167.35) + \dot{m} c_p (1323 - 1018.28) = 560.4 \dot{m} c_p$$

The turbine work output of the basic air-standard cycle is computed from the data obtained in worked example 10.7 as

$$\dot{W}_b = \dot{m} c_p (T_3 - T_{4s})$$

$$\dot{W}_b = \dot{m} c_p (1423 - 898.48) = 524.52 \dot{m} c_p$$

We notice that the rate of work output of the reheat cycle is about 6.8 percent larger than that of the basic cycle. However, the efficiency of the reheat cycle is 34.9% while the efficiency of the basic cycle is 36.8%.

Example 10.12 In an air-standard gas turbine cycle with reheat, the compressor is driven by the high-pressure turbine, while the low-pressure turbine delivers 9 MW of power to drive an electric generator. Air enters the compressor at 100 kPa and 15°C, and the pressure ratio of the compressor is 6. The air temperature at entry to the high-pressure turbine is 820°C. The temperature of the air leaving the reheater is 800°C. The efficiency of the compressor is 0.85 and the efficiency of each turbine is 0.8. Calculate (i) the mass rate of flow of air, (ii) the total heat input, and (iii) the thermal efficiency.

Solution In this cycle, the entire work output of the high-pressure turbine is consumed by the compressor as shown in Fig. E10.12. The work delivered by the low-pressure turbine is used to run the electric generator. For the isentropic compression in the compressor

$$T_{2s} = T_1 r_c^{(\gamma-1)/\gamma} = 288 \times 6^{0.2857} = 480.52 \text{ K}$$

where the second subscript s denotes quantities under ideal conditions. For the non-deal adiabatic compressor

$$T_2 = T_1 + (T_{2s} - T_1)/\eta_c$$

$$T_2 = 288 + (480.52 - 288)/0.85 = 514.49 \text{ K}$$

Rate of work input to the compressor is

$$\dot{W}_c = \dot{m}c_p(T_2 - T_1) = 226.49\dot{m}c_p$$

The work output of the high-pressure turbine is also equal to the work input to the compressor. Therefore

$$\dot{W}_{th} = \dot{m}c_p(T_3 - T_4) = \dot{W}_c$$

$$\dot{m}c_p(1093 - T_4) = 226.49\dot{m}c_p$$

Hence $$T_4 = 866.51 \text{ K}$$

For the isentropic expansion in the high-pressure turbine

$$T_{4s} = T_3/r_{th}^{(\gamma-1)/\gamma} = 1093/r_{th}^{0.2857} \qquad (E10.12.1)$$

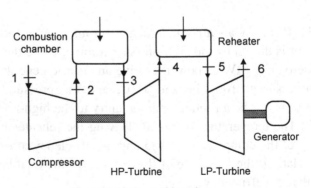

Fig. E10.12 Gas turbine plant

where r_{th} is the pressure ratio of the high pressure turbine. For the non-ideal adiabatic expansion in the high-pressure turbine

$$(T_3 - T_{4s})\eta_{th} = (T_3 - T_4)$$

$$T_{4s} = 1093 - (1093 - 866.51)/0.8 = 809.88 \text{ K}$$

Substituting in Eq. (E10.12.1) we have

$$r_{th} = 2.855 = P_3/P_4 \qquad \text{(E10.12.2)}$$

Now $P_3 = 600$ kPa. Therefore from Eq. (E10.12.2), $P_4 = 210.16$ kPa

The pressure ratio of the low-pressure turbine is

$$r_{tl} = P_4/P_6 = 210.16/100 = 2.1$$

For the isentropic expansion in the low-pressure turbine

$$T_{6s} = T_5/r_{tl}^{(\gamma-1)/\gamma} = 1073/r_{tl}^{0.2857} = 867.85 \text{ K}$$

For the non-ideal adiabatic expansion in the low-pressure turbine

$$(T_5 - T_{6s})\eta_{tl} = (T_5 - T_6)$$

Therefore $\qquad T_6 = 1073 - (1073 - 867.86) \times 0.8 = 908.89 \text{ K}$

The rate of work output of the low-pressure turbine is

$$\dot{W}_{tl} = \dot{m}c_p(T_5 - T_6) = \dot{m}c_p\eta_{tl}(T_5 - T_{6s})$$

$$\dot{W}_{tl} = 0.8 \times (1073 - 867.85)\dot{m}c_p = 9000 \text{ kW}$$

Hence the air flow rate, $\qquad \dot{m} = 54.56 \text{ kg s}^{-1}$

The total rate of heat input is

$$\dot{Q}_{in} = \dot{m}c_p(T_3 - T_2) + \dot{m}c_p(T_5 - T_4)$$

$$\dot{Q}_{in} = 54.56 \times 1.005 \times [(1093 - 514.49) + (1073 - 908.89)]$$

Hence $\qquad \dot{Q}_{in} = 40719.9 \text{ kW}$

The thermal efficiency of the plant is

$$\eta = \frac{\dot{W}_{net}}{\dot{Q}_{in}} = \frac{9000}{40719.9} = 22\%$$

Example 10.13 The highest and lowest temperatures of a Brayton cycle are T_3 and T_1 respectively, and the pressure ratio is r. Obtain expressions for the heat supply and rejection temperatures of a Carnot cycle with the same efficiency as the Brayton cycle.

Solution The T-s diagram of the air-standard Brayton cycle is shown in Fig. E10.13. The heat supplied in the constant pressure process 2-3 of the Brayton cycle is obtained by applying the SFEE.

$$Q_{23} = \dot{m}\int_2^3 dh \qquad (E10.13.1)$$

Using the 'T-ds' equation (Eq. (7.17)) and the relation, $h = u + Pv$ we obtain the thermodynamic relation

$$dh = Tds - vdP \qquad (E10.13.2)$$

For a constant pressure process, $dP = 0$

Therefore from Eq. (E10.13.2) $dh = Tds$.

Substituting in Eq. (E10.13.1)

$$Q_{23} = \dot{m}\int_2^3 Tds \qquad (E10.13.3)$$

Let the heat supply temperature of the equivalent Carnot cycle be T_h. If the heat input of the Carnot cycle is equal to that of the Brayton cycle then

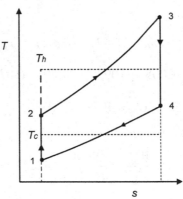

Fig. E10.13 *T-sdiagram*

$$Q_{23} = \dot{m}\int_{2}^{3} T ds = \dot{m}T_h(s_3 - s_2)$$

Therefore

$$T_h = \frac{\int_{2}^{3} T ds}{s_3 - s_2} \qquad (E10.13.4)$$

Similarly, for the same heat rejection rate

$$Q_{41} = \dot{m}\int_{1}^{4} T ds = \dot{m}T_c(s_4 - s_1)$$

Therefore the equivalent low temperature is

$$T_c = \frac{\int_{1}^{4} T ds}{s_4 - s_1} \qquad (E10.13.5)$$

The thermal efficiency of the Brayton cycle is

$$\eta = 1 - \frac{Q_{41}}{Q_{23}} = 1 - \frac{T_c(s_4 - s_1)}{T_h(s_3 - s_2)} = 1 - \frac{T_c}{T_h} \qquad (E10.13.6)$$

which is equal to the 'equivalent' Carnot efficiency.

For the isentropic processes 1-2 and 3-4

$$T_2 = T_1 r^{(\gamma-1)/\gamma} \quad \text{and} \quad T_4 = T_3 r^{-(\gamma-1)/\gamma} \qquad (E10.13.7)$$

For the constant pressure processes 2-3 and 4-1

$$s_3 - s_2 = c_p \ln(T_3/T_2) = c_p \ln[T_3/T_1 r^{(\gamma-1)/\gamma}]$$

$$s_4 - s_1 = c_p \ln(T_4/T_1) = c_p \ln[T_3/T_1 r^{(\gamma-1)/\gamma}]$$

Using the four relations above we can express the equivalent temperatures of the Carnot cycle given by Eqs. (E10.13.4) and (E10.13.5) in the form:

$$T_h = \frac{\int_{2}^{3} T ds}{s_3 - s_2} = \frac{c_p(T_3 - T_2)}{c_p \ln[T_3/T_1 r^{(\gamma-1)/\gamma}]} = \frac{[T_3 - T_1 r^{(\gamma-1)/\gamma}]}{\ln[T_3/T_1 r^{(\gamma-1)/\gamma}]}$$

$$T_c = \frac{\int_1^4 T ds}{s_4 - s_1} = \frac{c_p(T_4 - T_1)}{c_p \ln[T_3/T_1 r^{(\gamma-1)/\gamma}]} = \frac{[T_3 - T_1 r^{(\gamma-1)/\gamma}]}{r^{(\gamma-1)/\gamma} \ln[T_3/T_1 r^{(\gamma-1)/\gamma}]}$$

Notice that by substituting the above equivalent temperatures in Eq. (E10.13.6) we can recover the expression for the efficiency of the air-standard Brayton cycle obtained earlier as Eq. (10.23). From the *T-s* diagram in Fig. E10.13 it is clear that T_h and T_c are the heights of the rectangles with areas equal to the areas under the constant pressure curves 2-3 and 1-4 respectively.

Example 10.14 Figure E10.14 shows a gas turbine plant including a regenerator with an effectiveness of 0.7. The high-pressure turbine drives the compressor while the low-pressure turbine delivers a net work output of 150 kW. The air entering the compressor is at 100 kPa and 25°C. The efficiency of the compressor is 0.8 and the pressure ratio is 4. The efficiency of each of the two turbines is 0.87. Air enters the low-pressure turbine at 667°C. Each air stream experiences a pressure drop of 2 percent in the regenerator. The pressure of the air leaving the regenerator at 7 is 100 kPa. Calculate (i) the mass flow rate of air and (ii) the thermal efficiency of the cycle.

Solution The pressure at point 7 is 100 kPa. Now the pressure drop

$$\Delta P_{67} = P_6 - P_7 = 0.02 P_6$$

Therefore $P_6 = 102$ kPa.

The pressure at 2 is $P_2 = 400$ kPa.

The pressure drop $\Delta P_{23} = P_2 - P_3 = 0.02 P_2$

Therefore $P_3 = 392$ kPa

For the isentropic compression in the compressor

$$T_{2s} = T_1 r_c^{(\gamma-1)/\gamma} = 298 \times 4^{0.2857} = 442.8 \text{ K}$$

where the second subscript *s* denotes quantities under ideal conditions.

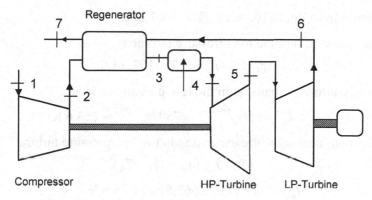

Fig. E10.14 Gas turbine plant

For the non-deal adiabatic compressor

$$T_2 = T_1 + (T_{2s} - T_1)/\eta_c$$

$$T_2 = 298 + (442.8 - 298)/0.8 - 479.0 \text{ K}$$

Rate of work input to the compressor is

$$\dot{W}_c = \dot{m}c_p(T_2 - T_1) = 181\dot{m}c_p$$

The work output of the high-pressure turbine is equal to the work input to the compressor. Therefore

$$\dot{W}_{th} = \dot{m}c_p(T_4 - T_5) = \dot{W}_c = 181\dot{m}c_p$$

For the isentropic expansion in the high-pressure turbine

$$T_{5s} = T_4/r_{th}^{(\gamma-1)/\gamma} = 940/r_{th}^{0.2857} \qquad (E10.14.1)$$

where r_{th} is the pressure ratio of the high pressure turbine. For the non-ideal adiabatic expansion in the high-pressure turbine

$$(T_4 - T_{5s})\eta_{th} = (T_4 - T_5)$$

$$T_{5s} = 940 - 181/0.87 = 731.95 \text{ K}$$

$$T_5 = 940 - 181 = 759 \text{ K}$$

Substituting in Eq. (E10.14.1) for T_{5s} we have

$$r_{th} = 2.4 = P_4/P_5 \qquad (E10.14.2)$$

Now $$P_4 = P_3 = 392 \text{ kPa}.$$

Therefore from Eq. (E10.14.2), $P_5 = 163.3$ kPa.

The pressure ratio of the low-pressure turbine is

$$r_{tl} = P_5 / P_6 = 163.3/102 = 1.6$$

For the isentropic expansion in the low-pressure turbine

$$T_{6s} = T_5 / r_{tl}^{(\gamma-1)/\gamma} = 759 / r_{tl}^{0.2857} = 663.6 \text{ K}$$

For the non-ideal adiabatic expansion in the low-pressure turbine

$$(T_5 - T_{6s})\eta_{tl} = (T_5 - T_6)$$

Hence $T_6 = 759 - (759 - 663.6) \times 0.87 = 676.0$ K

The effectiveness of the regenerator is given by

$$T_3 - T_2 = \varepsilon_{reg}(T_6 - T_2) = 0.7 \times (676 - 479) = 137.9$$

Hence $T_3 = 479 + 137.9 = 616.9$ K

The work output of the low-pressure turbine is

$$\dot{W}_{tl} = \dot{m}c_p(T_5 - T_6) = 150 \text{ kW}$$

Hence $(759 - 676)\dot{m}c_p = 150$ kW

Therefore the air flow rate, $\dot{m} = 1.8$ kg s^{-1}

The total rate of heat input is

$$\dot{Q}_{in} = \dot{m}c_p(T_4 - T_3)$$

$$\dot{Q}_{in} = 1.8 \times 1.005 \times (940 - 616.9) = 584.48 \text{ kW}$$

The thermal efficiency is

$$\eta = \frac{\dot{W}_{net}}{\dot{Q}_{in}} = \frac{150}{584.48} = 25.6\%$$

Notice that in this worked example we have taken into consideration several non-ideal conditions in the gas turbine plant including the pressure drops in the regenerator, the effectiveness of the regenerator, and the irreversibilities in the compressor and the turbines.

Example 10.15 Air enters the compressor of a gas turbine plant, used for propulsion, at 100 kPa and 20°C. The pressure ratio of the compressor is 5. At entry to the turbine the temperature is 880°C. The work output of the turbine is just sufficient to supply the work required by the compressor. The air leaving the turbine expands to a pressure of 100 kPa in an ideal reversible adiabatic nozzle. Calculate (i) the pressure ratio of the turbine, and (ii) the velocity of the air leaving the nozzle.

Solution A schematic diagram of a gas turbine cycle of a jet engine is shown in Fig. 10.14. The *T-s* diagram of the cycle is depicted in Fig. E10.15. We shall assume that the processes in the compressor and turbine are ideal. The work output of the turbine is just sufficient to operate the compressor.

For the isentropic compression in the compressor

$$T_2 = T_1 r_c^{(\gamma-1)/\gamma} = 293 \times 5^{0.2857} = 464.05 \text{ K}$$

Rate of work input to the compressor is

$$\dot{W}_c = \dot{m}c_p(T_2 - T_1) = 171.05\dot{m}c_p$$

The work output of the turbine is also equal to the work input to the compressor. Therefore

$$\dot{W}_t = \dot{m}c_p(T_3 - T_4) = \dot{W}_c$$

$$\dot{m}c_p(1153 - T_4) = \dot{W}_c = 171.05\dot{m}c_p$$

Hence $\qquad\qquad T_4 = 981.95 \text{ K}$

For the isentropic expansion in the turbine

$$T_4 = T_3 / r_t^{(\gamma-1)/\gamma} = 1153 / r_t^{0.2857} = 981.95 \text{ K}$$

Therefore the pressure ratio of the turbine is

$$r_t = 1.754 = P_3 / P_4 = 500 / P_4.$$

Hence $\qquad\qquad P_4 = 285 \text{ kPa}$

Since the expansion in the nozzle is isentropic, $s_4 = s_5$.

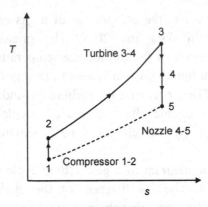

Fig. E10.15 *T-s* diagram for gas turbine with nozzle

Using the expression for the change in entropy of an ideal gas we have

$$S_5 - S_4 = \dot{m}c_p \ln\left(\frac{T_5}{T_4}\right) - \dot{m}R\ln\left(\frac{P_5}{P_4}\right) = 0$$

Rearranging the above equation and substituting numerical values we obtain

$$\frac{T_5}{T_4} = \left(\frac{P_4}{P_5}\right)^{-(\gamma-1)/\gamma} = \left(\frac{285}{100}\right)^{-0.2857} = 0.741$$

Therefore $T_4 = 728$ K.

Apply the SFEE to the nozzle neglecting the kinetic energy at the entrance and the changes in potential energy. Thus

$$c_p T_4 = \left(c_p T_5 + V_5^2/2\right)$$

$$V_5^2 = 2 \times 1.005 \times 10^3 (981.95 - 728)$$

Therefore the velocity at the exit, $V_5 = 714.4$ ms^{-1}

Problems

P10.1 The compression ratio of an air-standard Otto cycle is 8 and the heat input is 2300 kJ kg^{-1}. At the beginning of the compression stroke the conditions of the air are 100 kPa and 15°C. Calculate (i) the maximum temperature of the cycle, (ii) the net work output, and (iii) the thermal efficiency. [*Answers*: (i) 3592°C, (ii) 1299 kJ kg⁻¹, (iii) 56.5%]

P10.2 The compression ratio of an air-standard Diesel cycle is 14. The conditions of the air at the beginning of the compression stroke are 100 kPa and 18°C. The maximum temperature of the cycle is 2200°C. Calculate (i) the net work output, (ii) the heat input, and (iii) the thermal efficiency of the cycle. [*Answers*: (i) 900.5 kJ kg⁻¹, (ii) 1645 kJ kg⁻¹, (iii) 54.7%]

P10.3 The stroke volume of an engine operating on the dual cycle is 0.015 m³ and the compression ratio is 7. The conditions of the air at the beginning of the compression stroke are 100 kPa and 20°C. At the end of the constant volume heating process the pressure of the air is 5700 kPa. The constant pressure heating process occurs over 3 percent of the stroke volume. Calculate (i) the work output of the cycle, and (ii) the thermal efficiency. [*Answers*: (i) 905 kJ kg⁻¹, (ii) 53.7%]

P10.4 The pressure ratio of a simple Brayton cycle is 5. The conditions of the air at entry to the compressor are 100 kPa and 20°C. The efficiencies of the compressor and turbine are 0.8 and 0.82 respectively. The temperature of the air entering the turbine is 780°C. Calculate (i) the work input to the compressor, (ii) the work output of the turbine, and (iii) the thermal efficiency. [*Answers*: (i) 214.9 kJ kg⁻¹, (ii) 319.9 kJ kg⁻¹, (iii) 19%]

P10.5 A simple Brayton cycle has a pressure ratio of 5. The cycle includes a regenerator with an effectiveness of 0.75. The conditions of the air at entry to the compressor are 100 kPa and 20°C. The efficiencies of the compressor and turbine are 0.8 and 0.85 respectively. The temperature of the air entering the turbine is 800°C. Calculate (i) the work input to the compressor, (ii) the work output of the turbine, and (iii) the thermal efficiency.
[*Answers*: (i) 214.9 kJ kg⁻¹, (ii) 337.85 kJ kg⁻¹, (iii) 31.1%]

P10.6 The conditions of the air at entry to the compressor of a gas-turbine cycle are 100 kPa and 15°C. The pressure ratio of the compressor is 6. The air is heated to 900°C before it enters the high pressure turbine, whose work output is just sufficient to drive the compressor. The air is reheated to 850°C before entry to the low-pressure turbine. The efficiency of the compressor is 0.8, and the efficiency of each turbine is 0.85. Calculate (i) the pressure ratios of the two turbines, (ii) the heat input to the reheater, and (iii) the thermal efficiency. [*Answers*: (i) 2.63, 2.28, (ii) 191.6 kJ kg^{-1}, (iii) 24%]

P10.7 A gas-turbine cycle has a two-stage compressor and a two-stage turbine. Air enters the low-pressure compressor at 100 kPa and 15°C. The air is cooled to 20°C before it enters the high-pressure compressor. The pressure ratio of each compressor is 2. At entry to the high-pressure turbine the temperature of the air is 815°C. Before entering the low-pressure turbine the air is heated to 800°C in the reheater. The pressure ratio of each turbine is 2. Assume that the compressor and turbine are ideal. Calculate (i) the work input to the compressor, (ii) the work output of the turbine, and (iii) the thermal efficiency. [*Answers*: (i) 127.9 kJ kg^{-1}, (ii) 390.2 kJ kg^{-1}, (iii) 28.6%]

P10.8 The pressure ratio of an ideal Brayton cycle is 2. The lowest and highest temperatures of the cycle are 20°C and 880°C. Calculate (i) the heat supply and rejection temperatures of a Carnot cycle that has the same efficiency as the ideal Brayton cycle, and (ii) the thermal efficiency of the cycle. [*Answers*: (i) 679 K, 557 K, (ii) 18%]

P10.9 A gas-turbine cycle has a two-stage compressor and a two-stage turbine. Air enters the low-pressure compressor at 100 kPa and 15°C. The air is cooled to 20°C in an intercooler. The pressure ratio of each compressor is 2. At entry to the high-pressure turbine the temperature of the air is 815°C. The air is heated to 800°C in the reheater. The pressure ratio of each turbine is 2. The efficiency of each compressor is 0.8, and the efficiency of each turbine is 0.85. Calculate (i) the work input to the compressor, (ii) the work output of the turbine, and (iii) the thermal efficiency. [*Answers*: (i) 159 kJ kg^{-1}, (ii) 331.7 kJ kg^{-1}, (iii) 19.8%]

P10.10 The lowest and highest temperatures of an ideal Brayton cycle are 20°C and 900°C. Calculate (i) the pressure ratio that maximizes the net work output of the cycle, (ii) the maximum work output, and (iii) the thermal efficiency when the net work output is a maximum.
[*Answers*: (i) 11.33, (ii) 295 kJ kg^{-1}, (iii) 50%]

P10.11 The conditions of the air entering the compressor of a gas turbine-jet engine are 100 kPa and 16°C. The pressure ratio of the compressor is 5 and the temperature of the air entering the turbine is 1050°C. The output of the turbine is just equal to the work required by the compressor. The efficiencies of the compressor and turbine are 0.8 and 0.85 respectively. The air leaving the turbine expands isentropically in the nozzle to a pressure of 100 kPa. Calculate (i) the pressure at the nozzle inlet, and (ii) the velocity of the air leaving the nozzle.
[*Answers*: (i) 241.7 kPa, (ii) 705.8 ms^{-1}]

References

1   Bejan, Adrian, *Advanced Engineering Thermodynamics*, John Wiley & Sons, New York, 1988.
2.   Cravalho, E.G. and J.L. Smith, Jr., *Engineering Thermodynamics*, Pitman, Boston, MA, 1981.
3.   Jones, J.B. and G.A. Hawkins, *Engineering Thermodynamics*, John Wiley & Sons, Inc., New York, 1986.
4.   Reynolds, William C. and Henry C. Perkins, *Engineering Thermodynamics*, 2nd edition, McGraw-Hill, Inc., New York, 1977.
5.   Van Wylen, Gordon J. and Richard E. Sonntag, *Fundamentals of Classical Thermodynamics*, 3rd edition, John Wiley & Sons, Inc., New York, 1985.

Chapter 11

Refrigeration Cycles

The main function of refrigeration systems or reversed heat engines is to transfer heat continuously from a low temperature region to a high temperature region using energy from an external source. In the more common applications of refrigeration, like food preservation and air conditioning, the aim of the refrigeration system is to maintain the temperature of the cold space below the local ambient temperature. The same system, however, may be used to heat the high temperature region by using the heat extracted from the cold region. It is then called a heat pump.

Most household refrigerators operate on the vapor compression refrigeration cycle using mechanical work or electricity as the input. Vapor absorption systems, usually operated with a heat input, are being used in air conditioning systems. In this chapter we shall analyze several refrigeration cycles used in practical refrigeration systems including vapor compression cycles and vapor absorption cycles.

11.1 The Reversed-Carnot Cycle Using a Vapor

In chapter 6 we discussed the operation of the reversed Carnot heat engine cycle and derived an expression for its coefficient of performance (COP) (see worked example 7.3). We noted that the Carnot cycle has the highest COP of any cycle operating between given heat absorption and heat rejection temperatures. In this section we shall briefly review the operation of a reversed Carnot cycle using a vapor as the working fluid.

A schematic of the closed system that transfers heat from a low temperature region to a high temperature region is depicted in

Fig. 11.1(a). The *T-s* diagram of the reversed Carnot cycle in which all processes occur within the liquid-vapor region is shown in Fig. 11.1(b).

The working fluid enters the condenser as a saturated vapor at 1 and rejects heat to the high temperature region isothermally before exiting at 2 as a saturated liquid. The liquid undergoes an isentropic expansion 2-3 in an expander or turbine to produce a work output. During the evaporation process 3-4 the wet vapor absorbs heat from the low temperature region isothermally. Finally, the wet vapor is compressed from 4 to 1 in an isentropic process to complete the cycle. The work output of the expander (2-3) supplies a fraction of the work required by the compressor (4-1) while the rest is supplied from an external source.

We now apply the SFEE to each of the steady-flow processes of the cycle, neglecting the kinetic and potential energy of the fluid, to obtain the following expressions.

For the heat rejection process 1-2 in the condenser:

$$\dot{Q}_{12} = \dot{m}(h_1 - h_2) \tag{11.1}$$

where $\dot{m}$ is the steady mass flow rate of the working fluid.

For the heat absorption process 3-4 in the evaporator:

$$\dot{Q}_{34} = \dot{m}(h_4 - h_3) \tag{11.2}$$

Applying the first law to the cycle, the net work input becomes

$$\dot{W}_{net} = \dot{Q}_{12} - \dot{Q}_{34} \tag{11.3}$$

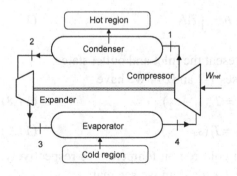

Fig. 11.1(a) Reversed-Carnot cycle

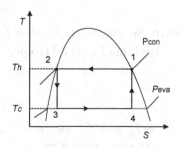

Fig. 11.1(b) *T-s* diagram

The coefficient of performance of the *refrigeration* cycle 1-2-3-4, whose purpose it is to absorb heat from the *cold* region, is given by

$$COP_r = \frac{\dot{Q}_{34}}{\dot{W}_{net}}$$
(11.4)

If the main purpose of the cycle is to supply heat to the hot region, then the cycle 1-2-3-4 is called a *heat pump* cycle, and its coefficient of performance is given by

$$COP_{hp} = \frac{\dot{Q}_{12}}{\dot{W}_{net}}$$

Manipulating Eqs. (11.1) to (11.4) we obtain

$$COP_r = \frac{(h_4 - h_3)}{(h_1 - h_2) - (h_4 - h_3)}$$
(11.5)

We relate the enthalpy changes in Eq. (11.5) to entropy changes by integrating the general thermodynamic property relation

$$dh = Tds - vdP$$
(11.6)

Note that the above relation follows directly from the 'T-ds' equation [Eq. (7.17)] when we substitute for u from the relation, $u = (h - Pv)$.

The pressure and temperature during the phase change processes 1-2 and 3-4 are constant. Therefore $dP = 0$ and $T = T_{sat}$, the saturation temperature.

Integrating Eq. (11.6) we have

$$h_o - h_i = \int_i^o Tds$$
(11.7)

where the subscripts i and o represent the inlet and outlet states.

Applying Eq. (11.7) to processes 1-2 and 3-4 we have

$$h_1 - h_2 = T_h(s_1 - s_2)$$
(11.8)

$$h_4 - h_3 = T_c(s_4 - s_3)$$
(11.9)

where T_h and T_c are the hot and cold region temperatures respectively. From the rectangular shape of the T-s diagram we see that

$$s_1 - s_2 = s_4 - s_3$$

Substituting in Eq. (11.5) from Eqs. (11.8) and (11.9) we obtain the familiar expression for the COP of the reversed Carnot refrigeration cycle,

$$COP_r = \frac{T_c}{T_h - T_c} \tag{11.10}$$

For the corresponding heat pump cycle we have

$$COP_{hp} = \frac{T_h}{T_h - T_c} = 1 + COP_r$$

11.2 The Vapor Compression Cycle

A number of practical difficulties have prevented the construction of a refrigerating system that operates on the reversed Carnot cycle using a vapor as the working fluid. The modifications done to the cycle shown in Fig. 11.1(b) to overcome these difficulties are shown in the *T-s* diagram depicted in Fig. 11.2(a).

Although, the isothermal processes 1-2 and 3-4 of the Carnot cycle may be well approximated in practice, the development of a practical compressor to carry out the *wet compression* process 4-1 poses several challenges. When a reciprocating compressor is used, liquid refrigerant may be trapped in the head of the cylinder by the piston which could result in valve damage. Ideally, there should only be dry vapor at the end of the compression. However, this may not be the case in practice

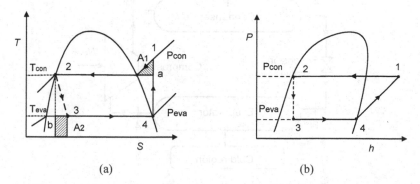

(a)   (b)

Fig. 11.2 Ideal vapor compression cycle. (a) *T-s* diagram, (b) *P-h* diagram

because of the finite time required for droplet evaporation. Moreover, the drops of liquid refrigerant may carry away lubricant from the walls of the cylinder, accelerating wear. Due to these reasons, the compression process in actual cycles is carried out in the *dry* vapor region (4-1) as shown in Fig. 11.2(a).

The second difference between the reversed Carnot cycle shown in Fig. 11.1(a), and the cycle shown in Fig. 11.2(a), is in the expansion process 2-3. The Carnot cycle requires the expansion to occur isentropically in the vapor-liquid mixture region. There are several difficulties when a wet vapor drives an expansion engine, which include the effect of droplets, and the carryover of lubricants. Moreover, the work output of the expander or turbine is only a small fraction of the work required by the compressor. For these reasons the expansion of the working fluid from the condenser pressure to the evaporator pressure is carried out in an *expansion valve*. This involves an irreversible throttling process for which the enthalpy is constant. Therefore, we have indicated the expansion process 2-3 by a broken-line in the *T-s* diagram in Fig. 11.2(a).

Most practical refrigeration systems operate on the revised cycle 1-2-3-4 shown in Fig. 11.2(a), which is called the *standard vapor-compression* cycle. A schematic diagram of the corresponding vapor compression refrigeration system is shown in Fig. 11.3.

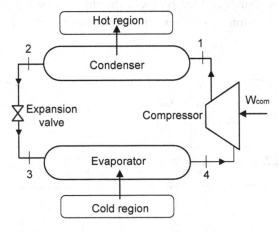

Fig. 11.3 Vapor compression refrigeration system

It is interesting to compare the performance of the reversed Carnot cycle 4-a-2-b with the vapor compression cycle 4-1-2-3 shown in Fig. 11.2(a). From Eq. (11.7) we observe that the area under a constant pressure line in the *T-s* diagram is equal to the difference in enthalpy of the states represented by the end points of the line.

Applying this condition we obtain the following relations between the heat interactions of the two cycles:

$$q_{12} = h_1 - h_2 = q_{c,out} + A_1 \qquad (11.11)$$

$$q_{34} = h_4 - h_3 = q_{c,in} - A_2 \qquad (11.12)$$

where $q_{c,out}$ and $q_{c,in}$ are respectively the heat rejected and absorbed per unit mass in the reversed Carnot cycle 4-a-2-b. A_1 and A_2 are the shaded areas shown in Fig. 11.2(a).

Applying the first law to the closed cycle 1-2-3-4 we have

$$w_{14} = q_{12} - q_{34} \qquad (11.13)$$

Substituting from Eqs. (11.11) and (11.12) in (11.13) we obtain

$$w_{14} = (q_{c,out} - q_{c,in}) + A_1 + A_2 = w_c + A_1 + A_2 \qquad (11.14)$$

The COP of the vapor compression cycle is given by

$$COP_r = q_{34} / w_{14} = (q_{c,in} - A_2)/(w_c + A_1 + A_2) \qquad (11.15)$$

The COP of the reversed Carnot cycle 4-a-2-b is

$$COP_c = q_{c,in} / w_c$$

Therefore

$$\frac{COP_r}{COP_c} = \frac{1 - A_2 / q_{c,in}}{1 + (A_1 + A_2)/w_c} \qquad (11.16)$$

We notice that the ratio of the COPs of the two cycles is a function of the areas A_1 and A_2. The area A_1, sometimes called the 'superheat horn', represents the additional work input required per unit mass due to superheating in the vapor compression cycle. The area A_2 represents the loss in refrigerating effect due to the use of a throttle valve instead of the isentropic expansion. From Eq. (11.14), we could also interpret area A_2 as the work lost due to throttling. For actual refrigerants, these areas depend on the shape of the saturation lines in the *T-s* diagram. Therefore we could use the magnitudes of these areas to compare graphically the impact of the properties of different refrigerants on the COP of the vapor compression cycle.

The pressure-enthalpy (*P-h*) diagram shown Fig. 11.2(b) is a common graphical means of representing refrigeration properties and is used widely for the analysis of vapor compression cycles. The heat rejection process, 1-2, the expansion process, 2-3 and the heat absorption process, 3-4 form three sides of a trapezium as seen in Fig. 11.2(b). Since the magnitudes of these heat interactions are related to the enthalpy differences, they can be read-off directly from the *P-h* chart of the refrigerant. Usually the *P-h* chart also includes constant temperature, constant entropy, and constant specific volume lines enabling these quantities to be obtained directly from the chart. However, unlike the *P-v* and *T-s* diagrams, the area of the *P-h* diagram does not have a special physical significance.

11.2.1 *Analysis of the vapor compression cycle*

We now apply the SFEE to each of the steady-flow processes of the vapor compression system shown in Fig. 11.3. Neglecting the kinetic and potential energy of the fluid, we obtain the following expressions.
For the heat rejection process 1-2 in the condenser:

$$\dot{Q}_{12} = \dot{m}(h_1 - h_2) \tag{11.17}$$

where $\dot{m}$ is the steady mass flow rate of the working fluid.
For the heat absorption process 3-4 in the evaporator:

$$\dot{Q}_{34} = \dot{m}(h_4 - h_3) \tag{11.18}$$

Note that $\dot{Q}_{34}$ is also called the *cooling capacity* of the cycle.

Treating the expansion process 2-3 through the valve as an adiabatic throttling process we have

$$h_2 = h_3 \tag{11.19}$$

Applying the first law to the cycle, the net work input becomes

$$\dot{W}_{net} = \dot{Q}_{12} - \dot{Q}_{34} \tag{11.20}$$

The coefficient of performance of the *refrigeration* cycle 1-2-3-4, whose purpose is to absorb heat from the *cold* region, is given by

$$COP_r = \frac{\dot{Q}_{34}}{\dot{W}_{net}}$$ (11.21)

Substituting from Eqs. (11.17) to (11.19) in Eq. (11.21) we obtain

$$COP_r = \frac{(h_4 - h_3)}{(h_1 - h_2) - (h_4 - h_3)}$$ (11.22)

$$COP_r = \frac{(h_4 - h_3)}{(h_1 - h_4)}$$ (11.23)

An additional design parameter of considerable practical importance is the theoretical compressor swept volume per unit time, which is called the compressor displacement. This is given by

$$\dot{V}_{com} = \dot{m}v_4$$

where v_4 is the specific volume of the refrigerant at entry to the compressor. The compressor displacement is a measure of the size of the compressor.

11.2.2 *Actual vapor compression cycle*

There are several differences between the ideal vapor compression cycle, a-b-c-d, and the actual cycle 1-2-3-4 shown in the *T-s* diagram in Fig. 11.4. Some of the differences are unavoidable while others are intentional. The pressure drops in the condenser and the evaporator are inevitable. These pressure drops tend to increase the pressure difference across the compressor, which in turn increases the required work input.

In the ideal cycle the liquid entering the expansion valve at b is just saturated. In actual practice, however, sub-cooling the liquid slightly to 2 is found to be advantageous. In the case of the common form of expansion valve, known as the capillary tube, the presence of any vapor at the entrance to the tube could cause flow blockage. Sub-cooling ensures that only liquid enters the expansion valve. It is also desirable to slightly superheat the vapor at 4 to ensure that no liquid drops are present in the vapor entering the compressor.

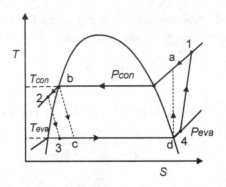

Fig. 11.4 *T-s* diagram of the actual vapor compression cycle

In some practical refrigeration systems superheating of the vapor from d-4 is carried out in a counter-flow heat exchanger using the saturated liquid leaving the condenser at b. The liquid leaving the heat exchanger is sub-cooled to state 2. Applying the SFEE to the heat exchanger we have

$$h_4 - h_d = h_b - h_2$$

It is clear that the heat absorbed by the refrigerant per unit mass in the evaporator is increased from $(h_d - h_c)$ to $(h_d - h_3)$ due to the sub-cooling of the liquid. However, when vapor at 4 enters the compressor, the refrigerant flow rate is decreased due to the lower vapor density of the superheated vapor at 4 compared to saturated vapor density at d. Although the overall thermodynamic advantage of including a heat exchanger in the cycle is marginal, the practical reasons mentioned above justify its use. In worked example 11.6 we illustrate the analysis of a vapor compression system including a heat exchanger.

The final important difference between the actual and ideal cycles is that the compression process in the actual cycle is not isentropic as assumed in the ideal cycle. This could be readily accounted for by assuming the compression process to be adiabatic and defining the isentropic efficiency of the compressor as the ratio of the isentropic work input to the actual work input.

The various differences mentioned above relate mainly to the internal processes of the vapor compression cycle. In addition there are two important external factors that affect the performance of the cycle.

For heat to flow from the cold region to the working fluid in the evaporator, and from the working fluid to the hot region in the condenser, there has to be finite temperature differences (see Fig. E11.7). These temperature differences constitute external irreversibilities that diminish the COP of the cycle as we observed in our studies related to the second law in chapter 6. A common practical situation is considered in worked example 11.7.

11.3 Modifications to the Vapor Compression Cycle

The single-stage vapor compression cycle considered thus far is an efficient method of refrigeration when the evaporating temperatures are relatively high. As the evaporating temperature is reduced, the power required by the compressor increases and the COP of the cycle decreases. Moreover, the compressor displacement, which is a measure of the size of the compressor, and the discharge temperature increase as the evaporator temperature is reduced. The use of a multi-stage compressor with intercooling can mitigate some of these detrimental effects at low evaporator temperatures.

11.3.1 *Two-stage compression with flash inter-cooling*

Figure 11.5(a) shows an arrangement with two compressors and a flash-intercooler. The *P-h* diagram of the system is depicted in Fig. 11.5(b).

The saturated liquid leaving the condenser at 2 is first expanded to an intermediate pressure and enters the flash intercooler at state 3. The liquid from intercooler at state 6 is then expanded through the expansion valve 2 to the evaporator pressure. The low-pressure vapor leaving the evaporator at state 8 is compressed by the low-pressure compressor to state 5 before entering the intercooler. The expansion valve 1 also functions as a float-valve to maintain a constant liquid level in the intercooler. Ideally, the vapor entering the high-pressure compressor at 4 is a saturated vapor at the intermediate pressure which is also the pressure of the intercooler. Finally, the vapor is compressed to state 1 and enters the condenser to complete the cycle.

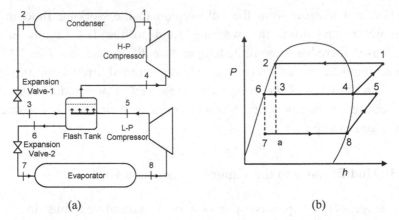

(a)   (b)

Fig. 11.5 Two-stage compression with flash-intercooling. (a) Schematic diagram, (b) *P-h* diagram

If a single compressor was used with the same condenser and evaporator pressures the flow through the expansion valve would take place from 2 to a. We observe that during this expansion, the fraction of vapor in the expanding mixture, usually called flash-gas, increases progressively. Unfortunately, in the evaporator the flash-gas makes no contribution to the heat absorption process, but work is needed to recompress the gas back to the condenser pressure to complete the cycle.

Ideally, if the flash-gas in the expanding mixture was to be removed continuously, the compressor work input would be significantly reduced. The single flash tank in the system shown in Fig. 11.5 facilitates the removal of a part of the flash-gas generated. Generally, the reduction in compressor work due to the flash-intercooler depends on the type refrigerant used in the cycle. We also notice that the higher liquid fraction at 7 increases the refrigeration capacity of the system. A vapor compression refrigeration system including a flash-intercooler is analyzed in worked example 11.9.

11.3.2 *Two-stage compression with two evaporators*

Refrigeration systems that have two evaporators operating at different temperatures are common in industrial applications. For instance, in a

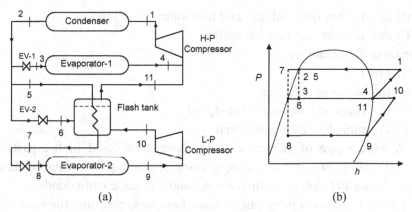

Fig. 11.6 Two-stage compression with two evaporators. (a) Schematic diagram, (b) *P-h* diagram

food processing plant, a low-temperature evaporator may be used to rapidly freeze the food, while an evaporator at a higher temperature may hold the frozen food. The system shown is Fig. 11.6 is an arrangement with two evaporators and a flash-intercooler. The *P-h* diagram shown in Fig. 11.6(b) shows the various state points. The application of a vapor compression refrigeration system with two evaporators is illustrated in worked example 11.10.

11.4 Refrigerants for Vapor Compression Systems

The following are some of the desirable characteristics of a substance that could be used as a refrigerant in a practical vapor compression system:

(i) a positive evaporating pressure to prevent leakage of ambient air into the system,

(ii) a relatively high critical pressure to enable operation in the liquid-vapor region of the phase diagram,

(iii) a low freezing point to operate at low evaporating temperatures,

(iv) a high latent enthalpy of evaporation to increase the heat absorbed per unit mass,

(v) thermo-physical properties that facilitate high heat transfer,

(vi) inertness and stability,

(vii) satisfactory oil solubility and low water solubility,
(viii) non-toxicity and non-irritability,
(ix) non-flammability,
(x) low-cost,
(xi) easy leakage detection,
(xii) low ozone-depletion potential, and
(xiii) low global-warming potential.

A wide range of substances have been considered in the past as possible refrigerants for use in vapor compression systems. These include halocarbons, azeotropes, hydrocarbons, and inorganic compounds.

(i) Halocarbons contain one or more halogens, chlorine, fluorine, and bromine. The names and chemical formulae of the more common halo-carbon refrigerants are: Refrigerant 11 or R-11 ($CFCl_3$), Refrigerant 12 or R-12 (CF_2Cl_2), Refrigerant 22 or R-22 ($CHClF_2$), Refrigerant 134a or R-134a ($CH_2F - CF_3$).

During the early days following their introduction, halocarbons were hailed as excellent refrigerants with most of the desirable characteristics mentioned above (i to xi). Unfortunately, in recent years a number of the widely used halocarbons have been found to cause ozone layer depletion in the outer atmosphere and also contribute to global warming. As a consequence their use as refrigerants has been discontinued by inter-national agreement. In the search for environmentally safe halocarbon refrigerants, R134a and R22 have emerged as the more satisfactory substances.

(ii) Azeotropes are mixtures of substances that evaporate and condense at a single temperature. Refrigerants with desirable properties have been produced by using existing refrigerants to form azeotropic mixtures. For example, refrigerant, R-502 that is used in industrial and commercial refrigeration systems is an azeotropic mixture of R-22 and R-115.

(iii) Hydrocarbons such as ethane (R-50), methane (R-170) and propane (R-290) have been used as refrigerants in industrial applications. where the safety procedures needed to deal with their flammability can be implemented satisfactorily.

(iv) Of the inorganic substances, ammonia has been widely used as a refrigerant in industrial applications. It has most of the desirable characteristics mentioned above. However, ammonia mixes easily with water and is a strong irritant. Therefore it has not been used widely in domestic applications.

11.5 The Vapor Absorption Cycle

The vapor absorption refrigeration cycle operates with heat as the main input energy form compared to the vapor compression system which requires work to drive the compressor. In Figs. 11.7(a) and (b) we have shown the two cycles side-by-side for easy comparison. The main difference is that the compressor in (a) is replaced by the unit within the broken-line boundary in (b), consisting of the absorber, the liquid pump, the generator, and the pressure reducing valve (PRV). As in the vapor compression cycle the heat absorption from the cold space occurs in the evaporator. The refrigerant vapor leaving the evaporator then enters the absorber where it is absorbed in a liquid called the absorbent. The absorbent solution, rich in refrigerant, is then pumped to the generator which is at the condenser pressure. In the generator the solution is heated using an external heat input to boil-off the refrigerant. The vapor is condensed in the condenser and returned to the evaporator through the expansion valve as in the vapor compression cycle.

In the meanwhile, the solution weak in refrigerant, flows back from the generator to the absorber through a pressure reducing valve (PRV) to complete the cycle. If the absorption of vapor occurs adiabatically in

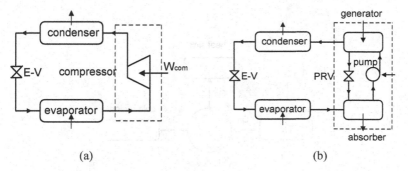

(a)   (b)

Fig. 11.7 Comparison of compression and absorption cycles, (a) vapor compression cycle, (b) vapor absorption cycle

the absorber, then the heat of absorption would increase the solution temperature, impeding further vapor absorption. Therefore the absorber is cooled using an external source of cold fluid. The main advantage of the absorption cycle is that the work input to the solution pump is only small fraction of the work required by the compressor in the vapor compression cycle. Moreover, for the absorption system the required heat input in the generator may be provided with a gas burner, a waste-heat stream or solar energy.

11.5.1 *The three-heat-reservoir-model*

Before we analyze the vapor absorption cycle with actual working fluids it is instructive to obtain an expression for the COP of the ideal cycle based on the three-heat-reservoir model. We observe from Fig. 11.7(b) that the absorption refrigerator may be modeled as a cyclic device exchanging heat with three reservoirs.

The refrigerated cold space is the low temperature reservoir at temperature T_c while the generator constitutes the high temperature reservoir at T_h. Since the condenser and the absorber are usually cooled by rejecting heat to the atmosphere, we assume it to be the third reservoir at an intermediate temperature T_o.

The work input to the liquid pump is assumed to be negligible. The ideal absorption system is the cyclic device where all the processes including the heat interactions with the three reservoirs are reversible. A schematic of this ideal absorption cycle is shown in Fig. 11.8.

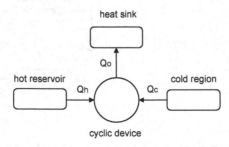

Fig. 11.8 Three-heat-reservoir model of the absorption cycle

Applying the first law to the cyclic system operating in a steady manner we have

$$Q_c + Q_h - Q_o = 0 \qquad (11.24)$$

We now apply the second law in the form of the entropy balance equation. The entropy change and storage in the cyclic device are zero. The entropy changes occur only in the three reservoirs. Therefore

$$\frac{Q_o}{T_o} - \frac{Q_c}{T_c} - \frac{Q_h}{T_h} = 0 \qquad (11.25)$$

The COP of the absorption cycle is defined as the ratio of the heat absorbed from the cold reservoir to the heat supplied by the hot reservoir.

$$COP_a = \frac{Q_c}{Q_h} \qquad (11.26)$$

Manipulating Eqs. (11.24) to (11.26) we obtain

$$COP_a = \left(\frac{T_h - T_o}{T_h}\right)\left(\frac{T_c}{T_o - T_c}\right) \qquad (11.27)$$

The right-hand-side of Eq. (11.27) may be interpreted as the product of the efficiency of a Carnot engine operating between T_h and T_o and the COP of a Carnot refrigerator operating between T_c and T_o.

11.5.2 *Analysis of the actual absorption cycle*

The most common refrigerant-absorbant fluid pairs used in actual absorption systems are water-lithium bromide (LiBr) and ammonia-water. Water-LiBr is used mainly in air conditioning applications where the evaporator temperatures are above the freezing temperature of water. These systems are easier to analyze because LiBr remains in the liquid state during entire operation and pure water vapor acts as the refrigerant. In ammonia-water systems, used in low-temperature applications, both the liquid phase and the vapor phase contain ammonia and water. In the following section we shall consider the conditions of equilibrium for water-LiBr mixtures.

11.5.3 *Equilibrium of water-LiBr mixtures*

Consider the absorption refrigeration cycle shown in Fig. 11.7(b). The condenser and the generator are at different temperatures but ideally at the same pressure. Therefore the LiBr-water solution in the generator is in equilibrium with the pure water and vapor in the condenser. Similarly, the solution in the absorber is at the same pressure as the water vapor in the evaporator although their temperatures are different.

We shall discuss these conditions of equilibrium by referring to the piston-cylinder apparatus shown in Figs. 11.9(a) and (b). In Fig. 11.9(a) pure water and vapor are in equilibrium under the constant pressure applied by the fixed load on the piston. We noted in chapter 2 that when vapor is in equilibrium with pure water, the temperature is a unique function of the applied pressure.

Figure 11.9(b) shows an identical piston-cylinder arrangement which now contains a solution of pure water and the salt, LiBr in equilibrium

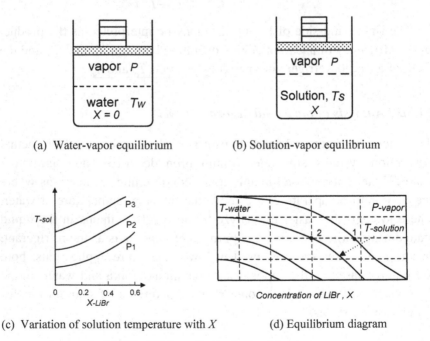

(a) Water-vapor equilibrium   (b) Solution-vapor equilibrium

(c) Variation of solution temperature with X (d) Equilibrium diagram

Fig. 11.9 Equilibrium of water-LiBr mixtures

with pure water vapor above it. The same pressure is applied on the system by the load on the piston. Unlike pure water the solution has an additional variable property called the concentration X, defined as the mass of LiBr per unit mass of solution, that affects the equilibrium. It is found that for the same pressure P, the vapor can be in equilibrium with the solution for different combinations of the solution temperature T_s and the concentration X.

In Fig. 11.9(c), the solution temperature is plotted against the concentration for three different values of the applied pressure P. We notice that when $X = 0$, the temperature is equal to the saturation temperature of pure water at the applied pressure. As the concentration of the solution increases, the equilibrium temperature increases monotonically.

The equilibrium relation between the solution temperature, the solution concentration and the water temperature may be represented in the compact-form shown graphically in Fig. 11.9(d). The vertical axis on the left is the saturation temperature of pure water and the corresponding saturation vapor pressure is indicated in the vertical axis on the right hand side. This data may also be obtained from the steam tables. The horizontal axis gives the mass concentration, X of LiBr in the solution. The family of curves represent different solution temperatures.

Note that the states of the solution indicated by 1 and 2 in Fig. 11.9(d) are both in equilibrium with the same state of pure water. The actual equilibrium diagram for LiBr-water is available in [5]. The specific enthalpy of the solution is a function of the solution concentration and temperature. This variation is presented in a compact graphical form available in [5]. The absorption cycle can be analyzed in a straightforward manner by applying the mass balance equation and the SFEE to each component, as we shall illustrate in the worked examples 11.12 and 11.13. The property data required are obtained directly from the graphs in [5].

11.6 The Air-Standard Refrigeration Cycle

The air-standard refrigeration cycle is essentially the reverse of the Brayton cycle. The main applications of the cycle are in the liquefaction of gases and the cooling of aircraft cabins.

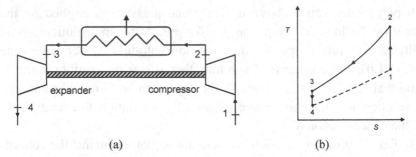

Fig. 11.10 Air cycle refrigeration: (a) Schematic diagram, (b) *T-s* diagram

A simplified form of the air standard refrigeration system used for the cooling of aircraft cabins is shown in Fig. 11.10. It is an open cycle where ambient air enters the compressor at 1 and is compressed isentropically to state 2. Constant pressure heat rejection from 2 to 3 brings the temperature ideally close to the ambient temperature. The air expands isentropically in the expander or turbine to a low temperature before it is ducted to the cabin. A fraction of the power required by the compressor is supplied by the turbine. The heat interactions in the reversed Brayton cycle occur over a range of temperatures whereas in the vapor compression cycle the evaporation and condensation processes occur at constant temperatures. Therefore the external heat transfer irreversibility could be larger when air is used as the working fluid in the refrigeration cycle compared to a vapor.

11.6.1 *The air-standard refrigeration cycle with a heat exchanger*

An air-cycle refrigeration system incorporating a counter-flow heat exchanger is shown in Fig. 11.11. With this arrangement it is possible to achieve very low temperatures of the refrigerated space. In the ideal cycle, air enters the compressor at 1 and is compressed to 2. It then rejects heat to a heat sink at T_h and cools to state a. In the heat exchanger, the air cools further to state 3 by transferring heat to the air stream leaving the refrigerated space at b.

In the turbine the air undergoes an isentropic expansion from 3 to 4. The cold air leaving the turbine (expander) at 4 absorbs heat from the

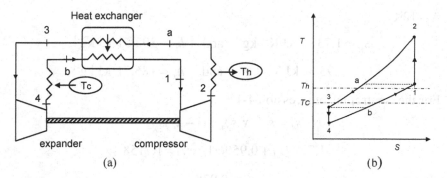

Fig. 11.11 Air cycle refrigeration system with heat exchanger: (a) Schematic diagram, (b) *T-s* diagram.

refrigerated space at temperature T_c. In the heat exchanger the cold air entering at b is heated to state 1 at constant pressure before entering the compressor. We notice from Fig. 11.11(b) that for the same heat sink temperature T_h and the same pressure ratio, the inclusion of the heat exchanger helps to lower the air temperature at state 4 significantly. Two air-cycle refrigeration systems are analyzed in worked examples 11.14 and 11.15.

11.7 Worked Examples

Example 11.1 The evaporating and condensing temperatures of a Carnot refrigeration cycle using refrigerant-134a as the working fluid are -10°C and 25°C respectively. The refrigerant flow rate is 0.15 kg s⁻¹. Calculate (i) the cooling capacity, (ii) the net work input, and (iii) the COP of the cycle.

Solution The *T-s* diagram of a reversed Carnot cycle operating with a vapor is shown in Fig. 11.1(b). We obtain the following properties of R134a from tabulated data in [4].

At 25°C,

$$s_g = 1.7158 \text{ kJ K}^{-1} \text{ kg}^{-1} \quad \text{and} \quad h_g = 412.23 \text{ kJ kg}^{-1},$$

$$s_f = 1.1198 \text{ kJ K}^{-1} \text{ kg}^{-1} \quad \text{and} \quad h_f = 234.52 \text{ kJ kg}^{-1}$$

At -10°C,

$$s_g = 1.7327 \text{ kJ K}^{-1} \text{ kg}^{-1} \quad \text{and} \quad h_g = 392.51 \text{ kJ kg}^{-1},$$

$$s_f = 0.9506 \text{ kJ K}^{-1} \text{ kg}^{-1} \quad \text{and} \quad h_f = 186.71 \text{ kJ kg}^{-1}$$

For the isentropic compression 4-1

$$s_4 = s_1 = x_4 s_{g4} + (1 - x_4) s_{f4}$$

$$1.7327 x_4 + 0.9506(1 - x_4) = 1.7158$$

Hence $x_4 = 0.9784$

$$h_4 = x_4 h_{g4} + (1 - x_4) h_{f4}$$

$$h_4 = 0.9784 \times 392.51 + (1 - 0.9784) \times 186.71 = 388.06 \text{ kJ kg}^{-1}$$

For the isentropic expansion 2-3

$$s_2 = s_3 = x_3 s_{g3} + (1 - x_3) s_{f3}$$

$$1.7327 x_3 + 0.9506(1 - x_3) = 1.1198$$

Hence $x_3 = 0.2163$

$$h_3 = x_3 h_{g3} + (1 - x_3) h_{f3}$$

$$h_3 = 0.2163 \times 392.51 + (1 - 0.2163) \times 186.71 = 231.24 \text{ kJ kg}^{-1}$$

The refrigeration capacity is

$$\dot{Q}_{34} = \dot{m}_r (h_4 - h_3)$$

$$\dot{Q}_{34} = 0.15 \times (388.06 - 231.24) = 23.52 \text{ kW.}$$

The net work input to the compressor is

$$\dot{W}_{net} = \dot{W}_{41} - \dot{W}_{23} = \dot{m}_r (h_1 - h_4) - \dot{m}_r (h_2 - h_3)$$

$$\dot{W}_{net} = 0.15 \times (412.23 - 388.06) - 0.15 \times (234.52 - 231.24)$$

Hence $\dot{W}_{net} = 3.1335 \text{ kW}$

The COP of the cycle is given by

$$COP = \dot{Q}_{34} / \dot{W}_{net} = 23.52 / 3.1335 = 7.51$$

The general expression for the COP of a Carnot refrigerator gives

$$COP = T_c / (T_h - T_c) = (273 - 10)/(25 + 10) = 7.51$$

Example 11.2 A standard vapor compression cycle using R134a as the working fluid has a condensing temperature of 30°C and an evaporating temperature of -5°C. The refrigeration capacity is 10 kW. Calculate (i) the refrigerant flow rate, (ii) the compressor work input, (iii) the volume flow rate of refrigerant at the compressor inlet, and (iv) the COP.

Solution The *T-s* and *P-h* diagrams of a standard vapor compression cycle are shown in Fig. 11.2(a) and (b) respectively. We obtain the following data for R134a from [4].

At -5°C, $s_4 = 1.7294$ kJ K^{-1} kg^{-1}, $h_4 = 395.49$ kJ kg^{-1}

At 30°C, $h_2 = 241.69$ kJ kg^{-1}.

For the isentropic compression 4-1, $s_1 = s_4 = 1.7294$ kJ K^{-1} kg^{-1}.

From the data in [4], it follows that the vapor is superheated at state 1. To calculate the enthalpy at state 1 we extract the following data from the tables in [4]:

s	h
1.7142	414.74
1.7294	h_1 (?)
1.7482	425.21

By linear interpolation, $h_1 = 419.42$ kJ kg^{-1}
For the throttling process 2-3

$$h_3 = h_2 = 241.69 \text{ kJ kg}^{-1}$$

The refrigeration capacity is

$$\dot{Q}_{34} = \dot{m}_r(h_4 - h_3)$$
$$\dot{m}_r(395.49 - 241.69) = 10 \text{kW}$$

Hence the refrigerant flow rate is $\dot{m}_r = 0.065$ kg s^{-1}
The work input to the compressor is

$$\dot{W}_{41} = \dot{m}_r(h_1 - h_4)$$
$$\dot{W}_{41} = 0.065 \times (419.42 - 395.49) = 1.555 \text{ kW}$$

From the data in [4] the specific volume at 4 is, $v_4 = 0.08273$ m³ kg^{-1}
The volume flow rate of the refrigerant at 4 is

$$\dot{V}_4 = v_4 \dot{m}_r = 0.08273 \times 0.065 = 5.5377 \times 10^{-3} \text{ m}^3 \text{ s}^{-1}$$

The COP of the cycle is

$$COP = \dot{Q}_{34}/\dot{W}_{41} = 10/1.555 = 6.43$$

For comparison we calculate the COP of the reversed Carnot cycle operating at the same condensation and evaporation temperatures as

$$COP = T_c/(T_h' - T_c) = (273 - 5)/(30 + 5) = 7.66$$

Example 11.3 The evaporator and condenser temperatures of a standard vapor compression cycle using R134a as the refrigerant are -5°C and 30°C respectively. The isentropic efficiency of the adiabatic compressor is 0.8, and the refrigeration capacity is 10 kW. Calculate (i) the refrigerant flow rate, (ii) the compressor work input, and (iii) the COP.

Solution The *T-s* diagram of the vapor compression cycle is shown in Fig. E11.3. For purposes of comparison we use the same condensing and evaporating temperatures as in worked example 11.2. For the irreversible adiabatic compression process 4-1 we define the isentropic efficiency as

$$\eta_i = \frac{w_{iso}}{w_{act}} = \frac{h_a - h_4}{h_1 - h_4}$$

In worked example 11.2 we obtained the data for the isentropic compression 4-a, which we shall use here. Substituting in the above equation we have

$$\frac{w_{iso}}{w_{act}} = \frac{419.42 - 395.49}{h_1 - h_4} = 0.8$$

Hence $\qquad h_1 - h_4 = 29.91 \text{ kJ kg}^{-1}$

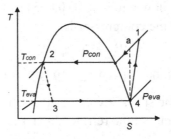

Fig. E11.3 *T-s* diagram

Now the cooling capacity and the refrigerant flow rate are the same as in example 11.2 because the state points 2, 3 and 4 are unchanged. The work input to the compressor is

$$\dot{W}_{41} = \dot{m}_r(h_1 - h_4) = 0.065 \times 29.91 = 1.944 \text{ kW}$$

The COP of the cycle is

$$COP = \dot{Q}_{34} / \dot{W}_{41} = 10/1.944 = 5.14$$

As expected, the COP is decreased due to the irreversibility of the compressor.

Example 11.4 A vapor compression cycle using R134a as the refrigerant operates with evaporating and condensing temperatures of -5°C and 30°C. Calculate (i) the Carnot-cycle compression work, (ii) the excess work due to the 'superheat horn', (iii) the loss of work due to throttling, and (iv) the loss in refrigeration capacity due to throttling.

Solution The work loss due to the 'superheat horn' and the loss in refrigeration capacity due to throttling are indicated by the shaded areas of the *T-s* diagram shown in Fig. E11.4. Now the area A_1 = area (1cde)-area (acde). We showed earlier that the area under a constant pressure line on the *T-s* diagram is equal to the change in enthalpy [Eq. (11.7)].

Therefore area, $1\text{cde} = h_1 - h_c$, and the rectangular area, acde $= T_c(s_a - s_c)$. Note that the evaporating and condensing temperatures of this example are the same as those in worked example 11.2. Hence

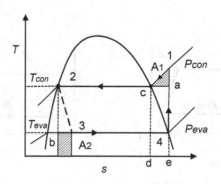

Fig. E11.4 *T-s* diagram

$$h_1 = 419.42 \text{ kJ kg}^{-1}, \qquad\qquad h_c = 414.74 \text{ kJ kg}^{-1},$$

$$s_a = s_4 = 1.7294 \text{ kJ K}^{-1} \text{ kg}^{-1}, \qquad s_c = 1.7142 \text{ kJ K}^{-1} \text{ kg}^{-1},$$

$$T_c = 273 + 30 = 303 \text{ K}.$$

Now $$A_1 = (h_1 - h_c) - T_c(s_a - s_c)$$

$$A_1 = (419.42 - 414.74) - 303 \times (1.7294 - 1.7142) = 0.0744 \text{ kJ kg}^{-1}$$

For the isentropic expansion 2-b, $s_2 = s_b$. Therefore

$$1.1434 = 0.9754(1 - x_b) + 1.7294 x_b, \qquad \text{Hence } x_b = 0.2228$$

$$h_b = 193.32(1 - x_b) + 395.49 x_b = 238.37 \text{ kJ kg}^{-1}$$

The loss of refrigeration capacity due to throttling is given by A_2 where

$$A_2 = h_3 - h_b = h_2 - h_b$$

$$A_2 = 241.69 - 238.37 = 3.326 \text{ kJ kg}^{-1}$$

From Eq. (11.14), the loss of work due to throttling is also equal to A_2.

Example 11.5 The evaporator and condenser temperatures of a vapor compression cycle using R134a as the refrigerant are -5°C and 30°C. The refrigerant leaving the condenser is subcooled to 25°C while the vapor entering the compressor is superheated to 5°C. The cooling capacity is 10 kW. Calculate (i) the refrigerant flow rate, (ii) the compressor work input, and (iii) the COP.

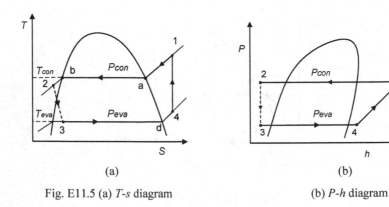

Fig. E11.5 (a) *T-s* diagram   (b) *P-h* diagram

Solution The *T-s* and *P-h* diagrams of the cycle are depicted in Figs. E11.5(a) and (b) respectively. We obtain the following properties of R-134a from the tables in [4].

For the vapor at state 4 that is superheated by 10°C,

$$s_4 = 1.7614 \text{ kJ K}^{-1} \text{ kg}^{-1} \quad \text{and} \quad h_4 = 404.25 \text{ kJ kg}^{-1}.$$

For the sub-cooled liquid at 2, we ignore the effect of pressure and obtain the saturated liquid enthalpy at 25°C as, $h_2 = 234.52 \text{ kJ kg}^{-1}$.

Now for the isentropic compression 4-1, $s_1 = s_4$.

The vapor at 1 is superheated. To obtain the enthalpy at 1 we extract the following data from the table in [4]

s	h
1.7482	425.21
1.7614	h_1 (?)
1.7803	435.44

By linear interpolation, $h_1 = 429.42 \text{ kJ kg}^{-1}$

For the throttling process 2-3,

$$h_3 = h_2 = 234.52 \text{ kJ kg}^{-1}$$

The refrigeration capacity is

$$\dot{Q}_{34} = \dot{m}_r (h_4 - h_3)$$
$$\dot{m}_r (404.25 - 234.52) = 10 \text{ kW}$$

Hence the refrigerant flow rate is, $\dot{m}_r = 0.0589 \text{ kg s}^{-1}$

The work input to the compressor is

$$\dot{W}_{41} = \dot{m}_r (h_1 - h_4)$$
$$\dot{W}_{41} = 0.0589 \times (429.42 - 404.25) = 1.4825 \text{ kW}$$

The COP of the cycle is

$$COP = \dot{Q}_{34} / \dot{W}_{41} = 10 / 1.4825 = 6.75$$

Example 11.6 A vapor compression system using R134a as the refrigerant includes a heat exchanger, as shown in Fig. E11.6(a), to superheat the vapor leaving the evaporator using the saturated liquid leaving the condenser. The condensing and evaporating temperatures are 30°C and −5°C respectively. The vapor leaving the evaporator is heated to 5°C. The compression process is isentropic. Calculate the COP of the cycle with and without the heat exchanger.

Solution The schematic diagram and the *P-h* diagram of the cycle are shown in Figs. E11.6(a) and (b) respectively. We obtain the following data for R134a from the tables in [4].

For the vapor at state 4 that is superheated by 10°C,

$$s_4 = 1.7614 \text{ kJ K}^{-1} \text{ kg}^{-1} \quad \text{and} \quad h_4 = 404.25 \text{ kJ kg}^{-1}$$

For the saturated liquid state, a and the saturated vapor state, b,

$$h_a = 241.69 \text{ kJ kg}^{-1} \quad \text{and} \quad h_b = 395.49 \text{ kJ kg}^{-1}$$

Apply the SFEE to the heat exchanger, neglecting the kinetic and potential energy of the fluid. Thus

$$\dot{m}_r(h_a - h_2) = \dot{m}_r(h_4 - h_b)$$

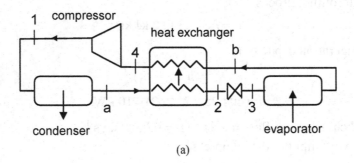

(a)

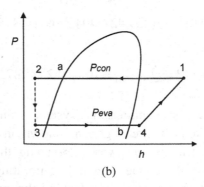

(b)

Fig. E11.6 (a) Refrigeration cycle with a heat exchange, (b) *P-h* diagram

Substituting numerical values in the above equation we have

$$h_2 = 232.93 \text{ kJ kg}^{-1}$$

For the throttling process,

$$h_3 = h_2 = 232.93 \text{ kJ kg}^{-1}$$

The refrigeration effect per unit mass is

$$q_r = h_b - h_2 = 395.49 - 232.93 = 162.56 \text{ kJ kg}^{-1}.$$

Note that the energy needed for superheating the refrigerant from b to 4 is supplied internally, and therefore does not contribute to the refrigeration effect as in example 11.5.

The work input to the compressor per unit mass is

$$w_c = h_1 - h_4 = 429.42 - 404.25 = 25.17 \text{ kJ kg}^{-1}$$

The COP is given by

$$COP = q_r/w_c = 162.56/25.17 = 6.46$$

In worked example 11.2, the COP of the cycle with the same conditions, but without the heat exchanger, was obtained as 6.43. We note that the increase in the COP due to the inclusion of the heat exchanger is marginal. However, the practical advantages of the heat exchanger are: (a) only liquid enters the expansion valve, and (b) the vapor entering the compressor has no liquid phase.

Example 11.7 A standard vapor compression cycle using R134a as the working fluid is used to produce chilled-water in an air conditioning plant. The condensing and evaporating temperatures of the cycle are 35°C and 5°C respectively. The chilled-water enters the evaporator at 20°C and leaves at 10°C. The flow rate of chilled-water is 0.2 kg s^{-1}. The condenser is cooled with water that enters at 23°C and leaves at 29°C. Calculate (i) the flow rate of refrigerant in the cycle, and (ii) the flow rate of condenser cooling water.

Solution Figure E11.7 shows the *T-s* diagram of the vapor compression cycle where the distribution of chilled-water and condenser cooling water temperatures are also indicated. The chilled-water flowing counter to the

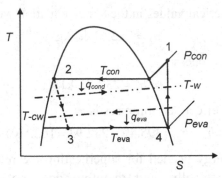

Fig. E11.7 T-s diagram

refrigerant in the evaporator is cooled by the evaporating refrigerant. In the condenser the cooling water is heated due to the heat absorption from the condensing refrigerant.

We obtain the following data for R134a from [4]:

At 5°C, $s_4 = 1.7238$ kJ K^{-1} kg^{-1}, $h_4 = 401.33$ kJ kg^{-1}

At 35°C, $h_2 = 248.98$ kJ kg^{-1}

For the throttling process, 2-3,

$$h_3 = h_2 = 248.98 \text{ kJ kg}^{-1}$$

Applying the SFEE to the evaporator we have

$$\dot{m}_r(h_4 - h_3) = \dot{m}_{cw}c_w(T_{cwi} - T_{cwo})$$

The specific heat capacity of water is 4.2 kJ K^{-1} kg^{-1}. Substituting numerical values in the above equation

$$\dot{m}_r(401.33 - 248.98) = 0.2 \times 4.2 \times (20 - 10)$$

Therefore the refrigerant flow rate is, $\dot{m}_r = 0.055$ kg s^{-1}.

For the isentropic compression 4-1, $s_1 = s_4 = 1.7238$ kJ K^{-1} kg^{-1}.

The vapor is superheated at state 1. To calculate the enthalpy at state 1 we extract the following data from the tables in [4]:

s	h
1.7126	417.14
1.7238	h_1 (?)
1.7470	427.93

By linear interpolation, $h_1 = 420.65$ kJ kg^{-1}

Applying the SFEE to the condenser we have

$$\dot{m}_r(h_1 - h_2) = \dot{m}_w c_w (T_{wo} - T_{wi})$$

$$0.055 \times (420.65 - 248.98) = \dot{m}_w \times 4.2 \times (29 - 23)$$

Therefore the condenser cooling water flow rate is, $\dot{m}_w = 0.3746\,\text{kg s}^{-1}$.

Example 11.8 A heat pump operates on the vapor compression cycle using R134a as the working fluid. The refrigerant flow rate is 0.15 kg s⁻¹. The evaporating and condensing temperatures are 10°C and 60°C respectively. The heat pump supplies heat at a constant rate to a tank of water of mass 1000 kg and specific heat capacity 4.2 kJ K⁻¹ kg⁻¹. (i) Calculate the time required to heat the water from 20°C to 40°C. (ii) If the heat pump is replaced with an electrical resistance heater that uses the same rate of energy input as the compressor of the heat pump, calculate the time taken to heat the water from 20°C to 40°C.

Solution The heat pump cycle is identical to the standard vapor compression cycle shown in Fig. 11.3. The heat rejected in the condenser is used to heat the water in the tank. We obtain the following data for R134a from the tables in [4]:

At 10°C, $\quad s_4 = 1.7215\,\text{kJ K}^{-1}\,\text{kg}^{-1}, \quad h_4 = 404.16\,\text{kJ kg}^{-1}$

At 60°C, $\quad h_2 = 287.15\,\text{kJ kg}^{-1}$

For the isentropic compression 4-1, $s_1 = s_4 = 1.7238\,\text{kJ K}^{-1}\,\text{kg}^{-1}$.

The vapor is superheated at state 1. To calculate the enthalpy at state 1 we extract the following data from the tables in [4]:

s	h
1.7026	426.69
1.7215	h_1 (?)
1.7412	439.77

By linear interpolation, $\quad h_1 = 433.09\,\text{kJ kg}^{-1}$

For the throttling process 2-3

$$h_3 = h_2 = 241.69\,\text{kJ kg}^{-1}$$

The rate of heat rejection in the condenser is

$$\dot{Q}_{12} = \dot{m}_r(h_1 - h_2) = 0.15 \times (433.09 - 287.15) = 21.89\ \text{kW}.$$

Apply the first law to the water tank, neglecting any heat losses and assuming that the rate of heat input is constant. Thus we have

$$M_w c_w (dT_w / dt) = \dot{Q}_{12} = 21.89 \text{ kW}$$

Integrating the above equation

$$M_w c_w (T_{wf} - T_{wi}) = 21.89\tau$$

The time to heat the water in the tank is

$$\tau = 1000 \times 4.2 \times (40 - 20) / 21.89 = 1.066 \text{ hours}$$

The work input rate in the compressor is

$$\dot{W}_{41} = \dot{m}_r (h_1 - h_4) = 0.15 \times (433.09 - 404.16) = 4.339 \text{ kW}$$

When a resistance heater is used, the time taken to heat the water is

$$\tau = 1000 \times 4.2 \times (40 - 20) / 4.339 = 5.38 \text{ hours.}$$

The heat pump consumes 4.63 kWh while the resistance heater requires an energy input of 23.3 kWh. We note that the heat pump is a more energy efficient method of heating compared to the use of a resistance heater with the same rate of electrical energy input.

Example 11.9 A vapor compression cycle with a flash-intercooler using ammonia as the working fluid is shown in Fig. E11.9. The condensing temperature is 30°C. The liquid entering the expansion valve, EV-2 is at -2°C and the vapor leaving the evaporator is at -30°C. The temperature of the fluid in the flash-tank is -4°C. Calculate the COP of the system with and without the intercooler.

Solution Let the refrigerant flow rates through the low-pressure compressor and high-pressure compressor be $\dot{m}_2$ and $(\dot{m}_1 + \dot{m}_2)$ respectively as indicated in Fig. E11.9. Applying the SFEE to the flash-tank we obtain

$$\dot{m}_2 (h_2 - h_3) = \dot{m}_1 (h_8 - h_7) \qquad (E11.9.1)$$

The following data for ammonia are obtained from the tables in [4].

At 30°C, $h_2 = 323.1 \text{ kJ kg}^{-1}$. Ignoring the effect of pressure, the sub-cooled liquid enthalpy at 3 is $h_3 = 170.69 \text{ kJ kg}^{-1}$.

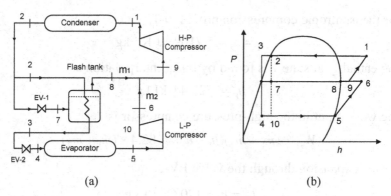

(a)   (b)

Fig. E11.9 Refrigeration system with flash intercooling, (a) Schematic diagram, (b) *P-h* diagram

At -4°C, $\qquad\qquad h_8 = 1439.9$ kJ kg^{-1}

For the throttling process 2-7, $h_7 = h_2 = 323.1$ kJ kg^{-1}

Substituting numerical values in Eq. (E11.9.1) we have

$$\dot{m}_2 = 7.328\dot{m}_1 \qquad\qquad (E11.9.2)$$

For the isentropic compression process 5-6,

$$s_6 = s_5 = 5.785 \text{ kJ K}^{-1} \text{ kg}^{-1}$$

At 6, the vapor is superheated at a pressure of 3.691 bar. The saturation temperature at 3.691 bar is -4°C.

By linear interpolation, the superheated vapor enthalpy at state 6,

$$h_6 = 1554.68 \text{ kJ kg}^{-1}$$

The work input in the low-pressure compressor is

$$\dot{W}_{56} = \dot{m}_2(h_6 - h_5) = (1554.68 - 1405.6)\dot{m}_2 = 149.08\dot{m}_2$$

Applying the SFEE to the mixing process at the junction 6-8-9 we have

$$\dot{m}_1 h_8 + \dot{m}_2 h_6 = (\dot{m}_1 + \dot{m}_2)h_9$$

Substituting numerical values and the condition in Eq. (E11.9.2) in the above equation

$$h_9 = (1439.9 + 7.328 \times 1554.68)/(1 + 7.328) = 1541.0 \text{ kJ kg}^{-1}$$

The entropy at state 9 is obtained by linear interpolation as

$$s_9 = 5.7169 \text{ kJ K}^{-1} \text{ kg}^{-1}$$

For the isentropic compression process 9-1,

$$s_1 = s_9 = 5.7169 \text{ kJ K}^{-1} \text{ kg}^{-1}$$

The enthalpy at state 1 is found by linear interpolation as

$$h_1 = 1723.44 \text{ kJ kg}^{-1}$$

The work input in the high-pressure compressor is

$$\dot{W}_{91} = (\dot{m}_1 + \dot{m}_2)(h_1 - h_9) = 182.4 \times (\dot{m}_1 + \dot{m}_2)$$

For the expansion through the valve EV-2,

$$h_4 = h_3 = 170.69 \text{ kJ kg}^{-1}$$

The refrigeration effect in the evaporator is

$$\dot{Q}_{45} = \dot{m}_2(h_5 - h_4) = (1405.6 - 170.69)\dot{m}_2 = 1234.9\dot{m}_2$$

The COP of the cycle is

$$COP = \frac{\dot{Q}_{45}}{\dot{W}_{56} + \dot{W}_{91}}$$

Substitute the relevant expressions in the above equation, and use the condition in Eq. (E11.9.2) to obtain

$$COP = \frac{1234.9\dot{m}_2}{149.08\dot{m}_2 + 182.4(\dot{m}_1 + \dot{m}_2)} = 3.46$$

(ii) Consider the standard vapor compression cycle without the intercooler. The condensing and evaporating temperatures are 30°C and -30°C respectively. Then the expansion process in the valve is 2-10, as indicated in Fig. E11.9(b).

Using the procedure in worked example 11.2 we obtain the following: The refrigeration effect,

$$q_{34} = 1082.5 \text{ kJ kg}^{-1}$$

The work input in the compressor,

$$w_{41} = 343.48 \text{ kJ kg}^{-1}$$

Therefore the COP of the standard vapor compression cycle using a single compressor is 3.15. We note that the use of flash inter-cooling improves the COP of the cycle by about 10 percent.

Example 11.10 Figure E11.10 depicts a vapor compression system with two evaporators and an inter-cooler. The working fluid is ammonia. The cooling capacity of the low-temperature evaporator operating at $-40°C$ is 250 kW. The high-temperature evaporator at $0°C$ has a cooling capacity of 150 kW. The vapor leaving the two evaporates is in a saturated state. The temperature of the fluid in the flash-tank is $0°C$. Dry saturated vapor enters the low-pressure compressor and the high-pressure compressor at $-40°C$ and $0°C$ respectively. The condensing temperature is $35°C$. Calculate (i) the work input to the two compressors, and (ii) the COP of the cycle.

Solution Let the refrigerant flow rates through the low-temperature evaporator and the high-temperature evaporator be $\dot{m}_1$ and $\dot{m}_2$ respectively, as indicated in Fig. E11.10(a). The flow rate through the expansion valve, EV-2 is $\dot{m}_3$. The various state points are also indicated in the *P-h* diagram depicted in Fig. 11.10(b). The following data are obtained from the tables in [4]:

At 35°C, $\qquad\qquad\qquad h_2 = 347.4 \text{ kJ kg}^{-1}$.

At 0°C, $\qquad h_6 = 181.2 \text{ kJ kg}^{-1}, \qquad h_4 = h_{10} = 1444.4 \text{ kJk g}^{-1}$

At -40°C, $\qquad\qquad\qquad h_8 = 1390 \text{ kJ kg}^{-1}$

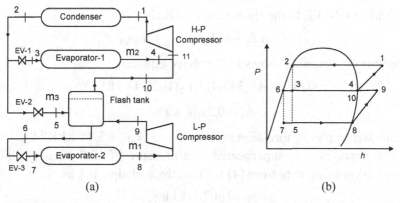

Fig. E11.10 Refrigeration system with two evaporators, (a) Schematic diagram, (b) *P-h* diagram

For the throttling processes 2-3, 2-5 and 6-7:

$$h_3 = h_2 = 347.4 \text{ kJ kg}^{-1}$$

$$h_5 = h_2 = 347.4 \text{ kJ kg}^{-1}$$

$$h_7 = h_6 = 181.2 \text{ kJ kg}^{-1}$$

For the low-temperature evaporator:

$$\dot{Q}_{78} = \dot{m}_1(h_8 - h_7)$$

$$250 = \dot{m}_1(1390 - 181.2)$$

$$\dot{m}_1 = 0.2068 \text{ kg s}^{-1}$$

For the high-temperature evaporator:

$$\dot{Q}_{34} = \dot{m}_2(h_4 - h_3)$$

$$150 = \dot{m}_2(1444.4 - 347.4)$$

$$\dot{m}_2 = 0.1367 \text{ kg s}^{-1}$$

For the isentropic compression process 8-9,

$$s_9 = s_8 = 5.962 \text{ kJ K}^{-1} \text{ kg}^{-1}$$

The vapor at 9 is superheated at a pressure of 4.295 bar. By linear interpolation using data from [4], we find the enthalpy at 9 as

$$h_9 = 1639.4 \text{ kJ kg}^{-1}$$

Applying the SFEE to the flash-tank we obtain

$$\dot{m}_3 h_5 + \dot{m}_1 h_9 = \dot{m}_1 h_6 + \dot{m}_3 h_{10}$$

Substituting numerical values in the above equation

$$\dot{m}_3(1444.4 - 347.4) = \dot{m}_1(1639.4 - 181.2)$$

$$\dot{m}_3 = 0.2748 \text{ kg s}^{-1}$$

For the isentropic compression process 4-1, $s_1 = s_4 = 5.34 \text{ kJ K}^{-1} \text{kg}^{-1}$
   The vapor at 1 is superheated at a pressure of 13.3 bar. By linear interpolation using data from [4] we find the enthalpy at 1 as

$$h_1 = 1607.3 \text{ kJ kg}^{-1}$$

The work input to the low-pressure compressor is

$$\dot{W}_{89} = \dot{m}_1(h_9 - h_8)$$

$$\dot{W}_{89} = 0.2068 \times (1639.4 - 1390) = 51.58 \text{ kW}$$

The work input to the high-pressure compressor is

$$\dot{W}_{41} = (\dot{m}_2 + \dot{m}_3)(h_1 - h_4)$$

$$\dot{W}_{41} = (0.2748 + 0.1367) \times (1607.3 - 1444.4) = 67.03 \text{ kW}$$

$$\dot{W}_{tot} = \dot{W}_{89} + \dot{W}_{41} = 51.58 + 67.03 = 118.6 \text{ kW}$$

Total work input is 118.6 kW.

The COP of the cycle is

$$COP = \frac{\dot{Q}_{34} + \dot{Q}_{78}}{\dot{W}_{89} + \dot{W}_{41}} = \frac{150 + 250}{51.58 + 67.03} = 3.37$$

Example 11.11 An ideal three-heat-reservoir absorption cycle extracts heat at the rate 6 kW from a refrigerated space at 10°C. The heat source temperature is 90°C. Heat is rejected to the ambient reservoir at 25°C. (i) Calculate the COP of the cycle, and the rate of heat supply from the heat source. (ii) If there is a temperature difference of 6°C between each reservoir and the working fluid of the internally reversible cyclic device, calculate the COP of the cycle and the entropy production rate.

Solution The three-heat-reservoir model of the absorption cycle is shown in Fig. E11.11. The temperatures of the refrigerated space, the generator and the heat sink are denoted by T_c, T_h, and T_o respectively.

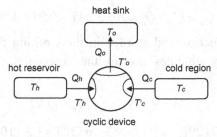

Fig. E11.11 Three-heat-reservoir model

The corresponding temperatures of the working fluid of the cycle are T_c', T_h', and T_o' respectively.

(i) Consider the case when there is no temperature difference between the working fluid and the reservoirs. The COP is given by Eq. (11.27) as

$$COP_a = \left(\frac{T_h - T_o}{T_h}\right)\left(\frac{T_c}{T_o - T_c}\right)$$

$$COP_a = \left(\frac{363 - 298}{363}\right)\left(\frac{283}{298 - 283}\right) = 3.38$$

$$COP_a = \dot{Q}_c / \dot{Q}_h = 6/\dot{Q}_h = 3.38$$

The heat input from the hot reservoir is, $\dot{Q}_h = 1.775\,\text{kW}$.

(ii) The temperatures of the working fluid when it interacts with the different reservoirs are:

$$T_c' = (283 - 6) = 277\,\text{K}, \qquad T_h' = (363 - 6) = 357\,\text{K}$$

$$T_o' = (298 + 6) = 304\,\text{K}.$$

The cycle undergone by the working fluid is *internally reversible*. Therefore the COP is given by

$$COP_a' = \left(\frac{T_h' - T_o'}{T_h'}\right)\left(\frac{T_c'}{T_o' - T_c'}\right)$$

$$COP_a' = \left(\frac{357 - 304}{357}\right)\left(\frac{277}{304 - 277}\right) = 1.52$$

The heat input from the reservoir is, $\dot{Q}_h = 6/1.52 = 3.947\ \text{kW}$

Applying the first law to the cycle

$$\dot{Q}_o = \dot{Q}_c + \dot{Q}_h = 6 + 3.947 = 9.947\ \text{kW}$$

The entropy production and storage in the working fluid are both zero. The only entropy changes are in the reservoirs. The total entropy production rate is

$$\dot{\sigma} = \dot{Q}_o / T_o - \dot{Q}_h / T_h - \dot{Q}_c / T_c$$

$$\dot{\sigma} = 9.947/298 - 3.947/363 - 6/283 = 1.3 \times 10^{-3}\ \text{kW K}^{-1}$$

We note that the entropy production is due to the irreversible heat transfer between the reservoirs and the working fluid, which has a significant effect on the COP of the cycle.

Example 11.12 A simple vapor absorption cooling system using water as the refrigerant and lithium-bromide as the absorbent is depicted in Fig. E11.12. The temperatures of the evaporator, the absorber, the condenser, and the generator are 10°C, 25°C, 35°C and 90°C respectively. The solution flow rate through the liquid pump is 0.5 kg s⁻¹. Calculate (i) the refrigerant flow rate, (ii) the cooling capacity, (iii) the heat rejected in the condenser, and (iv) the COP of the cycle.

Solution Let the flow rate of refrigerant (water) through the condenser be $\dot{m}_1$ and the flow rate of solution leaving the generator be $\dot{m}_7$. Now the saturated solution at 25°C, leaving the absorber at 5 is in equilibrium with the pure water vapor in the evaporator at 10°C.

From the equilibrium diagram for water-LiBr in [5] we obtain the concentration of the solution at 5 as $X_5 = 0.451$, (kg of LiBr per kg of solution). The saturated solution at 90°C, leaving the generator at 7, is in equilibrium with the pure water at 35°C in the condenser. From the equilibrium diagram we obtain the concentration at 7 as, $X_7 = 0.653$.

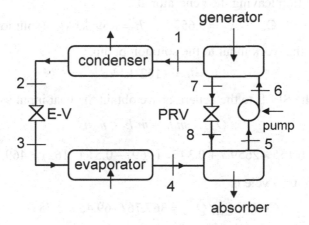

Fig. E11.12 Vapor absorption cycle

Applying the equation of mass conservation of LiBr for the generator

$$\dot{m}_7 X_7 = \dot{m}_6 X_6$$

$$\dot{m}_7 = 0.5 \times 0.451 / 0.653 = 0.345 \quad \text{kg s}^{-1}$$

Applying the equation of total mass conservation for the generator

$$\dot{m}_1 = \dot{m}_6 - \dot{m}_7 = 0.5 - 0.345 = 0.155 \quad \text{kg s}^{-1}$$

We obtain the following data for pure water from the steam tables in [4]:
For the saturated liquid at 35°C at 2, $h_2 = 146.55$ kJ kg^{-1}.
For the saturated vapor at 10°C at 4, $h_4 = 2519.2$ kJ kg^{-1}.
For saturated vapor at 90°C at 1, $h_1 = 2659.7$ kJ kg^{-1}.
For the throttling process, 2-3,

$$h_3 = h_2 = 146.55 \quad \text{kJ kg}^{-1}$$

The refrigeration capacity of the evaporator is

$$\dot{Q}_{34} = \dot{m}_1 (h_4 - h_3) = 0.155 \times (2519.2 - 146.55) = 367.76 \quad \text{kW}$$

The heat rejection rate in the condenser is

$$\dot{Q}_{12} = \dot{m}_1 (h_1 - h_2) = 0.155 \times (2659.7 - 146.55) = 389.5 \quad \text{kW}$$

We obtain the enthalpy of the water-LiBr solution from the chart in [5].
   For the solution leaving the absorber at 5,

$$T_5 = 25°C, \quad X_5 = 0.451, \qquad h_5 = -162 \quad \text{kJ kg}^{-1} \text{ (solution)}.$$

For the solution leaving the generator at 7

$$T_7 = 90°C, \qquad X_7 = 0.653, \qquad h_7 = -69 \quad \text{kJ kg}^{-1} \text{ (solution)}.$$

Neglecting the work input to the solution pump,

$$h_6 = h_5 = -162 \quad \text{kJ kg}^{-1}.$$

Applying the SFEE to the generator we obtain the heat input as

$$\dot{Q}_{gen} = \dot{m}_1 h_1 + \dot{m}_7 h_7 - \dot{m}_6 h_6$$

$$\dot{Q}_{gen} = 0.155 \times 2659.7 + 0.345 \times (-69) - 0.5 \times (-162) = 469.45 \, \text{kW}$$

The COP of the cycle is

$$COP = \dot{Q}_{34} / \dot{Q}_{gen} = 367.76 / 469.45 = 0.783$$

Example 11.13 A water-LiBr vapor absorption system incorporates a heat exchanger in which the weak liquid flowing from the generator heats the strong liquid leaving the pump to 50°C as shown in Fig. E11.13. The temperatures of the evaporator, the absorber, the condenser, and the generator are 10°C, 25°C, 35°C and 90°C respectively. The solution flow rate through the liquid pump is 0.5 kg s⁻¹. Calculate (i) the heat input in the generator, (ii) the heat rejected in the absorber, and (iii) the COP of the cycle.

Solution The purpose of the heat exchanger, shown in Fig. E11.13, is to heat the solution entering the generator using the hot solution leaving it. This reduces the heat input required in the generator. The basic parameters for this example are the same as those for worked example 11.12. Therefore we shall use the appropriate fluid properties obtained in the earlier example.

The enthalpy of the solution at 9, where the temperature and concentration are 50°C and 0.451 respectively, is obtained from the chart in [5] as $h_9 = -103$ kJ kg⁻¹ (solution).

Applying the SFEE to the generator we obtain the heat input as

$$\dot{Q}_{gen} = \dot{m}_1 h_1 + \dot{m}_7 h_7 - \dot{m}_6 h_9$$

$$\dot{Q}_{gen} = 0.155 \times 2659.7 + 0.345 \times (-69) - 0.5 \times (-103) = 439.95 \text{ kW}$$

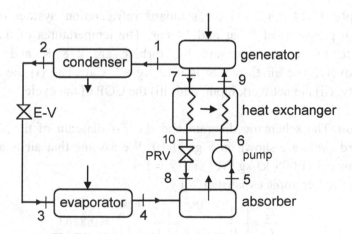

Fig. E11.13 Absorption cycle with heat exchanger

The COP of the cycle is

$$COP = \dot{Q}_{34} / \dot{Q}_{gen} = 367.76 / 439.95 = 0.836$$

We note that the inclusion of the heat exchanger increases the COP of the cycle by about 6.8 percent.

Neglecting the work input to the solution pump,

$$h_6 = h_5 = -162 \text{ kJ kg}^{-1}$$

Applying the SFEE to the heat exchanger we have

$$\dot{m}_7(h_7 - h_{10}) = \dot{m}_5(h_9 - h_6)$$

Substituting numerical values in the above equation

$$0.345 \times (-69 - h_{10}) = 0.5 \times [-103 - (-162)]$$

$$h_{10} = -154.5 \text{ kJ kg}^{-1}$$

For the throttling process 10-8,

$$h_8 = h_{10} = -154.5 \text{ kJ kg}^{-1}$$

Applying the SFEE to the absorber we obtain the heat rejection rate as

$$\dot{Q}_{abs} = \dot{m}_7 h_8 + \dot{m}_1 h_4 - \dot{m}_5 h_5$$

$$\dot{Q}_{abs} = 0.155 \times 2519.2 + 0.345 \times (-154.5) - 0.5 \times (-162)$$

$$\dot{Q}_{abs} = 418.17 \text{ kW}$$

Example 11.14 An ideal air standard refrigeration system operates between pressures of 7 bar and 14 bar. The temperatures of the air at entry to the compressor and the turbine are 278 K and 308 K respectively. The air flow rate is 0.12 kg s^{-1}. Calculate (i) the cooling capacity, (ii) the net work input, and (iii) the COP of the cycle.

Solution The schematic diagram and the *T-s* diagram of the ideal air standard cycle are shown in Fig. 11.10. We assume that air is an ideal gas with $c_p = 1.005$ kJ kg^{-1} K^{-1} and $\gamma = 1.4$.

For the isentropic expansion 3-4

$$\frac{T_4}{T_3} = \left(\frac{P_4}{P_3}\right)^{(\gamma-1)/\gamma} = \left(\frac{7}{14}\right)^{0.2857} = 0.8203$$

Therefore $T_4 = 252.7$ K.

For the isentropic compression 1-2

$$\frac{T_2}{T_1} = \left(\frac{P_2}{P_1}\right)^{(\gamma-1)/\gamma} = \left(\frac{14}{7}\right)^{0.2857} = 1.219$$

Therefore $T_2 = 338.9$ K.

The cooling capacity is

$$\dot{Q}_{41} = \dot{m}c_p(T_1 - T_4)$$

$$\dot{Q}_{41} = 0.12 \times 1.005 \times (278 - 252.7) = 3.05 \text{ kW}.$$

The heat rejection rate is

$$\dot{Q}_{23} = \dot{m}c_p(T_2 - T_3)$$

$$\dot{Q}_{23} = 0.12 \times 1.005 \times (338.9 - 308) = 3.726 \text{ kW}.$$

Applying the first law to the closed cycle, the net work input is

$$\dot{W}_{net} = \dot{Q}_{23} - \dot{Q}_{41} = 3.726 - 3.05 = 0.676 \text{ kW}$$

The COP of the refrigeration cycle is

$$COP = \dot{Q}_{41}/\dot{W}_{net} = 3.05/0.676 = 4.5$$

Example 11.15 The air standard refrigeration cycle shown in Fig. E11.15 incorporates a heat exchanger to cool the air before it enters the turbine by transferring heat to the air entering the compressor. The cycle operates between the pressures of 1 bar and 13 bar with an air flow rate of 0.2 kg s^{-1}. The following air temperatures are given: $T_a = 292$ K, $T_b = 222$ K and $T_1 = 284$ K. Calculate (i) the cooling capacity, (ii) the net work input, (iii) the COP of the cycle, and (iv) the entropy production rate in the heat exchanger.

Solution We assume that air is an ideal gas with

$$c_p = 1.005 \text{ kJ K}^{-1} \text{ kg}^{-1} \quad \text{and} \quad \gamma = 1.4.$$

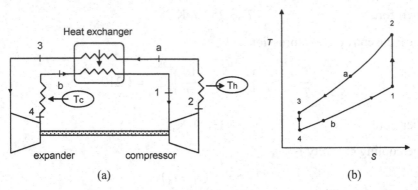

Fig. E11.15 Air standard refrigeration cycle: (a) Schematic diagram, (b) *T-s* diagram

Applying the SFEE to the heat exchanger we have

$$\dot{m}c_p(T_a - T_3) = \dot{m}c_p(T_1 - T_b)$$

$$T_3 = 292 - (284 - 222) = 230 \text{ K}$$

For the isentropic expansion 3-4

$$\frac{T_4}{T_3} = \left(\frac{P_4}{P_3}\right)^{(\gamma-1)/\gamma} = \left(\frac{1}{13}\right)^{0.2857} = 0.4805$$

Therefore $T_4 = 110.5$ K.

For the isentropic compression 1-2

$$\frac{T_2}{T_1} = \left(\frac{P_2}{P_1}\right)^{(\gamma-1)/\gamma} = \left(\frac{13}{1}\right)^{0.2857} = 2.081$$

Therefore $T_2 = 591.0$ K.

The cooling capacity is

$$\dot{Q}_{4b} = \dot{m}c_p(T_b - T_4)$$

$$\dot{Q}_{4b} = 0.2 \times 1.005 \times (222 - 110.5) = 22.4 \text{ kW}.$$

The heat rejection rate is

$$\dot{Q}_{2a} = \dot{m}c_p(T_2 - T_a)$$

$$\dot{Q}_{2a} = 0.2 \times 1.005 \times (591.0 - 292.0) = 60.1 \text{ kW}.$$

Applying the first law to the closed cycle, the net work input is

$$\dot{W}_{net} = \dot{Q}_{2a} - \dot{Q}_{4b} = 60.1 - 22.4 = 37.7 \ \text{kW}$$

The COP of the refrigeration cycle is

$$COP = \dot{Q}_{4b} / \dot{W}_{net} = 22.4 / 37.7 = 0.59$$

Assume that the pressure drops of the air streams through the heat exchanger are negligible. Using the expression for the change in entropy of an ideal gas [Eq. (7.25)], we obtain the entropy production in the heat exchanger as

$$\dot{\sigma} = \dot{m}c_p \ln\left(\frac{T_1}{T_b}\right) - \dot{m}c_p \ln\left(\frac{T_a}{T_3}\right)$$

$$\dot{\sigma} = 0.2 \times 1.005 \times \ln\left(\frac{284}{222}\right) - 0.2 \times 1.005 \times \ln\left(\frac{292}{230}\right)$$

$$\dot{\sigma} = 1.53 \times 10^{-3} \ \text{kW K}^{-1}$$

Note that all processes in the cycle are *internally* reversible except the process in the heat exchanger where heat is transferred across a finite temperature difference.

Problems

P11.1 A Carnot refrigeration cycle absorbs heat at −10°C and rejects heat at 50°C. (a) If the heat rejection rate is 25 kW, calculate the cooling capacity, work input, and the COP of the refrigeration cycle. (b) If the cycle operates as a heat pump calculate the COP of the cycle. (c) If there is a temperature difference of 5°C between the working fluid of the cycle and the two reservoirs at −10°C and 50°C, calculate the COP of the refrigeration cycle, and the rate of entropy production. [*Answers*: (a) 20.35 kW, 4.65 kW, 4.38, (b) 5.38, (c) 3.68, 0.00275 kW K^{-1}]

P11.2 A heat pump operates on the reversed Carnot cycle with R134a as the working fluid. The condenser and evaporator temperatures are 50°C and 10°C respectively. The refrigerant flow rate is 0.2 kg s^{-1}. Calculate

(i) the rate of heat input to the heat pump, (ii) the net work input, and (iii) the COP. [*Answers*: (i) 26.6 kW, (ii) 3.76 kW, (iii) 8.075]

P11.3 The condensing and evaporating temperatures of a standard vapor compression cycle using R134a are 5°C and 35°C. The cooling capacity is 8 kW. Calculate (i) the refrigerant flow rate, (ii) the net work input to the compressor, and (iii) the COP of the cycle. [*Answers*: (i) 0.0525 kg s⁻¹, (ii) 1.0144 kW, (iii) 7.89]

P11.4 A standard vapor compression cycle using R134a has a non-ideal, adiabatic compressor with an efficiency of 0.8. The condensing and evaporating temperatures are 40°C and 0°C respectively. The refrigerant flow rate is 0.25 kg s⁻¹. Calculate (i) the cooling capacity, (ii) the work input to the compressor, (iii) the COP of the cycle, and (iv) the entropy production rate in the compressor. [*Answers*: (i) 35.5 kW, (ii) 8.09 kW, (iii) 4.39, (iv) 0.0049 kW K⁻¹]

P11.5 In a vapor compression cycle, the vapor condenses at 40°C and is sub-cooled to 35°C before entering the expansion valve. The liquid evaporates at a temperature of −5°C and it is superheated to 5°C at entry to the compressor. The refrigerant is R134a. The cooling capacity is 10 kW, and the compression process is isentropic. Calculate (i) the refrigerant flow rate, (ii) the work input, and (iii) the COP. [*Answers*: (i) 0.0644 kg s⁻¹, (ii) 2.02 kW, (iii) 4.95]

P11.6 The condensing and evaporating temperatures of a vapor compression cycle using R134a are 40°C and −5°C respectively. The cycle incorporates a heat exchanger that uses the saturated liquid leaving the condenser to superheat the saturated vapor leaving the evaporator to 5°C. The refrigerant flow rate is 0.2 kg s⁻¹. The compression process is isentropic. Calculate (i) the cooling capacity, (ii) the work input to the compressor, and (iii) the COP.
[*Answers*: (i) 31.33 kW, (ii) 6.27 kW, (iii) 5]

P11.7 A vapor compression system with two evaporators and a flash inter-cooler using ammonia as the refrigerant is shown in Fig. 11.6(a).

The refrigerant flow rate through the low-temperature evaporator operating at $-30°C$ is 0.25 kg s^{-1}. The high-temperature evaporator at $-4°C$ has a refrigerant flow rate of 0.15 kg s^{-1}. The vapor leaving the two evaporators is in a saturated state. The temperature of the fluid in the flash-tank is $-4°C$. Dry saturated vapor enters the low-pressure compressor and the high-pressure compressor at $-30°C$ and $-4°C$ respectively. The liquid entering the expansion valve EV-2 is subcooled to $20°C$. The condensing temperature is $40°C$. Calculate (i) the cooling capacities of the two evaporators, (ii) the work input to the two compressors, and (iii) the COP of the cycle. [*Answers*: (i) 160.6 kW, 282.7 kW, (ii) 37.17 kW, 94.5 kW, (iii) 3.36]

P11.8 The temperatures of the three heat reservoirs representing the heat source, the refrigerated space and the heat sink of an ideal three-heat-reservoir absorption cycle are $100°C$, $10°C$ and $35°C$ respectively. The heat interaction between each reservoir and the working fluid of the internally reversible cycle requires a temperature difference of $5°C$. The cooling capacity is 10 kW. Calculate (i) the COP of the cycle, (ii) the rate of heat supply from the heat source, and (iii) the rate of entropy production. [*Answers*: (i) 1.187, (ii) 8.423 kW, (iii) 0.0019 kW K^{-1}]

P11.9 A water-LiBr vapor absorption system incorporates a heat exchanger as shown in Fig. E11.13. The temperatures of the evaporator, the absorber, the condenser, and the generator are $10°C$, $25°C$, $40°C$ and $100°C$ respectively. The strong liquid leaving the pump is heated to $50°C$ in the heat exchanger. The refrigerant flow rate through the condenser is 0.16 kg s^{-1}. Calculate (i) the heat input in the generator, (ii) the heat rejected in the absorber, and (iii) the COP of the cycle. [*Answers*: (i) 462.2 kW, (ii) 437.2 kW, (iii) 0.81]

P11.10 A regenerative heat exchanger is incorporated in an air standard refrigeration cycle as shown in Fig. 11.11. The pressure ratio of the cycle is 12 and the cooling capacity is 20 kW.

The following air temperatures are given: $T_a = 22\,°C$, $T_b = -48\,°C$ and $T_1 = 14\,°C$. Calculate (i) the air flow rate, (ii) the net work input, (iii) the COP of the cycle, and (iv) the entropy production rate in the heat exchanger.

[*Answers*: (i) 0.18 kg s^{-1}, (ii) 32.2 kW, (iii) 0.62, (iv) 0.001347 kW K^{-1}]

References

1. Jones, J.B. and G.A. Hawkins, *Engineering Thermodynamics*, John Wiley & Sons, Inc., New York, 1986.
2. Kuehn, Thomas H., Ramsey, James W. and Threlkeld, James L., *Thermal Environmental Engineering*, 3rd edition, Prentice-Hall, Inc., New Jersey, 1998.
3. Reynolds, William C. and Henry C. Perkins, *Engineering Thermodynamics*, 2nd edition, McGraw-Hill, Inc., New York, 1977.
4. Rogers, G.F.C. and Mayhew, Y.R., *Thermodynamic and Transport Properties of Fluids*, 5th edition, Blackwell, Oxford, U.K. 1998.
5. Stoecker, Wilbert F. and Jones, Jerold W., *Refrigeration and Air Conditioning*, International Edition, McGraw-Hill Book Company, London, 1982.
6. Van Wylen, Gordon J. and Richard E. Sonntag, *Fundamentals of Classical Thermodynamics*, 3rd edition, John Wiley & Sons, Inc., New York, 1985.

Chapter 12

Gas and Gas-Vapor Mixtures

In chapter 2 we studied the properties of a pure substance which is homogeneous and of fixed chemical composition. In the absence of chemical reactions, a homogeneous mixture of pure substances may also be treated as a pure substance. For example, in our work thus far we have considered air, which is a mixture of several gases, as a pure substance. The properties of a mixture of pure substances could be determined by knowing the values of the corresponding properties of the various constituents of the mixture.

There are numerous engineering applications where we need to deal with mixtures of pure substances. For instance, in air conditioning calculations, moist air is usually treated as a pure substance, and in combustion analysis, the mixture of exhaust products is often considered as a pure substance. In this chapter we shall develop some general methods for the computation of the properties of mixtures of gases, and gas-vapor mixtures.

12.1 Mixtures of Gases

In this section we shall introduce several quantities and rules that enable us to characterize a mixture of gases. These, in general, are applicable mainly to ideal gases but a few may also be applied to real gases under certain conditions. For mixtures of ideal gases it is possible to obtain simple expressions to compute properties like the internal energy, the enthalpy and the entropy of the mixture, in terms of the properties of the individual constituents.

577

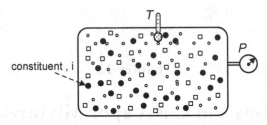

Fig. 12.1 Mixture of gases

12.1.1 *Mass-fraction and mole-fraction*

A mixture of gases confined to a rigid vessel is shown schematically in Fig. 12.1. The total pressure and the temperature of the mixture, assumed uniform in the vessel, are P and T respectively. Let the masses of the r constituent gases be m_1, m_2, ... m_i ..... m_i. Then the total mass is given by

$$m = \sum_{i=1}^{r} m_i \qquad (12.1)$$

where the subscript i denotes the ith constituent gas in the mixture.

The *mass fraction*, f_i of the ith constituent is defined as,

$$f_i = m_i / m$$

We recall from chapter 2 that a *mole* of gas is the amount of gas numerically equal to its molecular weight. Therefore the number of moles, N_i of the ith constituent in the vessel in Fig. 12.1 may be expressed as,

$$N_i = m_i / M_i$$

where M_i is the molecular mass of the ith constituent gas. The total number of moles is given by

$$N = \sum_{i=1}^{r} N_i \qquad (12.2)$$

The *mole fraction* of the ith constituent is defined as,

$$x_i = N_i / N$$

We define the *mean molecular mass*, $\overline{M}$ of the mixture by the equation

$$\overline{M}N = m \tag{12.3}$$

The following relation is obtained by manipulating the above equations:

$$\overline{M} = \sum_{i=1}^{r} \frac{m_i}{N} = \sum_{i=1}^{r} \frac{N_i M_i}{N} = \sum_{i=1}^{r} x_i M_i \tag{12.4}$$

We note the important summation rules:

$$\sum_{i=1}^{r} x_i = 1 \quad \text{and} \quad \sum_{i=1}^{r} f_i = 1 \tag{12.5}$$

12.1.2 *Partial pressure and partial volume*

The partial pressure of the ith constituent in a gas mixture is defined as

$$P_i = x_i P \tag{12.6}$$

where P is the total pressure of the mixture. Taking the summation of both sides of Eq. (12.6) and applying the summation rule, Eq. (12.5), we obtain

$$P = \sum_{i=1}^{r} P_i \tag{12.7}$$

Equations (12.6) and (12.7) are applicable both to real gases and ideal gases.

From a microscopic view point, we could interpret the partial pressure, P_i as the contribution of the ith constituent gas to the total average normal force per unit area at the boundary of the vessel. The important underlying assumption is that the contribution of each constituent to the normal force is *independent* of the presence of the other gases.

The partial volume of the ith constituent of a mixture of gases is defined by

$$V_i = x_i V \tag{12.8}$$

where V is the volume of the gas mixture. It is not possible to provide a physical interpretation for the partial volume that is applicable to any gas. For a mixture of *ideal gases*, however, the partial volume has a physical meaning which we shall discuss in the next section.

12.1.3 *Dalton's rule for ideal gas mixtures*

The constituents of a mixture of *ideal gases* are said to be *independent* if the properties of any constituent is not influenced by the presence of the other gases. This is an idealization that applies to mixtures of gases and liquids that are not too dense. It follows that the mixture of independent *ideal* gases is also an *ideal* gas.

Applying the equation of state in its *molar-form* to each constituent, and the mixture as a whole, we obtain the following relations,

$$P_i V = N_i \overline{R} T \quad (i = 1, r) \tag{12.9}$$

$$PV = N \overline{R} T \tag{12.10}$$

where $\overline{R}$ is the universal gas constant. It follows from Eq. (12.9) that P_i is the pressure of the *ideal gas i* when it occupies the mixture volume V alone at a temperature T. Taking the summation over i of both sides of Eq. (12.9), and comparing with Eq. (12.10), we obtain

$$P = \sum_{1}^{r} P_i \tag{12.11}$$

This relationship, known as Dalton's rule, may be restated as: the pressure of a mixture of ideal gases equals the sum of the pressures of its constituents if each existed alone at the volume and temperature of the mixture. We note that Dalton's rule is strictly true only for ideal gas mixtures. However, it may be applied to real gas mixtures under certain conditions of pressure and temperature.

From Eqs. (12.9) and (12.10) we have

$$P_i = P \left(\frac{N_i}{N} \right) = x_i P \tag{12.12}$$

Comparing Eq. (12.12) with our *general* definition of partial pressure, given by Eq. (12.6), it follows that the *partial pressure* of an ideal gas in a mixture of ideal gases is the pressure it would exert if it existed alone at the volume and temperature of the mixture.

12.1.4 *Amagat-Leduc rule for ideal gas mixtures*

As we stated earlier, the mixture of *independent* ideal gases may be treated as an ideal gas. Applying the equation of state to the mixture we have

$$PV = N\overline{R}T = \overline{R}T\sum_{i=1}^{r} N_i \qquad (12.13)$$

Hence
$$V = \sum_{i=1}^{r}\left(N_i\overline{R}T/P\right) = \sum_{i=1}^{r} V_i \qquad (12.14)$$

We interpret the volume V_i in Eq. (12.14) as the volume of the ith constituent gas when it exists alone at the pressure, P and temperature, T of the mixture. Therefore it follows from Eq. (12.14) that the volume of a mixture of ideal gases is equal to the sum of the volumes of the constituent gases if each existed alone at the pressure and temperature of the mixture. This statement, known as the *Amagat-Leduc rule*, is strictly true only for ideal gas mixtures. However, it may be valid for real gas mixtures under certain conditions of pressure and temperature.

Applying the ideal gas equation to the ith constituent when it exists at the pressure, P and temperature, T of the mixture we have

$$PV_i = N_i\overline{R}T \qquad (12.15)$$

From Eqs. (12.15) and (12.10) we obtain

$$V_i = \left(N_i/N\right)V = x_iV \qquad (12.16)$$

Comparing Eq. (12.16) with our general definition of partial volume, given by Eq. (12.8), we conclude that for *ideal gas* mixtures the partial volume of a constituent is the volume when it exists alone at the pressure and temperature of the mixture.

12.1.5 *Properties of ideal gas mixtures*

For a mixture of ideal gases that is in equilibrium, the temperature, T is the same for each constituent. The pressure of a mixture of *independent* ideal gases is equal to the sum of the pressures of the constituents if each existed alone at the volume and temperature of the mixture. The volume V_i of each constituent of a mixture of ideal gases is equal to the volume of the mixture because each constituent is free to occupy the entire volume.

For a mixture of ideal gases that are *independent*, the extensive properties, like the internal energy u, the enthalpy h, and the entropy s of each constituent, are not influenced by the presence of the rest of the constituents. Since each constituent gas is a pure substance we invoke the *two-property rule* with T and V as the independent properties, which are the same for all constituents of the mixture. We recall that according to the two-property rule, two independent properties are sufficient to completely determine the state of a simple pure substance. The mixture properties per unit mass u, h and s can then be expressed in terms of the corresponding constituent properties per unit mass u_i, h_i and s_i in the following form:

$$mu = \sum_{i=1}^{r} m_i u_i(T,V) \tag{12.17}$$

$$mh = \sum_{i=1}^{r} m_i h_i(T,V) \tag{12.18}$$

$$ms = \sum_{i=1}^{r} m_i s_i(T,V) \tag{12.19}$$

In a mixture of ideal gases, the specific heat capacities c_{vi} and c_{pi} of the ith constituent is a function of temperature only. Therefore by differentiating Eqs. (12.17) and (12.18) with respect to temperature we obtain the mean specific heat capacities as:

$$\bar{c}_v = (du/dT) = \sum_{i=1}^{r} f_i(du_i/dT) = \sum_{i=1}^{r} f_i c_{vi} \tag{12.20}$$

$$\bar{c}_p = (dh/dT) = \sum_{i=1}^{r} f_i(dh_i/dT) = \sum_{i=1}^{r} f_i c_{pi} \tag{12.21}$$

where f_i is the mass fraction of the ith constituent.

The above mean properties per unit mass were all obtained on a mass-basis. However, a similar procedure may be used to obtain the properties on a molar-basis which are more convenient for certain computational applications.

Consider the ideal gas equation of state of the ith constituent on a mass basis.

$$P_iV = m_iR_iT \tag{12.22}$$

where R_i is the gas constant of the ith constituent.

Taking the summation of Eq. (12.22) over the r constituents we have

$$PV = T\sum_{i=1}^{r}m_iR_i \tag{12.23}$$

We define the mean gas constant, R_m of the mixture by the following equation

$$PV = mR_mT \tag{12.24}$$

From Eqs. (12.23) and (12.24) we obtain the mean gas constant of the mixture as

$$R_m = \sum_{i=1}^{r}f_iR_i \tag{12.25}$$

We shall illustrate the application of the various property relations derived above in worked examples 12.1 to 12.6.

12.1.6 *Real gas mixtures*

In our work thus far, we have used the ideal gas equation widely in solving most of the worked examples, with a few exceptions in Chapter 2, where the van der Waals equation of state was applied. Under certain conditions of temperature and pressure the behavior of a real gas is better represented by an equation of state like the van der Waals equation that is more complex than the ideal gas equation.

Consider a mixture of real gases in which each constituent gas follows the van der Waals equation

$$P_i(T,V) = \overline{R}T/(v_i - b_i) - a_i/v_i^2 \tag{12.26}$$

where v_i is the molar specific volume and a_i and b_i are constants of the van der Waals equation for the ith constituent of the mixture. In order to determine the total pressure of the mixture assume that Dalton's rule is valid for the real gas mixture. Therefore

$$P = \sum_{i=1}^{r} P_i(T,V) \tag{12.27}$$

Now the molar specific volume, v_i of the ith constituent can be expressed in the form

$$v_i = V/N_i = V/(x_i N) = v_m / x_i \tag{12.28}$$

where v_m is the molar specific volume of the mixture and x_i is the mole fraction.

Applying Dalton's rule, expressed by Eq. (12.27), together with Eqs. (12.26) and (12.28) we obtain the following expression for the total pressure of the mixture:

$$P = \sum_{i}^{r} P_i = \sum_{1}^{r} \left[\frac{\overline{R}Tx_i}{(v_m - x_i b_i)} - \frac{a_i x_i^2}{v_m^2} \right] \tag{12.29}$$

A somewhat different approach to determine the total pressure of a real gas mixture is to assume the validity of the Amagat-Leduc rule, instead of Dalton's rule. These different assumptions usually give different numerical results for the total pressure. An application of the above relations is given in worked example 12.7.

12.2 Mixtures of Ideal Gases and Vapors

There are numerous processes of practical engineering importance, especially in air conditioning and drying, that involve mixtures of gases and vapors. When subjected to certain processes, gas mixtures and mixtures of gases and vapors behave quite differently. In the case of a mixture of gas and vapor, the vapor could change phase during the process whereas a mixture of gases remains in a gaseous state during the process.

For example, when ambient air passes through the cooling coil of an air conditioner, some of the water vapor in the air condenses to water on the surface of the coil. The air delivered to the conditioned space after passing through the cooling coil is therefore much drier than the original ambient air supplied to the cooling coil. The reverse process occurs in a dryer where the air leaving the dryer becomes more moist due to the evaporation of water from the material being dried. In order to analyze these practical situations we need to develop a set of parameters to characterize a mixture of gases and vapors, and also determine the relevant thermodynamic properties of the mixture.

12.2.1 *Mixtures of air and water vapor*

The main focus of our discussion in this section is ambient air, which is a mixture of water vapor in a superheated state, and dry air. We shall refer to air that is free of any moisture as *dry air*. The pressure of water vapor in typical ambient air is relatively low. For instance, at 25°C the pressure of water vapor in air that is fully saturated with water vapor is about 3.2 kPa (saturation pressure of water vapor at 25°C from the steam tables) compared with a mixture pressure of about 100 kPa. Under these dilute conditions we shall assume that water vapor and dry air in the ambient behave as ideal gases. Furthermore, we shall treat them as *independent pure substances* where the properties of water vapor are not influenced by the presence of air. These assumptions are found to be reasonable for most practical calculations involving atmospheric air.

In order to discuss some of the physical processes involving gas-vapor mixtures, we consider the piston cylinder arrangement shown in Fig. 12.2(a). Initially, the cylinder contains *moist* ambient air. The initial state of the *water vapor* in the mixture is indicated by A in the T-v diagram for water, depicted in Fig 12.2(b). In view of the aforementioned assumptions concerning ambient air, we may apply Dalton's rule to express the pressure of the mixture in the form,

$$P = P_a + P_v$$

where P_a and P_v are the partial pressures of the air and the water vapor respectively.

Imagine a quasi-static process in which the air is cooled at a *constant mixture pressure* which is maintained by the fixed weight of the piston.

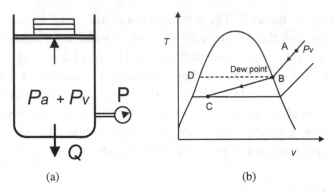

Fig. 12.2 (a) Piston-cylinder set-up with moist air, (b) T-v diagram for vapor

Since the mole fractions of air and water vapor are constant during the process, the partial pressure of the water vapor, $P_v = x_v P$, also remains constant. Therefore the cooling process follows a constant pressure line on the T-v diagram for the vapor. However, when the state of the vapor reaches the point B on the saturated vapor curve in Fig. 12.2(b), the first liquid drops appear in the cylinder, and condensation of the water vapor just begins. The temperature at state B, where condensation just begins, is called the *dew point* of the air, and the air is then referred to as *saturated air*.

As the cooling process continues more vapor would condense to water. Therefore, the mass of water vapor and consequently the mole fraction of water vapor in the mixture decrease. We note that the liquid water produced by condensation has a relatively small volume and therefore its effect on the air-vapor mixture in the cylinder is negligible.

As condensation continues, the partial pressure of the vapor decreases due to the reduced mole fraction of the vapor. Since vapor is in a saturated state during condensation, its temperature, and the temperature of the mixture, both decrease as shown by the line BC in Fig. 12.2(b). Because the mixture pressure is maintained constant by the fixed weight of the piston, by Dalton's rule, the partial pressure of the air increases to compensate for the reduced vapor pressure. It is noteworthy that in the absence of air in the cylinder the state of the pure vapor during condensation would follow the constant temperature line BD. We note that the presence of air significantly alters the overall behavior of the vapor during condensation. We shall consider the analysis of air-vapor mixtures subjected to constant volume and isothermal processes in the worked examples.

12.2.2 *Relative humidity and humidity ratio*

We now introduce two parameters called the *relative humidity* and the *humidity ratio* that are commonly used to characterize the state of a mixture of dry air and water vapor. The humidity ratio is sometimes referred to as the *specific humidity*.

The following assumptions are made in obtaining analytical expressions for these parameters. (i) The air and vapor phases can be treated as a mixture of independent ideal gases. (ii) The conditions of equilibrium between the liquid and the vapor in the mixture are not influenced by the presence of air. (iii) There are no physical and chemical interactions between the liquid phase and the gaseous phase. When applied to air-water mixtures, these assumptions allow us to obtain the properties of the vapor from the steam table at the mixture temperature and the *partial* pressure of the vapor in the mixture.

The *relative humidity*, ϕ is defined as the ratio of the partial pressure of the vapor P_v in the mixture to the saturation pressure of the vapor at the mixture temperature, $P_g(T)$. For air-water mixtures, the latter pressure is obtained directly from the steam table. Therefore

$$\phi = \frac{P_v}{P_g(T)} \tag{12.30}$$

The *humidity ratio*, ω is defined as the mass of water vapor in a given volume of mixture to the mass of dry air in the same volume.

Hence
$$\omega = \frac{\Delta m_v}{\Delta m_a} \tag{12.31}$$

where Δm_v and Δm_a are masses of vapor and air respectively in a volume ΔV.

Also
$$\omega = \frac{\Delta m_v / \Delta V}{\Delta m_a / \Delta V} = \frac{\rho_v}{\rho_a} \tag{12.32}$$

In the above equation, ρ_v and ρ_a are the densities of vapor and dry air respectively.

Applying the density-form of the ideal gas equation to the vapor and air we have

$$P_v = \rho_v R_v T \qquad (12.33)$$

$$P_a = \rho_a R_a T \qquad (12.34)$$

where R_v and R_a are the respective gas constants of vapor and air. Substituting from Eqs. (12.33) and (12.34) in Eq. (12.32) we obtain

$$\omega = \frac{\rho_v}{\rho_a} = \frac{P_v R_a}{P_a R_v} \qquad (12.35)$$

Applying Dalton's rule
$$P = P_v + P_a \qquad (12.36)$$

Therefore
$$\omega = \frac{P_v R_a}{(P - P_v) R_v} \qquad (12.37)$$

From Eqs. (12.30) and (12.37) it follows that

$$\omega = \frac{R_a \phi P_g(T)}{R_v [P - \phi P_g(T)]} \qquad (12.38)$$

The gas constants for dry air and water vapor may be expressed in terms of the universal gas constant and their molecular masses as:

$$R_a = \overline{R} / M_a = \overline{R} / 28.96$$

$$R_v = \overline{R} / M_v = \overline{R} / 18 \qquad (12.39)$$

Substituting in Eq. (12.38) from Eq. (12.39) we have

$$\omega = \frac{0.622 \phi P_g(T)}{P - \phi P_g(T)} \qquad (12.40)$$

The above equation is a very useful relation for computations involving ambient air, and it forms the basis of the *psychrometric chart*.

12.2.3 *The psychrometric chart*

The mixture temperature or the *dry-bulb* temperature T of air, represented along the horizontal axis, is the main independent variable of the *psychrometric* chart shown in Fig. 12.3. The humidity ratio is represented along the vertical axis.

We outline briefly the important steps in the construction of the psychrometric chart for a *fixed ambient pressure (mixture pressure)*, P, say 1.01 bar. Select a typical value of the dry-bulb temperature, say 25°C and obtain from the steam tables [4] the saturation pressure at 25°C as, $P_g(25) = 0.03166$ bar. The first family of curves that we propose to plot is for constant values of the relative humidity ϕ. The curve for saturated air, with $\phi = 100\%$ is the outermost curve of this family. Substituting the above values of P, P_g and ϕ in Eq. (12.40) we have $\omega = 0.0201$ *kg per kg of dry air*. The point A is plotted with the coordinates (25°C, 0.0201). The procedure is repeated for different values of the dry bulb temperature to generate the complete curve for $\phi = 100\%$.

Now consider a different, *fixed* value of the relative humidity, say $\phi = 40\%$. Considering the same temperature of 25°C, we substitute the new values in Eq. (12.40) to obtain $\omega = 0.007898$ *kg per kg of dry air*. The point B in Fig. 12.3 is plotted with the coordinates (25°C, 0.007898). By repeating the above procedure we are able to generate a family of curves for different values of the relative humidity. An important point to note is that a typical psychrometric chart, like the one included in [5], is for a fixed ambient pressure, usually about 1.01 bar. However, the procedure outlined above could be used to develop a chart at any other pressure.

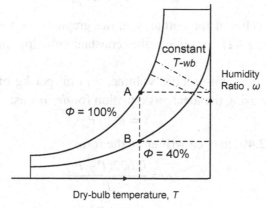

Fig. 12.3 Psychrometric chart

It is possible to include additional properties of moist air in the psychrometric chart. We shall now outline the procedure to plot the constant-enthalpy and constant-specific volume lines in the chart. Now the enthalpy of a mixture of dry air and water vapor is equal to the sum of the enthalpy of dry air and the enthalpy of the superheated water vapor. Therefore

$$h = c_{pa}T + \omega h_g \qquad (12.41)$$

The units of h are kJ kg^{-1} of *dry* air.

In the above equation c_{pa} is the specific heat capacity at constant pressure of *dry* air. For the typical temperature range of about 0°C to 50°C of the psychrometric chart, c_{pa} varies from about 1.006 to 1.009 kJ K^{-1} kg^{-1}. Therefore a constant value of 1.00 kJ K^{-1} kg^{-1} may be used in Eq. (12.41) without causing significant errors in the computed quantities.

Now the water vapor in the ambient is in a superheated state at a low pressure. We observe from the data in the steam tables [4] that under low pressures, the enthalpy of superheated steam is approximately the same as the enthalpy of saturated steam at the same *temperature*. Therefore in Eq. (12.41) we shall use the saturated vapor enthalpy $h_g(T)$ from the steam tables.

In order to generate a family of constant-enthalpy lines we rewrite Eq. (12.41) as

$$\omega = (h - c_{pa}T)/h_g(T) \qquad (12.42)$$

For a *fixed* value of the enthalpy h, the graph of ω versus T is plotted using Eq. (12.42) to obtain the constant-enthalpy lines sketched in Fig. 12.4.

Lines of constant-specific volume, v in m^3 per kg of *dry air* may be produced by using the ideal gas equation for *dry air* as:

$$v = R_a T / P_a \qquad (12.43)$$

Now Eq. (12.40) may be written in the form

$$\omega = \frac{0.622(P - P_a)}{P_a} \qquad (12.44)$$

where P_a is the pressure of *dry* air and P is the total pressure.

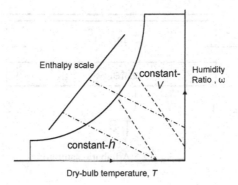

Fig. 12.4 Constant enthalpy and specific volume lines

From Eqs. (12.43) and (12.44) we have

$$\omega = 0.622\left(\frac{Pv}{R_aT(K)} - 1\right) \tag{12.45}$$

For fixed values of v and P, ω is plotted against T using Eq. (12.45) to produce the lines of constant specific volume sketched in Fig. 12.4.

12.2.4 *Adiabatic saturation and wet-bulb temperature*

The schematic diagram in Fig. 12.5 depicts a device that brings a stream of air passing steadily through it to a saturated state by a process known as *adiabatic saturation*. The walls of the device are perfectly insulated and the air flows over the surface of a pool of water, whose level is maintained constant by a steady supply of make-up water.

Let the air temperature and humidity ratio at the inlet section 1 and the exit section 2 be T_1, ω_1 and T_2, ω_2 respectively. The constant mass flow rate of *dry* air is $\dot{m}_a$ and the mass flow rate of make-up water at temperature T_3 is $\dot{m}_w$.

Applying the mass balance equation for water flowing through the control volume 1-3-2 we obtain

$$\dot{m}_w + \dot{m}_a\omega_1 = \dot{m}_a\omega_2 \tag{12.46}$$

Applying the SFEE, neglecting the kinetic and potential energy of the air, we have

$$\dot{m}_w h_{f3} + \dot{m}_a h_{a1} = \dot{m}_a h_{a2} \tag{12.47}$$

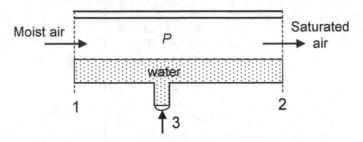

Fig. 12.5 Adiabatic saturator

From Eqs. (12.46) and (12.47) we obtain

$$h_{a1} + h_{f3}(\omega_2 - \omega_1) = h_{a2} \qquad (12.48)$$

Substituting for the enthalpy of moist air from Eq. (12.41) in Eq. (12.48)

$$c_{pa}T_1 + \omega_1 h_{g1} + h_{f3}(\omega_2 - \omega_1) = c_{pa}T_2 + \omega_2 h_{g2} \qquad (12.49)$$

Hence
$$\omega_1 = \frac{\omega_2(h_{g2} - h_{f3}) - c_{pa}(T_1 - T_2)}{h_{g1} - h_{f3}} \qquad (12.50)$$

Now h_{a2} and ω_2 are the enthalpy and humidity ratio respectively of saturated air at section 2. Since the relative humidity at section 2 is given by, $\phi_2 = 100\%$, we observe from Eqs. (12.40) and (12.41) that both ω_2 and h_{a2} are functions *only* of the temperature T_2.

We now introduce the important assumption that, under *ideal conditions*, the make-up water temperature T_3 is also equal to T_2, the exit air temperature. Substituting this condition in Eq. (12.50) we have

$$\omega_1 = \frac{\omega_2(h_{g2} - h_{f2}) - c_{pa}(T_1 - T_2)}{h_{g1} - h_{f2}} \qquad (12.51)$$

We note from Eq. (12.51) that, when *ideal conditions prevail*, air entering with different combinations T_1, ω_1 will have the same exit saturation temperature T_2. The temperature T_2, which is therefore a property of the inlet air stream, is called the *thermodynamic wet-bulb temperature* (TWT) of the air.

By choosing different values of T_2, we can construct a family of constant–TWT lines in the psychrometic chart. The main steps of the procedure are briefly outlined below. Select an air inlet temperature T_1

and a TWT, T_2. Obtain the following data from the steam tables [4]: the saturation vapor enthalpy, h_{g1} at T_1, the saturation vapor enthalpy, h_{g2}, the saturation vapor pressure, P_{g2} and the saturation liquid enthalpy, h_{f2} at T_2. Calculate the humidity ratio, ω_2 of the exit air using Eq. (12.40) with $\phi_2 = 100\%$. Substitute these values in Eq. (12.51) to obtain the humidity ratio, ω_1 of the inlet air stream and locate the point (T_1, ω_1) on the psychrometric chart. By repeating the procedure for different values of T_1 we construct the constant-TWT line for T_2. Two such constant wet-bulb temperature lines are sketched in Fig. 12.3.

The practical usefulness of the TWT lines is immediately evident because by measuring the TWT and the dry-bulb temperature of a sample of air we can locate its state on the psychrometric chart. Moreover, the TWT provides an indirect method of determining the relative humidity of air. The instrument commonly used to measure the wet-bulb temperature is called the *wet-bulb thermometer.*

A practical arrangement based on thermometry to measure the wet-bulb temperature is shown in Fig. 12.6. In its basic form the wet-bulb thermometer consists of an ordinary liquid-in-glass thermometer whose bulb is covered with a wet porous wick that is continuously supplied with water. The air whose humidity is to be determined is blown over the wick usually with aid of a fan. The temperature indicated by the thermometer under steady conditions is called the *wet-bulb temperature,* T_{wb}.

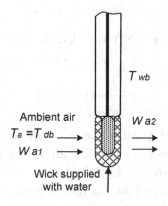

Fig. 12.6 Wet-bulb thermometer

The ideal conditions of adiabatic saturation assumed in the development of the thermodynamic wet bulb temperature (TWT) as a property may not be realized with the actual wet-bulb thermometer shown in Fig. 12.6. A complete analysis of the difference between TWT and T_{wb} is beyond the scope of this book. However, we shall briefly outline the main steps of a simplified analysis to make a comparison between the two types of wet-bulb temperatures.

Applying the SFEE to the control volume surrounding the wet wick in Fig. 12.6 we have

$$A_w h_c (T_a - T_{wb}) = A_w h_m h_{fg} (\omega_{a2} - \omega_{a1}) \qquad (12.52)$$

The left hand side of Eq. (12.52) is the heat gained by the wick from the air due to forced convection and the right hand side is the energy absorbed by the water vapor evaporating from the wick. Note that we have neglected all other forms of energy interactions that usually affect the reading of a thermometer, such as radiation and conduction. The terms h_c and h_m are respectively the heat transfer coefficient and the mass transfer coefficient, and A_w is the area of the control surface just outside the wick. Rearranging Eq. (12.52) we have

$$\omega_{a1} = \omega_{a2} - (h_c / h_m h_{fg})(T_a - T_{wb}) \qquad (12.53)$$

We may rewrite Eq. (12.51) in the following form by neglecting the variation of, $(h_g - h_f) = h_{fg}$ over the narrow temperature range of practical interest. Hence

$$\omega_1 = \omega_2 - (c_{pa} / h_{fg})(T_1 - T_2) \qquad (12.54)$$

From Eqs. (12.53) and (12.54) we observe that T_2 may be equal to T_{wb} if

$$\left(\frac{h_c}{h_m c_{pa}} \right) = Le = 1 \qquad (12.55)$$

The non-dimensional quantity Le is called the *Lewis number*. Even though the Lewis number for ambient air in the temperature range of 10 to 60°C, is about 0.86, the TWT included in the psychrometric chart is found to be nearly equal to the temperature, T_{wb} measured by the wet-bulb thermometer. In contrast, for most other mixtures of gases and vapors the equality of the two wet-bulb temperatures may not hold, and therefore serious errors may result if they are assumed to be equal.

12.3 Processes of Air-Vapor Mixtures

A number of processes of importance in air conditioning systems involve mixtures of air and water vapor. These include, heating, cooling, dehumidification and evaporative cooling which are sometimes called psychrometric processes. In this section we analyze some of these psychrometric processes.

12.3.1 *Cooling, dehumidification and heating*

A typical arrangement for cooling, dehumidification, and reheating of air in an air conditioning system is depicted in Fig. 12.7(a). Air is drawn into the duct through a filter by the suction pressure of the fan. It then passes over the cooling coil, where its temperature is lowered and some of the water vapor is removed by condensation.

A typical cooling coil consists of several parallel rows of horizontal tubes through which either refrigerant or chilled-water flows. In the diagram the first two rows remove sensible heat from the air to lower its temperature. In this section of the coil the temperature of the air is above the dew-point of the incoming air, and therefore condensation of moisture does not occur. The process is indicated in the psychrometric chart in Fig. 12.7(b) by the line 1-2, along which the humidity ratio of the air is constant.

Condensation commences as the air enters the third row of tubes where its temperature is ideally equal to the dew-point. As the air passes over the rest of the rows of tubes, it is dehumidified by the continuous condensation of water vapor. During this process the air remains saturated and therefore its state follows the saturation curve from 2 to 3 in the psychrometric chart. The air enters the heating section at 3 and passes over a heating coil where its temperature is increased without any change of the moisture content. In the psychrometric chart, the heating process is represented by the line 3-4 along which the humidity ratio is constant.

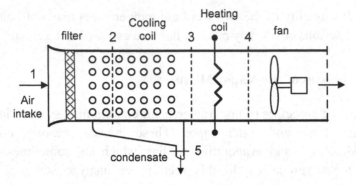

Fig. 12.7(a) System for cooling, dehumidification and heating of ambient air

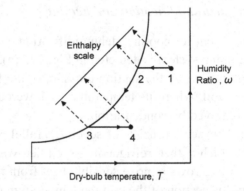

Fig. 12.7(b) Psychrometric processes for system shown in Fig. 12.7(a)

We now apply the mass conservation equation and the SFEE to the different sections of the air conditioning arrangement. The kinetic and potential energy of the air, any heat interactions with the surroundings, and the work input to the fan are neglected.

For sensible cooling of air in section 1-2 we have:

$$\dot{m}_a \omega_1 = \dot{m}_a \omega_2 \qquad (12.56)$$

and

$$\dot{Q}_{12} = \dot{m}_a (h_1 - h_2) \qquad (12.57)$$

where $\dot{m}_a$ is the mass flow rate of dry air and $\dot{Q}_{12}$ is the heat flow rate from the air to the cooling fluid in the tubes.

For the cooling and dehumidifying section 2-3:

$$\dot{m}_{con} = \dot{m}_a \omega_2 - \dot{m}_a \omega_3 \qquad (12.58)$$

and
$$\dot{Q}_{23} = \dot{m}_a h_2 - \dot{m}_a h_3 - \dot{m}_{con} h_5 \qquad (12.59)$$

where $\dot{m}_{con}$ is the rate at which the condensed water flows out of the system at 5 and h_5 is its specific enthalpy. The heat removal rate is $\dot{Q}_{23}$.

For the heating section 3-4:
$$\dot{m}_a \omega_3 = \dot{m}_a \omega_4 \qquad (12.60)$$

and
$$\dot{Q}_{34} = \dot{m}_a (h_4 - h_3) \qquad (12.61)$$

where $\dot{Q}_{34}$ is the rate of heat input.

The various humidity ratios and enthalpies of the air may be calculated using Eqs. (12.40) and (12.41) respectively. Alternatively, these quantities may be read-off directly from the psychrometric chart [5], depending on the information available on the conditions of the air at the different sections. Applications of air conditioning systems are given in worked examples 12.13 and 12.14.

12.3.2 Evaporative cooling

In hot and dry climates, evaporative cooling can be a means of lowering the temperature of air instead of passing it through a cooling coil. An evaporative cooler used for this purpose is shown in Fig. 12.8. The air is drawn by the suction pressure of the fan into the duct through a wet screen as shown in Fig. 12.8(a). The porous structure of the screen increases the area of contact between the water present inside it and the air passing through. The screen is kept moist by a steady supply of water at section 3. As in the adiabatic saturation process, the water in the screen evaporates into the air. The air cools as a result of the sensible heat transfer to the water that provides the latter with the latent heat of vaporization.

Applying the water-mass balance equation to a control volume between sections 1 and 2 we obtain
$$\dot{m}_{w3} + \dot{m}_a \omega_1 = \dot{m}_a \omega_2 \qquad (12.62)$$

Applying the SFEE, neglecting the kinetic and potential energy of the air, and the work input to the fan we have
$$\dot{m}_{w3} h_{w3} + \dot{m}_a h_{a1} = \dot{m}_a h_{a2} \qquad (12.63)$$

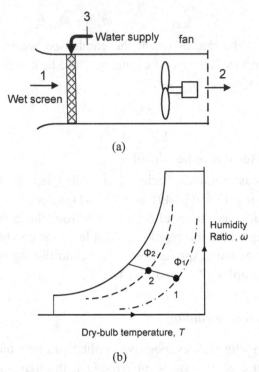

(a)

(b)

Fig. 12.8(a) Evaporative cooling system, (b) Psychrometric-process path

Substitute for the air enthalpy from Eq. (12.41) in Eq. (12.63) and solve the resulting equation simultaneously with Eq. (12.62). Hence we obtain

$$\omega_2 = \frac{\omega_1(h_{g1} - h_{w3}) + c_{pa}(T_1 - T_2)}{h_{g2} - h_{w3}} \qquad (12.64)$$

Since the temperature range of practical interest in evaporative cooling applications is small, we could neglect variations in the enthalpy difference $(h_g - h_{w3})$ at sections 1 and 2. Therefore Eq. (12.64) may be written in the approximate form:

$$\omega_2 = \omega_1 + \left(c_{pa}/h_{fg}\right)(T_1 - T_2) \qquad (12.65)$$

We note from Eqs. (12.65) and (12.51) that the outlet condition at 2 follows closely the TWT-line but the air at the exit is not necessarily saturated as in the case of an adiabatic saturator.

12.3.3 *Cooling towers*

Cooling towers are used in steam power plants, central air conditioning systems, and process industries to reject heat to the atmosphere. Usually water is the primary coolant that absorbs heat directly from the process heat load. For example, the condensers of power plants and vapor compression refrigeration systems are cooled by circulating water through them.

A simplified schematic diagram of a counter-flow cooling tower is depicted in Fig. 12.9. The hot water from the process heat load is sprayed onto a packing or filling placed within the walls of the cooling tower with the aid of a series of nozzles in a rotating arm. A fan at the top of the cooling tower draws ambient air through vents at the bottom of the outer wall. The water drips down as a thin film over the solid structure of the packing while the air flows up through pores of the packing. The moisture content of the air increases progressively due to the evaporation of water from the film. The latent heat of vaporization is provided mainly by the internal energy of water, which cools the water as a consequence.

The water carried away by the air is balanced by a steady supply of make-up water that maintains a constant water flow rate through the loop between the cooling tower and the process heat load.

We analyze the processes in the cooling tower by applying the mass balance equation and the SFEE to a control volume surrounding it, with water flow ports at 1, 3, 5, and air flow ports at 2 and 4. The work input to the fan, and any heat interactions with the surroundings are neglected. Let the dry air flow rate be $\dot{m}_a$ and the water flow rate be $\dot{m}_w$.

The mass balance equation for water is

$$\dot{m}_{w1} + \dot{m}_{w5} + \dot{m}_{a2}\omega_2 = \dot{m}_{w3} + \dot{m}_{a4}\omega_4 \qquad (12.66)$$

Mass balance equation for *dry* air is

$$\dot{m}_{a2} = \dot{m}_{a4} \qquad (12.67)$$

Apply the SFEE to a control volume surrounding the cooling tower, neglecting kinetic and potential energy changes, and the work input to the fan. Hence we obtain

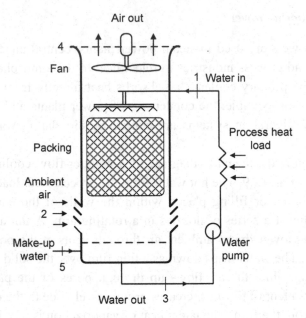

Fig. 12.9 Cooling tower

$$\dot{m}_{a2}h_{a2} + \dot{m}_{w1}h_{w1} + \dot{m}_{w5}h_{w5} = \dot{m}_{a4}h_{a4} + \dot{m}_{w3}h_{w3} \qquad (12.68)$$

where the subscripts a and w represent air and water respectively.

For the process heat load loop we have

$$\dot{m}_{w1} = \dot{m}_{w3} \qquad (12.69)$$

and

$$\dot{Q}_l = \dot{m}_{w3}(h_{w1} - h_{w3}) \qquad (12.70)$$

where $\dot{Q}_l$ is the process heat input rate.

The main advantage of the cooling tower is that ideally we could expect to cool the hot water from the process to the *wet-bulb* temperature of the ambient air. In contrast, with direct cooling using ambient air, we could only cool the water to the *dry-bulb* temperature of the air, at best. In actual cooling towers, the difference between the outlet temperature of the water at 3 and the wet-bulb temperature, $(T_3 - T_{wb})$ is called the *approach* of the cooling tower. Attaining the lowest possible approach is an important design goal for cooling towers.

12.4 Worked Examples

Example 12.1 Derive the following relations for the mass fraction f_i and the mole fraction x_i of a mixture of gases.

$$f_i = \frac{x_i M_i}{\sum\limits_{j=1}^{r} x_j M_j} \quad \text{and} \quad x_i = \frac{f_i / M_i}{\sum\limits_{j=1}^{r} f_j / M_j}$$

Solution Applying the definition of mass fraction

$$f_i = \frac{m_i}{\sum\limits_{i=1}^{r} m_i} = \frac{N_i M_i}{\sum\limits_{i=1}^{r} N_i M_i} = \frac{N_i M_i / N}{\sum\limits_{i=1}^{r} N_i M_i / N} = \frac{x_i M_i}{\sum\limits_{i=1}^{r} x_i M_i}$$

Applying the definition of mole fraction

$$x_i = \frac{N_i}{\sum\limits_{i=1}^{r} N_i} = \frac{m_i / M_i}{\sum\limits_{i=1}^{r} m_i / M_i} = \frac{m_i / mM_i}{\sum\limits_{i=1}^{r} m_i / mM_i} = \frac{f_i / M_i}{\sum\limits_{i=1}^{r} f_i / M_i}$$

Example 12.2 The volumetric analysis of a gaseous mixture at 110 kPa is as follows: $CO_2 = 23\%$, $CO = 9\%$, $O_2 = 6\%$, $N_2 = 62\%$. Calculate (i) the mass fraction of the gases in the mixture, (ii) the partial pressures of the gases, (iii) the mean molecular weight, (iv) the mean gas constant of the mixture, and (v) the mass of 20 m³ of the gas at 200 kPa and 30°C.

Solution For a mixture of ideal gases, it follows from Eq. (12.8) that $x_i = V_i / V$. We present the relevant data in the following table:

I	x_i	M_i	$M_i x_i$	f_i	$P_i = x_i P$ (kPa)
CO_2	0.23	44	10.12	0.3170	25.3
CO	0.09	28	2.52	0.07894	9.9
O_2	0.06	32	1.92	0.06015	6.6
N_2	0.62	28	17.36	0.54385	68.2

The mass fractions tabulated above are obtained using the equation

$$f_i = M_i x_i / \sum M_i x_i$$

The mean molecular weight using Eq. (12.4) is

$$\overline{M} = \sum_{i=1}^{r} x_i M_i = 31.92 \text{ kg kmol}^{-1}$$

The mean gas constant using Eq. (12.25) is

$$R_m = \sum_{i=1}^{r} f_i R_i = \sum_{i=1}^{r} \overline{R}(f_i / M_i) = 0.26 \text{ kJ kg}^{-1} \text{ K}^{-1}$$

Applying the ideal gas equation for the gas mixture we have

$$PV = mR_m T$$

Therefore $m = 200 \times 20 / 0.26 \times 303 = 50.77$ kg.

Example 12.3 The volumetric composition of a gaseous mixture at 100 kPa is as follows: $CO_2 = 30\%$, $O_2 = 8\%$, $N_2 = 62\%$. The specific heat capacities at constant volume (kJ kg^{-1} K^{-1}) of the gases are: $CO_2 = 0.85$, $O_2 = 0.92$, $N_2 = 1.04$. Calculate the mean values of c_v and c_p of the mixture.

Solution For a mixture of ideal gases, it follows from Eq. (12.8) that $x_i = V_i / V$. We present the relevant data in the following table:

i	x_i	M_i	$M_i x_i$	f_i	c_{pi}	c_{vi}
CO_2	0.30	44	13.2	0.39850	0.85	0.661
O_2	0.08	32	2.56	0.07729	0.92	0.660
N_2	0.62	28	17.36	0.5241	1.04	0.743

The mass fractions tabulated above are obtained using the equation

$$f_i = [M_i x_i / \sum M_i x_i]$$

The mean specific heat capacities are given by Eqs. (12.20) and (12.21) as:

$$\overline{c}_p = \sum_{1}^{r} f_i c_{pi} = 0.9548 \text{ kJ kg}^{-1} \text{ K}^{-1}$$

$$\overline{c}_v = \sum_{1}^{r} f_i (c_{pi} - \overline{R} / M_i) = 0.7038 \text{ kJ kg}^{-1} \text{ K}^{-1}$$

Example 12.4 A mixture of gases has the following volumetric composition at 1 bar and 30°C: $CO_2 = 30\%$, $CO = 20\%$, $N_2 = 50\%$. The mixture is compressed in an isentropic process to a pressure of 4 bar. The specific heat capacities at constant pressure, c_p (kJ kg^{-1} K^{-1}) of the gases are: $CO_2 = 0.846$, $CO = 1.04$, $N_2 = 1.04$. Calculate the (i) the final temperature, and (ii) the change in internal energy.

Solution We first calculate the mean specific heat capacity and the mean gas constant using the methods given in worked example 12.3. The relevant quantities are tabulated below:

i	x_i	M_i	$M_i x_i$	f_i	c_{pi}
CO_2	0.3	44	13.2	0.402	0.846
CO	0.2	28	5.6	0.1707	1.04
N_2	0.5	28	14.0	0.4268	1.04

The mean specific heat capacity is given by Eq. (12.21) as:

$$\bar{c}_p = \sum_1^r f_i c_{pi} = 0.9651 \text{ kJ kg}^{-1} \text{ K}^{-1}$$

The mean gas constant is given by Eq. (12.25) as:

$$R_m = \sum_1^r R(f_i / M_i) = 0.2534 \text{ kJ kg}^{-1} \text{ K}^{-1}$$

Using Eq. (12.19), the change in specific entropy of the mixture of ideal gases for a process from an initial state i to a final state f is

$$\bar{s}_f - \bar{s}_i = \sum_1^r f_i c_{vi} \ln(T_f / T_i) + f_i R_i \ln(V_f / V_i)$$

$$\bar{s}_f - \bar{s}_i = \bar{c}_v \ln(T_f / T_i) + R_m \ln(V_f / V_i) \qquad (E12.4.1)$$

For an isentropic process, $\bar{s}_f - \bar{s}_i = 0$. Therefore it follows from Eq. (E12.4.1) that

$$T_f V_f^{R_m / \bar{c}_v} = T_i V_i^{R_m / \bar{c}_v} \qquad (E12.4.2)$$

Applying the ideal gas equation to the mixture at the initial and final states we have

$$P_iV_i/T_i = P_fV_f/T_f \qquad \text{(E12.4.3)}$$

From Eqs. (E12.4.2) and (E12.4.3) we obtain the familiar relation for an isentropic process, now applicable to mixtures, as:

$$T_f/P_f^{R_m/\bar{c}_p} = T_i/P_i^{R_m/\bar{c}_p} \qquad \text{(E12.4.4)}$$

Applying Eq. (E12.4.4) the final temperature is obtained as:

$$T_f = 303 \times (4/1)^{0.2624} = 436 \text{ K}$$

The change in internal energy per unit mass is

$$\Delta\bar{u} = \sum_1^r f_i c_{vi}(T_f - T_i) = \bar{c}_v(T_f - T_i)$$

$$\Delta\bar{u} = (\bar{c}_p - R_m)(T_f - T_i) = 0.7118 \times 133 = 94.7 \text{ kJ kg}^{-1}$$

Example 12.5 A stream of nitrogen at 1.3 bar and 250°C is mixed with a stream of hydrogen at 1.3 bar and 40°C in a steady-flow adiabatic process. The ratio of the mass flow rates of nitrogen to hydrogen is 2.5. The pressure after mixing is 1.15 bar. Calculate (i) the final temperature of the mixture, and (ii) the net entropy change per unit mass of mixture. The specific heat capacities at constant pressure, c_p (kJ kg^{-1} K^{-1}) of the gases are: H_2 = 14.5, N_2 = 1.04.

Solution The specific heat capacity of the mixture is

$$\bar{c}_p = (f_1 c_{p1} + f_2 c_{p2})$$

$$\bar{c}_p = 14.5 \times 1/3.5 + 1.04 \times 2.5/3.5 = 4.886 \text{ kJ kg}^{-1} \text{ K}^{-1}$$

where subscripts 1 and 2 denote hydrogen and nitrogen respectively.
The gas constants for hydrogen and nitrogen are:

$$R_1 = 8.3145/2 = 4.157 \text{ kJ kg}^{-1} \text{ K}^{-1}$$

$$R_2 = 8.3145/28 = 0.297 \text{ kJ kg}^{-1} \text{ K}^{-1}$$

Applying the SFEE to the adiabatic mixing process we have

$$\dot{m}_1 h_1 + \dot{m}_2 h_2 = (\dot{m}_1 + \dot{m}_2)h_3$$

$$\dot{m}_1 c_{p1} T_1 + \dot{m}_2 c_{p2} T_2 = (\dot{m}_1 + \dot{m}_2)\bar{c}_p T_3$$

$$\dot{m}_1 \times 14.5 \times 40 + 2.5\dot{m}_1 \times 1.04 \times 250 = 3.5\dot{m}_1 \times 4.886T_3$$

where $\dot{m}_1$ is the mass flow rate of hydrogen.

Hence the temperature after mixing is, $T_3 = 71.9\,°C$

The mole fractions of hydrogen and nitrogen in the mixture are:

$$x_1 = \frac{1/2}{1/2 + 2.5/28} = 0.8485$$

$$x_2 - \frac{2.5/28}{1/2 + 2.5/28} = 0.1515$$

Therefore the partial pressures of the two gases after mixing are:

$$P_{m1} = 1.15 \times 0.8485 = 0.9757 \ \text{bar}$$

$$P_{m2} = 1.15 \times 0.1515 = 0.1742 \ \text{bar}$$

The changes in entropy of the two ideal gases are given by

$$\Delta S_1 = \dot{m}_1 \left[c_{p1} \ln(T_3/T_1) - R_1 \ln(P_{m1}/P_1) \right]$$

$$\Delta S_1 = \dot{m}_1 \left[14.5 \ln(344.9/313) - 4.157 \ln(0.9757/1.3) \right] = 2.60\dot{m}_1$$

and
$$\Delta S_2 = 2.5\dot{m}_1 \left[c_{p2} \ln(T_3/T_2) - R_2 \ln(P_{m2}/P_2) \right]$$

$$\Delta S_2 = 2.5\dot{m}_1 \left[1.04 \ln(344.9/523) - 0.297 \ln(0.1742/1.3) \right]$$

$$\Delta S_2 = 0.41\dot{m}_1$$

The net entropy change per unit mass of mixture is

$$\Delta S_{net} = (2.6\dot{m}_1 + 0.41\dot{m}_1)/3.5\dot{m}_1 = 0.86 \ \text{kJ K}^{-1} \text{kg}^{-1}$$

which is the entropy production per unit mass due to the mixing irreversibility.

Example 12.6 A rigid pressure vessel of volume 0.3 m³ contains oxygen at 350 kPa and 25°C. The vessel is connected through a valve to a pipe carrying nitrogen at a constant pressure and temperature of 700 kPa and 150°C respectively. The valve is opened, and nitrogen flows into the vessel in an adiabatic process until the pressure in the vessel reaches 600 kPa. The valve is then closed. Calculate (i) the temperature of the gas mixture in the tank, and (ii) the entropy production. The specific heat capacities, c_p (kJ kg⁻¹ K⁻¹) of the gases are: $O_2 = 0.92$ and $N_2 = 1.04$.

Solution Let the subscripts 1 and 2 denote quantities related to oxygen and nitrogen respectively. The initial and final states are identified by i and f respectively. Let the partial pressures of oxygen and nitrogen in the final state be P_{1f} and P_{2f}. The final equilibrium temperature is T_f. Now the mass of nitrogen that flows into the vessel is equal to the final mass of nitrogen in the vessel.

Therefore applying the ideal gas equation to the final mass of nitrogen in the vessel we obtain

$$\Delta m_{in} = P_{2f}V/R_2T_f \qquad (E12.6.1)$$

The mass of oxygen in the vessel is constant. Applying the ideal gas equation to the oxygen we have

$$P_{1f}V/R_1T_f = P_{1i}V/R_1T_{i1}$$

$$P_{1f} = 350T_f/298 = 1.174T_f \qquad (E12.6.2)$$

Applying Dalton's rule to the final state

$$P_{2f} = 600 - P_{1f} = 600 - 1.174T_f \qquad (E12.6.3)$$

Integrating the energy equation from the initial to the final states of the filling process we have

$$E_f - E_i = \Delta m_{in}h_o \qquad (E12.6.4)$$

where E is the internal energy of the contents of the vessel and h_o is the constant enthalpy of the nitrogen in the pipe.

Applying the ideal gas equation to the two gases, and substituting the resulting relations in Eq. (E12.6.4) we obtain

$$\left(\frac{P_{1f}V}{R_1T_f}\right)c_{v1}T_f + \left(\frac{P_{2f}V}{R_2T_f}\right)c_{v2}T_f - \left(\frac{P_{1i}V}{R_1T_{1i}}\right)c_{v1}T_{1i} = \Delta m_{in}c_{p2}T_o \qquad (E12.6.5)$$

The following data pertaining to oxygen and nitrogen are used in the numerical calculation:

$$R_1 = \overline{R}/32 = 0.2598 \text{ kJ kg}^{-1}\text{ K}^{-1}, \quad R_2 = \overline{R}/28 = 0.2969 \text{ kJ kg}^{-1}\text{ K}^{-1},$$

$$c_{v1} = c_{p1} - R_1 = 0.66 \text{ kJ kg}^{-1}\text{ K}^{-1}, \quad c_{v2} = c_{p2} - R_2 = 0.743 \text{ kJ kg}^{-1}\text{ K}^{-1}$$

We now substitute in Eq. (E12.6.5) from Eqs. (E12.6.1) and (E12.6.3). After manipulating the resulting equation and substituting the numerical data we obtain the following simplified form

$$2.982T_f + 2.502(600 - 1.174T_f) - 889.145$$

$$= 1481.7(600 - 1.174T_f)/T_f$$

$$0.04465T_f^2 + 2351.5T_f - 1481.7 \times 600 = 0$$

The positive root of the above quadratic equation gives the final temperature as $T_f = 375.4$ K.

We apply the entropy balance equation to the filling process to obtain

$$\sigma = S_{vf} - S_{vi} - \Delta m_{in} s_o$$

$$\sigma = \left(\frac{P_1 V}{R_1 T_{1i}}\right) c_{v1} \ln\left(\frac{T_f}{T_{1i}}\right) + \left(\frac{P_{2f} V}{R_2 T_f}\right)\left(c_{p2} \ln\left(\frac{T_f}{T_o}\right) - R_2 \ln\left(\frac{P_{2f}}{P_{2i}}\right)\right)$$

In the above equation the first term is the increase in entropy of the oxygen which has a constant volume but undergoes an increase in temperature. The second term is the increase in entropy of nitrogen in the vessel due to the change in pressure and temperature. Substituting numerical values in the above equation we have

$$\sigma = 1.356 \times 0.66 \ln\left(\frac{375.4}{298}\right) + 0.4287\left(1.04 \ln\left(\frac{375.4}{423}\right) - 0.2969 \ln\left(\frac{159.28}{700}\right)\right)$$

$$\sigma = 0.2066 + 0.1352 = 0.3418 \text{ kJ K}^{-1}$$

The entropy production is due to the irreversible mixing of the two gases.

Example 12.7 The volume of a mixture of carbon dioxide and nitrogen at 100 kPa and 20°C is 4 m³. The mass fractions of carbon dioxide and nitrogen are 0.55 and 0.45 respectively. The mixture is compressed to a final volume and temperature of 0.02 m³ and 40°C. In the initial state the gases may be treated as ideal gases. In the final compressed state assume that the van der Waals equation of state applies. The constants a [bar (m³ kmol⁻¹)²] and b [m³ kmol⁻¹] in the van der Waals equation for carbon dioxide and nitrogen are respectively, 3.643, 0.0427 and 1.361, 0.0385. Calculate the final pressure of the gas.

Solution The mole fractions of carbon dioxide and nitrogen are:

$$x_1 = \frac{0.55/44}{0.55/44 + 0.45/28} = 0.4375$$

$$x_2 = \frac{0.45/28}{0.55/44 + 0.45/28} = 0.5627$$

Applying the ideal gas equation to the initial state

$$P_i V_i = N\overline{R}T_i$$

Hence $\quad N = 100 \times 4/(8.3145 \times 293) = 0.164$

which is the total number of kmols.

Although the ideal gas equation is not applicable to the final state, we shall assume that Dalton's rule is still valid. Applying Eq. (12.29) we obtain the final pressure as

$$P = \sum_i^2 P_i = \sum_1^2 \left[\frac{\overline{R}Tx_i}{(v_m - x_i b_i)} - \frac{a_i x_i^2}{v_m^2} \right] \qquad \text{(E12.7.1)}$$

The molar specific volume in the final state is

$$v_m = 0.02/0.164 = 0.1219 \ \text{m}^3 \ \text{kmol}^{-1}$$

Substituting in Eq. (E12.7.1) we have

$$P_f = 8.3145 \times 313 \times \left(\frac{0.4375}{0.1219 - 0.4375 \times 0.0427} + \frac{0.563}{0.1219 - 0.563 \times 0.0385} \right)$$

$$-\left(\frac{364.3 \times 0.4375^2}{0.1219^2} + \frac{136.1 \times 0.563^2}{0.1219^2} \right) = 180.5 \ \text{bar}$$

For comparison, we now apply the ideal gas equation to the final state

$$P_f V_f = N\overline{R}T_f$$

$$P_f = 0.164 \times 8.3145 \times 313/0.02 = 213.4 \ \text{bar}$$

There is a difference of about 18% in the final pressure predicted by the van der Waals equation and the ideal gas equation. The ideal gas equation is generally not accurate at high pressures above the critical pressure.

Example 12.8 A piston-cylinder arrangement contains a mixture of dry air and water vapor at 55°C and 135 kPa. The initial relative humidity and volume are 40% and 0.055 m³ respectively. The mixture undergoes a constant pressure cooling process. (a) Calculate the temperature at which the condensation of water just begins, (b) If the cooling is continued until the temperature is 16°C, calculate (i) the partial pressure of the vapor, (ii) the mass of water condensed, and (iii) the total heat removed in the cooling process.

Solution From the steam tables in [4] the saturation vapor pressure at 55°C is 15.74 kPa. The relative humidity in the initial state is 40%. The humidity ratio in the initial state is obtained using Eq. (12.40) as:

$$\omega_1 = \frac{0.622\phi_1 P_{g1}(T_1)}{P - \phi_1 P_{g1}(T_1)} = \frac{0.622 \times 0.4 \times 15.74}{135 - 15.74 \times 0.4} = 0.0304$$

The partial pressure of the vapor in the initial state is

$$P_{v1} = \phi_1 P_{g1} = 0.4 \times 15.74 = 6.296 \text{ kPa}.$$

During the constant pressure process the vapor pressure remains constant until condensation just begins when the vapor pressure,

$$P_{v2} = P_{g2} = 6.296 \text{ kPa}.$$

From the steam table [4] we obtain the saturation temperature at 6.296 kPa as 37.04°C. This is the dew-point of the air in the cylinder.

Further cooling of the air to 16°C results in more condensation. The air is saturated during this cooling process with a relative humidity of 100%. The vapor pressure at 16°C is obtained from the steam table as 1.817 kPa. We calculate the humidity ratio using Eq. (12.40).

$$\omega_2 = \frac{0.622 P_{g2}(T_2)}{P - P_{g2}(T_2)} = \frac{0.622 \times 1.817}{135 - 1.817} = 0.00848$$

The constant mass of dry air in the cylinder is obtained by applying the ideal gas equation to the initial state.

$$P_{a1} V_1 = m_a R_a T_1$$

$$m_a = \frac{0.055 \times (135 - 6.296) \times 28.96}{8.3145 \times (273 + 55)} = 0.07516 \text{ kg}$$

The total mass of water condensed is

$$m_{con} = m_a(\omega_1 - \omega_2)$$

$$m_{con} = 0.07516 \times (0.0304 - 0.00848) = 0.00164 \text{ kg}$$

We calculate the enthalpy of moist air using Eq. (12.41).

$$h = c_{pa}T + \omega h_g$$

For the initial state, the saturated vapor enthalpy, h_g at 55°C, is obtained from the steam table [4] as 2600.3 kJ kg^{-1}. Hence for moist air

$$h_1 = 1.00 \times 55 + 0.0304 \times 2600.3 = 134.05 \text{ kJ kg}^{-1}$$

In the final state, the saturated enthalpy h_g at 16°C is 2530.2 kJ kg^{-1}, and

$$h_2 = 1.00 \times 16 + 0.00848 \times 2530.2 = 37.45 \text{ kJ kg}^{-1}$$

The enthalpy of the condensed liquid at 16°C is $h_f = 67.1$ kJ kg^{-1}

Applying the first law to the constant pressure process we have

$$Q_{12} = H_2 - H_1 = m_a h_2 + m_{con} h_f - m_a h_1$$

where H is the total enthalpy.

$$Q_{12} = 0.07516 \times (37.45 - 134.05) + 67.1 \times 0.00164 = -7.15 \text{ kJ}$$

The negative sign indicates that heat is removed from the system. We should not use the psychrometric chart to solve this example, because the chart in [5] is for a pressure of 101.3 kPa whereas the pressure here is 135 kPa.

Example 12.9 A rigid vessel of volume 0.28 m^3 contains a mixture of dry air and water vapor whose temperature, pressure and relative humidity are 27°C, 100 kPa and 60% respectively. (a) If the vessel is cooled, calculate the temperature at which condensation of water vapor just begins. (b) If the air is cooled by another 5°C, calculate (i) the mass of water condensed, and (ii) the total heat removed.

Solution From the steam table [4] the saturation vapor pressure at 27°C is 3.564 kPa. The relative humidity in the initial state is 60%. The humidity ratio in the initial state is obtained using Eq. (12.40) as

$$\omega_1 = \frac{0.622\phi_1 P_{g1}(T_1)}{P - \phi_1 P_{g1}(T_1)} = \frac{0.622 \times 0.6 \times 3.564}{100 - 3.564 \times 0.6} = 0.01359$$

The moist air undergoes a constant volume process. Applying the ideal gas equation to the air at the initial state, 1 and the state 2, when condensation just begins, we obtain

$$P_{a1} / P_{a2} = (T_1 + 273)/(T_2 + 273) \qquad \text{(E12.9.1)}$$

Now $\qquad P_{a1} = 100 - 0.6 \times 3.564 = 97.86$ kPa.

Substituting in Eq. (E12.9.1)

$$97.86 / P_{a2} = 300/(T_2 + 273)$$

$$P_{a2} = 0.3262(T_2 + 273) \qquad \text{(E12.9.2)}$$

During the constant volume process, the humidity ratio remains constant until condensation begins at state 2. Therefore

$$\omega_2 = \frac{0.622 P_{g2}}{P_{a2}} = 0.01359 \qquad \text{(E12.9.3)}$$

where P_{g2} is the saturation vapor pressure at 2.

Substituting from Eq. (E12.9.2) in Eq. (E12.9.3)

$$\frac{0.622 P_{g2}(T_2)}{0.3262(T_2 + 273)} = 0.01359$$

$$P_{g2}(T_2) = 0.007127(T_2 + 273) \qquad \text{(E12.9.4)}$$

We obtain the temperature T_2 at which condensation just begins, by solving Eq. (E12.9.4) using a trial-and-error approach. Make an initial guess for T_2 and obtain $P_{g2}(T_2)$ from the steam table [4]. Check whether Eq. (E12.9.4) is satisfied. After a few iterations we obtain $T_2 = 18°C$.

In the second process, 2-3 the cooling is continued until the temperature, $T_3 = 18 - 5 = 13°C$. During this process the air remains saturated with vapor. From the steam tables [4], the vapor pressure at 13°C is 1.497 kPa.

Applying the ideal gas equation to the dry air at state 1 and state 3 we have

$$P_{a3} / P_{a1} = (T_3 + 273)/(T_1 + 273) = 286/300$$

$$P_{a3} = 97.86 \times 286/300 = 93.29 \text{ kPa}$$

The humidity ratio at state 3 is

$$\omega_3 = \frac{0.622 P_{g3}}{P_{a3}} = \frac{0.622 \times 1.497}{93.29} = 0.00998$$

The mass of dry air in the vessel is

$$m_a = P_1 V_1 / R_a T_1 = 97.86 \times 0.28 / (0.287 \times 300) = 0.318 \text{ kg}$$

The mass of water condensed is

$$m_{con} = m_a (\omega_1 - \omega_3)$$

$$m_{con} = 0.318 \times (0.01359 - 0.00998) = 0.001148 \text{ kg}$$

The enthalpy of air at state 1 and state 3 are:

$$h_1 = c_{pa} T_1 + \omega_1 h_{g1} = 27 + 0.01359 \times 2550.3 = 61.66 \text{ kJ kg}^{-1}$$

$$h_3 = c_{pa} T_3 + \omega_3 h_{g3} = 13 + 0.00998 \times 2524.8 = 38.19 \text{ kJ kg}^{-1}$$

The internal energy of the condensed liquid at 13°C is $u_{f3} = 55 \text{ kJ kg}^{-1}$.

The total internal energy at state 1 and state 3 are obtained using the expression

$$U = H - PV$$

Hence

$$U_1 = m_a h_1 - P_1 V$$

$$U_3 = m_{con} u_{f3} + m_a h_3 - P_3 V$$

Applying the first law to the closed system of the rigid vessel we have

$$Q_{13} = U_3 - U_1$$

Therefore

$$Q_{13} = m_{con} u_{f3} + m_a (h_3 - h_1) + V(P_1 - P_3)$$

$$Q_{13} = 0.001148 \times 55 + 0.318 \times (38.19 - 61.66) + 0.28 \times (100 - 94.787)$$

$$Q_{13} = -5.936 \text{ kJ}$$

The negative sign signifies that heat is removed from the system.

Example 12.10 The molar composition of the gases present in a sample of exhaust products from a combustion process has been determined by gas analysis as:

$$CO_2 = 0.051, \quad H_2O = 0.18, \quad O_2 = 0.03, \quad N_2 = 0.738$$

The pressure and temperature of the gas mixture are 104 kPa and 180°C. (i) If the mixture is cooled at constant pressure, calculate the temperature at which condensation of water vapor just begins. (ii) If the mixture is cooled to 30°C at constant pressure, calculate the mass fraction of the water present originally in the mixture that is condensed.

Solution The partial pressure of water vapor in the mixture is

$$P_w = x_w P = 0.18 \times 104 = 18.72 \text{ kPa}$$

Condensation just begins at the temperature when the saturation vapor pressure is 18.72 kPa. From the steam tables [4] we obtain the saturation temperature as 58.56°C. This temperature is the dew-point of the exhaust gases and it is an important parameter of combustion processes.

When the temperature is 30°C, the saturation vapor pressure is 4.242 kPa. Therefore the mole fraction of the water vapor in the mixture is

$$x_w = P_w / P = 4.242 / 104 = 0.04079$$

Now the ratio of the moles of vapor to the moles of the gases in the mixture is given by

$$N_w / N_g = x_w / (1 - x_w)$$

Applying the above equation to the original mixture and the mixture at 30°C we have

$$N_w / N_g = 0.18 / (1 - 0.18) = 0.2195$$

$$N'_w / N_g = 0.04079 / (1 - 0.04079) = 0.0425$$

Hence $\quad N'_w / N_w = 0.0425 / 0.2195 = 0.1936$

Therefore the fraction of water condensed is given by

$$(N_w - N'_w) / N_w = 0.806$$

Example 12.11 Air enters a compressor at 30°C, 101 kPa and 60% relative humidity. The air is compressed to 400 kPa before entering an intercooler. Calculate the lowest temperature to which the air may be cooled in the intercooler if condensation is to be prevented.

Solution The humidity ratio of the air entering the compressor is given by

$$\omega_1 = \frac{0.622\phi_1 P_{g1}(T_1)}{P - \phi_1 P_{g1}(T_1)} = \frac{0.622 \times 0.6 \times 4.242}{101 - 4.242 \times 0.6} = 0.016079$$

The humidity ratio of the air remains constant during the compression process. Let the air be cooled in the intercooler to a temperature at which condensation just begins. The air is then in a saturated state at a pressure of 400 kPa. Therefore

$$\omega_2 = \frac{0.622 \times 1.0 \times P_{g2}(T_2)}{400 - 1.0 \times P_{g2}(T_2)} = 0.016079$$

Hence $P_{g2} = 10.08$ kPa. The corresponding saturation vapor temperature from the steam table [4] is about 46°C. Therefore if the air is cooled to a temperature below 46°C in the intercooler, condensation of water will occur.

Example 12.12 A psychrometric chart for a mixture of air and water is to be developed for a pressure of 80 kPa. To illustrate the procedure for constructing the following lines: (i) the constant relative humidity line at $\phi = 60\%$, (ii) the constant enthalpy line at $h = 50$ kg per kg of dry air, (iii) the constant specific volume line at 1.1 m³ per kg of dry air, and (iv) the constant thermodynamic wet-bulb temperature line at 15°C, obtain the humidity ratio, ω for a dry-bulb temperature of 20°C, so that (20°C, ω) is one point on the constant property line.

Solution At 20°C the saturation vapor pressure from the steam tables [4] is 2.337 kPa. (i) To calculate the humidity ratio at $\phi = 60\%$ and $P = 80$ kPa we use Eq. (12.40)

$$\omega = \frac{0.622\phi P_g(T)}{P - \phi P_g(T)} = \frac{0.622 \times 0.6 \times 2.337}{80 - 0.6 \times 2.337} = 0.01109$$

(ii) The line of constant enthalpy at $h = 50$ kJ kg⁻¹ is constructed using the Eq. (12.42),

$$\omega = (h - c_{pa}T)/h_g(T)$$

From the steam table [4], the saturation vapor enthalpy at 20°C is 2537.6 kJ kg^{-1}. The line of constant enthalpy for $h = 50$ kJ kg^{-1} is obtained by substituting in the above equation.

$$\omega = (50 - 1.0 \times 20) / 2537.6 = 0.01182$$

(iii) The line of constant specific volume is given by Eq. (12.45). For $v = 1.1$ we obtain

$$\omega = 0.622 \left(\frac{Pv}{R_a T(K)} - 1 \right) = 0.622 \left(\frac{80 \times 1.1}{0.287 \times 293} - 1 \right) = 0.0289$$

(iv) The line of constant wet-bulb temperature is given by Eq. (12.51),

$$\omega_1 = \frac{\omega_2 (h_{g2} - h_{f2}) - c_{pa}(T_1 - T_2)}{h_{g1} - h_{f2}} \tag{E12.12.1}$$

At the given wet-bulb temperature, $T_{wb} = 15$°C, we obtain the following data from the steam tables [4].

The saturation vapor enthalpy, $h_{g2} = 2528.4$ kJ kg^{-1}

The saturation liquid enthalpy, $h_{f2} = 62.9$ kJ kg^{-1}

The saturation pressure, $P_{g2} = 1.704$ kPa

At 20°C, $h_{g1} = 2537.6$ kJ kg^{-1}.

We then calculate ω_2 for saturated air at 15°C using Eq. (12.40) as

$$\omega_2 = \frac{0.622 P_{g2}}{P - P_{g2}} = \frac{0.622 \times 1.704}{80 - 1.704} = 0.01354$$

Finally, substituting numerical values in Eq. (E12.12.1) we have

$$\omega_1 = \frac{0.01354 \times (2528.4 - 62.9) - 1.0(20 - 15)}{2537.6 - 62.9} = 0.01147$$

In the above calculations we have illustrated how to obtain one point on each constant property line in the psychrometric chart. In order to plot smooth curves we need to repeat these calculations by taking different values of the dry bulb temperature. Then by varying the value of the particular property we generate the family of constant property lines.

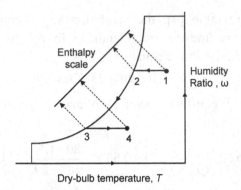

Fig. E12.13 Psychrometric chart

Example 12.13 The air conditioning system shown in Fig. 12.7(a) is used to supply a space with air at 20°C and 30% relative humidity. The ambient air at the intake to the system is at 30°C and 45% relative humidity. The pressure in the system is 101.3 kPa and the dry air flow rate is 0.9 kg s⁻¹. Calculate (i) the dew-point of the ambient air, (ii) the rate of moisture removal by the cooling coil, (iii) the heat removal rate of the cooling coil, and (iv) the heat supply rate of the heating coil.

Solution We refer to the schematic diagram in Fig. 12.7(a) and the corresponding psychrometric chart shown in Fig. E12.13. The data required for the solution could be obtained directly from the psychrometric chart in [5] that is developed for a pressure of 101.3 kPa.

However, for the purpose of illustrating their application we shall use the various equations derived in Sec. 12.2 to obtain the relevant data.

At the inlet section 1, $T_1 = 30°C$, $\phi_1 = 45\%$.

From the steam tables [4], $P_{g1} = 4.242$ kPa and $h_{g1} = 2555.7$ kJ kg⁻¹.

At the exit section 4, $T_4 = 20°C$, $\phi_4 = 30\%$.

From the steam tables [4], $P_{g4} = 2.337$ kPa and $h_{g4} = 2537.6$ kJ kg⁻¹.

We first calculate the following properties that are needed for the analysis.

$$\omega_1 = \frac{0.622\phi_1 P_{g1}}{P - \phi_1 P_{g2}} = \frac{0.622 \times 0.45 \times 4.242}{101.3 - 0.45 \times 4.242} = 0.011946$$

$$\omega_4 = \frac{0.622\phi_4 P_{g4}}{P - \phi_4 P_{g4}} = \frac{0.622 \times 0.3 \times 2.337}{101.3 - 0.3 \times 2.337} = 0.004335$$

$$h_1 = c_{pa}T_1 + \omega_1 h_{g1} = 1 \times 30 + 0.011946 \times 2555.7 = 60.53 \text{ kJ kg}^{-1}$$

$$h_4 = c_{pa}T_4 + \omega_4 h_{g4} = 1 \times 20 + 0.004335 \times 2537.6 = 31.0 \text{ kJ kg}^{-1}$$

At the dew-point temperature the air becomes just saturated as indicated by state 2 in Fig. E12.13. Moreover, during the sensible cooling process 1-2, the humidity ratio of the air remains constant because no moisture is condensed during this process.

Hence $\qquad\qquad \omega_2 = \omega_1 = 0.011946$

Applying Eq. (12.40) we have

$$\frac{0.622 P_{g2}}{101.3 - P_{g2}} = 0.011946$$

Hence $\qquad\qquad P_{g2} = 1.9088 \text{ kPa.}$

From the steam tables [4], the saturation temperature corresponding to the pressure of 1.9088 kPa is 16.77°C, which is the dew-point of the ambient air.

Applying the mass balance equation for water between sections 1 and 4 we obtain the rate of condensation as

$$\dot{m}_{con} = \dot{m}_a(\omega_1 - \omega_4)$$

$$\dot{m}_{con} = 0.9 \times (0.011946 - 0.004335) = 0.00685 \text{ kg s}^{-1}$$

Now during the heating process 3-4, the humidity ratio is constant and therefore $\omega_3 = \omega_4$. Also, the air leaving the cooling coil at section 3 is saturated as shown in Fig. E12.13 and therefore $\phi_3 = 100\%$. Applying Eq. (12.40) to state 3 we have

$$\omega_3 = \frac{0.622 P_{g3}}{101.3 - P_{g3}} = 0.004335$$

Hence $\qquad\qquad P_{g3} = 0.701 \text{ kPa.}$

From the steam tables [4], the saturation temperature corresponding to a pressure of 0.701 kPa is 2.08°C and $h_{g3} = 2504.5 \text{ kJ kg}^{-1}$.

The enthalpy at section 3 is given by

$$h_3 = c_{pa}T_3 + \omega_3 h_{g3} = 1 \times 2.08 + 0.004335 \times 2504.5 = 12.94 \text{ kJ kg}^{-1}$$

Assume that the condensate is at the mean temperature of the air, which is $0.5(T_2 + T_3) = 9°C$. Hence from the steam tables, $h_{f5} = 37.8 \text{ kJ kg}^{-1}$.

Applying the SFEE to a control volume between sections 1 and 3 we obtain

$$\dot{m}_a h_1 = \dot{m}_a h_3 + \dot{m}_{con} h_{f5} + \dot{Q}_{cool}$$

$$\dot{Q}_{cool} = 0.9 \times (60.53 - 12.94) - 0.00685 \times 37.8$$

$$\dot{Q}_{cool} = 42.83 - 0.25 = 42.58 \text{ kW}$$

We note that the enthalpy carried away by the condensate is negligible.

Applying the SFEE to a control volume between sections 3 and 4 we have

$$\dot{m}_a h_4 = \dot{m}_a h_3 + \dot{Q}_{heat}$$

$$\dot{Q}_{heat} = 0.9 \times (31 - 12.94) = 16.25 \text{ kW}$$

Example 12.14 Ambient air at 30°C dry-bulb temperature, 25°C wet-bulb temperature and pressure 101 kPa enters a window air conditioner of a room. Saturated air and the condensed water both leave the air conditioner at 14°C. The mass flow rate of air at entry is 0.5 kg s^{-1}. Using data from the psychrometric chart, calculate (i) the dry air flow rate, (ii) the dew-point of the air, (iii) the rate of removal of moisture, and (iv) the cooling capacity of the air conditioner.

Solution The psychrometric chart in Fig. E12.14 shows the cooling process undergone by the air. The air entry condition, 1 is given by the intersection of the 25°C- constant wet-bulb temperature line and the vertical line passing through the 30°C-dry-bulb temperature.

The horizontal line 1-2 indicates the sensible cooling of the air until at condition 2 on the saturation line, the air attains the dew-point temperature. We read the dew-point temperature as 23.2°C. There is no condensation during the process 1-2. Moisture is removed from the air in

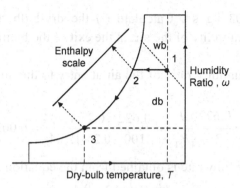

Fig. E12.14 Psychrometric chart

the cooling and dehumidification process 2-3. The given exit dry-bulb temperature of 14°C of the air enables us to locate condition 3 on the saturation line. The following properties are read directly from the chart in [5]:

$$\omega_1 = 0.018, \ h_1 = 76 \ \text{kJ kg}^{-1}, \ \omega_3 = 0.0098, \ h_3 = 39 \ \text{kJ kg}^{-1}$$

Now the dry air flow rate is

$$\dot{m}_a = \dot{m}/(1+\omega_1) = 0.5/1.018 = 0.491 \,\text{kg s}^{-1}$$

The rate of moisture removal is

$$\dot{m}_w = \dot{m}_a(\omega_1 - \omega_3) = 0.491 \times (0.018 - 0.0098) = 0.00402 \ \text{kg s}^{-1}$$

The enthalpy of the condensate at the mean temperature is 76 kJ kg^{-1}.

Apply the SFEE to the control volume surrounding the air conditioner. The cooling capacity is given by

$$\dot{Q}_c = \dot{m}_a(h_1 - h_3) - \dot{m}_w h_{con}$$

$$\dot{Q}_c = 0.491 \times (76 - 39) - 76 \times 0.00402 = 17.9 \,\text{kW}$$

Note that if the enthalpy of the condensate is neglected, the cooling capacity is 18.1 kW.

Example 12.15 Ambient air entering a steam humidifier at 100 kPa, 15°C and 20% relative humidity flows through it at a constant pressure. The mass flow rate of air at entry is 0.4 kg s^{-1}. In the humidifier, saturated steam at a pressure of 100 kPa is injected into the air stream at the rate of 0.003 kg s^{-1}. Calculate (i) the dry-bulb temperature, and (ii) the relative humidity of the air, at the exit of the humidifier.

Solution The humidity ratio of the air at entry to the humidifier is given by Eq. (12.40).

$$\omega_1 = \frac{0.622\phi_1 P_{g1}}{P - \phi_1 P_{g1}} = \frac{0.622 \times 0.2 \times 1.704}{100 - 0.2 \times 1.704} = 0.002127$$

The dry air mass flow rate is obtained from the equation

$$\dot{m}_a(1 + \omega_1) = 0.4$$

Hence $\dot{m}_a = 0.399$ kg s^{-1}

Consider a control volume surrounding the humidifier. Applying the mass balance equation for water we have

$$\dot{m}_a(\omega_2 - \omega_1) = \dot{m}_{steam} = 0.003 \text{ kg s}^{-1}$$

$$\omega_2 = 0.002127 + 0.003/0.399 = 0.009646$$

Applying the SFEE to the control volume surrounding the humidifier we obtain

$$\dot{m}_a h_1 + h_{gs}\dot{m}_{steam} = \dot{m}_a h_2 \qquad \text{(E12.15.1)}$$

We use Eq. (12.41) to obtain the air enthalpies as

$$h_1 = c_{pa}T_1 + \omega_1 h_{g1} = 1 \times 15 + 0.002127 \times 2528.4 = 20.378 \text{ kJ kg}^{-1}$$

and $h_2 = c_{pa}T_2 + \omega_2 h_{g2} = T_2 + 0.00964h_{g2}$

The saturated steam enthalpy at 100 kPa is $h_{gs} = 2675$ kJ kg^{-1}
   Substituting in Eq. (E12.15.1) we have

$$20.378 + 2675 \times 0.003/0.399 = T_2 + 0.009646h_{g2}(T_2)$$

$$T_2 + 0.009646h_{g2}(T_2) = 40.49$$

A trial-and-error method is used to solve the above equation by making an initial guess of the exit temperature T_2. The enthalpy $h_{g2}(T_2)$ is then obtained from the steam tables [4]. The iterative procedure is continued until the equation is satisfied. The exit dry-bulb temperature thus obtained is about 16°C. Applying Eq. (12.40)

$$\omega_2 = \frac{0.622\phi_2 P_{g2}}{P - \phi_2 P_{g2}} = \frac{0.622 \times 1.817\phi_2}{100 - 1.817\phi_2} = 0.009646$$

Hence we obtain the exit relative humidity as, $\phi_2 = 84\%$.

Example 12.16 In an air conditioning system outdoor air at 32°C and 50% relative humidity is mixed with return air at 26°C and relative humidity 64%. The air flow rates of the outdoor air stream and the return air stream are 1.8 kg s⁻¹ and 3.0 kg s⁻¹ respectively. The constant pressure in the adiabatic mixing section is 101 kPa. Calculate (i) the enthalpy of the mixture, (ii) the humidity ratio of the mixture, and (iii) the dry-bulb temperature of the mixture.

Solution We first calculate the following properties needed for the analysis of the mixing process. Let the subscripts 1, 2 and 3 denote quantities related to the outdoor air, the return air, and the mixed air respectively. Applying Eq. (12.40) we obtain

$$\omega_1 = \frac{0.622\phi_1 P_{g1}}{P - \phi_1 P_{g1}} = \frac{0.622 \times 0.5 \times 4.754}{101 - 0.5 \times 4.754} = 0.01499$$

$$\omega_2 = \frac{0.622\phi_2 P_{g2}}{P - \phi_2 P_{g2}} = \frac{0.622 \times 0.64 \times 3.36}{101 - 0.64 \times 3.36} = 0.01353$$

The dry air flow rates of the two streams are:

$$\dot{m}_{a1} = \dot{m}_1 /(1 + \omega_1) = 1.8/1.01499 = 1.773 \text{ kg s}^{-1}$$

$$\dot{m}_{a2} = \dot{m}_2 /(1 + \omega_2) = 3.0/1.01353 = 2.96 \text{ kJ kg}^{-1}$$

Using Eq. (12.41) we obtain the enthalpies as:

$$h_1 = c_{pa}T_1 + \omega_1 h_{g1} = 1 \times 32 + 0.01499 \times 2559.3 = 70.36 \text{ kJ kg}^{-1}$$

$$h_2 = c_{pa}T_2 + \omega_2 h_{g2} = 1 \times 26 + 0.01353 \times 2548.4 = 60.48 \text{ kJ kg}^{-1}$$

Consider a control volume surrounding the mixing section. Applying the mass balance equations for dry air and water we obtain

$$\dot{m}_{a1} + \dot{m}_{a2} = \dot{m}_{a3} \tag{E12.16.1}$$

$$\dot{m}_{a1}\omega_1 + \dot{m}_{a2}\omega_2 = \dot{m}_{a3}\omega_3 \tag{E12.16.2}$$

Eliminating $\dot{m}_{a3}$ between Eqs. (E12.16.1) and (E12.16.2) and substituting numerical values in the resulting equation

$$\omega_3 = (0.01499 \times 1.773 + 0.01353 \times 2.96)/4.733 = 0.014076$$

Applying the SFEE to the adiabatic control volume surrounding the mixing section we have

$$\dot{m}_{a1}h_1 + \dot{m}_{a2}h_2 = \dot{m}_{a3}h_3 \qquad (E12.16.3)$$

Eliminating $\dot{m}_{a3}$ between Eqs. (E12.16.1) and (E12.16.3) and substituting numerical values in the resulting equation we obtain

$$h_3 = (70.36 \times 1.773 + 60.48 \times 2.96)/4.733 = 64.18$$

$$h_3 = c_{pa}T_3 + \omega_3 h_{g3}(T_3) = 64.18$$

Hence $\qquad\qquad T_3 + 0.014076 h_{g3}(T_3) = 64.18$

A trial-and-error method is used to solve the above equation by first guessing a value for T_3. The enthalpy, $h_{g3}(T_3)$ is then obtained from the steam tables [4]. The value of T_3 is adjusted until the equation is satisfied. After a few iterations the temperature is obtained as, $T_3 = 28.1°C$.

Example 12.17 A steady stream of warm, dry ambient air at 38°C and 10% relative humidity is cooled to 24°C in an evaporative cooler by spraying cooling water at 10°C. The constant pressure in the cooler is 90 kPa. The air flow rate at the entrance to the cooler is 0.5 kg s⁻¹. Calculate the relative humidity of the air leaving the cooler.

Solution Using Eq. (12.40), the humidity ratio at the entrance is

$$\omega_1 = \frac{0.622\phi_1 P_{g1}}{P - \phi_1 P_{g1}} = \frac{0.622 \times 0.1 \times 6.624}{90 - 0.1 \times 6.624} = 0.004612$$

The mass flow rate of dry air is

$$\dot{m}_a = \dot{m}/(1 + \omega_1) = 0.5/1.004612 = 0.498 \text{ kg s}^{-1}$$

Using Eq. (12.41) we obtain the enthalpies of the air as:

$$h_1 = c_{pa}T_1 + \omega_1 h_{g1} = 1 \times 38 + 0.004612 \times 2570.1 = 49.85 \text{ kJ kg}^{-1}$$

$$h_2 = c_{pa}T_2 + \omega_2 h_{g2} = 1 \times 24 + \omega_2 \times 2544.8$$

The enthalpy of cooling water at 10°C is $h_w = 42$ kJ kg^{-1}.

Consider a control volume surrounding the evaporative cooler. Applying the mass balance equation for water

$$\dot{m}_w = \dot{m}_a(\omega_2 - \omega_1) \tag{E12.17.1}$$

Applying the SFEE to the control volume we have

$$\dot{m}_a h_1 + \dot{m}_w h_w = \dot{m}_a h_2 \tag{E12.17.2}$$

where 1 and 2 are the entrance and exit sections of the cooler. Substituting numerical values in Eqs. (E12.17.1) and (E12.17.2) and eliminating $\dot{m}_w$ we obtain

$$49.85 + 42(\omega_2 - 0.004612) = 24 + 2544.8\omega_2$$

Hence $\omega_2 = 0.01025$.

To find the relative humidity we use Eq. (12.40),

$$\omega_2 = \frac{0.622\phi_2 P_{g2}}{P - \phi_2 P_{g2}} = \frac{0.622 \times \phi_2 \times 2.982}{90 - \phi_2 \times 2.982} = 0.01025$$

Solving the above equation we obtain the relative humidity of the air at the outlet as $\phi_2 = 49\%$.

Example 12.18 The dry-bulb and wet-bulb temperatures of two streams of air A and B are respectively A: (32°C, 23.6°C) and B: (14°C, 10°C). The mass flow rates of the two air streams A and B are 0.5 and 0.3 kg s^{-1} respectively. The two air streams are mixed adiabatically in a steady flow process at a constant pressure of 101.3 kPa. Using the psyhrometric chart determine the dry-bulb temperature and the relative humidity of the mixed air stream.

Solution The psychrometric chart in Fig. E12.18 shows the states 1 and 2 of the air streams A and B. The state points are located by the intersection of the wet-bulb temperature line and the dry-bulb temperature line for each stream. The following properties are read directly from the chart in [5]:

$$\omega_1 = 0.015, \ h_1 = 71 \text{ kJ kg}^{-1}, \ \omega_2 = 0.006, \ h_2 = 29.5 \text{ kJ kg}^{-1}.$$

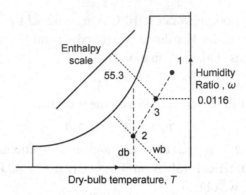

Fig. E12.18 Psychrometric chart

Now the dry air flow rates are:

$$\dot{m}_{a1} = \dot{m}_1/(1+\omega_1) = 0.5/1.015 = 0.4926 \text{ kg s}^{-1}$$

$$\dot{m}_{a2} = \dot{m}_2/(1+\omega_2) = 0.3/1.006 = 0.298 \text{ kg s}^{-1}$$

Consider an adiabatic control volume surrounding the mixing section. Applying the mass balance equations for dry air and water we have

$$\dot{m}_{a3} = \dot{m}_{a1} + \dot{m}_{a2}$$

$$\dot{m}_{a1}\omega_1 + \dot{m}_{a2}\omega_2 = \dot{m}_{a3}\omega_3$$

where 3 denotes the mixed air stream. Eliminate $\dot{m}_{a3}$ between the above equations to obtain an expression for ω_3.

Substituting numerical values in the two equations above, and eliminating $\dot{m}_{a3}$ we obtain

$$\omega_3 = (0.015 \times 0.4926 + 0.006 \times 0.298)/0.7908 = 0.0116$$

Applying the SFEE to the adiabatic control volume surrounding the mixing section we have

$$\dot{m}_{a1}h_1 + \dot{m}_{a2}h_2 = \dot{m}_{a3}h_3$$

Substituting numerical values in the above equation

$$h_3 = (71 \times 0.4926 + 29.5 \times 0.298)/0.7908 = 55.34 \text{ kJ kg}^{-1}$$

The state point 3 is found by locating the point of intersection of the constant humidity ratio line of, $\omega_3 = 0.0116$ and the constant enthalpy line of, $h_3 = 55.34$ kJ kg^{-1} as indicated in Fig. E12.18. Thus from the psychrometric chart the dry-bulb temperature and relative humidity at state 3 are 26°C and 55% respectively.

Example 12.19 A wet and dry bulb psychrometer is placed in an air stream at a pressure of 80 kPa. The readings of the dry-bulb thermometer and the wet-bulb thermometer are 30°C and 22°C respectively. Calculate the relative humidity of the air stream.

Solution The following data are obtained from the steam table [4]:
At the wet-bulb temperature, $T_2 = 22°C$,

$$P_{g2} = 2.642 \text{ kPa}, \quad h_{f2} = 92.2 \text{ kJ kg}^{-1}, \quad h_{g2} = 2541.2 \text{ kJ kg}^{-1}$$

At the dry-bulb temperature, $T_1 = 30\,°C$,

$$P_{g1} = 4.242 \text{ kPa} \quad \text{and} \quad h_{g1} = 2555.7 \text{ kJ kg}^{-1}.$$

The relative humidity of air at the wet-bulb temperature is 100%. The humidity ratio at the wet-bulb temperature, T_2 is given by Eq. (12.40) as:

$$\omega_2 = \frac{0.622 P_{g2}}{P - P_{g2}} = \frac{0.622 \times 2.642}{80 - 2.642} = 0.02124$$

Assume that the wet-bulb temperature indicated by the wet-bulb thermometer is equal to the thermodynamic wet-bulb temperature (TWT). Then the humidity ratio of the air stream is given by Eq. (12.51) as:

$$\omega_1 = \frac{\omega_2(h_{g2} - h_{h2}) - c_{pa}(T_1 - T_2)}{h_{g1} - h_{f2}}$$

Substituting numerical values in the above equation we have

$$\omega_1 = \frac{0.02124 \times (2541.2 - 92.2) - 1.0 \times (30 - 22)}{2555.7 - 92.2} = 0.01787$$

We use Eq. (12.40) to obtain the relative humidity. Hence

$$\omega_1 = \frac{0.622 \phi_1 P_{g1}(T_1)}{P - \phi_1 P_{g1}(T_1)} = \frac{0.622 \times \phi_1 \times 4.242}{80 - \phi_1 \times 4.242} = 0.01787$$

Therefore the relative humidity, $\phi_1 = 52.7\%$. The psychrometric chart in [5] is for a pressure of 101.3 kPa, and therefore it should not be used to solve this example, where the pressure is 80 kPa.

Example 12.20 Cooling water from the condenser of an air conditioning system enters a counter-flow cooling tower at 35°C and leaves at 22°C. Ambient air at 20°C dry-bulb temperature and 60% relative humidity flows into the cooling tower at the bottom. Saturated air leaves at the top with a temperature of 30°C. Make-up water is supplied to the cooling tower at 25°C. The cooling tower operates at a constant pressure of 100 kPa. For unit mass flow rate of cooling water, calculate the mass flow rate of air.

Solution We refer to the schematic diagram of the cooling tower depicted in Fig. 12.9. The following assumptions are made in applying the conservation equations of mass and energy: (i) heat interactions with the surroundings are negligible, (ii) the work input to the fan and the water pump are negligible, and (iii) the changes in kinetic and potential energy of the air and water streams are small.

The following data are obtained from the steam tables [4]:
For inlet air at temperature, $T_2 = 20°C$,

$$P_{g2} = 2.337 \text{ kPa}, \qquad h_{g2} = 2537.6 \text{ kJ kg}^{-1}$$

For exit air at temperature, $T_4 = 30°C$,

$$P_{g4} = 4.242 \text{ k Pa}, \qquad h_{g4} = 2555.7 \text{ kJ kg}^{-1}$$

For the inlet water at 35°C, $h_{f1} = 146.5 \text{ kJkg}^{-1}$

For the outlet water at 22°C, $h_{f3} = 92.2 \text{ kJkg}^{-1}$

For make-up water at 25°C, $h_{f5} = 104.8 \text{ kJ kg}^{-1}$

The humidity ratio of the inlet air is given by Eq. (12.40) as:

$$\omega_2 = \frac{0.622\phi_2 P_{g2}(T_2)}{P - \phi_2 P_{g2}(T_2)} = \frac{0.622 \times 0.6 \times 2.337}{100 - 0.6 \times 2.337} = 0.008845$$

The humidity ratio of the exit air is given by Eq. (12.40) as:

$$\omega_4 = \frac{0.622 P_{g4}}{P - P_{g4}} = \frac{0.622 \times 4.242}{100 - 4.242} = 0.02755$$

Consider a control volume surrounding the cooling tower (see Fig. 12.9). Applying the mass balance equations for dry air and water we have

$$\dot{m}_{a2} = \dot{m}_{a4} = \dot{m}_a$$

$$\dot{m}_{w5} = \dot{m}_a(\omega_2 - \omega_1) = \dot{m}_a(0.02755 - 0.008845) = 0.0187\dot{m}_a$$

Since the water flow rate in the cooling-load loop is constant

$$\dot{m}_{w1} = \dot{m}_{w3} = \dot{m}_w$$

Applying the SFEE to the control volume we obtain

$$\dot{m}_a h_{a2} + \dot{m}_{w5} h_{w5} + \dot{m}_w h_{w1} = \dot{m}_a h_{a4} + \dot{m}_w h_{w3}$$

The enthalpies of inlet and exit air are given by Eq. (12.41) as:

$$h_{a2} = c_{pa}T_2 + \omega_2 h_{g2} = 1 \times 20 + 0.008845 \times 2537.6 = 42.45 \text{ kJ kg}^{-1}$$

$$h_{a4} = c_{pa}T_4 + \omega_4 h_{g4} = 1 \times 30 + 0.02755 \times 2555.7 = 100.41 \text{ kJ kg}^{-1}$$

Substituting numerical values in the SFEE above we obtain

$$42.45\dot{m}_a + 0.0187\dot{m}_a \times 104.8 + 146.5\dot{m}_w = 100.41\dot{m}_a + 92.9\dot{m}_w$$

Hence

$$56\dot{m}_a = 53.6\dot{m}_w$$

The mass flow rate of moist air is

$$\dot{m}_2 = \dot{m}_{a2}(1 + \omega_2) = \dot{m}_a \times 1.008845$$

Hence

$$\dot{m}_2 / \dot{m}_w = 1.008845 \times 53.6 / 56 = 0.966$$

We note that some of the properties of moist air required for this example could have been obtained directly from the psychrometric chart.

Problems

P12.1 A mixture of gases with the following mass fractions is at 100 kPa and 35°C: $CO_2 = 0.4$, $CO = 0.18$, $N_2 = 0.42$. The specific heat capacities at constant pressure c_p (kJ kg^{-1} K^{-1}) of the gases are: $CO_2 = 0.846$, $CO = 1.04$, $N_2 = 1.04$. Calculate (i) the mole fractions of the gases, (ii) the partial pressures, (iii) the mean gas constant, (iv) the mean specific heat capacities of the mixture, and (v) the mass of 0.4 m^3 of the mixture. [*Answers*: (i) 0.298, 0.2107, 0.491, (ii) 29.8, 21.07, 49.1 kPa, (iii) 0.254 kJ kg^{-1} K^{-1}, (iv) 0.962, 0.708 kJ kg^{-1} K^{-1}, (v) 0.51 kg]

P12.2 A rigid vessel of volume 0.22 m^3 contains oxygen at 70 kPa and 40°C. Carbon monoxide from a high pressure source is forced into the vessel until the pressure and temperature of the mixture are 180 kPa and 80°C. Calculate (i) the partial pressures of the gases in the vessel, and (ii) the mass fractions of the gases. [*Answers*: (i) 78.95, 101.05 kPa, (ii) 0.47, 0.528]

P12.3 An insulated pipe with an initially-closed valve connects a rigid vessel A containing 0.4 m^3 of helium at 7 bar and 15°C to vessel B containing 0.3 m^3 of nitrogen at 3.5 bar and 70°C. The valve is now opened, and the gases in A and B mix in an adiabatic process. The specific heat capacities at constant volume of helium and nitrogen are 3.1 kJ kg^{-1} K^{-1} and 0.74 kJ kg^{-1} K^{-1} respectively. Calculate (i) the pressure and temperature of the gases after mixing, and (ii) the entropy production in the process. [*Answers*: (i) 560 kPa, 306.95 K, (ii) 0.81 kJ K^{-1}]

P12.4 A well-insulated rigid tank of volume 0.08 m^3 contains air at 110 kPa and 15°C. It is connected to a pipe carrying helium at 350 kPa and 50°C through a valve. The valve is opened and helium flows into the tank until the pressure is 200 kPa. The valve is then closed. Calculate (i) the final equilibrium temperature of the gas mixture in the tank, and (ii) the total entropy production. [*Answers*: (i) 346.7 K, (ii) 0.043 kJ K^{-1}]

P12.5 Air with a dry-bulb temperature of 30°C and relative humidity 70% enters a cooling and dehumidifying coil with a mass flow rate of 0.12 kg s^{-1}. The constant pressure of the air is 80 kPa. Saturated air leaves the coil with a temperature of 14°C. Calculate (i) the rate of moisture removal, and (ii) the cooling capacity of the coil. [*Answers*: (i) 0.001314 kg s^{-1}, (ii) 5.19 kW]

P12.6 A wet-bulb thermometer placed in a stream of air with a dry-bulb temperature of 46°C and a pressure of 80 kPa indicates a temperature of 26°C. Calculate (i) the relative humidity of the air, and (ii) the specific volume of the air. [*Answers*: (i) 23.3%, (ii) 1.179 m^3 kg^{-1}]

P12.7 A stream of ambient air at 20°C dry-bulb temperature and 65% relative humidity is mixed adiabatically with a stream of return air from a room at 30°C and relative humidity 80%.

The mass flow rates of ambient air and return air are 0.3 kg s^{-1} and 0.7 kg s^{-1} respectively. The pressure in the mixing section is 101 kPa. Calculate (i) the relative humidity, and (ii) the dry-bulb temperature of the mixed air stream. [*Answers*: (i) 79.3%, (ii) 27°C]

P12.8 Air with a mass flow rate of 0.4 kg s^{-1} enters an evaporative cooler at 34°C dry-bulb temperature and 15% relative humidity. The air leaves the cooler at 24°C dry-bulb temperature. The pressure inside the cooler is 90 kPa. The cooling water sprayed into the air is supplied at 10°C. Calculate (i) the relative humidity of the leaving air, and (ii) the rate at which cooling water is supplied. [*Answers*: (i) 45.8%, (ii) 0.0016 kg s^{-1}]

P12.9 Air enters a steam humidifier at the rate of 0.4 kg s^{-1} with a dry-bulb temperature of 20°C and a relative humidity of 30%. Steam at 100 kPa is sprayed into the air. The pressure inside the humidifier is 100 kPa. The air leaves with a dry-bulb temperature of 21°C. Calculate (i) the relative humidity of the air leaving the humidifier, and (ii) the steam flow rate. [*Answers*: (i) 74%, (ii) 0.0029 kg s^{-1}]

P12.10 Warm water, pumped to a cooling tower at the rate of 1.1 kg s^{-1}, is cooled from 40°C to 25°C. The conditions of the air entering the cooling tower are 18°C dry-bulb temperature and 50% relative humidity. The air leaves in a saturated state with a temperature of 28°C. The pressure is constant at 100 kPa. Make-up water is supplied to the cooling tower at 16°C. Calculate (i) the mass flow rate of air, and (ii) the rate of flow of make-up water. [*Answers*: (i) 1.268 kg s^{-1}, (ii) 0.0226 kg s^{-1}]

References

1. Jones, J.B. and G.A. Hawkins, *Engineering Thermodynamics*, John Wiley & Sons, Inc., New York, 1986.
2. Kuehn, Thomas H., Ramsey, James W. and Threlkeld, James L., *Thermal Environmental Engineering*, 3rd edition, Prentice-Hall, Inc., New Jersey, 1998.
3. Reynolds, William C. and Henry C. Perkins, *Engineering Thermodynamics*, 2nd edition, McGraw-Hill, Inc., New York, 1977.
4. Rogers, G.F.C. and Mayhew, Y.R., *Thermodynamic and Transport Properties of Fluids*, 5th edition, Blackwell, Oxford, U.K. 1998.
5. Stoecker, Wilbert F. and Jones, Jerold W., *Refrigeration and Air Conditioning*, International Edition, McGraw-Hill Book Company, London, 1982.
6. Van Wylen, Gordon J. and Richard E. Sonntag, *Fundamentals of Classical Thermodynamics*, 3rd edition, John Wiley & Sons, Inc., New York, 1985.

Chapter 13

Reactive Mixtures

In chapter 12 we considered the properties of mixtures of gases and vapors, that are pure substances. We shall now extend this study to mixtures that undergo chemical reactions where the chemical constituents of the system change during the process. The main focus of this chapter will be on combustion processes. These processes are used to provide high temperature thermal energy sources for numerous engineering applications. We have already discussed some of these applications in earlier chapters of this book.

In the steam power plant, for example, it is the enthalpy of the combustion gases from the furnace that supply the energy to evaporate the water in the boiler (Fig. 9.2). The combustion of fuel and air in internal combustion engines generates the high pressure and temperature in the gas mixture needed during the power stroke of the cycle (Fig. 10.1). In gas turbine plants, (Fig. 10.7) the hot gases leaving the combustion chamber provide the energy to drive the turbine. In this chapter we shall apply the mass balance equation, the first law, and the second law to combustion processes.

13.1 Chemical Reactions of Fuels

A fuel is a substance that is capable of undergoing a combustion reaction in the presence of oxygen. Common fuels include: (i) solids, like coal and wood, (ii) liquids, like gasoline, diesel, kerosene and palm oil, and (iii) gases, like natural gas, coal gas and producer gas. Most of the liquid fuels are hydrocarbons and these are obtained by refining 'crude' oil, which occurs naturally. In a typical combustion process, the hydrocarbon fuel reacts with oxygen contained in the air supplied to the combustion chamber.

13.1.1 *Mass balance for a combustion reaction*

During the chemical reaction between a fuel and oxygen, the mass of each element remains the same. The mass conservation equation for the process is usually represented by the chemical equation for the reaction, which ensures the mass balance of each element. We shall now write the chemical equation for three typical combustion reactions.

$$CO + 1/2\ O_2 = CO_2 \tag{13.1}$$

The molar-form of the mass balance for the above chemical reaction may be stated as: one mole of carbon monoxide reacts with half a mole of oxygen to form one mole of carbon dioxide. The mass conservation equation may also be expressed in terms of the molar masses in the form: 28 kg of carbon monoxide reacts with 16 kg of oxygen to form 44 kg of carbon dioxide.

The combustion of hydrogen can be represented by the chemical equation:

$$H_2 + 1/2\ O_2 = H_2O \tag{13.2}$$

Similarly, for the combustion of methane, we have

$$CH_4 + 2\ O_2 = CO_2 + 2\ H_2O \tag{13.3}$$

In the above reaction the products of combustion contain both carbon dioxide and water. Depending on the temperature and pressure of the products, the water can be in the vapor, liquid or solid phase. The chemical transformation that occurs during combustion may be represented schematically, as depicted in Fig. 13.1.

In most practical applications, the oxygen required for combustion is supplied in the form of air and not as pure oxygen. Air is a mixture of

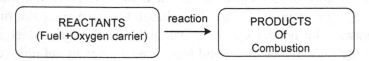

Fig. 13.1 Transformations during a combustion process

oxygen, nitrogen and argon with respective mole fractions of 21 percent, 78 percent, and 1 percent. In the absence of dissociation reactions, nitrogen does not take part in the combustion reaction. Argon and nitrogen leave as constituents of the products of combustion, usually at the same temperature as the other products of the chemical reaction. For engineering calculations, we neglect the presence of argon and assume that air is a mixture of 21 percent oxygen and 79 percent nitrogen by volume. Therefore the mole-ratio of nitrogen to oxygen is $79/21 = 3.76$.

When oxygen is introduced in the form of air, the chemical reactions represented by Eqs. (13.1) to (13.3) need to be amended to include the presence of nitrogen in the combustion system. We illustrate this by referring to Eq. (13.3) which now takes the form:

$$CH_4 + 2\,O_2 + 2 \times 3.76\,N_2 = CO_2 + 2\,H_2O + 2 \times 3.76\,N_2 \quad (13.4)$$

Note that in Eq. (13.4), 3.76 moles of nitrogen per mole of oxygen are included on both sides of the chemical equation.

The minimum quantity of air required for the complete oxidation of all the elements in a fuel, that are capable of undergoing combustion, is called the *'theoretical air'* or *'stoichiometric air'*. The coefficients in the chemical equation representing such a reaction are called *'stoichiometric coefficients'*. In Eq. (13.3) these coefficients are respectively, (1,2) and (1,2) for the reactants and products.

If the air supplied to the combustion system exceeds the theoretical air required for complete combustion of the fuel, then the *excess air* will be included in the products of combustion. For example, when the air supplied is 1.5 times the theoretical air, Eq. (13.4) has to be written in the form

$$CH_4 + 1.5 \times 2\,O_2 + 1.5 \times 2 \times 3.76\,N_2$$

$$= CO_2 + 2\,H_2O + 1.5 \times 2 \times 3.76\,N_2 + 0.5 \times 2 \times O_2 \quad (13.5)$$

Excess air is usually supplied to ensure *complete* combustion of the fuel. In internal combustion engines, the excess air is about 25 percent, while for gas turbines it could be up to 150 percent.

When the oxygen supplied for the combustion of a hydrocarbon fuel is less than that required for stoichiometric combustion, then the products may contain some fuel, and gases like carbon monoxide. The combustion is then said to be *incomplete*.

Another important parameter of a combustion reaction is the *air-fuel ratio* (AFR) which can be determined from the chemical equation. It is defined as the ratio of the quantity of air to the quantity of fuel, which could be expressed either in the molar-form or the mass-form. For example, the AFR for the stoichiometeric combustion of methane, can be obtained from Eq. (13.4). On a molar-basis, we have

$$(AFR)_{molar} = (2 + 2 \times 3.76)/1 = 9.52$$

moles of air per mole of fuel.

The AFR on a mass-basis may be expressed as:

$$(AFR)_{mass} = (2 + 2 \times 3.76) \times 29/(1 \times 16) = 17.2$$

kg of air per kg of fuel.

The application of the mass balance equation to combustion processes is illustrated in worked examples 13.1 to 13.5.

13.2 Energy Balance for a Combustion Process

We now consider the application of the first law to combustion processes. For a closed system undergoing a process from an initial state 1 to a final state 2, the first law may be written in the form

$$Q_{12} = U_2 - U_1 + W_{12} \qquad (13.6)$$

Equation (13.6) has the same form whether or not a chemical reaction occurs within the system. The work interaction, W_{12} and the heat interaction, Q_{12} that occur at the boundary of the system, could be evaluated using the methods developed in earlier chapters of this book. However, the determination of the change in internal energy, (U_2-U_1) accompanying a chemical reaction, needs special consideration.

In all applications of the first law, considered in our work thus far, the chemical composition of the substances constituting a system, remained unchanged during a process. Since we were dealing with changes in internal energy of the same chemical species, the reference state for zero internal energy was selected arbitrarily, with no effect on the computed values of the internal energy change.

In contrast, during a combustion process, the chemical composition of the constituents of a system change from reactants to products. This change in composition is accompanied by an energy transformation, where the 'chemical internal energy' of the reactants is converted to 'thermal internal energy' of the products of combustion.

13.2.1 *Enthalpy and internal energy of formation*

In the absence of chemical reactions, the internal energy and enthalpy of pure substances may be tabulated using an arbitrarily selected reference state for the data. This procedure is adopted in preparing the tables of properties for water and steam. However, for a proper energy analysis of processes involving chemical transformations, a common reference state, that is applicable to all substances, has to be developed. For this purpose, we first select a *standard reference state* at a temperature T^o (25°C) and pressure P^o (1 bar). We then determine the change in enthalpy that occurs when a given substance is formed from elemental substances under standard conditions. By convention, the enthalpy of every elemental substance, which consists of only one kind of atom, is assigned a value of zero at the standard reference state. For example, the enthalpy of carbon is zero at the standard state, and so is the enthalpy of oxygen, because these are elemental substances.

The chemical equations of formation and the measured values of the molar enthalpy of formation, $\Delta \bar{h}^o$ for three substances, are listed below:

$$C(s) + 2\ H_2(g) = CH_4(g); \quad \Delta \bar{h}^o = -74870 \ \text{kJ kmol}^{-1}$$

$$C(s) + O_2(g) = CO_2(g); \quad \Delta \bar{h}^o = -393520 \ \text{kJ kmol}^{-1}$$

$$H_2(g) + 1/2\ O_2(g) = H_2O(l); \quad \Delta \bar{h}^o = -285830 \ \text{kJ kmol}^{-1}$$

where s, l and g denote solid, liquid and gas phases respectively. The negative sign of the enthalpy of formation means that the enthalpy of the substance is less than the enthalpy of the elements that formed it. The enthalpy of formation of a substance is determined by experimental measurements or computational procedures.

The relationship between the enthalpy of a substance and its enthalpy of formation may be illustrated by using the formation of water as follows:

$$\Delta \bar{h}_w^{\,o} = \bar{h}_{water} - \bar{h}_H - \bar{h}_O$$

Since the enthalpies of the elements hydrogen and oxygen are zero, the enthalpy of water at the standard state is equal to its enthalpy of formation.

Now consider the following combustion reaction between CH_4 and O_2. The chemical equation is

$$CH_4 + 2\,O_2 = CO_2 + 2\,H_2O(l)$$

The change in total molar enthalpy, $\Delta \bar{H}_R$ between the reactants and the products of the above reaction may be expressed as:

$$\Delta \bar{H}_R = \bar{h}^o{}_{CO_2} + 2\bar{h}^o{}_{H_2O} - \bar{h}^o{}_{CH_4} - 2h^o{}_{O_2}$$

Substituting the enthalpies of formation of the different constituents listed on page 635, we obtain the enthalpy of the reaction as:

$$\Delta \bar{H}^o{}_R = -393520 - 2\times285830 + 74870 - 2\times0$$

$$\Delta \bar{H}^o{}_R = -890310 \text{ kJ kmol}^{-1}$$

We note that the use of the enthalpies of formation makes possible the calculation of the enthalpy change between the reactants and products for complex reactions, from data on the individual constituents, as illustrated above.

The molar internal energy of formation, $\Delta \bar{u}^o$ under standard conditions (P^o, T^o), may be obtained from the above data on the enthalpy of formation, by applying the following relation based on the definition of enthalpy:

$$\Delta \bar{h}^o = \Delta \bar{u}^o + P^o \Delta \bar{v} \qquad (13.7)$$

The change in volume between the products and reactants, $\Delta \bar{v}$ in the above equation, may be obtained by applying the ideal gas equation to the reactants and products. Thus, we have

$$\Delta \bar{h}^o = \Delta \bar{u}^o + (n_p - n_r)\bar{R}T^o \qquad (13.8)$$

The number of moles of the products and reactants are denoted by n_p and n_r respectively. The volumes of constituents that are liquids and solids may be neglected in using the above equation.

13.2.2 *Internal energy and enthalpy of reactants and products*

Consider a combustion process in which reactants at an initial temperature T_1 are converted to products at a temperature T_2. The total internal energy of the reactant mixture may be expressed as:

$$U_1 = U^o_r + \Delta U_r(T^o \to T_1) \tag{13.9}$$

Similarly, the internal energy of the product mixture is

$$U_2 = U^o_p + \Delta U_p(T^o \to T_2) \tag{13.10}$$

Note that in Eq. (13.9) the first term on the right hand side is the sum of the internal energy of formation of all the reactants at the *standard reference state*. The second term is the total change in *thermal* internal energy when the temperature of the reactants change from the *standard reference state* temperature of T^o (25°C) to the initial temperature of the reactants, T_1. The corresponding terms in Eq. (13.10) have the same meaning when applied to the products. The curves in Fig. 13.2(a) are graphical representations of Eqs. (13.9) and (13.10), which are useful aids for analyzing combustion processes.

Consider the graphs depicted in Fig. 13.2(a) for the variation of the internal energy of the reactants and the products. As expected, the internal energies of the mixtures increase monotonically with temperature. The two graphs are displaced vertically due to the difference in the internal energy of formation of the reactants and the products at the standard

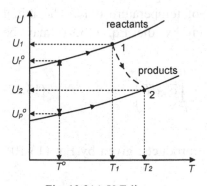

Fig. 13.2(a) U-T diagram

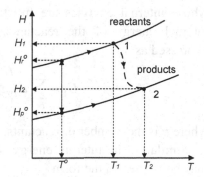

Fig. 13.2(b) H-T diagram

reference temperature T^o (25°C). For combustion reactions, the internal energy of formation of the reactants is usually higher. This difference in internal energy at the reference state could be attributed to the difference between the 'chemical binding energy' of the reactants and the products.

In Fig. 13.2(a), the initial and final states of the combustion process are indicated by 1 and 2 respectively. The relative positions of these two points are determined by Eq. (13.6), which is the first law. For example, consider an adiabatic combustion process that occurs at constant volume. From the first law it follows that for such a process, $U_1 = U_2$, and hence the 'path' 1-2 is a horizontal line. However, it is important to note that combustion processes are irreversible, and therefore, the intermediate points of the path 1-2 are not defined.

For the analysis of steady-flow combustion processes, we shall make use of the SFEE, and the graphs shown in Fig. 13.2(b) for the variation of the enthalpy of the reactants and products.

Consider the chemical reaction, represented by the general chemical equation

$$v_{r1}A_{r1} + v_{r2}A_{r2} + v_{r3}A_{r3} = v_{p1}A_{p1} + v_{p2}A_{p2} + v_{p3}A_{p3} \qquad (13.11)$$

where A_r and A_p are respectively the constituents of the reactants and the products. The corresponding stoichiometric coefficients are v_r and v_p.

Now if the gases in the reactants and products are treated as ideal, then we could express the thermal internal energy changes in terms of the specific heat capacities. We now write the expressions for the internal energy of the reactants and products, assuming that they are ideal gases whose internal energies are functions of temperature only. The initial internal energy of the reactants, given by the Eq. (13.9), may be expressed as

$$U_1 = \sum_{i=1}^{n} v_{ri} \left(\Delta \bar{u}_{ri}^{\,o} + \int_{T^o}^{T_1} \bar{c}_{vri} dT \right) \qquad (13.12)$$

where n is the number of reactants.

Similarly, the internal energy of the products, given by Eq. (13.10), may be written in the form

$$U_2 = \sum_{i=1}^{m} \upsilon_{pi} \left(\Delta \overline{u}_{pi}^{\,o} + \int_{T^o}^{T_2} \overline{c}_{vpi} dT \right) \qquad (13.13)$$

where m is the number of products.

The molar internal energy of formation, $\Delta \overline{u}^o$ and the molar specific heat capacities, $\overline{c}_{vi}$ for the various reactants and products are obtained from tabulated data sources. In applying the first law to a combustion process in a closed system we shall make use of Eqs. (13.6), (13.12) and (13.13).

13.2.3 Heats of reaction and heating values

Consider the control volume shown in Fig. 13.3 where fuel and air undergo combustion in a steady flow process. Assume that the reactants and the products are maintained at the standard conditions of 1 bar and 25°C. The heat output from the control volume *per mole of fuel* is $\overline{Q}^o$. We neglect changes in kinetic and potential energy and assume that the work interaction is zero. Applying the SFEE to the control volume we have

$$\overline{Q}^o = \overline{H}^o{}_r - \overline{H}^o{}_p \qquad (13.14)$$

where $\overline{H}^o{}_r$ and $\overline{H}^o{}_p$, are respectively, the total enthalpies of the reactants and products *per mole of fuel*. The heat output $\overline{Q}_o$ is *called the heat of reaction* or *heating value* of the fuel under *standard* conditions. Note that $\overline{Q}^o$ may be viewed as the heat that must transferred from the combustion system to maintain it at the standard temperature.

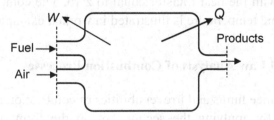

Fig. 13.3 Steady flow combustion process

If water is present in the products of combustion, then we need to consider effect of the state (phase) of water on the heating value. If the water is in the *liquid state* the heat transferred from the combustion system will include the latent heat of condensation of the water. The value of $\overline{Q}^o$ is then called the *higher heating value* (*HHV*). However, if the water is in the vapor state, the heat transferred is lower, and $\overline{Q}^o$ is then called the *lower heating value* (*LHV*). The difference between the higher and lower heating values is given by

$$HHV - LHV = \upsilon_w \overline{h}_{fg} \tag{13.15}$$

where υ_w is the number of moles of water in the products per mole of fuel and $\overline{h}_{fg}$ is the molar enthalpy of vaporization. Note that we have developed all the relations above on a molar-basis because most sources of data on chemical reactions tend to be presented on a molar-basis. It should be quite straightforward to express these relations on a mass-basis.

13.2.4 *Adiabatic flame temperature*

In some applications of combustion processes it is necessary to know the theoretically maximum temperature that the products may attain under adiabatic conditions. For examples, in gas turbines, it is important to control the maximum operating temperature to satisfy metallurgical limits of the components of the turbine. For a given fuel and specified reactant inlet conditions, the temperature reached by the products is a maximum when the reactant mixture is stoichiometric. The adiabatic flame temperature for steady flow combustion may be obtained from Eq. (13.14) with the heat transfer equal to zero. The computation of the adiabatic flame temperature is illustrated in worked example 13.8.

13.3 Second Law Analysis of Combustion Processes

The performance limits and irreversibilities of combustion processes may be obtained by applying the second law in the form of the entropy balance equation. However, this requires information on the change of entropy of a substance during a combustion process. As for the internal energy and the enthalpy, the reference state for entropy has to be chosen in such a manner that it applies to all the reactants as well as all the products.

The *third law of thermodynamics*, which enables us to assign a reference state for entropy, could be stated in a number of different forms. The statement of the third law that is relevant to the present work may be stated as follows: *the entropy of a perfect crystal is zero at the absolute zero of temperature.* We note that the third law provides a reference state for entropy that is applicable to all substances.

The absolute entropy of various substances, determined from experimental data and theoretical computations are available in tabular form. As for the enthalpy of formation, the data for absolute entropy are determined at the standard conditions of 1 bar and 25°C. The absolute molar entropy of a number of gases at different temperatures T, and a pressure of 1 bar are tabulated in [4]. Using the ideal gas assumption, the absolute entropy at other pressures may be computed. Hence

$$\bar{s}(T,P) = \bar{s}^o(T,P^o) - \bar{R}\ln\left(P/P^o\right) \qquad (13.16)$$

where $\bar{s}$ is the molar specific entropy at temperature T and pressure P. $\bar{s}^o$ is the molar specific entropy at the standard pressure P^o and temperature T. When the temperature range of interest is small, the ideal gas expression, with a constant specific heat capacity, may be used to compute the absolute entropy, from known values at the standard temperature and pressure. Hence

$$\bar{s}(T,P) = \bar{s}_o^{\,o}(T^o,P^o) + \bar{c}_p\ln\left(T/T^o\right) - \bar{R}\ln\left(P/P^o\right)$$

where $\bar{s}$ is the molar specific entropy, $\bar{s}_o^{\,o}$ is the molar specific entropy at the standard conditions (T^o, P^o).

We now apply the second law to the steady flow combustion system depicted in Fig. 13.3. Assume that the kinetic and potential energy changes are negligible. Let the control volume exchange heat with a single reservoir at temperature T_o.

Applying the SFEE to the control volume we obtain

$$\dot{Q}_{in} + \sum_R \dot{n}_{ri}\bar{h}_{ri} = \sum_P \dot{n}_{pi}\bar{h}_{pi} + \dot{W}_{out} \qquad (13.17)$$

where $\dot{n}_{ri}$ and $\dot{n}_{pi}$ are the molar flow rates of constituent i of the reactants and the products respectively.

The molar specific enthalpies of the constituents of the reactants and the products are given by

$$\overline{h}_{ri} = \Delta\overline{h}_{ri}^{\,o} + \Delta\overline{h}_{ri}\left(T^o \to T_r\right)$$

$$\overline{h}_{pi} = \Delta\overline{h}_{pi}^{\,o} + \Delta\overline{h}_{pi}\left(T^o \to T_p\right)$$

In the above equations the first term on the right hand side is the enthalpy at the standard state of 25°C. The second term is the change in enthalpy due to the difference between the standard temperature and the prevailing temperature. This change in enthalpy for a number of gases is tabulated in [4]. However, if the gases in the mixture are assumed to be ideal, then the latter change in enthalpy may be expressed in terms of the specific heat capacities.

Applying the second law to the control volume we have

$$\dot{\sigma} = \sum_P \dot{n}_{pi}\overline{s}_{pi} - \sum_R \dot{n}_{ri}\overline{s}_{ri} - \dot{Q}_{in}/T_o \qquad (13.18)$$

where $\dot{\sigma}$ is the entropy production rate. The subscripts p and r denote quantities related to the products and reactants respectively. The last term in Eq. (13.18) is the change in entropy of the reservoir due to heat transfer to the control volume. Worked examples 13.8–13.10 illustrate the application of the second law to combustion processes.

13.4 Chemical Equilibrium

In this section we develop the criterion of equilibrium for a chemically reacting closed system. The system is in thermal and mechanical equilibrium with the temperature and pressure uniform throughout the system. However, the system is not in equilibrium chemically. The chemical reactions occurring within the system tend to bring the system progressively toward equilibrium.

Consider an infinitesimal change of state of the system and the surroundings with which the system interacts reversibly. Applying the first law to the system

$$\delta Q = dU + PdV \tag{13.19}$$

Applying the second law, in the form of the principle of increase of entropy, we have

$$dS - \delta Q/T \geq 0 \tag{13.20}$$

In Eq. (13.20) we have assumed that heat transfer between the system and the surroundings, which is a reservoir at T, is reversible. From the two equations above we obtain the relation

$$TdS - dU - PdV \geq 0 \tag{13.21}$$

Note that Eq. (13.21) is a relationship between properties of a system and therefore it applies to any change of state of a system with uniform temperature and pressure. We now apply Eq. (13.21) to a closed system with additional restrictions to determine the manner in which the system proceeds towards an equilibrium state.

Consider a system undergoing a process at *constant volume and internal energy*. Therefore $dV = 0$ and $dU = 0$. From Eq. (13.21) it follows that $dS \geq 0$, which is the criterion of equilibrium for the closed system.

We now obtain a criterion which we shall, in the next section, apply to chemically reacting systems. Consider an infinitesimal process in which the *temperature and pressure* of the system are constant. From Eq. (13.21) it follows that

$$d(TS)_{P,T} - dU_{P,T} - d(PV)_{P,T} \geq 0 \tag{13.22}$$

We can express Eq. (13.22) in terms of the Gibbs free energy function, G which was introduced in chapter 7 as

$$G = U + PV - TS \tag{13.23}$$

It follows from Eqs. (13.22) and (13.23) that when the *pressure and temperature* during a process are constant,

$$dG_{P,T} \leq 0 \tag{13.24}$$

The above equation states that when a system attains equilibrium, under conditions of *constant temperature and pressure*, the Gibbs function is a minimum. For a chemically reacting system, for which the composition of the constituents is the independent variable, the variation of G is depicted Fig. 13.4.

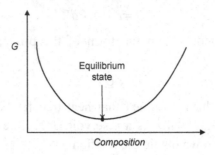

Fig. 13.4 Equilibrium condition under constant pressure and temperature

13.4.1 *Reactions in ideal-gas mixtures*

We now apply the above criterion to a reacting mixture of ideal gases to determine the equilibrium composition of the constituents. Consider the reaction represented by the chemical equation

$$v_1A_1 + v_2A_2 = v_3A_3 + v_4A_4 \qquad (13.25)$$

Let the number moles, the partial pressure, the molar specific enthalpy, and the absolute molar entropy, of the ith constituent in the mixture be respectively, n_i, P_i, $\overline{h}_i$ and $\overline{s}_i^{\,o}$. The Gibbs function of the mixture is given by

$$\overline{G}_m = \overline{H}_m - T\overline{S}_m \qquad (13.26)$$

where the subscript m denotes the properties of the mixture.

We substitute in Eq. (13.26), the expressions obtained earlier for the molar enthalpy and molar entropy of an ideal gas. Thus

$$\overline{G}_m = \sum_{i=1}^{4} n_i (\overline{h}_i - T\overline{s}_i) \qquad (13.27)$$

$$\overline{G}_m = \sum_{i=1}^{4} n_i \overline{h}_i - \sum_{i=1}^{4} n_i T\left[\overline{s}_i^{\,o}(T^o, P^o) + \overline{c}_{pi}\ln(T/T^o) - \overline{R}\ln(P_i/P^o)\right]$$

$$\overline{G}_m = \sum_{i=1}^{4} n_i F_i(T^o, T) + T\sum_{i=1}^{4} n_i \overline{R}\ln(P_i/P^o) \qquad (13.28)$$

where the function, $F_i(T^o, T) = \overline{h}_i - T\left[\overline{s}_i^{\,o} + \overline{c}_{pi}\ln(T/T^o)\right]$

Note that the function $F_i(T^o, T)$ depends on the properties of the ith constituent, but not on the number of moles of the ith constituent in the mixture.

Now consider an infinitesimal change in the number moles of the constituents of the mixture. For the mass balance and the chemical equation [Eq. (13.25)] to be satisfied

$$-dn_1 / \upsilon_1 = -dn_2 / \upsilon_2 = dn_3 / \upsilon_3 = dn_4 / \upsilon_4 = dn \qquad (13.29)$$

The infinitesimal change in the Gibbs function of the mixture, resulting from the change in composition, is obtained by differentiating Eq. (13.28). Thus

$$d\overline{G}_m = \sum_{i=1}^{4} F_i dn_i + T \sum_{i=1}^{4} [\overline{R} \ln(P_i / P^o) dn_i + n_i \overline{R} dP_i / P_i] \qquad (13.30)$$

Note that the change in composition of the constituents affects the partial pressures of the constituents because, $P_i = n_i P_m / N_m$, where N_m is the total number of moles and P_m is the total pressure.

We simplify the last term of the Eq. (13.30) as follows:

$$n_i dP_i / P_i = (x_i N_m) P_m dx_i / (P_m x_i) = N_m dx_i$$

where x_i is the mole fraction of the ith constituent. Hence Eq. (13.30) may be written in the form

$$d\overline{G}_m = \sum_{i=1}^{4} F_i dn_i + T \sum_{i=1}^{4} [\overline{R} \ln(P_i / P^o) dn_i + \overline{R} N_m dx_i] \qquad (13.31)$$

Note that after summation the last term of Eq. (13.31) becomes zero because the mole fractions satisfy the condition:

$$\sum_{i=1}^{4} x_i = 1$$

Hence Eq. (13.31) takes the form

$$d\overline{G}_m = \sum_{i=1}^{4} \left[F_i + T\overline{R} \ln(P_i / P^o) \right] dn_i \qquad (13.32)$$

Substituting from Eq. (13.29) in Eq. (13.32) and invoking the criterion of equilibrium expressed by Eq. (13.24), we have

$$\ln\left[\frac{(P_3/P^o)^{\nu_3}(P_4/P^o)^{\nu_4}}{(P_1/P^o)^{\nu_1}(P_2/P^o)^{\nu_2}}\right] = (F_1\nu_1 + F_2\nu_2 - F_3\nu_3 - F_4\nu_4)/\overline{R}T$$

The relationship for the partial pressures of the constituents at equilibrium is obtained from the above equation as

$$\left[\frac{(P_3/P^o)^{\nu_3}(P_4/P^o)^{\nu_4}}{(P_1/P^o)^{\nu_1}(P_2/P^o)^{\nu_2}}\right] = K_P(T) \qquad (13.33)$$

The function $K_P(T)$ is called the *equilibrium constant* for the reaction whose chemical equation is Eq. (13.25). It is a function of the temperature of the reacting mixture.

We can also express Eq. (13.33) in terms of the mole fractions of the constituents as

$$\left[\frac{x_3^{\nu_3}x_4^{\nu_4}}{x_1^{\nu_1}x_2^{\nu_2}}\right](P_m/P^o)^{(\nu_3+\nu_4-\nu_1-\nu_2)} = K_P(T)$$

The equilibrium constants, for a number of ideal-gas reactions, are tabulated as a function of temperature in [4].

13.4.2 *Dissociation*

The adiabatic flame temperature is the theoretically maximum temperature that can be attained in a steady-flow combustion system. In the calculation of this limiting temperature, we usually assume that the combustion reaction proceeds to completion. This means that when a hydrocarbon fuel is burned with oxygen, the products contain only carbon dioxide and water. However, under certain conditions of temperature and pressure, some of the products of combustion could undergo dissociation.

For example, if the temperature is sufficiently high, the CO_2 in the products of a reaction can dissociate into carbon monoxide and oxygen

$$CO_2 \leftrightarrow CO + 1/2\ O_2$$

The dissociation reaction is an endothermic reaction that absorbs energy from the products, thus lowering the temperature. Finally a state of dynamic equilibrium is established, with the reaction proceeding in both directions at the same rate. The fraction of CO_2 that undergoes dissociation is determined by the equilibrium constant, which in turn, is a function of the temperature. Additional examples of dissociation reactions are:

$$H_2O \leftrightarrow H_2 + 1/2\ O_2$$
$$N_2 \leftrightarrow 2\ N$$

Numerical calculations involving dissociation reactions and the equilibrium constant are presented in worked examples 13.14 to 13.16.

13.5 Worked Examples

Example 13.1 A fuel consists of a mixture of gases with the following volumetric composition:

$H_2 = 26\%$, $CH_4 = 38\%$, $CO = 26\%$, $N_2 = 6\%$, $CO_2 = 4\%$

Calculate (i) the AFR of a stoichiometric mixture of air and fuel, and (ii) the volumetric composition of the products of combustion, when the water produced is: (a) a vapor, and (b) a liquid.

Solution The fuel gases in the mixture are: CH_4, H_2 and CO. Consider 1 kmol of each constituent in the mixture and write the following chemical equations for the stoichiometric combustion reactions with air:

$$CH_4 + 2\ O_2 + 2 \times 3.76\ N_2 = CO_2 + 2\ H_2O + 2 \times 3.76\ N_2$$
$$H_2 + 1/2\ O_2 + (1/2 \times 3.76)\ N_2 = H_2O + (1/2 \times 3.76)\ N_2$$
$$CO + 1/2\ O_2 + (1/2 \times 3.76)\ N_2 = CO_2 + (1/2 \times 3.76)\ N_2$$

Now consider 1 kmol of the fuel mixture. The number of kmoles of air required for the complete combustion of all the constituent fuel gases is

$$N_A = (0.38 \times 2 + 0.26 \times 0.5 + 0.26 \times 0.5) \times 4.76 = 4.86$$

The mass of air required is

$$m_A = 4.86 \times 29 = 140.94\ \text{kg}$$

The mass of 1 kmol of fuel is

$$m_F = 0.26 \times 2 + 0.38 \times 16 + 0.26 \times 28 + 0.06 \times 28 + 0.04 \times 44$$

$$m_F = 17.32\ \text{kg}$$

(i) The air-fuel ratio by mass is

$$AFR = m_A / m_F = 140.94 / 17.32 = 8.13 \text{ kg air/kg fuel}$$

(ii) The number of kmoles of the various constituents in the products are obtained from the above chemical equations as:

$$CO_2: \ 0.38 + 0.26 + 0.04 = 0.68$$
$$H_2O \text{ (vapor)}: \ 2 \times 0.38 + 0.26 = 1.02$$
$$N_2: \ (0.38 \times 2 + 0.26 \times 1/2 + 0.26 \times 1/2) \times 3.76 + 0.06 = 3.895$$

Total number of product kmols is 5.595.

(a) The volumetric composition of the products, with water in the vapor state, is the same as the mole-ratio of the products. Hence

$$CO_2: H_2O: N_2 = 0.121 : 0.182 : 0.696$$

(b) When the water is in the liquid state, its volume is negligible, and therefore does not contribute to the mole-ratio of the products. Then the total number of kmols of products is: $0.68 + 3.895 = 4.575$. The volumetric composition is:

$$CO_2: N_2 = 0.149 : 0.85$$

Example 13.2 The fuel supplied to a furnace is a mixture of gases with the following volumetric composition:

$$C_4H_{10} = 22\%, \ CH_4 = 52\%, \ O_2 = 2\%, \ N_2 = 24\%$$

The pressure, temperature and relative humidity of the ambient air supplied for combustion are: 100 KPa, 30°C and 90% respectively. The volumetric composition of the dry exhaust gases is as follows:

$$CO_2 = 9.92\%, \ O_2 = 4.86\%, \ N_2 = 85.22\%$$

Calculate (i) the kmols of oxygen for stoichiometric combustion of a kmol of fuel, (ii) the excess air supplied, (iii) the partial pressure of the water vapor in the air supplied, and (iv) the kmols of water in the products of combustion.

Solution The combustible gases in the fuel-air mixture are C_4H_{10} and CH_4. Consider 1 kmol each of the above gases and write the following reaction equations:

$$CH_4 + 2\,O_2 + 2 \times 3.76\,N_2 = CO_2 + 2\,H_2O + 2 \times 3.76\,N_2$$

$$C_4H_{10} + 6.5\,O_2 + 6.5 \times 3.76\,N_2 = 4\,CO_2 + 5\,H_2O + 6.5 \times 3.76\,N_2$$

(i) From the two equations above we obtain the total number of kmols of O_2 for stoichiometric combustion of 1 kmol of fuel mixture as:

$$0.52 \times 2 + 0.22 \times 6.5 - 0.02 = 2.45$$

(ii) Let the fraction of excess air supplied be e. Then the number of kmols of O_2 and N_2 in the reacting mixture are, $2.45 \times (1 + e)$ and $2.45 \times (1 + e) \times 3.76$ respectively. The volumetric composition of the *dry* combustion products is

$$O_2 = 2.45e$$

$$N_2 = [0.24 + 2.45 \times 3.76(1 + e)]$$

and $$CO_2 = 0.52 + 4 \times 0.22$$

Let the total number of kmols of *dry* products be N_p. Then the composition of the products may be written as:

$$O_2 = 2.45e / N_p = 0.0486$$

$$N_2 = [0.24 + 2.45 \times 3.76(1 + e)] / N_p = 0.8522$$

$$CO_2 = [0.52 + 4 \times 0.22] / N_p = 0.0992$$

From the three equations above we obtain, $e = 0.28$ as the fraction of excess air.

The saturation vapor pressure of water at 30°C is 4.242 kPa. The partial pressure of the water vapor in the air supplied is

$$P_w = 0.9 \times 4.242 = 3.818 \text{ kPa.}$$

The number of kmols of water vapor per kmol of dry air in the air supplied to the furnace is given by

$$n_w / n_a = 0.9 \times 4.242 / (100 - 0.9 \times 4.242) = 0.0397$$

The number of kmols of water vapor in the products per kmol of fuel gas is

$$2 \times 0.52 + 5 \times 0.22 + 1.28 \times 2.45 \times 4.76 \times 0.0397 = 2.73$$

Total number of kmols of reactants is

$$1 + 1.28 \times 2.45 \times 4.76 \times 1.0397 = 16.52$$

The number of kmols of water vapor per kmol of reactants is

$$2.73/16.52 = 0.165$$

Example 13.3 A piston-cylinder set-up contains a stoichiometric mixture of C_2H_4 and air. The mass of C_2H_4 is 2×10^{-4} kg. The mixture undergoes a combustion process exchanging heat with the surroundings. The final pressure and temperature are 300 kPa and 300°C respectively. Calculate (i) the final volume of the products, and (ii) the dew-point of the products, if they are cooled at a constant pressure of 300 kPa,

Solution The chemical equation for the combustion reaction is

$$C_2H_4 + 3\ O_2 + 3 \times 3.76\ N_2 = 2\ CO_2 + 2\ H_2O + 3 \times 3.76\ N_2$$

The number of kmols of products per kmol of C_2H_4 is

$$(2 + 2 + 3 \times 3.76) = 15.28$$

The number of kmols of C_2H_4 in the reactants is

$$2 \times 10^{-4}/28 = 0.714 \times 10^{-5}$$

The total number of kmols in the products is

$$0.714 \times 10^{-5} \times 15.28 = 10.91 \times 10^{-5}$$

Apply the molar-form of the ideal gas equation to the products.

$$PV = N_p \overline{R} T$$

$$V = 10.91 \times 10^{-5} \times 8.3145 \times (273 + 300)/300 = 1.73 \times 10^{-3}\ \text{m}^3$$

The partial pressure of the water vapor in the products is

$$P_w = 300 \times 2/15.28 = 39.27\ \text{kPa}.$$

The saturation temperature of water at 39.27 kPa is 75.3°C, which is the dew-point of the water vapor in the products.

Example 13.4 The products of combustion of a hydrocarbon fuel with air has the following volumetric composition on a dry-basis: $CO_2 = 11\%$, $O_2 = 3\%$, $CO = 2\%$, $N_2 = 84\%$. (i) Calculate the AFR. (ii) If the total pressure of the products is 100 kPa, calculate the dew-point of the water vapor in the products.

Solution Let the chemical formula of the hydrocarbon fuel be C_nH_m.

The chemical equation for the combustion reaction is:

$$C_nH_m + x_1O_2 + x_1 \times 3.76\ N_2 = x_2CO_2 + x_3H_2O + x_4CO + x_5O_2 + x_6N_2$$

Consider 1 kmol of *dry* products of combustion. Using the given analysis of the dry products we have:

$$x_2 = 0.11,\ x_4 = 0.02,\ x_5 = 0.03\ \text{and}\ x_6 = 0.84.$$

Balancing the number of carbon atoms on both sides of the equation

$$n = x_2 + x_4 = 0.13$$

Balancing the number of nitrogen atoms on both sides of the equation

$$3.76x_1 = 0.84,\quad \text{Hence } x_1 = 0.2234$$

Balancing the number of oxygen atoms on both sides of the equation

$$2x_1 = 2x_2 + x_3 + x_4 + 2x_5 = 2 \times 0.2234$$

Hence $$x_3 = 0.1468$$

Balancing the number of hydrogen atoms on both sides of the equation

$$m = 2x_3 = 0.2936$$

Mass of fuel is, $$m_f = 12 \times 0.13 + 0.2936 \times 1 = 1.8536$$

Mass of air is, $$m_a = 0.2234 \times 4.76 \times 29 = 30.83$$

(i) The air to fuel ratio by mass is

$$AFR = 30.83 / 1.8536 = 16.63$$

The number of kmols of water vapor in the products is, $x_3 = 0.1468$
The total number of kmols of products is

$$N_p = x_2 + x_3 + x_4 + x_5 + x_6 = 1.1468$$

The partial pressure of the water vapor in the product mixture is

$$P_w = 100 \times 0.1468 / 1.1468 = 12.8\ \text{kPa}$$

The saturation temperature of water at 12.8 kPa is 51°C, which is the dew point of the water vapor in the products. If the products are cooled below this temperature at constant pressure, then condensation of the water would occur.

Example 13.5 A sample of coal has the following ultimate analysis: carbon - 70%, hydrogen - 6%, oxygen - 10%, sulfur - 1%, nitrogen - 2%, ash 11%. The coal undergoes combustion with 20% excess air at a pressure of 100 kPa. Calculate (i) the AFR, and (ii) the volumetric analysis of the products on a dry-basis.

Solution The combustible constituents, their reaction equations, and the mass balance are as follows:

$$C + O_2 = CO_2$$

$$12 \text{ kg} + 32 \text{ kg} = 44 \text{ kg}$$

$$2 H_2 + O_2 = 2 H_2O$$

$$4 \text{ kg} + 32 \text{ kg} = 36 \text{ kg}$$

$$S + O_2 = SO_2$$

$$32 \text{ kg} + 32 \text{ kg} = 64 \text{ kg}$$

From the above equations we obtain the oxygen required per kg of coal for stoichiometric combustion as:

$$0.7 \times 32/12 + 0.06 \times 32/4 + 0.01 \times 32/32 = 2.357 \text{ kg}$$

The oxygen content of the coal is 0.1. The excess air supplied is 20%. Therefore the required supply of oxygen is

$$1.2 \times (2.357 - 0.1) = 2.708 \text{ kg}.$$

The mass of nitrogen supplied with the oxygen is

$$2.708 \times 3.29 = 8.91 \text{ kg}.$$

The mass of air supplied is

$$2.708 + 8.91 = 11.62 \text{ kg}.$$

Hence the AFR is 11.62 kg of air per kg of coal.

Now consider the products of the combustion process. The total mass per kg of coal is

$$0.7 \times 44/12 + 0.06 \times 36/4 + 0.01 \times 64/32 = 3.127 \text{ kg}$$

The constituents in the reactant mixture, whose mass is unaffected by combustion are nitrogen in the coal, the nitrogen in the air supplied, the excess oxygen, and the ash. Their total mass is

$$0.02 + 8.91 + 0.4514 + 0.11 = 9.49 \text{ kg}$$

The total mass of the product mixture is

$$3.127 + 9.49 = 12.62 \text{ kg}$$

This agrees with the total mass of air and fuel which is

$$1.0 + 11.62 = 12.62 \text{ kg}$$

The number of kmols of the gases in the product mixture are as follows:

$$CO_2 \rightarrow 0.7 \times 44/(12 \times 44) = 0.05833$$

$$SO_2 \rightarrow 0.01 \times 64/(32 \times 64) = 0.0003125$$

$$O_2 \rightarrow 0.4514/32 = 0.0141$$

and

$$N_2 \rightarrow 8.91/28 = 0.32$$

Hence the volumetric analysis of the product gases is

$$CO_2 = 0.1485, \ SO_2 = 0.7955 \times 10^{-3}, \ O_2 = 0.0359, \ N_2 = 0.8146$$

Example 13.6 Calculate (i) the enthalpy of reaction, and (ii) the internal energy of reaction at 1 bar and 25°C for the combustion of propane, C_3H_8, when the water formed is: (a) a vapor, and (b) a liquid. The molar enthalpy of formation for CO_2, $H_2O(g)$, and C_3H_8 are respectively $-393510 \text{ kJ kmol}^{-1}$, $-241820 \text{ kJ kmol}^{-1}$ and $-104680 \text{ kJ kmol}^{-1}$.

Solution The chemical equation for the combustion process is

$$C_3H_8 + 5 \ O_2 = 3 \ CO_2 + 4 \ H_2O$$

(a) The enthalpy of reaction per kmol of C_3H_8 is given by

$$\Delta \bar{h}_R = 3\bar{h}_{CO_2} + 4\bar{h}_{H_2O} - \bar{h}_{C_3H_8} - 5\bar{h}_{O_2}$$

$$\Delta \bar{h}_R = -3 \times 393510 - 4 \times 241820 + 104680 - 5 \times 0 = -2043130 \text{ kJ kmol}^{-1}$$

We now write the following relations between the enthalpy and internal energy of the reactants and products under standard conditions:

$$\bar{h}_R = \bar{u}_R + n_p \bar{R} T^o$$

$$\bar{h}_P = \bar{u}_P + n_p \bar{R} T^o$$

Hence

$$\bar{h}_P - \bar{h}_R = (\bar{u}_P - \bar{u}_R) + (n_P - n_R)\bar{R} T^o$$

$$\Delta \bar{h}_R = \Delta \bar{u}_R + (n_P - n_R)\bar{R} T^o$$

Substituting numerical values in the above equation

$$-2043130 = \Delta\bar{u}_R + (7-6)\times 8.1345\times 298$$

Therefore the internal energy of reaction is

$$\Delta\bar{u}_R = -2045608 \text{ kJ kmol}^{-1} \text{ of fuel}$$

(b) The enthalpy of vaporization of water at 25°C is 2441.8 kJ kg^{-1}. The corresponding molar value is $2441.8\times 18 = 43952.4$ kJ kmol^{-1}.

When the 4 kmols of water in the products condense to liquid the enthalpy of reaction becomes

$$\Delta\bar{h}_R(l) = -2043130 - 4\times 43952.4 = -2218940 \text{ kJ kmol}^{-1}$$

The corresponding value of the internal energy of reaction is given by

$$-2218940 = \Delta\bar{u}_R(l) + (3-6)\times 8.1345\times 298$$

Note that the 4 kmols of water are not included in the total number of product kmols because of the negligible volume of liquid water.

Therefore the internal energy of reaction is

$$\Delta\bar{u}_R(l) = -2211507 \text{ kJ kmol}^{-1} \text{ of fuel}$$

Example 13.7 One kmole of C_3H_8 and 3 kmoles of oxygen undergo combustion in a constant volume process. The initial temperature and pressure of the reactants are 25°C and 1 bar respectively. The products of combustion are cooled to 316°C. (a) Calculate the heat transferred per kg of fuel. (b) If the products are cooled at constant volume, calculate the temperature at which condensation of water vapor just begins. The enthalpy of reaction of C_3H_8 is −2043130 kJ kmol^{-1}. The average molar specific heat capacity, $\bar{c}_v$ (kJ kmol^{-1} K^{-1}) for CO_2 and H_2O are 35.88 and 26.58 respectively.

Solution The chemical equation for the reaction is

$$C_3H_8 + 5\,O_2 = 3\,CO_2 + 4\,H_2O$$

The internal energy of combustion is obtained from the given value of the enthalpy of combustion using the relation

$$\Delta U^o = \Delta H^o - (n_P - n_R)\bar{R}T^o$$

$$\Delta U^o = -2043130 - (7-6)\times 8.3145\times 298 = 2045608 \text{ kJ kmol}^{-1}$$

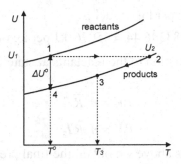

Fig. E13.7 U-T diagram

Initially, the reactants are at state 1 as shown in the U-T diagram in Fig. E13.7. Applying the first law to the constant volume, adiabatic combustion process, 1-2 we have

$$U_2 = U_1$$

Apply the first law to the constant volume cooling process 2-3.

Hence the heat removed is given by

$$Q_{23} = U_2 - U_3$$

From the two equations above it follows that

$$Q_{23} = U_1 - U_3 \qquad (E13.7.1)$$

For the combustion reaction occurring at the standard condition of 25°C we have

$$\Delta U^o = U_4 - U_1 \qquad (E13.7.2)$$

Adding Eqs. (E13.7.1) and (E13.7.2) we obtain the heat transfer as:

$$Q_{23} = (U_4 - U_3) - \Delta U^o$$

The above equation is now expressed in terms of the average specific heat capacities of the products as:

$$Q_{23} = -\Delta U^o - \sum_{i=1}^{2} n_{pi} \bar{c}_{pi} (T_3 - T_4)$$

Substituting numerical values

$$Q_{23} = 2045608 - (3 \times 35.88 + 4 \times 26.58) \times (589 - 298)$$

$$Q_{23} = 1983346 \text{ kJ per kmol of } C_3H_8$$

Hence the heat transfer per kg of fuel is

$$Q_{23} = 1983346/44 = 45076 \text{ kJ per kg of } C_3H_8$$

Apply the ideal gas equation to the reactants at state 1 and the products at state 3 to obtain

$$P_1V = n_R \overline{R} T_1$$

$$P_3V = n_P \overline{R} T_3$$

From the two equations above we obtain the final pressure as

$$P_3 = 7 \times 589/(6 \times 298) = 2.3 \text{ bar}$$

The partial pressure of water vapor in the products is

$$P_w = 2.3 \times 4/7 = 1.32 \text{ bar}$$

Condensation of water occurs at the saturation temperature of water corresponding to 1.32 bar, which from the steam tables [4], is 107.4°C.

Note that to evaluate the change in thermal internal energy we used the specific heat capacities at the mean temperature. This was possible because the temperature range of interest was specified. However, for the wide temperature changes encountered in combustion problems, the use of the mean values of the specific heat capacity could cause significant errors in the results.

It is instructive to recall the worked example 4.19, where we demonstrated that, the use of the mean specific heat capacities is equivalent to assuming a linear variation of the specific heat capacity with temperature, over the temperature range of interest.

In worked example 13.9, we shall use a more accurate technique, where the specific heat capacity data are fitted with cubic equations in terms of the temperature. The molar enthalpy at different temperatures of several gases, that are of importance in combustion processes, are tabulated in [4]. We shall use these data in worked examples 13.12 to 13.15.

Example 13.8 Calculate the internal energy of combustion of CH_4 at 200°C. The enthalpy of combustion of CH_4 at the standard temperature of 25°C is −802300 kJ kmol^{-1}. The average molar specific heat capacity, $\overline{c}_v$ (kJ kmol^{-1} K^{-1}) for the various constituents are as follows:

$$CH_4 = 31.45, \quad O_2 = 21.77, \quad CO_2 = 32.17, \quad H_2O = 25.89$$

Solution The chemical equation for the combustion reaction is

$$CH_4 + 2\,O_2 = CO_2 + 2\,H_2O$$

The internal energy of combustion is obtained from the given value of the enthalpy of combustion using the relation

$$\Delta U^o = \Delta H^o - (n_P - n_R)\bar{R}T^o$$

Substituting numerical values in the above equation we have

$$\Delta U^o = -802300 - (3-3)\times 8.3145 \times 298 = -802300 \text{ kJ kmol}^{-1}$$

The U-T diagram for the combustion process is shown in Fig. E13.8. The internal energies of combustion at the standard temperature of 25°C and at the given temperature of 200°C are indicated in the figure. The variation of the internal energy of the reactants and the products may be expressed in terms of the average specific heat capacities as follows:

$$U_{R2} = U_{R1} + \sum_{i=1}^{2} n_{ri}\bar{c}_{ri}(T_2 - T_1)$$

$$U_{P3} = U_{P4} + \sum_{i=1}^{2} n_{pi}\bar{c}_{pi}(T_3 - T_4)$$

From the two equations above we have

$$U_{P3} - U_{R2} = U_{P4} - U_{R1} + \sum_{i=1}^{2} n_{pi}\bar{c}_{pi}(T_3 - T_4) - \sum_{i=1}^{2} n_{ri}\bar{c}_{ri}(T_2 - T_1)$$

Since $T_1 = T_4$ and $T_2 = T_3$, the above equation may be simplified to the form

$$\Delta U_1 = \Delta U^o + \sum_{i=1}^{2} n_{pi}\bar{c}_{pi}(T_2 - T_1) - \sum_{i=1}^{2} n_{ri}\bar{c}_{ri}(T_2 - T_1)$$

where ΔU_1 is the internal energy of reaction at 200°C. Substituting numerical values we have

$$\Delta U_1 = -802300 + (32.17 + 25.89 \times 2 - 31.85 - 21.77 \times 2)(200 - 25)$$

The internal energy of combustion at 200°C is

$$\Delta U_1 = -800802 \text{ kJ kmol}^{-1}$$

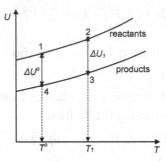

Fig. E13.8 *U-T* diagram

Example 13.9 Calculate the adiabatic flame temperature for the steady flow combustion of a stoichiometric mixture of CH_4 and air. The reactants are at 25°C. The enthalpy of combustion of CH_4 at the standard temperature of 25°C is −802300 kJ kmol⁻¹. The variation of the molar specific heat capacities of the constituents CO_2, H_2O, N_2 and O_2 with temperature, T (K), in the temperature range 300-1500 K, may be expressed in the form of the cubic equation:

$$\bar{c}_p = a_o + a_1 T + a_2 T^2 \ [\text{kJ kmol}^{-1} \ \text{K}^{-1}]$$

The coefficients are tabulated below:

Substance	a_o	$a_1 \times 10^3$	$a_2 \times 10^6$
CO_2	26.098	43.66	−14.89
H_2O	30.475	9.65	1.189
N_2	27.4	5.25	−0.0042
O_2	25.82	13.0	−3.877

Solution The chemical equation for the steady-flow combustion process is

$$CH_4 + 2\ O_2 + 7.52\ N_2 = CO_2 + 2\ H_2O + 7.52\ N_2$$

The *H-T* diagram is depicted in Fig. E13.9.

Apply the SFEE to the combustion process, neglecting changes in kinetic and potential energy. The work output is zero. For adiabatic combustion the heat transfer is also zero. Hence

$$H_1 = H_2 \qquad\qquad (E13.9.1)$$

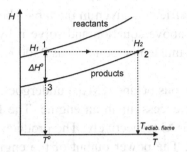

Fig. E13.9 *H-T* diagram

For the change in enthalpy of the products we have

$$H_2 = H_3 + \sum_{i=1}^{3} \int_{T^o}^{T_2} n_{pi} \bar{c}_{pi}(T) dT \qquad (E13.9.2)$$

where T_2 is the adiabatic flame temperature and n is the number of kmols. The subscript i denotes the three constituents CO_2, H_2O and N_2, in the products.

From Eqs. (E13.9.1) and (E13.9.2) it follows that

$$H_1 - H_3 = -\Delta H^o = \sum_{i=1}^{3} \int_{T^o}^{T_2} n_{pi} \bar{c}_{pi} dT$$

The specific heat capacities in the above equation are now represented by the cubic equations to obtain the following expression:

$$-\Delta H^o = \sum_{i=1}^{3} \int_{T^o}^{T_2} n_{pi} (a_{oi} + a_{1i}T + a_{2i}T^2) dT$$

Performing the integration we obtain

$$-\Delta H^o = \sum_{i=1}^{3} n_{pi} \left(a_{oi}T + a_{1i}T^2 / 2 + a_{2i}T^3 / 3 \right) + A$$

where A is the constant of integration.

Substituting numerical values

$$-\Delta H^o = \sum_{i=1}^{3} n_{pi} \left[a_{oi}(T_2 - 298) + a_{1i}(T_2^2 - 298^2)/2 + a_{2i}(T_2^3 - 298^3)/3 \right]$$

where $\quad\quad\quad\quad\quad \Delta H_o = -802300 \text{ kJ kmol}^{-1}$

We substitute the coefficients given in the table above for CO_2, H_2O and N_2, ($i = 1$, 3) in the above equation and solve it by trial-and-error. This gives the adiabatic flame temperature, T_2 as 2300 K.

Example 13.10 Gaseous octane (C_8H_{18}) undergoes steady-flow combustion with 50 percent excess air in an engine. The fuel and air enter the engine at 25°C and 65°C respectively. The products of combustion leave the engine at 450°C. The power output of the engine is 90 kW and the heat transfer from the engine is 200 kW. Calculate the fuel consumption rate of the engine. The enthalpy of combustion of octane at 1 bar and 25°C is -5116112 kJ kmol^{-1}.

The molar specific heat capacities, $\bar{c}_p$ (kJ kmol^{-1} K^{-1}) at the mean temperature of the reactants are: $O_2 = 29.5$, H_2O, $N_2 = 29.1$.

The specific heat capacities at the mean temperature of the products are:

$$O_2 = 31.1, \quad CO_2 = 44.6, \quad H_2O = 35.2, \quad N_2 = 29.6$$

Solution The chemical equation for the combustion process with 50% excess air is

$$C_8H_{18} + 1.5 \times 12.5\, O_2 + 1.5 \times 12.5 \times 3.76\, N_2$$
$$= 8\, CO_2 + 9\, H_2O + 1.5 \times 12.5 \times 3.76\, N_2 + 0.5 \times 12.5\, O_2$$

Note that the excess oxygen is included in the products of combustion, and we assume that the water is in the vapor state. The above equation may be written as:

$$C_8H_{18} + 18.75\, O_2 + 70.5\, N_2 = 8\, CO_2 + 9\, H_2O + 70.5\, N_2 + 6.25\, O_2$$

Apply the SFEE to the combustion process, neglecting changes in kinetic and potential energy.

$$\dot{Q}_{in}/\dot{n}_f + h^o{}_f + 18.75 \times (h^o{}_{O_2} + \bar{c}_{O_2}\Delta T_1) + 70.5 \times (h^o{}_{N_2} + \bar{c}_{N_2}\Delta T_1)$$
$$= \dot{W}_o/\dot{n}_f + 8 \times (h^o{}_{CO_2} + \bar{c}_{CO_2}\Delta T_2) + 9 \times (h^o{}_{H_2O} + \bar{c}_{H_2O}\Delta T_2)$$
$$+ 6.25 \times (h^o{}_{O_2} + \bar{c}_{O_2}\Delta T_2) + 70.5 \times (h^o{}_{N_2} + \bar{c}_{N_2}\Delta T_2)$$

In the above equation h^o is the enthalpy of formation under standard conditions, $\dot{n}_f$ is the molar flow rate of the fuel. From the given data:

$$\Delta T_1 = 40°C, \quad \Delta T_2 = 425°C, \quad \dot{Q}_{in} = -200 \text{ kW}, \quad \dot{W}_o = 90 \text{ kW}.$$

The molar specific heat capacities of the reactants and the products at the respective mean temperatures are listed above. The enthalpy of combustion is given by

$$\Delta H^o = (8 \times h^o{}_{CO_2} + 9 \times h_{H_2O}) - h^o{}_f = -5116112 \text{ kJ kmol}^{-1}$$

Note that the enthalpies of formation of the elements are zero.

We substitute these numerical data in the SFEE given above to obtain

$$-290 / \dot{n}_f = 2.954 \times 10^3 \times 425 - 2.6047 \times 10^3 \times 40 - 5116112$$

Hence
$$\dot{n}_f = 7.314 \times 10^{-5} \text{ kmol s}^{-1}$$

The fuel consumption rate is

$$\dot{m}_f = 7.314 \times 10^{-5} \times 114 = 8.34 \times 10^{-3} \text{ kg s}^{-1}$$

Example 13.11 A mixture of gaseous propane (C_3H_8) and air at an initial temperature of 25°C undergoes adiabatic combustion at constant volume. The products of combustion reach a temperature of 900°C. Calculate the air-fuel ratio of the mixture. The molar internal energy of reaction at 25°C is −2045608 kJ kmol^{-1}. The specific heat capacity, $\bar{c}_v$ (kJ kmol^{-1} K^{-1}) of the products at the mean temperature are:

$$O_2 = 25.2, \quad CO_2 = 42.2, \quad H_2O = 29.7, \quad N_2 = 22.8$$

Solution Assume that the reaction is complete and that the products do not undergo dissociation reactions. The water is assumed to be in the vapor state. Let the excess air fraction in the reactant mixture be e. The chemical equation for the reaction is:

$$C_3H_8 + 5(1 + e) O_2 + 18.8(1 + e) N_2$$
$$= 3 CO_2 + 4 H_2O + 5e O_2 + 18.8(1 + e) N_2$$

Apply the first law to the constant volume adiabatic process. The work done and heat transfer are both zero. Therefore the internal energy is constant as indicated in the *U-T* diagram (Fig. E13.11).

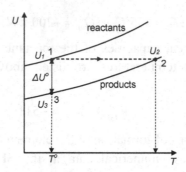

Fig. E13.11 *U-T* diagram

Hence $$U_1 = U_2$$

Now for the products, the change in internal energy is:

$$U_2 - U_3 = [3 \times 42.2 + 4 \times 29.7 + 5e \times 25.2 + 18.8(1 + e) \times 22.8](875)$$

For the standard conditions

$$U_3 - U_1 = \Delta U^o = -2045608 \text{ kJ kmol}^{-1}$$

From the three equations above we obtain

$$554.64e + 674.04 = 2045608 / 875$$

Hence the excess air fraction is, $e = 3$.

The mass of fuel per kmol is 44 kg. The mass of air is

$$4 \times 5 \times (32 + 3.76 \times 28) = 2745.6 \text{ kg.}$$

Hence the air-fuel-ratio is given by

$$AFR = 2745.6 / 44 = 62.4 \text{ kg/kg of fuel}$$

Example 13.12 The volumetric composition of the constituents of the products of a combustion process is as follows:

$$CO_2 = 0.095, \quad H_2O = 0.19, \quad N_2 = 0.715$$

The temperature and total pressure of the products are 400 K and 200 kPa respectively. Calculate the total entropy of the products per unit mass.

Solution The following values of the absolute molar entropy (kJ K^{-1} kmol^{-1}) at 1 bar and 400 K have been obtained from the table in [4].

$$CO_2 = 225.22, \qquad H_2O = 198.67, \qquad N_2 = 200.07$$

The partial pressures of the constituents are:

$$P_{CO_2} = 0.095 \times 2 = 0.19 \text{ bar}, \qquad P_{H_2O} = 0.19 \times 2 = 0.38 \text{ bar}$$

and

$$P_{N_2} = 0.715 \times 2 = 1.43 \text{ bar}$$

The absolute entropy (kJ K^{-1} kmol^{-1}) of the constituents at 400 K and partial pressure P is given by:

$$\bar{s}(T,P) = \bar{s}^o(T,1) - \bar{R}\ln(P/1)$$

Applying the above equation we obtain the following values of the molar entropy at 400 K:

$$\bar{s}_{CO_2} = 225.22 - 8.3145 \times \ln(0.19/1) = 239.03 \text{ kJ K}^{-1} \text{ kmol}^{-1}$$

$$\bar{s}_{H_2O} = 198.67 - 8.3145 \times \ln(0.38/1) = 206.71 \text{ kJ K}^{-1} \text{ kmol}^{-1}$$

$$\bar{s}_{N_2} = 200.07 - 8.3145 \times \ln(1.43/1) = 197.1 \text{ kJ K}^{-1} \text{ kmol}^{-1}$$

The total molar entropy of the product mixture per kmol is

$$\bar{s}_m = 0.095 \times 239.03 + 0.19 \times 206.71 + 0.715 \times 197.1 = 202.9$$

The mass of a kmol of products is

$$m_m = 0.095 \times 44 + 0.19 \times 18 + 0.715 \times 28 = 27.62 \text{ kg}$$

Hence the entropy per unit mass of products is

$$202.9/27.62 = 7.35 \text{ kJ K}^{-1} \text{ kg}^{-1}$$

Example 13.13 Determine the change in the molar Gibbs function at the standard conditions of 1 bar and 25°C for the reaction

$$CO + 1/2 \, O_2 = CO_2$$

Hence calculate the maximum reversible work per unit mass of carbon monoxide for a reaction occurring under standard conditions.

Solution The molar Gibbs function is defined as:

$$\overline{g}^{\,o} = \overline{h}^{\,o} - T^{o}\overline{s}^{\,o} \tag{E13.13.1}$$

The following data for the molar entropy (kJ K^{-1} kmol^{-1}) under standard conditions are obtained from the tables in [4]:

$$\overline{s}^{\,o}_{CO} = 197.54, \quad \overline{s}^{\,o}_{O_2} = 205.03, \quad \overline{s}^{\,o}_{CO_2} = 213.69$$

The molar enthalpy of the given reaction under standard conditions is [4]

$$\Delta\overline{h}^{\,o} = -282990 \text{ kJ kmol}^{-1}$$

The reactant gases and product gases are each at a pressure of 1 bar.

Applying Eq. (E13.13.1) to the reaction we obtain the change in the molar Gibbs function as:

$$\Delta\overline{g}^{\,o} = \left(\overline{h}^{\,o}_{CO_2} - \overline{h}^{\,o}_{CO} - 1/2\overline{h}^{\,o}_{O_2}\right) - T^{o}\left(\overline{s}^{\,o}_{CO_2} - \overline{s}^{\,o}_{CO} - 1/2\overline{s}^{\,o}_{O_2}\right)$$

$$\Delta\overline{g}^{\,o} = \Delta\overline{h}^{\,o} - T^{o}\left(\overline{s}^{\,o}_{CO_2} - \overline{s}^{\,o}_{CO} - 1/2\overline{s}^{\,o}_{O_2}\right)$$

Substituting numerical values we have

$$\Delta\overline{g}^{\,o} = -282990 - 298 \times \left(213.69 - 197.54 - 1/2 \times 205.03\right)$$

$$\Delta\overline{g}^{\,o} = -257253 \text{ kJ kmol}^{-1}$$

The change in Gibbs function per unit mass of carbon monoxide is

$$\Delta g = -257253/28 = -9188 \text{ kJ kg}^{-1}$$

The maximum reversible work is given by

$$w_{rev} = -\Delta g = 9188 \text{ kJ kg}^{-1}$$

Example 13.14 A stream of ethylene (C_2H_4) gas and a stream of air enter a steady flow combustion chamber at 25°C and 100 kPa. The mass flow rates are in stoichiometric proportions. The products of combustion leave at 25°C and 100 kPa. Heat is transferred from the combustion chamber to a reservoir at 25°C. Assume that the water in the products of combustion is in the vapor state. Calculate (i) the entropy change of the streams, (ii) the entropy production, and (iii) the irreversibility, if the dead-state temperature is 25°C. Molar enthalpy of reaction, $\Delta\overline{h}^{\,o} = -1323170$ kJ kmol^{-1}.

Solution The chemical equation for the combustion process is

$$C_2H_4 + 3\,O_2 + 3 \times 3.76\,N_2 = 2\,CO_2 + 2\,H_2O + 3 \times 3.76\,N_2$$

We first calculate the total molar entropy of the reactants and the products. The fuel and the air are each at a pressure of 1 bar while the product mixture is at a total pressure of 1 bar. The absolute entropy of the various constituents at 25°C and 1 bar are obtained from tabulated data in [4]. At entry to the combustion system the oxygen and the nitrogen in the air are at their partial pressures while the fuel is at 1 bar. In the product mixture, all the constituents are at their partial pressures. The difference in pressure is accounted for by using Eq. (13.16). Thus the molar entropies (kJ K^{-1} kmol^{-1}) of the constituents are as follows:
For the reactants at 25°C:

$$\bar{s}_{C_2H_4} = 219.22 \ \text{kJ K}^{-1} \text{kmol}^{-1}$$

$$\bar{s}_{O_2} = 205.03 - 8.3145\ln(0.21) = 218.0 \ \text{kJ K}^{-1} \text{kmol}^{-1}$$

$$\bar{s}_{N_2} = 191.5 - 8.3145\ln(0.79) = 193.46 \ \text{kJ K}^{-1} \text{kmol}^{-1}$$

For the products at 25°C:

$$\bar{s}_{CO_2} = 231.69 - 8.3145\ln(2/15.28) = 248.6 \ \text{kJ K}^{-1} \text{kmol}^{-1}$$

$$\bar{s}_{H_2O} = 188.72 - 8.3145\ln(2/15.28) = 205.6 \ \text{kJ K}^{-1} \text{kmol}^{-1}$$

$$\bar{s}_{N_2} = 191.5 - 8.3145\ln(11.28/15.28) = 194.0 \ \text{kJ K}^{-1} \text{kmol}^{-1}$$

Note that the total number of product moles is 15.28.
The total molar entropy of the reactants per kmol of fuel is:

$$\bar{s}_R = 219.22 + 3 \times 218.0 + 11.28 \times 193.46 = 3055.45 \ \text{kJ K}^{-1} \text{kmol}^{-1}$$

The total entropy of the products per kmol of fuel is:

$$\bar{s}_P = 2 \times 248.6 + 2 \times 205.6 + 11.28 \times 194.0 = 3096.7 \ \text{kJ K}^{-1} \text{kmol}^{-1}$$

Applying the first law, the heat transfer from the reservoir to the combustion system per kmol of fuel is:

$$Q_{in} = \bar{h}_P - \bar{h}_R = \Delta \bar{h}^o = -1323170 \ \text{kJ kmol}^{-1}$$

The change in entropy of the reservoir per kmol of fuel is:

$$\Delta \bar{s}_{res} = -Q_{in}/T_{res} = 1323170/298 = 4440.0 \ \text{kJ K}^{-1} \text{kmol}^{-1}$$

The entropy production per kmol of fuel is:

$$\sigma = \bar{s}_P - \bar{s}_R + \Delta \bar{s}_{res}$$

$$\sigma = 3096.7 - 3055.4 + 4440.0 = 4481.3 \text{ kJ K}^{-1}\text{ kmol}^{-1}$$

The irreversibility per kmol of fuel is:

$$I = \sigma T^o = 4481.3 \times 298 = 1335427.0 \text{ kJ kmol}^{-1}$$

Example 13.15 Butane gas (C_4H_{10}) at 25°C and 100 kPa is burned with 100 percent excess air at 25°C and 100 kPa in a steady flow process. The products of combustion leave at 1 bar and a temperature of 727°C. Heat is transferred from the combustion system to a reservoir at 60°C. Calculate (i) the entropy production, and (ii) the irreversibility, if the dead-state temperature is 25°C. The molar enthalpy of reaction for butane with oxygen under standard conditions is

$$\Delta H^o = -2657 \times 10^3 \text{ kJ kmol}^{-1}$$

Solution Consider the flow of 1 kmol of butane through the combustion system shown in Fig. E13.15(a). The chemical equation for the combustion process with 100 percent excess air is

$$C_4H_{10} + 13 \text{ } O_2 + 44.88 \text{ } N_2 = 4 \text{ } CO_2 + 5 \text{ } H_2O + 44.88 \text{ } N_2 + 6.5 \text{ } O_2$$

The *H-T* diagram for the combustion process is depicted in Fig. E13.15(b).

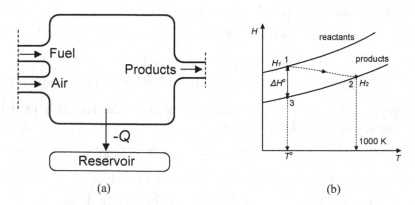

(a)   (b)

Fig. E13.15 (a) Steady-flow combustion system, (b) *H-T* diagram

Apply the first law to the control volume, neglecting changes in kinetic and potential energy. The work done is zero.

Hence
$$H_{R1} + Q = H_{P2}$$

For the reaction occurring under standard conditions
$$\Delta H^o = H_{P3} - H_{R1} = -2657 \times 10^3 \text{ kJ kmol}^{-1}$$

From the two equations above we have
$$Q = H_{P2} - H_{R1} = (H_{P2} - H_{P3}) + \Delta H^o = \Delta H_P - 2657 \times 10^3$$

The change in enthalpy of the product gases can be calculated using the values of enthalpy at different temperatures, tabulated in [4]. The reference state is at 25°C. From the tabulated data at 1000 K, the product temperature, we obtain
$$\Delta H_P = 4 \times 33405 + 5 \times 25978 + 44.88 \times 21460 + 6.5 \times 22707$$

$$\Delta H_p = 1374.23 \times 10^3 \text{ kJ kmol}^{-1}$$

Hence the heat transfer from the reservoir is:
$$Q = -2657 \times 10^3 + 1374.23 \times 10^3 = -1282.77 \times 10^3 \text{ kJ kmol}^{-1}$$

The change in entropy of the reservoir per kmol of fuel is:
$$\Delta \bar{s}_{res} = -Q/T_{res} = 1282.77 \times 10^3 /(273 + 60) = 3.852 \times 10^3$$

The fuel and air enter as separate streams each at a pressure of 1 bar and temperature 25°C. The product mixture leaves at a *total* pressure of 1 bar and temperature 1000 K. We obtain the values of the absolute entropy of the different constituents from [4] where the tabulated data are for 1 bar. The difference in pressure is accounted for by using Eq. (13.16).

For the reactants at 298 K:
$$\bar{s}_{C_4H_{10}} = 310.0 \text{ kJ K}^{-1} \text{ kmol}^{-1}$$

$$\bar{s}_{O_2} = 205.03 - 8.3145 \ln(0.21) = 218.0 \text{ kJ K}^{-1} \text{ kmol}^{-1}$$

$$\bar{s}_{N_2} = 191.5 - 8.3145 \ln(0.79) = 193.46 \text{ kJ K}^{-1} \text{ kmol}^{-1}$$

The total molar entropy of the reactants per kmol of fuel is:
$$\bar{s}_R = 310.0 + 13 \times 218.0 + 44.88 \times 193.46 = 11.826 \times 10^3$$

For the products at 1000 K:

$$\bar{s}_{CO_2} = 269.22 - 8.3145 \ln(4/60.38) = 291.79 \text{ kJ K}^{-1} \text{ kmol}^{-1}$$

$$\bar{s}_{H_2O} = 232.6 - 8.3145 \ln(5/60.38) = 253.3 \text{ kJ K}^{-1} \text{ kmol}^{-1}$$

$$\bar{s}_{O_2} = 243.48 - 8.3145 \ln(6.5/60.38) = 262.0 \text{ kJ K}^{-1} \text{ kmol}^{-1}$$

$$\bar{s}_{N_2} = 228.06 - 8.3145 \ln(44.88/60.38) = 230.52 \text{ kJ K}^{-1} \text{ kmol}^{-1}$$

The total entropy of the products per kmol of fuel is:

$$\bar{s}_P = 4 \times 291.79 + 5 \times 253.3 + 6.5 \times 262.0 + 44.88 \times 230.52$$

$$\bar{s}_P = 14.482 \times 10^3 \text{ kJ K}^{-1} \text{ kmol}^{-1}$$

The entropy production per kmol of fuel is:

$$\sigma = \bar{s}_P - \bar{s}_R + \Delta \bar{s}_{res}$$

$$\sigma = 14.482 \times 10^3 - 11.826 \times 10^3 + 3.852 \times 10^3 = 6.508 \times 10^3$$

The irreversibility per kmol of fuel is:

$$I = \sigma T^o = 6.508 \times 10^3 \times 298 = 1939.4 \times 10^3 \text{ kJ kmol}^{-1}$$

Example 13.16 The products of combustion of a stoichiometric mixture of carbon monoxide gas and oxygen are at a pressure of 100 kPa and a certain temperature. Analysis of the products reveals that 0.35 of each kmol of carbon dioxide has dissociated to carbon monoxide. (i) Determine the equilibrium constant for the temperature. (ii) Calculate the percentage dissociation when the products are at the same temperature but compressed to 1000 kPa.

Solution The chemical equation for stoichiometric combustion is

$$CO + 1/2 \, O_2 = CO_2$$

The chemical equation for the equilibrium mixture when a fraction $\alpha(0.35)$ of CO_2 has dissociated may be written as:

$$CO + 1/2 \, O_2 = (1 - \alpha) CO_2 + \alpha \, CO + (\alpha/2) O_2$$

The molar composition of the equilibrium mixture is:

$$CO_2 = 0.65, \qquad CO = 0.35, \qquad O_2 = 0.35/2$$

The total number of kmols is 1.175 and the total pressure is 1 bar.

Hence the partial pressures of the constituents in the equilibrium mixture are:

$$p_{CO_2} = 0.65/1.175 = 0.553 \text{ bar}, \quad p_{CO} = 0.35/1.175 = 0.298 \text{ bar}$$

and
$$p_{O_2} = 0.175/1.175 = 0.149 \text{ bar}$$

(i) The equilibrium constant for the reacting mixture is given by Eq. (13.33) as

$$K_P(T) = p_{CO_2} / (p_{CO}\sqrt{p_{O_2}})$$

$$K_P(T) = 0.553/(0.298 \times 0.149^{1/2}) = 4.81 \text{ bar}^{-1/2}$$

(ii) Let the degree of dissociation at 10 bar (1000 kPa) pressure be α. The total number of moles in the equilibrium mixture is:

$$N_m = 1 - \alpha + \alpha + \alpha/2 = 1 + \alpha/2$$

Hence the partial pressures of the constituents in the equilibrium mixture are:

$$p_{CO_2} = (1-\alpha) \times 10/N_m \text{ bar}, \quad p_{CO} = \alpha \times 10/N_m \text{ bar}$$

and
$$p_{O_2} = \alpha \times 10/(2N_m) \text{ bar}$$

Now the equilibrium constant, which is a function of temperature, is the same as in (i) above and equal to 4.81. Hence

$$K_P(T) = p_{CO_2} / (p_{CO}\sqrt{p_{O_2}}) = 4.81 \text{ bar}^{-1/2}$$

Substituting the expressions for the partial pressures in the above equation we have

$$(1-\alpha)(2+\alpha)^{1/2}/\alpha^{3/2} = 4.81 \times 10^{1/2}$$

$$230.4\alpha^3 + 3\alpha = 2$$

The solution of the above equation is obtained by trial-and-error as $\alpha = 0.185$. Hence at 10 bar and the same temperature the degree of dissociation is 0.185.

Example 13.17 A rigid vessel contains a mixture of CO and air at a pressure of 900 kPa and a temperature of 800 K. The mixture is 25 percent richer in fuel than the stoichiometric mixture. The mixture undergoes adiabatic combustion attaining a final equilibrium temperature of 3000 K. Calculate (i) the composition of the products of combustion, and (ii) the equilibrium constant for the reaction at 3000 K. The internal energy of reaction for stoichiometric combustion at the standard state is -281751 kJ kmol^{-1}.

Solution The chemical equation for the combustion of a mixture of air and 25 percent excess carbon monoxide is:

$$1.25 \; CO + 1/2 \; O_2 + (3.76/2) \; N_2 = 0.25 \; CO + CO_2 + (3.76/2) \; N_2$$

When a fraction α of CO_2 has dissociated, the above equation may written as:

$$1.25 \; CO + 1/2 \; O_2 + (3.76/2) \; N_2$$
$$= 0.25 \; CO + (1 - \alpha) \; CO_2 + \alpha \; CO + \alpha/2 \; O_2 + (3.76/2) \; N_2$$

The equilibrium molar composition of the products is as follows:

$$CO_2 = (1 - \alpha), \quad CO = (0.25 + \alpha), \quad O_2 = 0.5\alpha \quad \text{and} \quad N_2 = 1.88$$

The total number of moles of reactants is:

$$N_R = 1.25 + 0.5 + 1.88 = 3.63$$

The total number of moles of products is:

$$N_P = (1 - \alpha) + (0.25 + \alpha) + 0.5\alpha + 1.88 = 3.13 + 0.5\alpha$$

Apply the ideal gas equation to the reactants and the products. For the constant volume process we have

$$P_P / P_R = N_P T_P / (N_R T_R)$$

$$P_P / 9 = (3.13 + 0.5\alpha) \times 3000 / (3.63 \times 800)$$

$$P_P = 9 \times (3.13 + 0.5\alpha) \times 3000 / (3.63 \times 800) = 4.649 \times (6.26 + \alpha)$$

Apply the first law to the constant volume adiabatic combustion process to obtain

$$U_{R1} = U_{P2}$$

The adiabatic process 1-2 is also indicated in the U-T diagram in Fig. E13.17. Now for the change in internal energy of the products from the standard state:

$$U_{P2} = U^o{}_P + \Delta U_P (T^o \rightarrow 3000)$$

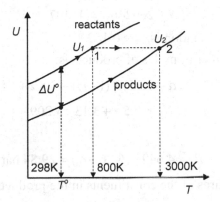

Fig. E13.17 *U-T* diagram for adiabatic combustion

For the change in internal energy of the reactants from the standard state:

$$U_{R1} = U^o_R + \Delta U_R(T^o \to 800)$$

From the three equations above it follows that

$$U^o_P - U^o_R = \Delta U^o = \Delta U_R(T^o \to 800) - \Delta U_P(T^o \to 3000)$$

The following values of the internal energy, $\bar{u}$ (kJ kmol^{-1}) of the constituents at 298 K, 800 K and 3000 K have been obtained from the tabulated data in [4].

T (K)	CO_2	CO	O_2	N_2
298	−2479	−2479	−2479	−2479
800	16164	8524	9189	8394
3000	127920	68598	73155	67795

Substituting these values in the above equation we have

$$-(1-\alpha)\times 281751 = 1.25\times(8524+2479) + 0.5\times(9189+2479)$$
$$+1.88\times(8394+2479) - (1-\alpha)(127920+2479)$$
$$-(\alpha+0.25)\times(68598+2479) - 0.5\alpha\times(73155+2479)$$
$$-1.88\times(67795+2479)$$

Note that on the left hand side of the above equation the internal energy of reaction is multiplied by $(1-\alpha)$ because only that fraction of CO reacts with oxygen. From the above equation we have

$$260246\alpha = 41497$$

The degree of dissociation is, $\alpha = 0.159$

(ii) The total number of moles of products is:

$$N_P = (\alpha + 0.25) + (1 - \alpha) + 0.5\alpha + 1.88$$

$$N_P = 0.5\alpha + 3.13 = 3.209$$

The total pressure is:

$$P_P = 4.649 \times (6.26 + \alpha) = 29.84 \text{ bar}$$

The partial pressures of the constituents in the products are as follows:

$$p_{CO_2} = (1 - \alpha)P_P / N_P = 7.82 \text{ bar}$$

$$p_{CO} = (\alpha + 0.25)P_P / N_P = 3.803 \text{ bar}$$

$$p_{O_2} = 0.5\alpha P_P / N_P = 0.739 \text{ bar}$$

The equilibrium constant at 3000 K is given by

$$K_P = (p_{CO})(p_{O_2})^{1/2} / p_{CO_2}$$

$$K_P = (3.803 \times 0.739^{1/2})/7.82 = 0.418 \text{ bar}^{1/2}$$

Example 13.18 Gaseous propane (C_3H_8) undergoes combustion with 20 percent excess air in a constant pressure process at 200 kPa. The equilibrium temperature of the products of combustion is 2800 K. The CO_2 and H_2O in the products undergo dissociation reactions. The products of combustion contain only CO_2, CO, O_2, H_2O, H_2 and N_2. Calculate (i) the degree of dissociation of CO_2 and (ii) the volumetric percentage of H_2 in the products.

The equilibrium constants at 2800 K are given by

$$P_{CO_2}/(P_{CO}P_{O_2}^{1/2}) = 6.6 \text{ bar}^{-1/2}$$

$$P_{CO_2}P_{H_2}/(P_{CO}P_{H_2O}) = 0.15$$

Solution The chemical equation for combustion with 20 percent excess air is:

$$C_3H_8 + 1.2 \times 5 \; O_2 + 1.2 \times 5 \times 3.76 \; N_2$$
$$= 3 \; CO_2 + 4 \; H_2O + 0.2 \times 5 \; O_2 + 1.2 \times 5 \times 3.76 \; N_2$$

The chemical equations for the two dissociation reactions are:

$$CO_2 \leftrightarrow CO + 1/2\ O_2$$
$$H_2O \leftrightarrow H_2 + 1/2\ O_2$$

Let the degree of dissociation of CO_2 and H_2O be α_1 and α_2 respectively. Then the product mixture will have the following composition:

$$CO_2 = 3(1-\alpha_1), \quad CO = 3\alpha_1, \quad H_2O = 4(1-\alpha_2), \quad H_2 = 4\alpha_2$$
$$O_2 = 0.2 \times 5 + 1.5\alpha_1 + 2\alpha_2 \quad \text{and} \quad N_2 = 1.2 \times 5 \times 3.76 = 22.56$$

Total number of product moles is:

$$N_P = 3(1-\alpha_1) + 3\alpha_1 + 4(1-\alpha_2) + 4\alpha_2 + 1 + 1.5\alpha_1 + 2\alpha_2 + 22.56$$
$$N_P = 1.5\alpha_1 + 2\alpha_2 + 30.56$$

Let the total pressure of the product mixture be P_m (2 bar)
The partial pressures of the constituents are:

$$P_{CO_2} = 3(1-\alpha_1)P_m / N_P, \quad P_{CO} = 3\alpha_1 P_m / N_P.$$
$$P_{H_2O} = 4(1-\alpha_2)P_m / N_P, \quad P_{H_2} = 4\alpha_2 P_m / N_P$$
$$P_{O_2} = (1 + 1.5\alpha_1 + 2\alpha_2)P_m / N_P$$

Substitute the above expressions for the partial pressures in the equilibrium equations for the two dissociation reactions. Thus

$$P_{CO_2} / (P_{CO}P_{O_2}^{1/2}) = 6.6 \text{ bar}^{-1/2}$$

$$\left(\frac{1-\alpha_1}{\alpha_1}\right)\left(\frac{30.56 + 1.5\alpha_1 + 2\alpha_2}{2(1 + 1.5\alpha_1 + 2\alpha_2)}\right)^{1/2} = 6.6 \qquad (E13.18.1)$$

and
$$P_{CO_2}P_{H_2} / (P_{CO}P_{H_2O}) = 0.15$$

$$\left(\frac{1-\alpha_1}{\alpha_1}\right)\left(\frac{\alpha_2}{1-\alpha_2}\right) = 0.15 \qquad (E13.18.2)$$

Equations (E13.18.1) and (E13.18.2) are solved simultaneously to obtain the degrees of dissociation as:

$$\alpha_1 = 0.32 \quad \text{and} \quad \alpha_2 = 0.0659$$

Note that a trial-and-error procedure has to be used in the solution.

The volumetric fraction of hydrogen in the products is:

$$n_{H_2} = 4\alpha_2 / N_P = 4\alpha_2 /(30.56+1.5\alpha_1 +2\alpha_2)$$

$$n_{H_2} = 4\times 0.0659/(30.56+1.5\times 0.32+2\times 0.0659) = 0.85\%$$

Problems

P13.1 Methane gas, (CH_4) undergoes combustion with air. The products of combustion have the following volumetric composition on a dry basis: 9.8 percent CO_2, 0.5 percent CO, and 2.8 percent O_2. Calculate (i) the percentage excess air, and (ii) the fuel-air ratio. [*Answers*: (i) 24%, (ii) 19.4]

P13.2 Octane (C_8H_{18}) is burned with 20% excess air in a combustion process. Calculate (i) the volumetric composition of the products, and (ii) the dew-point of the products, if the total pressure is 1.2 bar. [*Answers*: (i) CO_2 - 10.54%, O_2 - 3.3%, H_2O - 11.85%, N_2 - 74.3%, (ii) 52.9°C]

P13.3 A hydrocarbon fuel of unknown composition is burned with air. The products of combustion have the following volumetric composition: 8.5% - CO_2, 8.5% - O_2, 0.8% - CO, 82.2% - N_2. Calculate (i) the composition of the fuel, and (ii) the fuel-air ratio. [*Answers*: (i) m_c/m_H = 6.28, (ii) 23.3]

P13.4 In a steady-flow combustion chamber propane (C_3H_8) gas is burned with 20 percent excess air. The reactants are initially at 25°C. The products are cooled at a constant pressure of 1.85 bar to 627°C. Calculate (i) the heat removed per unit mass of fuel, and (ii) the dew-point of the product mixture. The enthalpy of reaction is −2043130 kJ kmol⁻¹. [*Answers*: 32810 kJ kg⁻¹, 64°C]

P13.5 Acetylene gas (C_2H_2) and air enter a steady-flow combustion system at 25°C. The products of combustion are cooled to 627°C by removing 28.6×10^3 kJ of heat per kg of fuel. The enthalpy of reaction is -1255618 kJ $kmol^{-1}$. Calculate (i) the air-fuel ratio, and (ii) the dew-point of the products if the total pressure is 2 bar. [*Answers*: 29.3, 40°C]

P13.6 Determine the adiabatic flame temperature for the steady-flow, stoichiometric combustion of butane (C_4H_{10}). The initial temperature of the reactants is 25°C. The enthalpy of reaction is -2657480 kJ $kmol^{-1}$. [*Answers*: 2500 K]

P13.7 A mixture of butane gas and 20 percent excess air enters a steady-flow combustion system at 1 bar and 25°C. The products of combustion are cooled by transferring heat to a reservoir at 25°C. The product mixture leaves the system at 1 bar and 25°C. The enthalpy of reaction is -1427860 kJ $kmol^{-1}$. Calculate (i) the entropy of the reactant and product streams per kg of fuel, (ii) the change in entropy of the reservoir, (iii) the entropy production, and (iv) the irreversibility, if the dead-state temperature is 25°C. [*Answers*: (i) 133.2, 137.2 kJ kg^{-1} K^{-1}, (ii) 153.7 kJ kg^{-1} K^{-1}, (iii) 157.7 kJ kg^{-1} K^{-1}, (iv) 47000 kJ kg^{-1}]

P13.8 Ethane (C_2H_6) gas undergoes adiabatic combustion with 30 percent excess air in a steady-flow process. The fuel and air streams enter separately at 1 bar and 25°C. The product stream leaves at 1 bar. The enthalpy of reaction is -2657480 kJ $kmol^{-1}$. Calculate (i) the temperature of the product stream, and (ii) the entropy production per kg of fuel. [*Answers*: (i) 1730°C, (ii) 54.6 kJ kg^{-1} K^{-1}]

P13.9 The equilibrium constant at 3300 K for the reaction
$$1/2\ O_2 + 1/2\ N_2 = NO$$
is 0.17. Obtain the equilibrium constants at 3300 K for the following reactions:

(i) $O_2 + N_2 = 2\ NO$ and (ii) $NO = 1/2\ N_2 + 1/2\ O_2$

[*Answers*: (i) 0.0289 (ii) 5.88]

P13.10 Carbon monoxide undergoes combustion with stoichiometric air. The temperature of the products of combustion is 2600 K. The equilibrium constant at 2600 K for the reaction

$$CO + 1/2\ O_2 = CO_2$$

is 16.4 bar$^{-1/2}$. Calculate the composition of the equilibrium mixture if the pressure is: (i) 1 bar and (ii) 6 bar. [*Answers*: (i) CO_2 = 75.7%, CO = 16.2%, O_2 = 8.1%, (ii) CO_2 = 91.6%, CO = 5.6%, O_2 = 2.8%]

References

1. Bejan, Adrian, *Advanced Engineering Thermodynamics*, John Wiley & Sons, New York, 1988.
2. Jones, J.B. and G.A. Hawkins, *Engineering Thermodynamics*, John Wiley & Sons, Inc., New York, 1986.
3. Reynolds, William C. and Henry C. Perkins, *Engineering Thermodynamics*, 2nd edition, McGraw-Hill, Inc., New York, 1977.
4. Rogers, G.F.C. and Mayhew, Y.R., *Thermodynamic and Transport Properties of Fluids*, 5th edition, Blackwell, Oxford, U.K. 1998.
5. Van Wylen, Gordon J. and Richard E. Sonntag, *Fundamentals of Classical Thermodynamics*, 3rd edition, John Wiley & Sons, Inc., New York, 1985.
6. Wark, Jr., Kenneth, *Advanced Thermodynamics for Engineers*, McGraw-Hill, Inc., New York, 1995.

Chapter 14

Thermodynamic Property Relations

In this chapter we shall develop some general relationships between the thermodynamic properties that were presented in earlier chapters. In chapter 1, pressure, volume and density were introduced as basic properties of a system, and the zeroth law was used to define temperature and clarify its measurement. The property called the internal energy followed from the first law in chapter 4. It was used to derived the properties, internal energy, enthalpy, and the specific heat capacity. The second law resulted in the property called entropy. From it we derived two additional properties, the Helmholtz function and the Gibbs function, which were related to the maximum work output of a system. We also used thermodynamic property tables [3] on a number of occasions to solve numerical problems.

Not all the properties mentioned above can be measured experimentally. Careful consideration of all these properties shows that only four properties are directly measurable. These are mass, volume, pressure and temperature. Other useful properties, such as the internal energy, the enthalpy and the entropy have to be calculated from measured experimental data.

Fortunately, using the laws of thermodynamics and the 'path independent' nature of properties we are able to derive a large number of general relations between the measurable properties and other useful properties mentioned above. In this chapter we shall develop these relations, and illustrate how they are used to prepare thermodynamics property tables [3] for pure substances.

Engineering Thermodynamic

14.1 Some Mathematical Relations

The development of property relations requires the extensive use of partial derivatives. Therefore we shall briefly review a few general relations between partial derivatives. Consider a variable z which is a continuous function of two variables x and y,

$$z = f(x, y)$$

Therefore
$$dz = \left(\frac{\partial z}{\partial x}\right)_y dx + \left(\frac{\partial z}{\partial y}\right)_x dy \tag{14.1}$$

The curves in Fig. 2.4 for a pure substance may be used to clarify the physical meaning of the partial derivatives in Eq. (14.1). If the variables z, x, y are respectively the pressure P, the volume v and the temperature T, then the partial derivative,

$$\left(\frac{\partial z}{\partial x}\right)_y = \left(\frac{\partial P}{\partial v}\right)_T$$

which is the slope of the *constant temperature* line JKLM in Fig. 2.4.

Similarly,
$$\left(\frac{\partial y}{\partial x}\right)_z = \left(\frac{\partial T}{\partial v}\right)_P$$

which is the slope of the *constant pressure* line ABCD.

We can also write Eq. (14.1) in the form

$$dz = a(x, y)dx + b(x, y)dy \tag{14.2}$$

where the partial derivatives are now expressed as :

$$a(x, y) = \left(\frac{\partial z}{\partial x}\right)_y \quad \text{and} \quad b(x, y) = \left(\frac{\partial z}{\partial y}\right)_x \tag{14.2a}$$

Recall that in chapter 2 we defined a *thermodynamic property* as a quantity that depends only on the state. Moreover, the change of a property during a process is independent of the process path (see worked example 1.4). If the variables x, y and z in Eq. (14.1) are properties of a system, then the above requirement of *path independence* leads to the following mathematical condition:

$$\left(\frac{\partial a}{\partial y}\right)_x = \left(\frac{\partial b}{\partial x}\right)_y \qquad (14.3)$$

This is also known as the *condition for an exact differential*.

In addition, the three *properties* x, y, z satisfy the following relations:

$$\left(\frac{\partial z}{\partial x}\right)_y = \frac{1}{\left(\dfrac{\partial x}{\partial z}\right)_y} \qquad (14.4)$$

which is called the *reciprocal relation*,

and

$$\left(\frac{\partial z}{\partial x}\right)_y \left(\frac{\partial x}{\partial y}\right)_z \left(\frac{\partial y}{\partial z}\right)_x = -1 \qquad (14.5)$$

which is called the *cyclic relation*.

For the sake of brevity we are not including the proofs of the relations (14.3) to (14.5). However, the proofs are available in most standard text books on thermodynamics, including Ref. [1, 2 and 5].

14.2 The Maxwell Relations

We shall now derive a set of relations between the properties P, v, T and s. These are called the *Maxwell relations* and they follow from the '*T-ds*' – *equation* (7.17) which we derived in section 7.1.3. When expressed in terms of the *specific* properties, Eq. (7.17) becomes

$$Tds = du + Pdv$$

Therefore

$$du = Tds - Pdv \qquad (14.6)$$

Applying condition (14.3) to Eq. (14.6) we obtain

$$\left(\frac{\partial T}{\partial v}\right)_s = -\left(\frac{\partial P}{\partial s}\right)_v \qquad (14.7)$$

A variation of the '*T-ds*' – equation involving enthalpy is obtained by substituting for du from the relation

$$dh = du + Pdv + vdP \qquad (14.8)$$

Hence we have

$$dh = Tds + vdP \qquad (14.9)$$

Applying condition (14.3) to Eq. (14.9) we obtain

$$\left(\frac{\partial T}{\partial P}\right)_s = \left(\frac{\partial v}{\partial s}\right)_P \tag{14.10}$$

We defined the Helmholtz function A in section 7.3.1. The Helmholtz function for unit mass a is given by

$$a = u - Ts \tag{14.11}$$

$$da = du - Tds - sdT \tag{14.12}$$

Substituting for du from Eq. (14.6) in Eq. (14.12) we obtain

$$da = -Pdv - sdT \tag{14.13}$$

Applying condition (14.3) to Eq. (14.13) we have

$$\left(\frac{\partial P}{\partial T}\right)_v = \left(\frac{\partial s}{\partial v}\right)_T \tag{14.14}$$

We defined the Gibbs function G in section 7.3.1. The Gibbs function for unit mass g is given by

$$g = h - Ts \tag{14.15}$$

Therefore

$$dg = dh - Tds - sdT \tag{14.16}$$

Substituting for dh from Eq. (14.9) in Eq. (14.16) we obtain

$$dg = vdP - sdT \tag{14.17}$$

Applying condition (14.3) to Eq. (14.17) we have

$$\left(\frac{\partial v}{\partial T}\right)_P = -\left(\frac{\partial s}{\partial P}\right)_T \tag{14.18}$$

The four property relations given by Eqs. (14.7), (14.10), (14.14) and (14.18) are called the *Maxwell relations* for a simple compressible substance. In later sections of this chapter we shall demonstrate the application of these relations in obtaining expressions for properties like u, h and s in terms of the measurable properties P, v and T.

We can derive a number of other useful relations by applying the condition given by Eq. (14.2a) to Eqs. (14.6), (14.9), (14.13) and (14.17) respectively. Hence we have

$$\left(\frac{\partial u}{\partial s}\right)_v = T \quad \text{and} \quad \left(\frac{\partial u}{\partial v}\right)_s = -P \qquad (14.19)$$

$$\left(\frac{\partial h}{\partial s}\right)_P = T \quad \text{and} \quad \left(\frac{\partial h}{\partial P}\right)_s = v \qquad (14.20)$$

$$\left(\frac{\partial a}{\partial v}\right)_T = -P \quad \text{and} \quad \left(\frac{\partial a}{\partial T}\right)_v = -s \qquad (14.21)$$

$$\left(\frac{\partial g}{\partial P}\right)_T = v \quad \text{and} \quad \left(\frac{\partial g}{\partial T}\right)_P = -s \qquad (4.22)$$

The Maxwell relations presented above are written for a simple compressible substance. However, similar relations may be readily obtained for substances involving other effects, such as magnetic and electrical work by appropriately modifying the compressible work term, Pdv in Eq. (14.6) [5].

14.3 Clapeyron Equation

The Clapeyron equation is an important property relation applicable to phase-change processes. It involves the slope of the saturation curve, the saturation temperature, the change in enthalpy, and the change in specific volume during a phase-change process. There are several different changes of phase that occur at constant temperature and pressure. We shall denote the properties of the two phases with the subscripts α and β respectively.

Consider the Maxwell relation given by Eq. (14.14),

$$\left(\frac{\partial P}{\partial T}\right)_v = \left(\frac{\partial s}{\partial v}\right)_T \qquad (14.23)$$

For a constant temperature phase change process the above equation may be written in the form

$$\left(\frac{dP}{dT}\right)_{sat} = \frac{s_\beta - s_\alpha}{v_\beta - v_\alpha} \qquad (14.24)$$

Now the change in enthalpy and the change entropy during a constant temperature phase change process are related by the equation

$$h_\beta - h_\alpha = T(s_\beta - s_\alpha)$$

Substituting for the change in entropy in Eq. (14.24) we have

$$\left(\frac{dP}{dT}\right)_{sat} = \frac{h_\beta - h_\alpha}{T(v_\beta - v_\alpha)} = \frac{h_{\alpha\beta}}{Tv_{\alpha\beta}} \tag{14.25}$$

We shall demonstrate the application of Eq. (14.25) to actual phase change processes in the worked examples to follow in this chapter.

14.4 Thermodynamics Relations Involving Internal Energy, Enthalpy, Entropy and Specific Heat Capacities

In this section we shall obtain expressions for the properties u, h, s, c_v and c_p in terms of the measurable properties P, v and T. In section 4.5.1 we defined the specific heat capacities as

$$c_v = \left(\frac{\partial u}{\partial T}\right)_v \quad \text{and} \quad c_p = \left(\frac{\partial h}{\partial T}\right)_P \tag{14.26}$$

From Eq. (14.6) we have

$$du = Tds - Pdv \tag{14.27}$$

For a *constant volume* process, Eq. (14.27) gives

$$c_v = \left(\frac{\partial u}{\partial T}\right)_v = T\left(\frac{\partial s}{\partial T}\right)_v \tag{14.28}$$

From Eq. (14.9) we have

$$dh = Tds + vdP \tag{14.29}$$

For a *constant pressure* process, Eq. (14.29) gives

$$c_p = \left(\frac{\partial h}{\partial T}\right)_P = T\left(\frac{\partial s}{\partial T}\right)_P \tag{14.30}$$

14.4.1 *Expression for the change in internal energy*

From the state principle (see section 4.4) it follows that

$$u = f(T,v) \tag{14.31}$$

$$du = \left(\frac{\partial u}{\partial T}\right)_v dT + \left(\frac{\partial u}{\partial v}\right)_T dv \qquad (14.32)$$

For an *isothermal* process Eq. (14.6) gives

$$\left(\frac{\partial u}{\partial v}\right)_T = T\left(\frac{\partial s}{\partial v}\right)_T - P \qquad (14.33)$$

Substituting the Maxwell relation, Eq. (14.14) in Eq. (14.33) we have

$$\left(\frac{\partial u}{\partial v}\right)_T = T\left(\frac{\partial P}{\partial T}\right)_v - P \qquad (14.34)$$

Substituting in Eq. (14.32) from Eqs. (14.28) and (14.34) we obtain

$$du = c_v dT + \left[T\left(\frac{\partial P}{\partial T}\right)_v - P\right]dv \qquad (14.35)$$

It follows from Eq. (14.35) that along a *constant volume* line

$$du_v = c_v dT,$$

and along a *constant temperature* line

$$du_T = \left[T\left(\frac{\partial P}{\partial T}\right)_v - P\right]dv$$

If the variation of c_v and the equation of state (P-v-T relation) for a substance are known then Eq. (14.35) may be integrated to obtain the change of internal energy for any process.

14.4.2 *Expression for the change in enthalpy*

From the state principle (see section 4.4) it follows that

$$u = f(T, P) \qquad (14.36)$$

$$dh = \left(\frac{\partial h}{\partial T}\right)_P dT + \left(\frac{\partial h}{\partial P}\right)_T dP \qquad (14.37)$$

For an *isothermal* process Eq. (14.9) gives

$$\left(\frac{\partial h}{\partial P}\right)_T = T\left(\frac{\partial s}{\partial P}\right)_T + v \qquad (14.38)$$

Substituting the Maxwell relation, Eq. (14.18) in Eq. (14.38) we have

$$\left(\frac{\partial h}{\partial P}\right)_T = v - T\left(\frac{\partial v}{\partial T}\right)_P \qquad (14.39)$$

Substituting in Eq. (14.37) from Eqs. (14.30) and (14.39) we obtain

$$dh = c_P dT + \left[v - T\left(\frac{\partial v}{\partial T}\right)_P\right]dP \qquad (14.40)$$

It follows from Eq. (14.40) that along a *constant pressure* line

$$dh_P = c_P dT,$$

and along a *constant temperature* line

$$dh_T = \left[v - T\left(\frac{\partial v}{\partial T}\right)_P\right]dP$$

If the variation of c_P and the equation of state (P-v-T relation) for a substance are known then Eq. (14.40) may be integrated to obtain the change of enthalpy for any process.

14.4.3 *Expression for the change in entropy*

From the state principle (see section 4.4) it follows that

$$s = f(T, P) \qquad (14.41)$$

$$ds = \left(\frac{\partial s}{\partial T}\right)_P dT + \left(\frac{\partial s}{\partial P}\right)_T dP \qquad (14.42)$$

Substituting in Eq. (14.42) from Eqs. (14.30) and (14.18) we have

$$ds = c_P \frac{dT}{T} - \left(\frac{\partial v}{\partial T}\right)_P dP \qquad (14.43)$$

Following a similar procedure we can derive an alternative expression for the change in entropy in terms of c_v. Hence we have

$$ds = c_v \frac{dT}{T} + \left(\frac{\partial P}{\partial T}\right)_v dv \qquad (14.44)$$

If the variation of c_P or c_v and the equation of state (P-v-T relation) for a substance are known then either Eq. (14.43) or Eq. (14.44) may be integrated to obtain the change of entropy for any process.

14.4.4 *Relations involving specific heat capacities*

In this section we shall obtain relations for the specific heat capacities in terms of the measurable properties P, v and T. Consider Eq. (14.43), which may be expressed as

$$ds = \left(\frac{c_P}{T}\right)dT - \left(\frac{\partial v}{\partial T}\right)_P dP$$

Apply condition (14.3) to the above equation. Hence we have

$$\left(\frac{\partial(c_P/T)}{\partial P}\right)_T = -\left[\frac{\partial}{\partial T}\left(\frac{\partial v}{\partial T}\right)_P\right]_P \qquad (14.45)$$

$$\left(\frac{\partial c_P}{\partial P}\right)_T = -T\left(\frac{\partial^2 v}{\partial T^2}\right)_P \qquad (14.46)$$

Following a similar procedure using Eq. (14.44), we obtain the following relation for the specific heat capacity c_v:

$$\left(\frac{\partial c_v}{\partial v}\right)_T = T\left(\frac{\partial^2 P}{\partial T^2}\right)_v \qquad (14.47)$$

We shall now obtain a relation for the difference in specific heat capacities, ($c_P - c_v$). From Eqs. (14.43) and (14.44) we have

$$c_v \frac{dT}{T} + \left(\frac{\partial P}{\partial T}\right)_v dv = c_P \frac{dT}{T} - \left(\frac{\partial v}{\partial T}\right)_P dP \qquad (14.48)$$

$$dT = \frac{T}{c_P - c_v}\left(\frac{\partial v}{\partial T}\right)_P dP + \frac{T}{c_P - c_v}\left(\frac{\partial P}{\partial T}\right)_v dv \qquad (14.49)$$

From the state principle (see section 4.4) it follows that

$$T = f(P, v) \qquad (14.50)$$

$$dT = \left(\frac{\partial T}{\partial P}\right)_v dP + \left(\frac{\partial T}{\partial v}\right)_P dv \tag{14.51}$$

Comparing Eqs. (14.49) and (14.51) we have

$$\left(\frac{\partial T}{\partial P}\right)_v = \frac{T}{c_P - c_v}\left(\frac{\partial v}{\partial T}\right)_P \tag{14.52}$$

$$\left(\frac{\partial T}{\partial v}\right)_P = \frac{T}{c_P - c_v}\left(\frac{\partial P}{\partial T}\right)_v \tag{14.53}$$

Both the above equations yield the result that

$$c_P - c_v = T\left(\frac{\partial v}{\partial T}\right)_P\left(\frac{\partial P}{\partial T}\right)_v \tag{14.54}$$

14.5 Volume Expansivity, Isothermal Compressibility, and Adiabatic Compressibility

The *volume expansivity* α is defined as the fractional change in volume per unit change in temperature at constant pressure. Therefore

$$\alpha = \frac{1}{v}\left(\frac{\partial v}{\partial T}\right)_P \tag{14.55}$$

The *isothermal compressibility* is defined as the fractional change in volume per unit pressure at constant temperature. Therefore

$$\beta_T = -\frac{1}{v}\left(\frac{\partial v}{\partial P}\right)_T \tag{14.56}$$

The *adiabatic compressibility* is defined as the fractional change in volume per unit pressure at constant entropy. Therefore

$$\beta_s = -\frac{1}{v}\left(\frac{\partial v}{\partial P}\right)_s \tag{14.57}$$

The volume expansivity, the isothermal compressibility, and the adiabatic compressibility are thermodynamics properties of a substance. For a simple compressible substance they are functions of two independent properties.

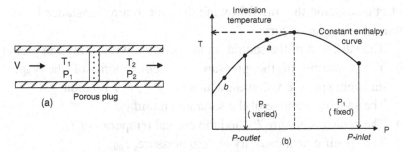

Fig. 14.1 (a) Throttling process (b) Temperature change during the process

14.6 The Joule-Thomson Coefficient

The Joule-Thomson coefficient (JTC) is defined as the change in temperature per unit change in pressure, when a fluid undergoes a throttling process at *constant enthalpy*. Therefore

$$\mu_{JT} = \left(\frac{\partial T}{\partial P}\right)_h \qquad (14.58)$$

The experimental arrangement to measure the JTC is depicted schematically in Fig. 14.1(a). The upstream pressure, P_1 is held fixed and the downstream pressure P_2 is varied. For the throttling process through the porous plug, the enthalpy is constant. The JTC is the slope of the constant enthalpy curve shown in Fig. 14.1(b), and it depends on the equation of state of the fluid. For the outlet condition a the gas is heated, while for b it is cooled. The temperature at which the slope of the curve changes from positive to negative is called the *inversion temperature*.

14.7 Developing Thermodynamic Property Tables

We made extensive use of thermodynamic property tables [3] in earlier chapters of this book. Not all the data available in such tables are directly measured. The main purpose of the present chapter is to develop the various thermodynamic relations required to compute the important tabulated properties used in thermodynamic analysis from measured quantities. We shall now outline the procedure to develop such a property table for a pure substance like water and ammonia.

Let us assume that the following data for a pure substance have been obtained in the laboratory.

(a) The variation of the saturation temperature with saturation pressure.

(b) The variation of the pressure and temperature of the vapor for different specific volumes, which is the P, v, T relation.

(c) The specific volume of the saturated liquid v_f.

(d) The critical pressure P_c, and the critical temperature T_c.

(e) The specific heat capacity at zero pressure, c_{p0}.

From the above data we can calculate a complete set of thermo-dynamic tables for the saturated liquid, the saturated vapor, and the superheated vapor. The first step is to obtain an equation for the variation of saturation vapor pressure with temperature (vapor-pressure curve), which may be obtained by a curve-fitting procedure. Hence we have

$$P_{sat} = F_1(T_{sat}) \qquad (14.59)$$

Worked example 14.1 gives a simple functional form for F_1.

The second step is to obtain the equation of state (P-v-T relation) of the vapor. Hence we have

$$P = F_2(v,T) \qquad (14.60)$$

There are several possible forms of the equation of state which may be selected. We shall discuss these in the next section.

Equation (14.60) will enable us to determine the specific volume of superheated vapor at different values of pressure and temperature. Using Eq. (14.59) and Eq. (14.60) simultaneously, we are able to calculate the specific volume, v_{sat} of the saturated vapor. The specific volume of saturated liquid has been measured.

The enthalpy of vaporization h_{fg} can be found from the Clapeyron equation (14.25), the LHS of which is obtained by differentiating Eq. (14.59). The entropy change during evaporation, s_{fg} is given by

$$s_{fg} = h_{fg} / T_{sat} \qquad (14.61)$$

The change of enthalpy of superheated vapor at any temperature and pressure is calculated by integrating Eq. (14.40) (see problem P14.15). The change of entropy of superheated vapor at any temperature and pressure is calculated by integrating Eq. (14.43) (see worked example 14.10 and problem 14.16). By following the above steps systematically we can prepare an entire data table of the type given in Ref. [3], using the measured quantities (a) to (e), listed above.

14.8 Equations of State for Gases

The equation of state of a simple compressible substance is its *P-v-T* relation. In sections 2.6 and 2.9 of chapter 2 we introduced the ideal gas equation, and the van der Waals equation of state. The latter has the form:

$$(P + a/v^2)(v - b) = \overline{R}T \tag{14.62}$$

We recall from worked examples 2.20 and 2.21 in chapter 2, that the constants of the van der Walls equation of state could be expressed in terms of the critical pressure P_c, and the critical temperature T_c of the fluid. Hence we have

$$v_c = 3\overline{R}T_c/(8P_c) = 3b$$

$$b = v_c/3 = \overline{R}T_c/8P_c \tag{14.62a}$$

$$a = 9\overline{R}T_c v_c/8 = 27\overline{R}^2 T_c^2/64P_c \tag{14.62b}$$

From the above relations it follows that the *compressibility factor* at the critical state is

$$Z_c = P_c v_c/(\overline{R}T_c) = 3/8 = 0.375$$

The experimentally determined values of Z_c for a large number of common gases are in the range of 0.25 to 0.3 [2].

The van der Walls equation of state may be written in terms of the dimensionless thermodynamic property called the *compressibility factor* $Z = Pv/(\overline{R}T)$. Hence we have

$$\left(\frac{Pv}{\overline{R}T} + \frac{27\overline{R}^2 T_c^2}{64P_c \overline{R}Tv} \right) \left(1 - \frac{\overline{R}T_c}{8P_c v} \right)_c = 1 \tag{14.63}$$

We now define three additional dimensionless parameters, the reduced pressure, the reduced volume, and the reduced temperature by

$$P^* = P/P_c, \qquad v^* = v/v_c \qquad \text{and} \qquad T^* = T/T_c$$

On substituting these parameters, Eq. (14.63) becomes

$$\left(Z + \frac{27 P^*}{64 T^{*2} Z} \right)\left(1 - \frac{P^*}{8 T^* Z} \right) = 1 \tag{14.64}$$

We can use Eq. (14.64) to correlate P-v-T data for real gases by expressing it in the general form

$$Z = Z(P^*, T^*) \tag{14.65}$$

When experimentally determined values of the compressibility factor Z for large number gases are plotted against P^*, for different values of T^*, they fall on a common set of curves. This fact is called the *principle of corresponding states*, and the resulting family of curves constitutes the *compressibility chart* [1,5]. This chart is useful for predicting the properties of substances for which more accurate P-v-T data are not available. At very low values of P^*, Z^* approaches unity, the value corresponding to the ideal-gas approximation.

A two-constant equation of state, which is more accurate then the van der Waals equation, was developed by Redlich and Kwong [1,2]. It may be expressed in the form:

$$P = \frac{\overline{R}T}{(v - b)} - \frac{a}{T^{1/2} v(v + b)} \tag{14.66}$$

The constants a and b are given by

$$a = 0.42748 \overline{R}^2 T_c^{5/2} / P_c \tag{14.66a}$$

$$b = 0.08664 \overline{R} T_c / P_c \tag{14.66b}$$

A number of empirical equations of state that fit experimental data better have been developed over the years. The best known of these is the Beattie-Bridgman equation [1,5] which may be expressed in the form:

$$P = \frac{RT(1 - \varepsilon)}{v^2}(v + B) - \frac{A}{v^2} \tag{14.67}$$

$$A = A_0(1 - a/v), \quad B = B_0(1 - b/v), \quad \varepsilon = c/(v T^3) \tag{14.67a}$$

The Beattie-Bridgeman equation can expressed as a polynomial in density, ρ $(1/v)$ and it is quite accurate for densities less than about 0.8

times the critical density. The constants in Eq. (14.67a) are tabulated in most standard textbooks in thermodynamics [1,3].

A more complex equation of state suitable for use at higher densities is the Benedict-Webb-Rubin equation. The eight constants of this equation for a number of gases are tabulated in Ref. [5].

14.9 Worked Examples

Example 14.1 If a saturated vapor in equilibrium behaves like an ideal gas show that

$$\frac{d(\ln P)}{dT} = \frac{h_{fg}}{RT^2}$$

Hence obtain an expression for the variation of saturation pressure with temperature.

Solution The Clapeyron equation, Eq. (14.25) is

$$\left(\frac{dP}{dT}\right)_{sat} = \frac{h_\beta - h_\alpha}{T(v_\beta - v_\alpha)} = \frac{h_{\alpha\beta}}{Tv_{\alpha\beta}}$$

For a liquid-vapor phase change process, $v_g \gg v_f$.

Assuming vapor to be an ideal gas, $Pv = RT$. Hence we have

$$\left(\frac{dP}{dT}\right)_{sat} = \frac{h_{fg}}{Tv} = \frac{Ph_{fg}}{RT^2}$$

Therefore

$$\left(\frac{d\ln P}{dT}\right) = \frac{h_{fg}}{RT^2}$$

Integrating the above equation we obtain

$$\ln P = -\frac{h_{fg}}{RT} + A = -\frac{C}{T} + A$$

where C and A are constants.

Example 14.2 Use the Clapeyron equation, and data for water from the steam table [3] to predict the change in saturation pressure between 50°C and 51°C.

Engineering Thermodynamic

Solution We obtain the following data from the steam tables [3].

$T\,(^\circ C)$	$h_g\,(kJ\,kg^{-1})$	h_f	$v_g\,(m^3\,kg^{-1})$	v_f
50	2591.4	209.3	12.04	0.0010121
51	2593.9	213.4	11.50	0.0010126
50.5 (mean)	2592.65	211.35	11.77	0.00101235

Substituting in the Clapeyron equation, Eq. (14.25) we obtain

$$\left(\frac{\Delta P}{51-50}\right)_{sat} = \frac{2592.65 - 211.35}{(50.5 + 273.15)(11.77 - 0.00101235)} = 0.625$$

Hence the change in saturation pressure is 0.625 kPa.

Example 14.3 Obtain the following relation for an isentropic process

$$\left(\frac{\partial T}{\partial P}\right)_s = \frac{Tv\alpha}{c_P}$$

Solution Applying the *cyclic relation*, Eq. (14.5) we have

$$\left(\frac{\partial T}{\partial P}\right)_s \left(\frac{\partial P}{\partial s}\right)_T \left(\frac{\partial s}{\partial T}\right)_P = -1 \qquad (E14.3.1)$$

$$c_P = T\left(\frac{\partial s}{\partial T}\right)_P \qquad (E14.3.2)$$

$$\alpha = \frac{1}{v}\left(\frac{\partial v}{\partial T}\right)_P \qquad (E14.3.3)$$

Using the Maxwell relation in Eq. (14.18), and Eq. (E14.3.3) we obtain

$$\left(\frac{\partial s}{\partial P}\right)_T = -\left(\frac{\partial v}{\partial T}\right)_P = -v\alpha \qquad (E14.3.4)$$

Applying the reciprocal relation, Eq. (14.4) to Eq. (E14.3.4) we have

$$\left(\frac{\partial P}{\partial s}\right)_T = -\frac{1}{v\alpha} \qquad (E14.3.5)$$

Substituting from Eq. (E14.3.2) and (E14.3.5) in Eq. (E14.3.1) we have

$$\left(\frac{\partial T}{\partial P}\right)_s = \frac{Tv\alpha}{c_P} \tag{E14.3.6}$$

Example 14.4 Show that the difference specific heat capacities

$$c_p - c_v = Tv\alpha^2/\beta_T$$

Solution We start with Eq. (14.54)

$$c_P - c_v = T\left(\frac{\partial v}{\partial T}\right)_P\left(\frac{\partial P}{\partial T}\right)_v \tag{E14.4.1}$$

Applying the *cyclic relation*, Eq. (14.5) we have

$$\left(\frac{\partial P}{\partial T}\right)_v\left(\frac{\partial T}{\partial v}\right)_P\left(\frac{\partial v}{\partial P}\right)_T = -1 \tag{E14.4.2}$$

Applying the *reciprocal relation*, Eq. (14.4) we have

$$\left(\frac{\partial T}{\partial v}\right)_P = \frac{1}{\left(\dfrac{\partial v}{\partial T}\right)_P} = \frac{1}{v\alpha} \tag{E14.4.3}$$

Manipulating Eqs. (E14.4.1) to (E14.4.3) we obtain

$$c_p - c_v = Tv\alpha^2/\beta_T$$

Example 14.5 Show that the ratio of specific heat capacities

$$\gamma = c_p/c_v = \beta_T/\beta_s$$

Solution Apply Eqs. (14.43) and (14.44) to a *constant entropy* process. Hence we have

$$\frac{c_P}{c_v} = \frac{-\left(\dfrac{\partial v}{\partial T}\right)_P dP_s}{\left(\dfrac{\partial P}{\partial T}\right)_v dv_s} = \frac{-\left(\dfrac{\partial v}{\partial T}\right)_P}{\left(\dfrac{\partial P}{\partial T}\right)_v\left(\dfrac{\partial v}{\partial P}\right)_s} \tag{E14.5.1}$$

Applying the *cyclic relation*, Eq. (14.5) we have

$$\left(\frac{\partial P}{\partial T}\right)_v \left(\frac{\partial T}{\partial v}\right)_P \left(\frac{\partial v}{\partial P}\right)_T = -1 \qquad (E14.2.2)$$

Manipulating Eqs. (E14.5.1) and (E14.5.2) we obtain

$$\gamma = c_p / c_v = \beta_T / \beta_s$$

Example 14.6 Derive the following relation for the Joule-Thomson coefficient

$$\mu_{JT} = \left(\frac{\partial T}{\partial P}\right)_h = \frac{1}{c_P}\left[T\left(\frac{\partial v}{\partial T}\right)_P - v\right]$$

Obtain an expression for μ_{JT} of an ideal gas.

Solution Applying the *cyclic relation*, Eq. (14.5) we have

$$\left(\frac{\partial T}{\partial P}\right)_h \left(\frac{\partial P}{\partial h}\right)_T \left(\frac{\partial h}{\partial T}\right)_P = -1 = \mu_{JT} c_P \left(\frac{\partial P}{\partial h}\right)_T \qquad (E14.6.1)$$

From Eq. (14.39)

$$\left(\frac{\partial h}{\partial P}\right)_T = v - T\left(\frac{\partial v}{\partial T}\right)_P \qquad (E14.6.2)$$

Substituting from Eq. (E14.6.2) in Eq. (E14.6.1) we have

$$\mu_{JT} = \left(\frac{\partial T}{\partial P}\right)_h = \frac{1}{c_P}\left[T\left(\frac{\partial v}{\partial T}\right)_P - v\right] \qquad (E14.6.3)$$

For an ideal gas, $Pv = RT$. Differentiating this equation of state and substituting in Eq. (E14.6.3) we obtain, $\mu_{JT} = 0$.

Example 14.7 A rigid vessel of volume 0.015 m³ contains 3.5 kg of nitrogen at 400°C. Use (i) the ideal gas equation, (ii) the van der waals equation, and (iii) the Redlich-Kwong equation to predict the pressure. For Nitrogen: $T_c = 126.2$ K, $P_c = 3390$ kPa.

Solution (i) Using the ideal gas equation we obtain
$$P = mR_nT/v = 3.5 \times 8.314 \times (273.15 + 400)/(28 \times 0.015) = 46.6$$

The pressure is 46.6 MPa.

(ii) We compute the constants a and b in van der Waals equation by substituting in Eqs. (14.62a) and (14.62b). Hence we have

$$a = 0.1746 \quad \text{and} \quad b = 0.0014$$

Substitute the given numerical data directly in van der Waals equation, (14.62) to obtain the pressure as 59.27 MPa

(iii) We compute the constants a and b in Redlich-Kwong (RK) equations by substituting in Eqs. (14.66a) and (14.66b). Hence we have

$$a = 1.9892 \quad \text{and} \quad b = 0.0009577$$

Substitute the given numerical data directly in the Redlich-Kwong equation, (14.66) to obtain the pressure as 56.65 MPa

Example 14.8 Use (i) the ideal gas equation, (ii) the van der waals equation, and (iii) the Redlich-Kwong equation to predict the specific volume of steam at 350°C and 5000 kPa. For water, $T_c = 647$ K, $P_c = 221.2$ kPa

Solution (i) Using the ideal gas equation we obtain

$$v = R_w T / P = 8.314 \times (273.15 + 350)/(18 \times 5000) = 0.057565$$

The specific volume is 0.057565 m³ kg⁻¹.

(ii) We first compute the constants a and b in van der Waals equation by substituting in Eqs. (14.62a) and (14.62b). Hence we have

$$a = 1.7054 \quad \text{and} \quad b = 0.0017$$

We rearrange van der Waals equation as a cubic equation in v:

$$v^3 - \left(b + \frac{RT}{P}\right)v^2 + \left(\frac{a}{P}\right)v - \left(\frac{ab}{P}\right) = 0$$

Substituting numerical values in the above equation we have

$$v^3 - 0.0593v^2 + 3.4107 \times 10^{-4} v - 5.7635 \times 10^{-7} = 0$$

We solve this cubic equation using the roots(c) function in the MATLAB software package. Hence we obtain the specific volume as 0.053 m³ kg⁻¹.

(iii) We compute the constants a and b in the Redlich-Kwong (RK) equation by substituting in Eqs. (14.66a) and (14.66b). Hence we have

$$a = 43.9681 \qquad \text{and} \qquad b = 0.0012$$

We rearrange the Redlich-Kwong equation as a cubic equation in v:

$$v^3 - \left(\frac{RT}{P}\right)v^2 - \left(b^2 + \frac{RTb}{P} - \frac{a}{PT^{1/2}}\right)v - \left(\frac{ab}{PT^{1/2}}\right) = 0$$

Substituting numerical values in the above equation we have

$$v^3 - 0.0576v^2 + 2.8347 \times 10^{-4}v - 4.1259 \times 10^{-7} = 0$$

We solve this cubic equation using the roots (c) function in the MATLAB software package. Hence we obtain the specific volume as 0.0523 m³ kg⁻¹.

Example 14.9 Obtain an expression for $(c_P - c_v)$ for a gas obeying the van der Waals equation of state.

Solution From Eq. (14.54) we have

$$c_P - c_v = T\left(\frac{\partial v}{\partial T}\right)_P \left(\frac{\partial P}{\partial T}\right)_v \qquad \text{(E14.9.1)}$$

Applying the *cyclic relation*, Eq. (14.5),

$$\left(\frac{\partial v}{\partial T}\right)_P \left(\frac{\partial T}{\partial P}\right)_v \left(\frac{\partial P}{\partial v}\right)_T = -1 \qquad \text{(E14.9.2)}$$

Applying the *reciprocal relation*, Eq. (14.4),

$$\left(\frac{\partial T}{\partial P}\right)_v = \frac{1}{\left(\dfrac{\partial P}{\partial T}\right)_v} \qquad \text{(E14.9.3)}$$

Substituting from Eqs. (E14.9.2) and (E14.9.3) in Eq. (E14.9.1) we have

$$c_P - c_v = -T\left[\left(\frac{\partial P}{\partial T}\right)_v\right]^2 \left[\left(\frac{\partial P}{\partial v}\right)_T\right]^{-1} \qquad \text{(E14.9.4)}$$

The van der Waals equation, (14.62) may be expressed in the form

$$P = RT/(v-b) - a/v^2 \qquad \text{(E14.9.5)}$$

The partial differentials in Eq. (E14.9.4) are obtained by differentiating Eq. (E14.9.5). Upon substitution in Eq. (E14.9.4) we obtain

$$c_P - c_v = R\left[1 - \frac{2a(v-b)^2}{RTv^3}\right]^{-1}$$

Example 14.10 Compute the entropy of superheated steam at 400°C and 4500 kPa given that the entropy of saturated water vapor at 1 kPa is 8.974 kJ kg^{-1} K^{-1}. Use the Redlich-Kwong equation of state. The following data are available: The variation of c_{p0} with temperature T (K) is given by [5]

$$c_{p0} = 1.8652 - 1.5963 \times 10^{-4} T + 6.528 \times 10^{-7} T^2$$

For water, $t_c = 374.15°C$ and $P_c = 22.120$ MPa.

Solution We apply Eq. (14.43) to calculate the change in entropy.

$$ds = c_P \frac{dT}{T} - \left(\frac{\partial v}{\partial T}\right)_P dP \qquad \text{(E14.10.1)}$$

Integration of Eq. (E14.10.1) is performed along two paths. These are: (i) a constant pressure path from 1 (1 kPa, 7°C) to 2 (1 kPa, 400°C), and (ii) a constant temperature path from 2 (1 kPa, 400°C) to 3 (4500 kPa, 400°C).
Along path (i) the integral becomes:

$$\Delta s_{1-2} = c_P \frac{dT}{T} = \int_{280}^{673} \left[1.8652 - 1.5963 \times 10^{-4} T + 6.528 \times 10^{-7} T^2\right]\frac{dT}{T}$$

$$\Delta s_{1-2} = 1.8652 \ln T - 1.5963 \times 10^{-4} T + 6.528 \times 10^{-7} T^2 / 2$$

evaluated from 280 K to 673 K.
Hence we have $(s_2 - s_1) = 1.6952$ kJ kg^{-1} K^{-1}.
Along path (ii) the integral at constant temperature is:

$$ds = -\left(\frac{\partial v}{\partial T}\right)_P dP \qquad \text{(E14.10.2)}$$

Applying the *cyclic relation*, Eq. (14.5) and the *reciprocal relation*, Eq. (14.4), we express Eq. (E14.10.2) in the form

$$ds = \left[\left(\frac{\partial P}{\partial T}\right)_v \left(\frac{\partial P}{\partial v}\right)_T\right]^{-1} dP \qquad (E14.10.3)$$

We differentiate the Redlich-Kwong equation (14.66) to obtain the partial differentials in Eq. (E14.10.3). Hence we have

$$\left(\frac{\partial P}{\partial T}\right)_v = \frac{R}{(v-b)} + \frac{a}{2T^{3/2}v(v+b)} = f_1(v,T) \qquad (E14.10.4)$$

$$\left(\frac{\partial P}{\partial v}\right)_T = -\frac{RT}{(v-b)^2} + \frac{a(2v+b)}{T^{1/2}(v^2+bv)^2} = f_2(v,T) \qquad (E14.10.5)$$

Substitute from Eqs. (E14.10.4) and (E14.10.5) in Eq. (E14.10.3) and integrate the resulting equation to obtain the entropy change $s_3 - s_2$.

$$\Delta s_{2-3} = \int_1^{4500} \frac{f_1(v,673)}{f_2(v,673)} dP \qquad (E14.10.6)$$

We integrate Eq. (E14.10.6) numerically, keeping the temperature constant at 673 K. At each pressure, P the volume, v is obtained by solving the Redlich-Kwong equation as was done in example 14.8. Hence we have $\Delta s_{2-3} = (s_3 - s_2) = -3.9795$ kJ kg^{-1} K^{-1}. Therefore the entropy at state 3 is

$$s_3 = 8.974 + 1.6952 - 3.9795 = 6.6897 \text{ kJ kg}^{-1} \text{ K}^{-1}.$$

Problems

P14.1 Determine the increase in melting point of ice, due to an increase of pressure of 100 kPa. Assume that at 0°C, $v_f = 0.0010$ m^3 kg^{-1} and $v_s = 0.001091$ m^3 kg^{-1}. The latent heat of fusion of ice is 334 kJ kg^{-1}. [*Answer*: 0.00744°C]

P14.2 Estimate the enthalpy of vaporization of refrigerant 134a at 30°C using the Clapeyron equation. Compare the result with the value in the data tables. [*Answer*: 173.4 kJ kg^{-1}]

P14.3 Show that for water at the triple-point, the slope of the saturation curve, $[dP/dT]_{sat}$ is larger for vaporization than for sublimation. Use data from the steam tables.

P14.4 The equation of state of a gas is

$$v = RT/P - a/T^3$$

where a is a constant.
(i) Obtain an expression for the exact differential dv.
(ii) Show that the condition in Eq. (14.3) is satisfied.
(iii) Obtain an expression for $(c_P - c_v)$ for the gas.

P14.5 Obtain expressions for the variation of c_v and c_P for (i) an ideal gas, (ii) a van der Waals' gas, and (iii) an incompressible substance.

P14.6 Obtain expressions for the change in (a) internal energy, (b) enthalpy, and (c) entropy of (i) an ideal gas, and (ii) a van der Waals' gas. Obtain the T-v relation for a van der Waals' gas, when it undergoes an isentropic process.

P14.7 Obtain expressions for (a) the expansion coefficient α, and (b) the isothermal compressibility β_T of (i) an ideal gas, and (ii) a van der Waals' gas.

P14.8 Using Eq. (E14.3.6) derive the T-v variation for an ideal gas undergoing an isentropic process.

P14.9 Show that

$$\left(\frac{\partial \alpha}{\partial P}\right)_T = -\left(\frac{\partial \beta_T}{\partial T}\right)_P$$

P14.10 Show that

$$\left(\frac{\partial P}{\partial T}\right)_s = \left(\frac{\gamma}{\gamma-1}\right)\left(\frac{\partial P}{\partial T}\right)_v$$

$$\left(\frac{\partial v}{\partial T}\right)_s = \left(\frac{1}{\gamma-1}\right)\left(\frac{\partial v}{\partial T}\right)_P$$

The equation of state of a gas is $P(v-b) = RT$, where b is a constant. Obtain the P-T relation for the gas, when it undergoes an isentropic process.

P14.11 Obtain an expression for the Joule-Thomson coefficient (JTC) of a van der Waals' gas in terms a, b, R, T, v and c_p. Show that the inversion temperature for a van der Waals' gas is given by

$$T = (2a/b\overline{R})(1-b/v)^2$$

Deduce the JTC for an ideal gas.

P14.12 Show that the van der Waals equation may be expressed in terms of the reduced coordinates as:

$$(P_r + 3/v_r^2)(v_r - 1/3) = 8T_r/3$$

If the reduced pressure and temperature are 0.75 and 1.0 respectively, calculate the compressibility factor, Z. [*Answer*: 0.7043]

P14.13 The molecular mass, the critical temperature and critical pressure of refrigerant-12 are 120.93, 112°C, and 4113 kPa respectively. Compute the specific volume of R-12 at 110°C and 2350 kPa using (i) the ideal gas equation, (ii) the van der Waals equation, and (iii) the Redlich-Kwong equation. [*Answer*: (i) 0.0112 m³ kg⁻¹ (ii) 0.0089, (iii) 0.0086]

P14.14 Calculate $(c_P - c_v)$ for water vapor at 7000 kPa and 300°C using the Redlich-Kwong equation. [*Answer*: 0.9215 kJ kg⁻¹ K⁻¹]

P14.15 Compute the enthalpy of superheated steam at 500°C and 3500 kPa taking the enthalpy of liquid water at 0°C as zero. Use the Redlich-Kwong equation of state. The following data are available:
(i) The variation of c_{p0} with temperature t, °C is given by [4]

$$c_{p0} = 1.8703 + 0.1968\times10^{-3}t + 0.6528\times10^{-6}t^2$$

(ii) The equation for the liquid-vapor saturation curve is [4]

$$\ln p = 19.335 - 5416T^{-1}$$

For water, t_c = 374.15°C and P_c = 22.12 MPa. [*Answer*: 3453 kJ kg^{-1}]

P14.16 Compute the entropy of superheated steam at 500°C and 4000 kPa taking the entropy of liquid water at 0°C as zero. Use the data for water given in problem 14.15. [*Answer*: 7.051 kJ K^{-1} kg^{-1}]

References

1. Jones, J.B. and G.A. Hawkins, *Engineering Thermodynamics*, John Wiley & Sons, Inc., New York, 1986.
2. Reynolds, William C. and Henry C. Perkins, *Engineering Thermodynamics*, 2nd edition, McGraw-Hill, Inc., New York, 1977.
3. Rogers, G.F.C. and Mayhew, Y.R., *Thermodynamic and Transport Properties of Fluids*, 5th edition, Blackwell, Oxford, U.K. 1998.
4. Stoecker, Wilbert F., *Design of Thermal Systems*, International Edition, McGraw Hill Book Company, London, 1989.
5. Van Wylen, Gordon J. and Richard E. Sonntag, *Fundamentals of Classical Thermodynamics*, 3rd edition, John Wiley & Sons, Inc., New York, 1985.
6. Zemansky, M., *Heat and Thermodynamics*, 5th edition, McGraw-Hill, Inc., New York, 1968.

Chapter 15

Statistical Interpretation of Entropy

15.1 Background

The new property called entropy arose as a consequence of the second law and it was defined by Eq. (7.10). This equation enables us to relate entropy to other measurable macroscopic properties like pressure, volume and temperature (see Sec. 7.1.4) of a system and thereby obtain its numerical value, which is useful for the analysis of engineering systems.

This development follows closely the analysis presented in chapter 4, to establish the existence of the property, internal energy due to the first law. Later in section 4.6.2 we used the kinetic theory of gases to explain the physical meaning of internal energy in terms of the microscopic properties of the constituent molecules of the gas. In a similar manner, we shall in this chapter seek a more generalized physical interpretation of entropy using the microscopic view point of matter. In a recent book [1], Arieh Ben-Naim has argued that the microscopic approach is the only satisfactory way to explain the physical meaning of the second law and entropy.

In several popular science books, written by well-known physicists [3,4,6], the meaning of entropy has been explained by referring to the notion of 'order' and 'disorder' as applied to everyday situations. For example, it is a common observation that a room or a work desk that is initially tidy, becomes untidy or disorganized as time goes by. This trend is sometimes thought to be a manifestation of the law of increase of entropy. Here the entropy is taken to be directly related to the number of ways of arranging the contents of the room, such as the furniture, or the items on the work desk. It is clear that there are many more arrangements of the contents of the room and the items on the desk that make the room or desk disorganized or disorderly, compared to the number of arrangements for which the room or desk can be described as organized or orderly. Therefore unless deliberate action is taken to tidy the room and the desk (non-isolated), they will progressively become more disorganized. Generally, we would also have more information about the contents of the room and the items on the desk when they are orderly compared to when they are disorderly. There are many other examples of this type [3,4] which have been cited to explain the meaning of entropy in a qualitative way.

The above examples, however, are over simplifications because the meaning of 'order and disorder' are subjective and are not easily quantifiable. Nevertheless, by using these everyday situations we can explain the physical meaning of entropy sometime better than by using the macroscopic-level definition, $\delta S = \delta Q / T$.

In the next section we shall develop a formal approach to apply of concepts of 'order and disorder' or 'organization and disorganization' to an assembly of atoms or molecules constituting a fixed quantity of matter. The disorganization or spreading of energy among the motions of atoms and molecules in an assembly is sometimes called '*thermal entropy*'. The disorganization of atoms and molecules due to spreading in space in mixtures of gases is called '*configurational entropy*'. In later sections of this chapter we shall develop simple physical models to quantify thermal entropy and configurational entropy.

15.2 Boltzmann Equation

Boltzmann was the first to develop a successful interpretation of entropy from a microscopic view point [2]. His work led to the development of an entirely new branch of thermodynamics called *statistical mechanics*. We shall briefly discuss the main concepts that form the basis of Boltzmann's microscopic approach to entropy.

The complete microscopic description of one gram of a gas needs about 10^{23} parameters, which includes the positions and velocities of all

the molecules of the gas at any instant. On the other hand, the thermodynamic description requires only 3 parameters P, v and T. However,
from the simple kinetic theory introduced in chapters 2 and 4 it is clear
that the properties like the pressure and temperature could be computed
by suitably averaging the properties of the individual molecules of the
gas. Statistical mechanics relies on this averaging process to develop the
definition of entropy from a microscopic view point.

The second important input to the statistical approach comes from
law of distribution of velocities among the molecules of a gas, which was
developed by Maxwell in 1859. Although the velocities of individual
molecules of a gas change frequently due to collisions, the average
distribution of velocities at equilibrium remains unchanged. Typical
velocity distributions of a gas at three different temperatures are depicted
in Fig. 15.1. We notice that as the temperature increases the velocity
distributions become flatter due to the increased spread of velocities of
the molecules. From the above velocity distributions we could also
compute the energy distributions of the molecules. In section 4.6.2 we
discussed the various modes of motion of molecules and the energies
associated with these modes. Like the macroscopic properties pressure
and temperature, the velocity and energy distributions of the molecules at
equilibrium are also time-independent, and are therefore characteristic
parameters of the system.

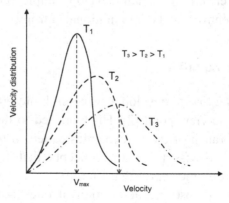

Fig. 15.1 Maxwell-Boltzmann velocity distribution in a gas at three different temperatures

Conceptually, the statistical approach provides a new perspective on temperature and thermodynamic equilibrium. In practical terms, using statistically averaged expressions it is possible to calculate typical thermodynamic properties such as the specific heat capacities, the internal energy or enthalpy, the entropy, and the Gibbs function.

The fundamental equation that links the macroscopic-level entropy, S to the parameters of atoms and molecules is the Boltzmann equation, which may expressed as:

$$S = k \ln W \qquad (15.1)$$

The Boltzmann constant, k [JK^{-1} per molecule] in Eq. (15.1) is given by

$$k = 1.38 \times 10^{-23} = \overline{R} / N_o \qquad (15.2)$$

where $\overline{R}$ is the universal gas constant and N_o is the Avogadro number.

The dimensionless quantity W in Eq. (15.1) is the *total number of microstates*, which is the number of arrangements of energy and matter in the system. These arrangements are distinguishable at the microscopic level but indistinguishable at the macroscopic level. In other words, the microstates or molecular arrangements have to be compatible with the prescribed macroscopic values of energy, volume and composition of the system. *A fundamental assumption in the statistical approach is that all microstates, compatible with a macroscopic description of a system, are equally probable.*

The direct relationship between entropy and the number of microscopic arrangements is already seen in Eq. (15.1) because as W increases the entropy also increases. Moreover, we shall demonstrate that the entropies of two systems, defined by Eq. (15.1), can be added to obtain the entropy of the composite system, which satisfies the additive condition of an extensive property.

Consider a system A consisting of two sub-systems B and C, whose entropies are S_B and S_C respectively. Since entropy is an extensive property

$$S_A = S_B + S_C \qquad (15.3)$$

Let the number of microscopic arrangements (microstates) of the constituents of B and C be W_B and W_C respectively. Because the microscopic arrangements in B and C are independent, any arrangement in B can be combined with W_C arrangements in C to obtain the number of arrangements W_A of the composite system A. Hence the *total* number of arrangements of A is the product of W_B and W_C. Therefore

$$W_A = W_B W_C \tag{15.4}$$

We obtain the entropy of the composite system A from Eqs. (15.1) and (15.4) as:

$$S_A = k \ln W_A = k \ln(W_B W_C) = k \ln W_B + k \ln W_C \tag{15.5}$$

Equation (15.5) shows that the entropy defined by Eq. (15.1) satisfies the additive condition [Eq. (15.3)] for an extensive property.

To count or calculate the microstates of a system we need an explicit mechanical or geometrical model for the system. In the next section we shall develop such a model to compute thermal entropy using Eq. (15.1).

15.3 Model for Thermal Entropy

In the present introductory treatment of the statistical approach we shall consider a mechanical model of a solid where the mean positions of the atoms are fixed.

The simplified model of a solid depicted in Fig. 15.2 consists of four atoms connected by springs, which represent the bonds between the atoms. Each mass-spring unit is called a *harmonic oscillator*. It should be noted that this model is a simplified version of the three-dimensional Einstein model of a crystal [2]. The main advantage of the simplified model is that we can count the number of arrangements, or microstates W of the system when the number of oscillators and the total energy of the system are specified.

Atoms   Springs

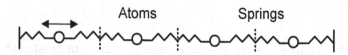

Fig. 15.2 Simplified microscopic model of a solid composed of four harmonic oscillators

By applying Newton's law to the mass-spring system or harmonic oscillator we obtain its classical frequency v as [3]:

$$v = \frac{1}{2\pi} \sqrt{\frac{k}{m}} \tag{15.6}$$

where m is the mass and k is the spring constant.

For the atomic-level harmonic oscillators shown in Fig.15.2 the possible energy levels are quantized according to quantum mechanics. The stationary states of such an oscillator have energies given by:

$$\varepsilon_n = nh v \tag{15.7}$$

where $h = 6.63 \times 10^{-34}$ (Js^{-1}) is Planck's constant. The quantum number n can take values 0, 1, 2, 3, ... In Eq. (15.7) the zero-point energy [2] of the ground-state ($n = 0$), $hv/2$ has been neglected, because the model is used mainly to illustrate the procedure for counting microstates.

Using the above result from quantum mechanics we are now able to determine the number of different ways in which a specified amount of energy can be shared by the four harmonic oscillators of our model. We will illustrate the counting procedure to determine the number of microstates for the different ways (macrostates) by which a total energy, $U = 6hv = 6\varepsilon$ can be shared between 4 oscillators.

Now each atom or oscillator can be in one of seven different energy states: 0, ε, 2ε, 3ε, 4ε, 5ε, and 6ε as shown schematically in Fig. 15.3. We need to determine what possible distributions of energy are consistent with the given total energy of 6ε. Let us assume the following general distribution:

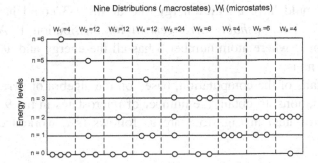

Fig. 15.3 Number of microstates for four-oscillator model of a solid

n_0 atoms have energy 0
n_1 atoms have energy ε
n_2 atoms have energy 2ε
n_3 atoms have energy 3ε (15.8)
n_4 atoms have energy 4ε
n_5 atoms have energy 5ε
n_6 atoms have energy 6ε

We require that

$$n_0 + n_1 + n_2 + n_3 + n_4 + n_5 + n_6 = 4 \qquad (15.9)$$

$$\varepsilon n_1 + 2\varepsilon n_2 + 3\varepsilon n_3 + 4\varepsilon n_4 + 5\varepsilon n_5 + 6\varepsilon n_6 = 6\varepsilon \qquad (15.10)$$

in order that the total number of atoms be 4 and their total energy 6ε.

If the above conditions are satisfied, the n's define a possible distribution of energy of the assembly. We need to: (i) write down all possible distributions (macrostates) of the type in Eq. (15.8) that satisfy the restrictions implied in Eqs. (15.9) and (15.10), and (ii) evaluate the different arrangements of the atoms (microstates) corresponding to each of these distributions.

The atomic distributions required in task (i) above are shown in Fig. 15.3. For example, in distribution 1, one atom has all the energy, 6ε and three atoms have zero energy. In distribution 3, one atom has 4ε, one atom has 2ε and two atoms have zero energy. In each case the total energy is 6ε.

In task (ii) above we need to work out the number of *distinguishable* ways the atoms numbered from 1 to 4 could be selected to satisfy the 9 distributions in Fig. 15.3. For example, in distribution 1, atom number 4 could have all the energy and atoms 1-3 could have none. This is one *distinguishable arrangement* for distribution 1. A second arrangement is where atom number 3 has all the energy and atoms 1, 2 and 4 have none.

The details of the computation, based on the algebra of permutations and combinations, to obtain the number of microstates in the 9 distributions (macrostates) in Fig. 15.3 are as follows (see worked examples 15.1-15.6):

$$W_1 = {}^4C_1 \times {}^3C_3 = 4 \times 1 = 4$$

$$W_2 = {}^4C_1 \times {}^3C_1 \times {}^2C_2 = 4 \times 3 \times 1 = 12$$

$$W_3 = {}^4C_1 \times {}^3C_1 \times {}^2C_2 = 4 \times 3 \times 1 = 12$$

$$W_4 = {}^4C_1 \times {}^3C_2 \times {}^1C_1 = 4 \times \frac{(3 \times 2)}{(2 \times 1)} \times 1 = 12$$

$$W_5 = {}^4C_1 \times {}^3C_1 \times {}^2C_1 \times {}^1C_1 = 4 \times 3 \times 2 \times 1 = 24$$

$$W_6 = {}^4C_2 \times {}^2C_2 = \frac{(4 \times 3)}{(2 \times 1)} \times 1 = 6$$

$$W_7 = {}^4C_1 \times {}^3C_3 = 4 \times 1 = 4$$

$$W_8 = {}^4C_2 \times {}^2C_2 = \frac{(4 \times 3)}{(2 \times 1)} \times 1 = 6$$

$$W_9 = {}^4C_3 \times {}^1C_1 = \frac{(4 \times 3 \times 2)}{(3 \times 2 \times 1)} \times 1 = 4$$

where ${}^nC_r$ is the number of combinations of n distinct things taken r at a time. Some general applications of combinatorial algebra are illustrated in the worked examples to follow in this chapter.

The general formula to calculate the number of arrangements (microstates) is [2]

$$W_N = \frac{N!}{n_0! \times n_1! \times n_2! \dots n_k!} \tag{15.12}$$

where N is the total number of atoms and n_1, n_2, ...n_k are number of atoms in each energy level as shown in Fig. 15.3.

The total number of microstates for the 9 distributions is obtained as 84, by adding the microstates in each distribution. A general formula to calculate the *total* number of microstates for N oscillators and a total energy of n units, when the number of quantum levels exceed n is [3]

$$W_{tot} = \frac{(N + n - 1)!}{(N - 1)! \times n!} \tag{15.13}$$

From the data in Fig. 15.3 we notice that distribution 5 has the largest number of microstates of 24. We stated earlier that a fundamental postulate of the statistical approach is that all microstates are equally probable. But due the larger number of microstates (24) in distribution 5, the system, left to itself, would be found most often in the state corresponding to distribution 5.

We used a simple model of a solid mainly to illustrate how the number of microstates or atomic arrangements are calculated from first principles. If we take the same set of energy levels for a real system with a much larger number of molecules, N (order of 10^{23}), and a correspondingly increased total energy, U we would find that the dominant distribution (largest W_i) will now have an overwhelmingly larger number of microstates compared to the rest of the distributions. This would correspond to the equilibrium state of the system because the system, left to itself, would be found most often in the dominant distribution. From Eq. (15.1) it follows that the entropy will then be a maximum.

The above procedure for computing the entropy can be generalized for real systems where the number of atoms N is very large (order 10^{23}) by making two important assumptions [6]. First we replace the *total* number of microstates W in Eq. (15.1) by W_{mp}, the number of microstates in the *dominant* distribution (largest W). We then use Stirling's approximation (see Eq. (15.16)) to evaluate $ln\ N!$ in the resulting expression [5,6]. Pursuing such as an analysis we could obtain most of the fundamental results used in statistical thermodynamics [5,6]. However, the detailed development of statistical thermodynamics is beyond the scope of this book.

15.4 Model for Configurational Entropy

Configurational entropy is defined as the disordering due to the dispersion of matter in space. A typical example could be the ideal adiabatic mixing of two liquids, initially maintained at the same pressure and temperature. We shall now develop a microscopic model to obtain the entropy change of ideal mixing processes by counting the microstates

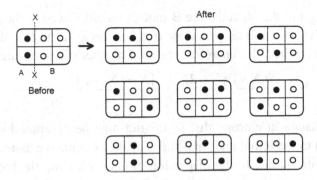

Fig.15.4 Cell model for the mixing of two molecules of type A and four molecules type B

of the system. The model is called a *cell-model* or a *lattice-model* because the volume occupied by the molecules is divided into small cells, each containing one molecule as shown in Fig. 15.4. The cells are merely the average volume the molecules occupy in space.

This is a geometrical model as distinct from the mechanical model used for thermal entropy. Initially 2 molecules of A indicated by black circles and 4 molecules of B indicated by white circles are in 6 cells, separated by a thin well-conducting diaphragm x-x as seen in Fig. 15.4. This is the initial equilibrium state of the system. We assume that the molecules of each type are indistinguishable. When the diaphragm is broken the two types of molecules mix. A *few* of the possible arrangements in the final state are shown on the right hand side of the figure. For the unmixed state, shown on the left hand side, the number of microstates is given by $W_i = 1$.

For the final state the first A molecule can be in 6 cells while the second can be in 5 cells. Therefore number of arrangements of the A molecules, if they were distinguishable, would be 30. Since they are indistinguishable we need to divide 30 by 2! to account for the fact that half the arrangements are repeated.

Now consider the general case when the number of A molecules are N_1, and the number of B molecules are N_2, and $N_1 + N_2 = N$. The first A molecule can be placed in N cells, the second in $(N-1)$ cells and N_1^{th} molecule in $(N-N_1 +1)$ cells. Since the molecules are indistinguishable, they can arranged among themselves in $N_1!$ ways. Once the A molecules are arranged in the N_1 cells, the B molecules are placed in the remaining $(N-N_1)$ cells. This can only be done one way because the B molecules are also indistinguishable. Therefore the total number of microstates is [3]:

$$W = \frac{N(N-1)(N-2)........[N-(N_1-1)]}{N_1!} = \frac{N!}{N_1! \times N_2!} \qquad (15.14)$$

The change in entropy due to mixing may be obtained by applying Eq. (15.1) to the initial and final states of the system. We assume that the thermal entropy due to the dispersion of energy among the molecules is unchanged. Note that by applying the first law to the macroscopic system we can deduce that the temperature of the system is unchanged under ideal mixing conditions.

Hence we have

$$\Delta S = k \ln(W/W_i) = k \ln \frac{N!}{N_1! \times N_2!} \qquad (15.15)$$

Now for real systems, the number of molecules are very large (order 10^{23}). For such cases we can use Stirling's approximation to express $\ln n!$ in the form

$$\ln n! = n \ln n - n \qquad (15.16)$$

Applying the approximation to the various terms in Eq. (15.15) we have

$$\frac{\Delta S}{k} = -N_1 \ln\left(\frac{N_1}{N}\right) - N_2 \ln\left(\frac{N_2}{N}\right) \qquad (15.17)$$

$$\Delta S = -NkX_1 \ln(X_1) - NkX_2 \ln(X_2) \qquad (15.18)$$

where X_1 and X_2 are the fractions of molecules A and B respectively.

We can also express Eq. (15.18) in terms of the Avogadro number, N_0 and the number of moles n_1 and n_2 of the two types of molecules. Hence we obtain the molar entropy change as:

$$\Delta S = -\bar{R}n_1 \ln\left(\frac{n_1}{n_1+n_2}\right) - \bar{R}n_2 \ln\left(\frac{n_2}{n_1+n_2}\right) \qquad (15.18)$$

where $\bar{R}$ is the universal gas constant. Note that Eq. (15.18) can be derived using the macroscopic definition of entropy.

15.5 Summary

At a microscopic level, entropy can be interpreted as the disordering or dispersal of energy among the molecules, and the dispersion of molecules in the space available. The former is called thermal entropy and the latter configurational entropy. These terms are useful for explaining the physical meaning of entropy in a descriptive manner. However, for most systems the two types of entropy are strongly coupled, and the configurational entropy can be obtained from the thermal entropy.

In the statistical approach to entropy, the main quantity to be determined is the number of ways of distributing the total energy among the molecules of the system. These are called the microstates, W of the system. The entropy S and the total number of microstates W are related by the Boltzmann equation. To evaluate W an explicit model is needed. For a solid, where the average positions of the molecules are fixed, we used a quantum mechanical harmonic oscillator model. For the ideal mixing processes, we used a geometric model called the cell model. Here the expression obtained for the entropy change due to ideal mixing agrees with that obtained using the macroscopic approach.

We shall now use the idea of 'disorder' or 'spread' to explain the changes in entropy that occur during a phase change process [2]. Consider a fixed quantity of saturated steam, initially at a pressure of 1 bar and temperature of 100°C (see Fig. 2.1). When cooled quasi-statically, the steam will condense to water at 100°C and its volume will decrease by a factor of about two thousand. The entropy of the water will decrease due to entropy transfer to the cooling medium. In statistical terms the entropy of the steam is due to the spread of the molecules in the actual space, and the spread of their velocities. Since the temperature is constant during condensation, the spread of the velocities of the molecules in steam is the same as that in water. However, the entropy due to the spread of the molecules in space (microstates) is much lower for water because of large reduction in volume during condensation. Hence the decrease in entropy can be attributed to the 'ordering' process represented by the volume change.

If we continue the cooling process, there is very little change in the volume, but the spread in molecular velocities diminishes as the temperature decreases (see Fig. 15.1). This results in a gradual decrease of entropy. When the water freezes at 0°C, the entropy of the water reduces further due to the crystalline structure of ice where the atoms are arranged in an orderly manner in arrays on a lattice. Therefore the continuous reduction of entropy during the phase change process from steam to ice can be attributed to the decrease in the spread of the molecules in actual space and the decrease in spread of the molecular velocities.

15.6 Worked Examples

Example 15.1 Find the number of ways in which the letters of the word 'optical' could be arranged so that the vowels are always next to each other.

Solution We treat the three vowels o,i,a as a single item. The word then becomes (oia)ptcl. The number of ways of arranging it is 5! (=120). But the three vowels can be arranged in 3! (=6) ways. Hence the total number of arrangements is given by 5! times 3! which is 720.

Example 15.2 How many words, with 2 vowels and 3 consonants, can be formed using 4 vowels and 7 consonants?

Solution Number of ways of selecting 2 vowels is $W_1 = {}^4C_2 = \dfrac{(4 \times 3)}{(2 \times 1)} = 6$

Number of ways of selecting 3 consonants is

$$W_2 = {}^7C_3 = \frac{(7 \times 6 \times 5)}{(3 \times 2 \times 1)} = 35$$

Hence the total number ways is: $W_3 = 6 \times 35 = 210$

The number of ways of arranging 5 letters among themselves is 5! (120).

Hence the total number words is: $W_{tot} = 210 \times 120 = 25200$

Example 15.3 A committee consisting of five persons, with at least 3 men, are to be selected from 7 men and 6 women. In how many ways can it be done?

Solution The choices for selecting the committee are as follows:
(i) 3 men and 2 women (*the order is unimportant*):

$$W_1 = {}^7C_3 \times {}^6C_2 = \frac{(7 \times 6 \times 5)}{(3 \times 2 \times 1)} \times \frac{(6 \times 5)}{(2 \times 1)} = 35 \times 15 = 525$$

(ii) 4 men and 1 women:

$$W_2 = {}^7C_4 \times {}^6C_1 = \frac{(7 \times 6 \times 5 \times 4)}{(4 \times 3 \times 2 \times 1)} \times 6 = 35 \times 6 = 210$$

(iii) 5 men and no women:

$$W_3 = {}^7C_5 = \frac{(7 \times 6 \times 5 \times 4 \times 3)}{(5 \times 4 \times 3 \times 2 \times 1)} = 21$$

Hence the total number of ways is $= 525 + 210 + 21 = 756$

Example 15.4 Twenty eight persons are to be divided into three groups having 4, 12 and 12 persons. In how many ways can this be done?

Solution The number of ways of dividing the N distinguishable items into r unequal groups (*the order is unimportant*) containing n_1, n_2, n_3 .....n_r such that

$$N = n_1 + n_2 + n_3 + ... + n_r$$

is given by

$$W = {}^NC_{n_1} \times {}^{(N-n_1)}C_{n_2} \times {}^{(N-n_1-n_2)}C_{n_3} \times {}^{(N-n_1-n_2-n_3)}C_{n4} \times ....$$

$$W = \frac{N!}{n_1! \times n_2! \times n_3! \times ......n_r!}$$

Using the above equation we obtain the number of ways as:

$$W = \frac{28!}{4! \times 12! \times 12!}$$

Since two of the groups are of the same size (12), we divide by 2! to avoid repetition. Hence the number of ways is $W/2!$.

Example 15.5 In how many ways can 6 distinguishable balls be placed in 8 distinguishable boxes if *no box* can contain more than one ball ?

Solution The first ball can be placed in any of the 8 boxes, the 2nd in any of the remaining 7 boxes, the 3rd in any of the remaining 6 boxes, and so on. Therefore the total number of ways is:

$$W_{tot} = 8 \times 7 \times 6 \times 5 \times 4 \times 3 = ^8P_6$$

Example 15.6 In how many ways can 8 distinguishable balls be placed in 6 distinguishable boxes if *any box* can contain more than one ball?

Solution The first ball can be placed in any of the 6 boxes, the 2nd also can be placed in any of the 6 boxes, the 3rd also can be in any of the 6 boxes, and so on. Hence the total number of ways is

$$W_{tot} = 6 \times 6 \times 6 \times 6 \times 6 \times 6 \times 6 \times 6 = 6^8$$

Example 15.7 A solid is modeled using 3 harmonic oscillators. Find the most probable macrostate and the total number of microstates when the total energy is (i) 2 units, (ii) 4 units.

Solution

$U = 2,$	$N= 3$			$U = 4,$	$N= 3$		
W_i	3	3		W_i	3	6	3	3
$\varepsilon = 2$	1	0		$\varepsilon = 4$	1	0	0	0
$\varepsilon = 1$	0	2		$\varepsilon = 3$	0	1	0	0
$\varepsilon = 0$	2	1		$\varepsilon = 2$	0	0	2	1
				$\varepsilon = 1$	0	1	0	2
				$\varepsilon = 0$	2	1	1	0

The results of the calculation of the number of microstates for the different macrostates (W_i) are summarized above. Equation (15.12), which was derived in example 15.4, was used to calculate the number of microstates.

Example 15.8 Find the number of microstates for a mixture of 3 atoms of argon and 3 atoms of xenon. Neglect thermal motion.

Solution The number of arrangements of the 3 argon atoms in the 6 locations is given by

$$W_{tot} = 6 \times 5 \times 4 = 120$$

Since the 3 atoms are indistinguishable, we need to divide by 3!. Hence the total number of microstates is $120/6 = 20$.

Example 15.9 Two identical bodies A and B are modeled using 4 harmonic oscillators. Initially A has 10 units of energy and B has 2 units of energy. The bodies are brought into contact adiabatically, and they achieve a final equilibrium state. Find the change in entropy for the process.

Solution Initially the identical bodies A and B have energies of 10 units and 2 units respectively. After the adiabatic heat interaction the bodies will have energies of 6 units each. We use Eq. (15.13) to calculate the total number microstates. Hence we have the following for the initial states of A and B:

$$W_{Ai} = \frac{(4+10-1)!}{(4-1)!10!} = 286, \; W_{Bi} = \frac{(4+2-1)!}{(4-1)!2!} = 10$$

For the final equilibrium states of A and B:

$$W_{Af} = \frac{(4+6-1)!}{(4-1)!6!} = 84, \; W_{Bf} = \frac{(4+6-1)!}{(4-1)!6!} = 84$$

The total initial entropy of A and B is

$$S_i = k \ln(W_{Ai} W_{Bi}) = k \ln(286 \times 10) = k \ln(2860)$$

The total final entropy of A and B is

$$S_f = k \ln(W_{Af} W_{Bf}) = k \ln(84 \times 84) = k \ln(7056)$$

Hence the change in entropy is

$$\Delta S = S_f - S_i = k \ln(7056/2860) = 0.903k$$

Example 15.10 A system has energy levels, ε of 0, 1, 2, 3 units. It contains 8 distinguishable particles (N) and has a total system energy of 6 (n) units. Find the thermodynamic probability of each of the macrostates.

Solution

P_i	0.0196	0.0392	0.0196	0.2353	0.1961	**0.2941**	0.1961
W_i	28	56	28	336	280	420	280
$\varepsilon = 3$	2	0	0	1	1	0	0
$\varepsilon = 2$	0	3	0	1	0	2	1
$\varepsilon = 1$	0	0	6	1	3	2	4
$\varepsilon = 0$	6	5	2	5	4	4	3

The results of the calculation of the number of microstates for the different macrostates (W_i) are summarized above. Equation (15.12), which was derived in example 15.4, was used to calculate the number of microstates in each macrostate. The total number of microstates, $W_{tot} = 1456$. The thermodynamics probability of a macrostate is, $P_i = W_i/W_{tot}$. The computed results are summarized in the table above.

Problems

P15.1 Find the number of ways in which the letters of the word 'corporation' could be arranged so that the vowels are always next to each other. [*Answer*: 50400]

P15.2 Four children, including at least one boy, are to be selected from a group of 6 boys and 4 girls. In how many ways can this be done? [*Answer*: 209]

P15.3 Twelve persons are to be divided into 2 groups having 3 people, and 3 groups having 2 people. In how many ways can this be done? [*Answer*: 138600]

P15.4 In how many ways can 3 distinguishable balls be put in 5 distinguishable boxes if no box can contain more than one ball? [*Answer*: 60]

P15.5 A coin is tossed 3 times. What is the total number of possible outcomes? [*Answer*: 8]

P15.6 Five distinguishable balls are put randomly in three boxes A,B and C. What is the probability that boxes A, B and C will have 3, 2 and 0 balls respectively. There is no limit on the number of balls that each box could contain. [*Answer*: 4.115%]

P15.7 In how many ways can 4 distinguishable balls be placed in 6 distinguishable boxes if, *any box* can contain more than one ball, and some boxes may have no balls? [*Answer*: 1296]

P15.8 In how many ways (W) can we put 10 distinguishable balls into three boxes A, B and C having 2, 3 and 5 balls respectively? [*Answer*: W = 2520]

P15.9 The boxes A, B, C in problem P15.8 above have 2, 2 and 3 separate compartments respectively. In how many ways (W_t) can the balls be put in the boxes, with the same distribution 2, 3 and 5, if there is no limit on the number of balls in any one compartment? [*Answer*: W_t = 7776W]

P15.10 A rectangular box contains N molecules. Each molecule can be on either half of the box with equal probability. What is the probability of all the molecules being on one half if the molecules are (i) indistinguishable, (ii) distinguishable? [*Answer*: (i) $1/N$, (ii) $1/2^N$]

P15.12 A system has energy levels 0, 1, 2, 3 units. The number of particles is 4. Find the total number of microstates if the total system energy is (i) 3 units, and (ii) 4 units. [*Answer*: (i) 20, (ii) 31]

P15.13 An ideal solution is formed by mixing 5 molecules of A and 10 molecules of B. Calculate the total number of microstates for the mixed and unmixed states. Assume that the thermal entropy is constant.
[*Answers*: 30030, 1]

P15.14 A solid is modeled using 7 harmonic oscillators. The available energy states are from 0 to 7ε. The total energy is 7ε units. Obtain (i) the total number of microstates of the system, (ii) the most probable macrostate and, (iii) the probability of the macrostates.
[*Answer*: (i) 1716, (ii) 420]

P15.15 Two identical bodies A and B are modeled using 6 harmonic oscillators. Initially A has 8 units of energy and B has 4 units of energy. The bodies are brought into contact adiabatically, and they achieve a final equilibrium state. Find the change in entropy for the process.
[*Answers*: $0.275k$]

P15.16 An ideal solution is formed by mixing 4 molecules of A and 12 molecules of B. Calculate the change in entropy. Assume that the thermal entropy is constant. [*Answers*: $7.507k$]

References

1. Arieh Ben-Naim, *Entropy Demystified: The Second Law Reduced to Plain Common Sense*, World Scientific Publishing Co., Singapore, 2008.
2. Dugdale, J.S., *Entropy and Low Temperature Physics*, Hutchinson and CO Ltd, London, 1966.
3. Greene, Brian, *The Fabric of the Cosmos*, Vintage Books, New York, 2005, Page: 151.
4. Hawking, Stephen W, *A Brief History of Time*, Bantam Books, New York, 1988, Page: 145.
5. Norman, Craig C, *Entropy Analysis: An Introduction to Chemical Thermodynamics*, Wiley-VCH, 1992.
6. Penrose, Roger, *The Emperor's New Mind*, Oxford University Press, Oxford, 1999, Page: 394.
7. Reynolds, William C. and Henry C. Perkins, *Engineering Thermodynamics*, 2nd edition, McGraw-Hill, Inc., New York, 1977.
8. Richard E. Sonntag and Van Wylen, Gordon J, *Introduction to Thermodynamics: Classical and Statistical,* 3rd edition, John Wiley & Sons, Inc., New York, 1991.

9. Tester, Jefferson W. and Michael Modell, *Thermodynamics and Its Applications*, 3rd edition, Prentice Hall PTR, New Jersey, 1996.
10. Wark, Jr., Kenneth, *Advanced Thermodynamics for Engineers*, McGraw-Hill, Inc., New York, 1995.
11. Zemansky, M., *Heat and Thermodynamics*, 5th edition, McGraw-Hill, Inc., New York, 1968.

Index

Printed in the United States
By Bookmasters